Matrix Models of String Theory

Matrix Models of String Theory

Badis Ydri
Annaba University, Annaba, Algeria

IOP Publishing, Bristol, UK

ISBN 978-0-7503-1726-9 (ebook)
ISBN 978-0-7503-1724-5 (print)
ISBN 978-0-7503-1725-2 (mobi)

DOI 10.1088/978-0-7503-1726-9

Version: 20181001

IOP Expanding Physics
ISSN 2053-2563 (online)
ISSN 2054-7315 (print)

British Library Cataloguing-in-Publication Data: A catalogue record for this book is available from the British Library.

Published by IOP Publishing, wholly owned by The Institute of Physics, London

IOP Publishing, Temple Circus, Temple Way, Bristol, BS1 6HG, UK

US Office: IOP Publishing, Inc., 190 North Independence Mall West, Suite 601, Philadelphia, PA 19106, USA

To my father for his continuous support throughout his life …

Saad Ydri

1943–2015

Also to my …

Nour

Contents

Author biography

Badis Ydri

Badis Ydri—currently a professor of theoretical particle physics, teaching at the Institute of Physics, Badji Mokhtar Annaba University, Algeria—received in 2001 his PhD from Syracuse University, New York, USA and in 2011 his Habilitation from Annaba University, Annaba, Algeria.

His doctoral work, titled 'Fuzzy Physics', was supervised by Professor A P Balachandran. Professor Ydri is a research associate at the Dublin Institute for Advanced Studies, Dublin, Ireland, and a regular ICTP associate at the Abdus Salam Center for Theoretical Physics, Trieste, Italy. His postdoctoral experience comprises a Marie Curie fellowship at Humboldt University Berlin, Germany, and a Hamilton fellowship at the Dublin Institute for Advanced Studies, Ireland.

His current research directions include: the gauge/gravity duality; the renormalization group method in matrix and noncommutative field theories; noncommutative and matrix field theory; emergent geometry, emergent gravity and emergent cosmology from matrix models.

Other interests include string theory, causal dynamical triangulation, Hořava–Lifshitz gravity, and supersymmetric gauge theory in four dimensions.

He has recently published three books. His hobbies include reading philosophic works and the history of science.

IOP Publishing

Matrix Models of String Theory

Badis Ydri

Chapter 1

Introduction

In string theory we are really dealing with a very complex conflation of several ideas and theories at once which we will attempt to list first below.

1. Gauge theory. The most famous prototype is Yang–Mills theory. The main symmetries playing a major role here besides local gauge invariance are supersymmetry and conformal invariance.
2. Theory of general relativity especially concerning black holes.
3. Supergravity. Especially in 11 and 10 dimensions.
4. Bosonic string theory. This exists only in 26 dimensions.

   The fundamental objects in gauge theory are fields and particles. The fundamental objects in string theory are strings (open and closed) while Dp-branes and the NS 5-branes are non-perturbative configurations in string theory. The Dp-branes are p-branes with Dirichlet boundary conditions. The p-branes are particles ($p = 0$), strings ($p = 1$), membranes ($p = 2$), etc. They really play the role of electric and magnetic charges in ordinary physics.

   The spectrum of string theory in Hilbert space is discrete with a mass gap and can thus be mapped one-to-one with elementary particle-like states in the target space (spacetime).
5. Superstring theories. They exist in 10 dimensions and they admit supergravity theories as low energy limits. They are:
   (a) Type I is a theory of open strings with $\mathcal{N} = 1$ supersymmetry. Only $SO(32)$ gauge charges are possible by the Green–Schwarz mechanism of anomaly cancellation. These charges are attached at the ends of the open strings by the Chan–Paton method. In this theory closed strings appear in the quantum theory as singlets under the gauge group.
   (b) Type IIA is a theory of closed strings with $\mathcal{N} = 2$ supersymmetry and where the two Majorana–Weyl spinors (or the corresponding two conserved supercharges) are of opposite chirality. There is no allowed gauge group.

   (c) Type IIB is a theory of closed strings with $\mathcal{N} = 2$ supersymmetry and where the two Majorana–Weyl spinors (or the corresponding two conserved supercharges) are of the same chirality. There is no allowed gauge group.
   (d) Heterotic $SO(32)$. This is a theory of closed strings with $\mathcal{N} = 1$ supersymmetry where the $SO(32)$ gauge charges are distributed on the closed strings.
   (e) Heterotic $E_8 \times E_8$. The same as above except that the allowed gauge group by the Green–Schwarz mechanism is $E_8 \times E_8$.

   The local diffeomorphism symmetry or reparametrization invariance of the worldsheet of the string plays a fundamental role. Only the above two local gauge groups $SO(32)$ and $E_8 \times E_8$ are allowed in string theory by demanding the principle of anomaly cancellation. Roughly, we can think of open strings as gauge theories and closed strings as gravity theories.

   These superstrings are connected to each other via an intricate web of dualities which are generalizations of the electric–magnetic duality present in electromagnetism if magnetic monopoles exist.

6. M-theory. Mostly unknown. We only know for sure that it exists in 11 dimensions and admits supergravity in 11 dimension as a low energy limit. Objects of M-theory include the supergraviton, the M2-brane and the M5-brane. The 11-dimensional supergravity contains membrane solutions.
7. Type IIB matrix model. This is the IKKT model which is the only known non-perturbative regularization of string theory.
8. M-(atrix) theory. In the context of string theory what we mean by M-(atrix) theory is the BFSS matrix quantum mechanics and its BMN pp-wave deformation. The BFSS model can be obtained as
   (a) the regularized and quantized 11-dimensional supermembrane theory, or as
   (b) the discrete light-cone quantization (DLCQ) of M-theory.
   (c) More simply, it is the dimensional reduction of supersymmetric Yang–Mills theory in 10 dimensions and as such it is a theory of D0-branes.
   (d) Or as the compactification of the IKKT matrix model on a circle. This in fact gives the finite temperature BFSS.

   Some authors have suggested that 'M-(atrix) theory is perhaps even more powerful than string theory' [1] in the sense that string theory gives only a first quantized theory in the target space whereas M-(atrix) theory gives possibly a second quantized theory.
9. Quantum gravity in two dimensions. The dynamical triangulation of quantum gravity in two dimensions and its matrix models are closely related to the IKKT ($D = 0$) and BFSS ($D = 1$) models.
10. The gauge/gravity duality. The idea that every nonabelian gauge theory has a dual description as a quantum theory of gravity is not just an idea but it is in fact the most important idea which came out of string theory and this idea is now a theory in its own right. M-(atrix) theory is one example but a more general scheme is given by the AdS/CFT correspondence.

In the following, we will present three supersymmetric matrix theories in dimensions $0+0$ (type IIB matrix model), $1+0$ (M-(atrix) theory), and $1+1$ (matrix string theory) which are of paramount importance to superstring theory and M-theory.

M-(atrix) theory is the matrix quantum mechanics model discovered by Banks, Fischler, Shenker and Susskind [2]. This is obtained by the reduction to one dimension of $\mathcal{N}=1$ super Yang–Mills theory in $D=10$ dimensions. This matrix quantum mechanics model describes the low energy dynamics of a system of N type IIA D0-branes [3]. Banks, Fischler, Shenker, and Susskind conjectured that the large N limit of this model describes precisely M-theory in the infinite momentum frame ($P_z \longrightarrow \infty$) in the uncompactified limit ($R_s \longrightarrow \infty$). We remark that the infinite momentum limit is equivalent to the light cone quantization only in the limit $N \longrightarrow \infty$.

This BFSS model can also be derived from compactification as follows.

We can start with the IKKT or type IIB matrix model discovered in 1996 by Ishibashi *et al* [4] which is the reduction to 0 dimension of $\mathcal{N}=1$ super Yang–Mills theory in D dimensions. Next, it is seen that the Euclidean IKKT model compactified on a circle gives the BFSS model at finite temperature [5].

We will also discuss the so-called matrix string theory of Dijkgraaf *et al* [6] which can be thought of as a matrix gauge theory in the same way that M-(atrix) theory or the BFSS model is a matrix quantum mechanics. Furthermore, in the same way that the Euclidean IKKT matrix model decompactified on a circle S^1 gives the BFSS model at finite temperature, M-(atrix) theory decompactified on a circle S^1 should give the above DVV matrix string theory [7].

These theories play a vital role in the emergence of geometry, gravity and cosmology, and in the description of gravitational instabilities such as the information loss problem and the black-hole/black-string transition. They are also featured prominently in the gauge/gravity duality which relates supersymmetric $U(N)$ gauge theories in $p+1$ dimensions and type II string theories in 10 dimensions around black p-brane solutions. These string models together with the techniques of random matrix theory, noncommutative geometry and lattice gauge theory provide a starting point for what we call 'computational string theory'.

This book is organized as follows. Part I contains a very short but detailed (standard) exposition of the essential material of string theory required for a grounded understanding of the supersymmetric matrix models presented in part II. Readers experienced with string theory may skip this part altogether.

Part II is the core of this book. It contains a comprehensive review which we believe is the first of its kind.

Chapter 6 contains a brief introduction to advanced string theory topics (T-duality and D-branes in particular) and some other related topics.

Chapter 7 deals mostly with the BFSS matrix quantum mechanics but also with the DVV matrix gauge theory and their physics. Two main applications are considered in detail: (1) the black-hole/black-string phase transition and its connection with the confinement/deconfinement phase transition [8] and (2) the black hole evaporation process and its relation to the gauge/gravity duality tests [9, 10].

Chapter 8 deals with the IKKT matrix model. The two selected applications here are the emergent matrix Yang–Mills cosmology from the Lorentzian IKKT matrix model and noncommutative/matrix emergent gravity from the Euclidean version.

In the first application, the (Lorentzian) type IIB matrix model allows us to study the real time dynamics of the emergence of (1 + 3)-dimensional Minkowski spacetime, the spontaneous symmetry breaking of $SO(9)$ down to $SO(3)$, as well as providing a mechanism for avoiding the Big Bang singularity, and allowing us to obtain the expansion (exponential at early times and power law, i.e. $\sqrt{t}$, at late times) of the Universe [11–18].

In the second application, the idea of emergent gravity from a noncommutative $U(1)$ gauge theory which was put forward first by Rivelles [19] is discussed. See also Yang [23–25]. A semi-classical derivation of Einstein equations from the IKKT model around fuzzy $\mathbf{S}^4$ background due to Steinacker is the main result discussed here [20–22].

Monte Carlo algorithms and methods used for matrix models of string theory are discussed in the appendix.

All illustrations in this review are only sketches of original Monte Carlo and numerical results. They were created by Dr Khaled Ramda and Dr Ahlam Rouag.

References

[1] Taylor W 2000 The M-(atrix) model of M theory *NATO Sci. Ser.* C **556** 91

[2] Banks T, Fischler W, Shenker S H and Susskind L 1997 M theory as a matrix model: a conjecture *Phys. Rev.* D **55** 5112

[3] Witten E 1996 Bound states of strings and p-branes *Nucl. Phys.* B **460** 335

[4] Ishibashi N, Kawai H, Kitazawa Y and Tsuchiya A 1997 A large N reduced model as superstring *Nucl. Phys.* B **498** 467

[5] Connes A, Douglas M R and Schwarz A S 1998 Noncommutative geometry and matrix theory: compactification on tori *J. High Energy Phys.* **9802** 003

[6] Dijkgraaf R, Verlinde E P and Verlinde H L 1997 Matrix string theory *Nucl. Phys.* B **500** 43

[7] Kawahara N and Nishimura J 2005 The large N reduction in matrix quantum mechanics: a bridge between BFSS and IKKT *J. High Energy Phys.* **0509** 040

[8] Kawahara N, Nishimura J and Takeuchi S 2007 Phase structure of matrix quantum mechanics at finite temperature *J. High Energy Phys.* **0710** 097

[9] Hyakutake Y 2014 Quantum near-horizon geometry of a black 0-brane *Progr. Theor. Exp. Phys.* **2014** 033B04

[10] Hanada M, Hyakutake Y, Ishiki G and Nishimura J 2014 Holographic description of quantum black hole on a computer *Science* **344** 882

[11] Kim S W, Nishimura J and Tsuchiya A 2012 Expanding (3+1)-dimensional universe from a Lorentzian matrix model for superstring theory in (9+1)-dimensions *Phys. Rev. Lett.* **108** 011601

[12] Kim S W, Nishimura J and Tsuchiya A 2012 Expanding universe as a classical solution in the Lorentzian matrix model for nonperturbative superstring theory *Phys. Rev.* D **86** 027901

[13] Nishimura J and Tsuchiya A 2013 Local field theory from the expanding universe at late times in the IIB matrix model *Progr. Theor. Exp. Phys.* **2013** 043B03

[14] Kim S W, Nishimura J and Tsuchiya A 2012 Late time behaviors of the expanding universe in the IIB matrix model *J. High Energy Phys.* **1210** 147

[15] Nishimura J and Tsuchiya A 2013 Realizing chiral fermions in the type IIB matrix model at finite N *J. High Energy Phys.* **1312** 002

[16] Ito Y, Kim S W, Koizuka Y, Nishimura J and Tsuchiya A 2014 A renormalization group method for studying the early universe in the Lorentzian IIB matrix model *Progr. Theor. Exp. Phys.* **2014** 083B01

[17] Aoki H, Nishimura J and Tsuchiya A 2014 Realizing three generations of the Standard Model fermions in the type IIB matrix model *J. High Energy Phys.* **1405** 131

[18] Ito Y, Nishimura J and Tsuchiya A 2015 Power-law expansion of the Universe from the bosonic Lorentzian type IIB matrix model *J. High Energy Phys.* **1511** 070

[19] Rivelles V O 2003 Noncommutative field theories and gravity *Phys. Lett.* B **558** 191

[20] Steinacker H 2010 Emergent geometry and gravity from matrix models: an introduction *Class. Quant. Grav* **27** 133001

[21] Steinacker H C 2016 Emergent gravity on covariant quantum spaces in the IKKT model *J. High Energy Phys.* **1612** 156

[22] Steinacker H 2007 Emergent gravity from noncommutative gauge theory *J. High Energy Phys.* **0712** 049

[23] Yang H S 2009 Emergent spacetime and the origin of gravity *J. High Energy Phys.* **0905** 012

[24] Yang H S 2009 Emergent gravity from noncommutative spacetime *Int. J. Mod. Phys.* A **24** 4473

[25] Yang H S and Sivakumar M 2010 Emergent gravity from quantized spacetime *Phys. Rev.* D **82** 045004

Part I

String theory

Chapter 2

Systematics of bosonic strings

The bosonic string is a standard topic in string theory which is discussed in great detail in every textbook on string theory. In this chapter we will mainly follow the first, and still the best, textbook [1].

2.1 Actions, symmetries and solutions

2.1.1 Polyakov action

We will consider a point particle and a string propagating in a D-dimensional curved spacetime, with Minkowski signature $(-, +, \ldots, +)$. We will employ the natural units $\hbar = c = 1$. The point particle sweeps out a 1-dimensional worldline, whereas the string sweeps out a 2-dimensional worldsheet, both in spacetime. The worldline of the point particle will be parametrized by a real number τ, while the worldsheet of the string will be parametrized by two real numbers τ and σ. Obviously, τ is timelike and σ is spacelike. The coordinates of the point particle and the string are given by

$$X^\mu = X^\mu(\tau), \quad \text{point particle}, \tag{2.1}$$

$$X^\mu = X^\mu(\tau, \sigma), \quad \text{string}. \tag{2.2}$$

We will define

$$\dot{X}^\mu = \frac{\partial X^\mu}{\partial \tau}, \quad X^{\mu'} = \frac{\partial X^\mu}{\partial \sigma}. \tag{2.3}$$

The action principle of the point particle is proportional to the invariant length of the worldline, viz

$$S = -m \int ds = -m \int d\tau \sqrt{-g_{\mu\nu}(X)\dot{X}^\mu \dot{X}^\nu} = -m \int d\tau \sqrt{-\dot{X}^2}. \tag{2.4}$$

doi:10.1088/978-0-7503-1726-9ch2

Similarly, the action principle of the string is proportional to the invariant area of the worldsheet, viz

$$S = -T\int d\mu_1 = -T\int d\tau d\sigma\sqrt{-\det G_{\alpha\beta}}. \tag{2.5}$$

The induced metric $G_{\alpha\beta}$, i.e. the metric on the worldsheet, is given by

$$G_{\alpha\beta} = g_{\mu\nu}(X)\partial_\alpha X^\mu \partial_\beta X^\nu. \tag{2.6}$$

Clearly, we have α, $\beta = 0{,}1$ and $\sigma^0 = \tau$ and $\sigma^1 = \sigma$. We compute

$$S = -T\int d\tau d\sigma\sqrt{(\dot{X}\cdot X')^2 - \dot{X}^2\cdot X'^2}. \tag{2.7}$$

The parameter T is the string tension and it is related to the so-called Regge slope by

$$T = \frac{1}{2\pi\alpha'}. \tag{2.8}$$

The so-called Nambu–Goto action (2.5) or (2.7) is quite difficult to quantize due to the square root. The point particle action (2.4) suffers from the same problem. This problem can be solved, for the point particle, by introducing an auxiliary coordinate $e(\tau)$ on the worldline, which is an einbein for the 1-dimensional geometry of the worldline, and re-expressing the action (2.4) in the equivalent form

$$S = \frac{1}{2}T\int d\tau(e^{-1}\dot{X}^2 - em^2). \tag{2.9}$$

Similarly, an equivalent action to (2.7), which does not involve the square root, and as a consequence, is easy to quantize, is the string sigma model action. We introduce the metric $h_{\alpha\beta}$ on the worldsheet, which is an auxiliary field, and re-express the action (2.7) in the equivalent form

$$S = -\frac{1}{2}T\int d^2\sigma\sqrt{-h}\,h^{\alpha\beta}g_{\mu\nu}(X)\partial_\alpha X^\mu\partial_\beta X^\nu. \tag{2.10}$$

This action is also called the Polyakov action, and it was discovered originally, by Brink, Di Vecchia and Howe, and by Deser and Zumino. In the above equation, we have used the notation

$$h = \det\ h_{\alpha\beta},\quad h^{\alpha\beta} = (h^{-1})_{\alpha\beta}. \tag{2.11}$$

Since there is no kinetic term for $h_{\alpha\beta}$, its equation of motion implies the vanishing of the worldsheet energy–momentum tensor $T_{\alpha\beta}$, that is

$$T_{\alpha\beta} = -\frac{2}{T}\frac{1}{\sqrt{-h}}\frac{\delta S}{\delta h^{\alpha\beta}} = 0. \tag{2.12}$$

We use the identities

$$\delta h = hh^{\alpha\beta}\delta h_{\alpha\beta} = -hh_{\alpha\beta}\delta h^{\alpha\beta}, \tag{2.13}$$

$$\delta\sqrt{-h} = -\frac{1}{2}\sqrt{-h}\,h_{\alpha\beta}\delta h^{\alpha\beta}. \tag{2.14}$$

We compute the equation of motion for $h_{\alpha\beta}$:

$$\begin{aligned}\delta S &= -\frac{1}{2}T\int d^2\sigma\sqrt{-h}\,\delta h^{\alpha\beta}\Big(g_{\mu\nu}\partial_\alpha X^\mu\partial_\beta X^\nu - \frac{1}{2}h_{\alpha\beta}h^{\gamma\delta}g_{\mu\nu}\partial_\gamma X^\mu\partial_\delta X^\nu\Big) = 0\\ &\Rightarrow T_{\alpha\beta} = g_{\mu\nu}\partial_\alpha X^\mu\partial_\beta X^\nu - \frac{1}{2}h_{\alpha\beta}h^{\gamma\delta}g_{\mu\nu}\partial_\gamma X^\mu\partial_\delta X^\nu = 0.\end{aligned} \tag{2.15}$$

Equivalently,

$$g_{\mu\nu}\partial_\alpha X^\mu\partial_\beta X^\nu = \frac{1}{2}h_{\alpha\beta}h^{\gamma\delta}g_{\mu\nu}\partial_\gamma X^\mu\partial_\delta X^\nu. \tag{2.16}$$

By taking the determinant of both sides of this equation we get:

$$\sqrt{-\det\ g_{\mu\nu}\partial_\alpha X^\mu\partial_\beta X^\nu} = \frac{1}{2}\sqrt{-h}\,h^{\gamma\delta}g_{\mu\nu}\partial_\gamma X^\mu\partial_\delta X^\nu. \tag{2.17}$$

By substitution into the Polyakov action we obtain the Nambu–Goto action.

2.1.2 Boundary conditions and symmetries

The equations of motion for X^μ in the case of a flat spacetime are:

$$\delta S = T\int d^2\sigma\partial_\alpha\Big(\sqrt{-h}\,h^{\alpha\beta}\partial_\beta X_\mu\Big)\delta X^\mu - T\int d^2\sigma\partial_\alpha\Big(\sqrt{-h}\,h^{\alpha\beta}\partial_\beta X^\mu\delta X_\mu\Big). \tag{2.18}$$

In the bulk the equations of motion are

$$\partial_\alpha\Big(\sqrt{-h}\,h^{\alpha\beta}\partial_\beta X_\mu\Big) = \sqrt{-h}\,\nabla^2 X_\mu = 0. \tag{2.19}$$

For worldsheets with boundary there is also a surface term:

$$\delta S = -T\int d\tau\sqrt{-h}\,\partial^\sigma X^\mu\delta X_\mu|_{\sigma=0}^{\sigma=l}. \tag{2.20}$$

This vanishes if:

- $$\partial^\sigma X^\mu(\tau, 0) = \partial^\sigma X^\mu(\tau, l) = 0. \tag{2.21}$$

  These are Neumann boundary conditions, i.e. the ends of the open string move freely in spacetime, and the component of the momentum normal to the boundary of the worldsheet vanishes.

- $$X^\mu(\tau, 0) = X^\mu(\tau, l), \quad h_{\alpha\beta}(\tau, 0) = h_{\alpha\beta}(\tau, l). \tag{2.22}$$

  The fields are periodic which corresponds to a closed string.

These are the only boundary conditions which are consistent with D-dimensional Poincaré invariance.

The Polyakov action (2.10) defines a 2-dimensional field theory on the string worldsheet. The worldsheets for open strings are rectangular surfaces while for closed strings they are cylinders. Thus, the string amplitudes for spacetime processes are given by matrix elements in the 2-dimensional quantum field theory.

The Polyakov action (2.10) has the following symmetries:

- D-dimensional Poincaré invariance:

$$X'^{\mu} = \Lambda^{\mu}_{\nu} X^{\nu} + a^{\mu}, \quad h'_{\alpha\beta} = h_{\alpha\beta}. \tag{2.23}$$

  These are global transformations. In other words, from the point of view of the worldsheet, Poincaré transformations are internal symmetries.

- Reparametrization or diffeomorphism invariance which are given by the 2-dimensional coordinate transformations:

$$\begin{aligned} \sigma^{\alpha} &\longrightarrow \sigma^{\alpha'} = f^{\alpha}(\sigma) \\ X^{\mu}(\tau, \sigma) &\longrightarrow X'^{\mu}(\tau', \sigma') = X^{\mu}(\sigma, \tau) \\ h^{\alpha\beta}(\tau, \sigma) &\longrightarrow h'^{\alpha\beta}(\tau', \sigma') \ : \ h_{\alpha\beta}(\tau, \sigma) = \frac{\partial f^{\gamma}}{\partial \sigma^{\alpha}} \frac{\partial f^{\delta}}{\partial \sigma^{\beta}} h'_{\gamma\delta}(\tau', \sigma'). \end{aligned} \tag{2.24}$$

  These are local transformations. These coordinate transformations and their inverses are infinitely differentiable, and hence, they are diffeomorphisms. As a consequence of diffeomorphism invariance, the energy–momentum tensor is conserved:

$$\nabla_{\alpha} T^{\alpha\beta} = 0. \tag{2.25}$$

  The corresponding infinitesimal transformations are:

$$\sigma \longrightarrow \sigma' = \sigma + \xi \tag{2.26}$$

$$\delta X^{\mu} = X'^{\mu}(\sigma) - X^{\mu}(\sigma) = -\xi^{\alpha} \partial_{\alpha} X^{\mu} \tag{2.27}$$

$$\delta h_{\alpha\beta} = h'_{\alpha\beta}(\sigma) - h_{\alpha\beta}(\sigma) = -\partial_{\beta} \xi^{\delta} \cdot h_{\alpha\delta} - \partial_{\alpha} \xi^{\delta} \cdot h_{\delta\beta} - \xi^{\delta} \partial_{\delta} h_{\alpha\beta} \tag{2.28}$$

$$\delta h^{\rho\gamma} = h'^{\rho\gamma}(\sigma) - h^{\rho\gamma}(\sigma) = \partial_{\beta} \xi^{\rho} \cdot h^{\beta\gamma} + \partial_{\alpha} \xi^{\gamma} \cdot h^{\rho\alpha} - \xi^{\delta} \partial_{\delta} h^{\rho\gamma} \tag{2.29}$$

$$\delta\sqrt{h} = -\partial_{\alpha}\left(\xi^{\alpha}\sqrt{h}\right). \tag{2.30}$$

- 2-dimensional Weyl transformations:

$$\begin{aligned} X^{\mu}(\tau, \sigma) &\longrightarrow X'^{\mu}(\tau, \sigma) = X^{\mu}(\sigma, \tau) \\ h_{\alpha\beta}(\tau, \sigma) &\longrightarrow h'_{\alpha\beta}(\tau, \sigma) = \exp(2\omega(\tau, \sigma)) h_{\alpha\beta}(\tau, \sigma). \end{aligned} \tag{2.31}$$

  These are also local transformations. As a consequence of this invariance, the energy–momentum tensor is traceless:

$$T^{\alpha}_{\alpha} = 0. \tag{2.32}$$

The infinitesimal Weyl scaling is given by

$$\delta h_{\alpha\beta} = 2\omega h_{\alpha\beta}. \tag{2.33}$$

We have therefore three local symmetries, two reparametrizations and one Weyl, which can be used to choose three independent elements of $h_{\alpha\beta}$ so that

$$h_{\alpha\beta} = \eta_{\alpha\beta} = \mathrm{diag}(-1, +1). \tag{2.34}$$

The Polyakov action becomes

$$S = -\frac{1}{2}T\int d^2\sigma g_{\mu\nu}(X)\partial_\alpha X^\mu \partial^\alpha X^\nu. \tag{2.35}$$

2.1.3 Closed string solutions

In flat spacetime, i.e. $g_{\mu\nu}(X) = \eta_{\mu\nu}$, the gauge-fixed Polyakov action becomes

$$S = \frac{1}{2}T\int d^2\sigma\left(\dot{X}^\mu \dot{X}_\mu - X'^\mu X'_\mu\right). \tag{2.36}$$

The equations of motion are:

$$\Box X^\mu = \left(\frac{\partial^2}{\partial\sigma^2} - \frac{\partial^2}{\partial\tau^2}\right)X^\mu = 0. \tag{2.37}$$

The general solution is

$$X^\mu = X_R^\mu(\sigma^-) + X_L^\mu(\sigma^+). \tag{2.38}$$

The function X_R^μ describes right-moving modes of the string, while the function X_L^μ describes left-moving modes.

The worldsheet lightcone coordinates are:

$$\sigma^\mp = \tau \mp \sigma. \tag{2.39}$$

The derivatives conjugate to $\sigma^\pm$ are defined by

$$\partial_\pm = \frac{1}{2}(\partial_\tau \mp \partial_\sigma). \tag{2.40}$$

The metric in worldsheet coordinates is

$$ds^2 = -h_{\alpha\beta}d\sigma^\alpha d\sigma^\beta = d\tau^2 - d\sigma^2 = d\sigma^- d\sigma^+. \tag{2.41}$$

Thus

$$\eta_{+-} = \eta_{-+} = -\frac{1}{2}, \quad \eta^{+-} = \eta^{-+} = -2. \tag{2.42}$$

The rest are zero. Hence, worldsheet indices are raised and lowered by the rule

$$U^+ = -2U_-, \quad U^- = -2U_+. \tag{2.43}$$

Since the metric on the worldsheet has been gauged fixed, the vanishing of the energy–momentum tensor, originating from the equations of motion of the worldsheet metric, must be imposed as a constraint, viz

$$T_{\alpha\beta} = g_{\mu\nu}\partial_\alpha X^\mu \partial_\beta X^\nu - \frac{1}{2}h_{\alpha\beta}h^{\gamma\delta}g_{\mu\nu}\partial_\gamma X^\mu \partial_\delta X^\nu = 0. \tag{2.44}$$

Equivalently,

$$\begin{aligned} T_{01} = T_{10} &= \dot{X}^\mu X'_\mu = \partial_+ X^\mu \partial_+ X_\mu - \partial_- X^\mu \partial_- X_\mu = 0 \\ T_{00} = T_{11} &= \frac{1}{2}\left(\dot{X}^\mu \dot{X}_\mu + X'^\mu X'_\mu\right) = \partial_+ X^\mu \partial_+ X_\mu - \partial_- X^\mu \partial_- X_\mu = 0. \end{aligned} \tag{2.45}$$

We use the identity

$$T_{\alpha\beta}d\sigma^\alpha d\sigma^\beta = \frac{1}{2}(T_{00} + T_{01})(d\sigma^+)^2 + \frac{1}{2}(T_{00} - T_{01})(d\sigma^-)^2. \tag{2.46}$$

Thus in lightcone coordinates we have:

$$\begin{aligned} T_{++} &= \partial_+ X^\mu \partial_+ X_\mu = \partial_+ X_L^\mu \partial_+ X_{L\mu} = 0 \\ T_{--} &= \partial_- X^\mu \partial_- X_\mu = \partial_- X_R^\mu \partial_- X_{R\mu} = 0. \end{aligned} \tag{2.47}$$

The tracelessness of the energy–momentum tensor $h^{\alpha\beta}T_{\alpha\beta} = 0$ becomes the statement that $T_{11} = T_{00}$ or equivalently $T_{+-} = T_{-+} = 0$.

We have, therefore, the constraints

$$\begin{aligned} T_{++} &= \partial_+ X_L^\mu \partial_+ X_{L\mu} = 0 \\ T_{--} &= \partial_- X_R^\mu \partial_- X_{R\mu} = 0. \end{aligned} \tag{2.48}$$

The closed string must satisfy the boundary condition

$$X^\mu(\tau, \sigma + \pi) = X^\mu(\tau, \sigma), \quad \partial^\sigma X^\mu(\tau, \sigma + \pi) = \partial^\sigma X^\mu(\tau, \sigma). \tag{2.49}$$

The most general solution is

$$X_R^\mu = \frac{1}{2}x^\mu + \frac{1}{2}l_s^2 p^\mu(\tau - \sigma) + \frac{i}{2}l_s \sum_{n\neq 0} \frac{1}{n}\alpha_n^\mu \exp(-2in(\tau - \sigma)) \tag{2.50}$$

$$X_L^\mu = \frac{1}{2}x^\mu + \frac{1}{2}l_s^2 p^\mu(\tau + \sigma) + \frac{i}{2}l_s \sum_{n\neq 0} \frac{1}{n}\tilde{\alpha}_n^\mu \exp(-2in(\tau + \sigma)). \tag{2.51}$$

The α_n^μ and $\alpha_{-n}^\mu = (\alpha_n^\mu)^*$ are oscillator coordinates, while x^μ and p^μ are the position and momentum of the string center of mass. The parameter l_s is the string length scale defined by

$$T = \frac{1}{2\pi\alpha'}, \quad \frac{1}{2}l_s^2 = \alpha', \quad \pi l_s^2 T = 1. \tag{2.52}$$

Thus

$$X^\mu = x^\mu + l_s^2 p^\mu \tau + \frac{i}{2} l_s \sum_{n\neq 0} \frac{1}{n} e^{-2in\tau}(\alpha_n^\mu e^{2in\sigma} + \tilde{\alpha}_n^\mu e^{-2in\sigma}). \tag{2.53}$$

We remark that the linear term in σ cancels, and thus, the boundary condition is clearly obeyed. From the action (2.36), the momentum conjugate associated to X_μ is

$$P_\mu = \frac{\partial L}{\partial \dot{X}_\mu} = T\dot{X}_\mu. \tag{2.54}$$

The classical Poisson brackets are therefore:

$$\begin{aligned} &[\dot{X}^\mu(\tau, \sigma), \dot{X}^\nu(\tau, \sigma')]_{P.B} = 0 \\ &[X^\mu(\tau, \sigma), X^\nu(\tau, \sigma')]_{P.B} = 0 \\ &[\dot{X}^\mu(\tau, \sigma), X^\nu(\tau, \sigma')]_{P.B} = \frac{1}{T}\eta^{\mu\nu}\delta(\sigma - \sigma'). \end{aligned} \tag{2.55}$$

A straightforward calculation shows that

$$[\alpha_n^\mu, x^\nu]_{P.B} = [\alpha_n^\mu, p^\nu]_{P.B} = [\tilde{\alpha}_n^\mu, x^\nu]_{P.B} = [\tilde{\alpha}_n^\mu, p^\nu]_{P.B} = 0 \tag{2.56}$$

$$[\alpha_n^\mu, \alpha_{n'}^\nu]_{P.B} = [\tilde{\alpha}_n^\mu, \tilde{\alpha}_{n'}^\nu]_{P.B} = in\delta_{n+n',0}\eta^{\mu\nu}, \quad [\alpha_n^\mu, \tilde{\alpha}_{n'}^\nu]_{P.B} = 0 \tag{2.57}$$

$$[p^\mu, x^\nu]_{P.B} = \eta^{\mu\nu}. \tag{2.58}$$

In the above calculation we use the Dirac delta function

$$\delta(\sigma - \sigma') = \frac{1}{\pi} \sum_{n=-\infty}^{+\infty} e^{2in(\sigma-\sigma')}. \tag{2.59}$$

This confirms that α_n^μ and $\tilde{\alpha}_n^\mu$ for $n \neq 0$ are two sets of independent harmonic oscillator coordinates. Equation (2.57) remains valid if we adopt the convention

$$\alpha_0^\mu = \tilde{\alpha}_0^\mu = \frac{1}{2} l_s p^\mu. \tag{2.60}$$

Further, we compute

$$\partial_- X_R^\mu = \dot{X}_R^\mu = l_s \sum_n \alpha_n^\mu e^{-2in(\tau-\sigma)} \tag{2.61}$$

$$\partial_+ X_L^\mu = \dot{X}_L^\mu = l_s \sum_n \tilde{\alpha}_n^\mu e^{-2in(\tau+\sigma)}. \tag{2.62}$$

Thus

$$T_{--} = \partial_- X^\mu_R \partial_-(X_\mu)_R \\ = 2l_s^2 \sum_m L_m e^{-2im(\tau-\sigma)} \tag{2.63}$$

$$T_{++} = \partial_+ X^\mu_L \partial_+(X_\mu)_L \\ = 2l_s^2 \sum_m \tilde{L}_m e^{-2im(\tau+\sigma)}. \tag{2.64}$$

The coefficients L_m and $\tilde{L}_m$ are called Virasoro generators, and they are given by

$$L_m = \frac{1}{2} \sum_n \alpha^\mu_{m-n} (\alpha_n)_\mu \tag{2.65}$$

$$\tilde{L}_m = \frac{1}{2} \sum_n \tilde{\alpha}^\mu_{m-n} (\tilde{\alpha}_n)_\mu. \tag{2.66}$$

For completeness, we compute the Hamiltonian

$$H = \int_0^\pi (\dot{X}_\mu P^\mu - \mathcal{L}) d\sigma \\ = \frac{T}{2} \int_0^\pi (\dot{X}_\mu \dot{X}^\mu + X'^\mu X'_\mu) d\sigma \\ = \sum_n (\alpha^\mu_n (\alpha_{-n})_\mu + \tilde{\alpha}^\mu_n (\tilde{\alpha}_{-n})_\mu). \tag{2.67}$$

We remark that

$$\frac{1}{2} H = L_0 + \tilde{L}_0. \tag{2.68}$$

The constraints become

$$L_m = 0, \quad \tilde{L}_m = 0, \quad 0, \pm 1, \pm 2, \ldots \tag{2.69}$$

Furthermore, the total momentum of the string is:

$$\mathcal{P}^\mu = T \int_0^\pi d\sigma \dot{X}^\mu \\ = T l_s \pi (\alpha^\mu_0 + \tilde{\alpha}^\mu_0) \\ = p^\mu. \tag{2.70}$$

Hence, the mass-shell condition is

$$M^2 = - \mathcal{P}^\mu \mathcal{P}_\mu \\ = - p^\mu p_\mu \\ = - \frac{2}{l_s^2} (\alpha^\mu_0 (\alpha_0)_\mu + \tilde{\alpha}^\mu_0 (\tilde{\alpha}_0) \mu). \tag{2.71}$$

But we have the constraints

$$L_0 = \sum_{n=1} \alpha^{\mu}_{-n}(\alpha_n)_{\mu} + \frac{1}{2}(\alpha^{\mu}_0)(\alpha_0)_{\mu} = 0 \tag{2.72}$$

$$\tilde{L}_0 = \sum_{n=1} \tilde{\alpha}^{\mu}_{-n}(\tilde{\alpha}_n)_{\mu} + \frac{1}{2}(\tilde{\alpha}^{\mu}_0)(\tilde{\alpha}_0)_{\mu} = 0. \tag{2.73}$$

Thus

$$M^2 = \frac{2}{\alpha'}\sum_{n=1}^{\infty}\left(\alpha^{\mu}_{-n}(\alpha_n)_{\mu} + \tilde{\alpha}^{\mu}_{-n}(\tilde{\alpha}_n)_{\mu}\right). \tag{2.74}$$

2.1.4 Open string solutions

The boundary condition in this case is

$$X'^{\mu}|_{\sigma=0} = X'^{\mu}|_{\sigma=\pi} = 0. \tag{2.75}$$

In other words, the normal derivative of X^{μ} must vanish at the string boundary. These are free boundary conditions which prevent momentum from flowing off the ends of the string. The general solution in this case is given by

$$X^{\mu} = x^{\mu} + l_s^2 p^{\mu}\tau + il_s \sum_{n\neq 0} \frac{1}{n}\alpha^{\mu}_n \exp(-in\tau)\cos n\sigma. \tag{2.76}$$

In other words, the left-moving and right-moving components have combined into standing waves.

In this case we compute

$$2\partial_- X^{\mu} = \dot{X}^{\mu} - X'^{\mu} = l_s \sum_n \alpha^{\mu}_n e^{-in(\tau-\sigma)} \tag{2.77}$$

$$2\partial_+ X^{\mu} = \dot{X}^{\mu} + X'^{\mu} = l_s \sum_n \alpha^{\mu}_n e^{-in(\tau+\sigma)}. \tag{2.78}$$

In the case of open strings we have the definition

$$\alpha^{\mu}_0 = l_s p^{\mu}. \tag{2.79}$$

We compute now

$$\begin{aligned} T_{--} &= \partial_- X^{\mu}_R \partial_-(X_{\mu})_R \\ &= \frac{1}{2} l_s^2 \sum_m L_m e^{-im(\tau-\sigma)} \end{aligned} \tag{2.80}$$

$$T_{++} = \partial_+ X_L^\mu \partial_+ (X_\mu)_L$$
$$= \frac{1}{2} l_s^2 \sum_m L_m e^{-im(\tau+\sigma)}. \tag{2.81}$$

The coefficients L_m are called Virasoro generators, and they are given by

$$L_m = \frac{1}{2} \sum_n \alpha_{m-n}^\mu (\alpha_n)_\mu. \tag{2.82}$$

Also, we compute the Hamiltonian

$$H = \int_0^\pi (\dot{X}_\mu P^\mu - \mathcal{L}) d\sigma$$
$$= \frac{T}{2} \int_0^\pi (\dot{X}_\mu \dot{X}^\mu + X'^\mu X'_\mu) d\sigma \tag{2.83}$$
$$= \frac{1}{2} \sum_n \alpha_n^\mu (\alpha_{-n})_\mu.$$

We remark that

$$H = L_0. \tag{2.84}$$

The constraints become

$$L_m = 0, \quad m = 0, \pm 1, \pm 2, \ldots \tag{2.85}$$

The mass relation for the open string is

$$M^2 = \frac{1}{\alpha'} \sum_{n=1}^\infty \alpha_{-n}^\mu (\alpha_n)_\mu. \tag{2.86}$$

2.2 Canonical quantization and Virasoro algebra

2.2.1 Canonical quantization

The worldsheet theory can be quantized by replacing Poisson brackets by commutators:

$$[,]_{P.B} \longrightarrow i[,]. \tag{2.87}$$

We obtain the fundamental commutation relations

$$[P^\mu(\tau, \sigma), P^\nu(\tau, \sigma')] = 0$$
$$[X^\mu(\tau, \sigma), X^\nu(\tau, \sigma')] = 0 \tag{2.88}$$
$$[P^\mu(\tau, \sigma), X^\nu(\tau, \sigma')] = -i\eta^{\mu\nu}\delta(\sigma - \sigma').$$

Equivalently, we obtain from (2.56), (2.57) and (2.58), the commutation relations

$$[\alpha_n^\mu, x^\nu] = [\alpha_n^\mu, p^\nu] = [\tilde{\alpha}_n^\mu, x^\nu] = [\tilde{\alpha}_n^\mu, p^\nu] = 0 \tag{2.89}$$

$$[\alpha_n^\mu, \alpha_{n'}^\nu] = [\tilde{\alpha}_n^\mu, \tilde{\alpha}_{n'}^\nu] = n\delta_{n+n',0}\eta^{\mu\nu}, \quad [\alpha_n^\mu, \quad \tilde{\alpha}_{n'}^\nu] = 0 \tag{2.90}$$

$$[p^\mu, x^\nu] = -i\eta^{\mu\nu}. \tag{2.91}$$

As before

$$\alpha_0^\mu = \tilde{\alpha}_0^\mu = \frac{1}{2} l_s p^\mu \text{ (closed)}, \quad \alpha_0^\mu = l_s p^\mu \text{ (open)}. \tag{2.92}$$

Define

$$a_m^\mu = \frac{1}{\sqrt{m}}\alpha_m^\mu, \quad a_m^{\mu+} = \frac{1}{\sqrt{m}}\alpha_{-m}^\mu, \quad m > 0. \tag{2.93}$$

Then we have

$$[a_m^\mu, a_n^{\nu+}] = [\tilde{a}_m^\mu, \tilde{a}_n^{\nu+}] = \eta^{\mu\nu}\delta_{m,n}, \quad m, n > 0. \tag{2.94}$$

This is the algebra of raising and lowering operators for quantum mechanical harmonic oscillators. The ground state $|0\rangle$ is defined by the condition

$$a_m^\mu|0\rangle = 0, \quad m > 0. \tag{2.95}$$

As it turns out, this does not determine the state of the string completely, since the oscillators can be in their ground state, but the string center of mass can still have a non-zero momentum k. A state with this property will be denoted by $|0, k\rangle$. The momentum carried by the state $|\phi\rangle$ defined by

$$|\phi\rangle = a_{m_1}^{\mu_1+}a_{m_2}^{\mu_2+}\ldots a_{m_n}^{\mu_n+}|0, k\rangle \tag{2.96}$$

is k since

$$p^\mu|\phi\rangle = k^\mu|\phi\rangle. \tag{2.97}$$

The commutators of the timelike operators a_0^m have a negative sign, i.e. $[a_m^0, a_m^{0+}] = -1$, and therefore a state of the form $a_m^{0+}|0\rangle$ has negative norm. Thus, the Fock space is not positive definite, since there exists negative norm states, which we call ghosts. These states would lead to violations of causality and unitarity. The physical space of allowed string states is a subspace of this Fock space. The states with an even number of timelike operators a_0^{m+} have positive norm, while those constructed with an odd number of timelike operators have negative norm.

In the quantum theory, there is a possible ambiguity in the ordering of the operators α_m and $\tilde{\alpha}_m$ in the equations defining L_m and $\tilde{L}_m$ given by equations (2.65) and (2.66) respectively. Since α_{m-n} commutes with α_n, unless $m = 0$, the only such ambiguity arises in the expression for L_0. Thus the normal ordered expression of L_0 given by

$$L_0 = \frac{1}{2}\alpha_0^\mu\alpha_{0\mu} + \sum_{n=1}^{\infty}\alpha_{-n}^\mu\alpha_{n\mu} \tag{2.98}$$

used in the classical calculation is only correct up to an arbitrary additive constant. The classical constraint $L_0 = 0$ must then be replaced by the condition

$$(L_0 - a)|\phi\rangle = 0. \tag{2.99}$$

where a is some constant which we will determine shortly.

For the closed string, we must impose

$$(L_0 - a)|\phi\rangle = (\tilde{L}_0 - a)|\phi\rangle = 0. \tag{2.100}$$

Obviously,

$$(L_0 - \tilde{L}_0)|\phi\rangle = 0. \tag{2.101}$$

This equation is called level-matching condition of the bosonic string. This is the only constraint that relates left- and right-moving modes.

The mass squared of the closed string becomes

$$\begin{aligned} M^2 &= -\frac{2}{\alpha'}\cdot\frac{1}{2}\left(\alpha_0^{\mu}\alpha_{0\mu} + \tilde{\alpha}_0^{\mu}\tilde{\alpha}_{0\mu}\right) \\ &= -\frac{2}{\alpha'}\cdot\left(-\sum_{n=1}\alpha_{-n}^{\mu}\alpha_{n\mu} - \sum_{n=1}\tilde{\alpha}_{-n}^{\mu}\tilde{\alpha}_{n\mu} + 2a\right). \end{aligned} \tag{2.102}$$

Equivalently,

$$\begin{aligned} \frac{\alpha'}{4}M^2 &= \sum_{n=1}\alpha_{-n}^{\mu}(\alpha_n)_{\mu} - a = N - a \\ &= \sum_{n=1}\tilde{\alpha}_{-n}^{\mu}\tilde{\alpha}_{n\mu} - a = \tilde{N} - a. \end{aligned} \tag{2.103}$$

The number operators are defined by

$$N = \sum_{n=1} n a_n^{\mu+}a_{n\mu}, \quad \tilde{N} = \sum_{n=1} n\tilde{a}_n^{\mu+}\tilde{a}_{n\mu}. \tag{2.104}$$

The ground state has $N = \tilde{N} = 0$ which gives

$$\frac{\alpha'}{4}M^2 = -a. \tag{2.105}$$

For the excited states we obtain

$$\frac{\alpha'}{4}M^2 = 1 - a,\ 2 - a,\ \dots. \tag{2.106}$$

A similar calculation gives for the open string the mass relation

$$\alpha' M^2 = \sum_{n=1}\alpha_{-n}^{\mu}(\alpha_n)_{\mu} - a = N - a. \tag{2.107}$$

The ground state has

$$\alpha' M^2 = -a. \tag{2.108}$$

Thus, the mass squared of the ground state for closed strings is four times that for open strings.

Besides the mass-shell conditions, we need also to impose the constraints $L_m = 0$. In the quantum theory, these constraints are replaced by the weaker requirements that the positive frequency Virasoro generators annihilate physical states, viz

$$L_m|\phi\rangle = 0, \quad m = 1, 2, \ldots \tag{2.109}$$

As it turns out, these conditions are sufficient to ensure that L_m, for both positive and negative m, have vanishing matrix elements between pairs of physical states, since $L_{-m} = L_m^+$, and thus

$$\langle\phi|L_m = 0, \quad m = -1, -2, \ldots \tag{2.110}$$

In summary, the physical subspace corresponds to the states satisfying the Virasoro conditions

$$(L_m - a\delta_{m,0})|\phi\rangle = 0, \quad m \geqslant 0. \tag{2.111}$$

These conditions are obviously, in one-to-one correspondence, with timelike oscillators, and as a consequence, their number is sufficient to give a positive definite subspace, i.e. to decouple the ghosts. However, a ghost-free spectrum is only possible for certain values of the constant a and the spacetime dimension D, as we will now show.

2.2.2 Virasoro algebra

We start by computing the classical Poisson bracket of two Virasoro generators, viz $[L_m, L_n]_{P.B}$. This Poisson bracket involves four equal terms, we find

$$[L_m, L_n]_{P.B} = i\sum_l (m - l)\alpha^\mu_{m+n-l}\alpha_{l\mu}. \tag{2.112}$$

Equivalently,

$$[L_m, L_n]_{P.B} = -i\sum_l (n - l)\alpha^\mu_{m+n-l}\alpha_{l\mu}. \tag{2.113}$$

Thus

$$[L_m, L_n]_{P.B} = \frac{i}{2}\sum_l (m - n)\alpha^\mu_{m+n-l}\alpha_{l\mu} = i(m - n)L_{m+n}. \tag{2.114}$$

This is called the Virasoro algebra.

This algebra is due to the fact that our gauge choice $h^{\alpha\beta} = \eta^{\alpha\beta}$ did not fully fix the reparametrization invariance. Indeed, any combined reparametrization ξ^α and Weyl scaling Λ for which

$$\partial^\alpha \xi^\beta + \partial^\beta \xi^\alpha = \Lambda \eta^{\alpha\beta} \tag{2.115}$$

preserves the gauge choice. This can be seen as follows. Recall that

$$\delta h_{\alpha\beta} = h'_{\alpha\beta} - h_{\alpha\beta} = -\partial_\beta \xi^\delta h_{\alpha\delta} - \partial_\alpha \xi^\delta \cdot h_{\delta\beta} - \xi^\delta \partial_\delta h_{\alpha\beta}. \tag{2.116}$$

If $h = \eta$, then

$$\delta h_{\alpha\beta} = h'_{\alpha\beta} - \eta_{\alpha\beta} = -\partial_\beta \xi_\alpha - \partial_\alpha \xi_\beta. \tag{2.117}$$

This can be undone by the Weyl scaling

$$\delta h_{\alpha\beta} = 2\omega h_{\alpha\beta} = 2\omega \eta_{\alpha\beta}, \quad 2\omega = \Lambda. \tag{2.118}$$

The condition (2.115) is solved by

$$\xi^\pm = \xi^0 \pm \xi^1, \quad \xi^+ = \xi^+(\sigma^+), \quad \xi^- = \xi^-(\sigma^-). \tag{2.119}$$

By analogy with the generators of the worldsheet reparametrization $\delta\sigma^\alpha = \xi^\alpha$, which are given by $V = \xi^\alpha \partial/\partial\sigma^\alpha$, the generators of the above residual gauge symmetries are given by

$$V^\pm = \frac{1}{2}\xi^\pm(\sigma^\pm)\frac{\partial}{\partial\sigma^\pm}. \tag{2.120}$$

These are the generators of the group of conformal transformations in 2-dimensional Minkowski spacetime. This is clearly an infinite dimensional group. Indeed, we can easily see that there exists an infinite number of conserved quantities given by

$$Q_f = \int d\sigma f(\sigma^+) T_{++}, \tag{2.121}$$

for any function f of σ^+. These charges are conserved, i.e. $dQ_f/d\tau = 0$, since $\partial_- T_{++} = 0$.

By replacing the Poisson bracket by the commutator $i[.., ..]$, the Virasoro algebra becomes

$$[L_m, L_n] = (m - n)L_{m+n}. \tag{2.122}$$

However, this cannot be totally correct since there exists a quantum correction. Indeed, we have obtained $[L_m, L_n]_{P.B}$ by steps that are mostly, but not all, valid at the quantum level. Thus, we redo the calculation as follows

$$\begin{aligned} [L_m, L_n]_{P.B} &= \frac{i}{4}\sum_{m'}\sum_{n'}\big[(m - m')\alpha_{m'\mu}\alpha^\mu_{n'}\delta_{m-m'+n-n'} + m'\alpha^\mu_{m-m'}\alpha_{n-n'\mu}\delta_{m'+n'} \\ &\quad + (m - m')\alpha_{m'\mu}\alpha^\mu_{n-n'}\delta_{m-m'+n'} + m'\alpha^\mu_{m-m'}\alpha_{n'\mu}\delta_{m'+n-n'}\big] \\ &= \frac{i}{2}\sum_{m'}\sum_{n'}\big[(m - m')\alpha_{m'\mu}\alpha^\mu_{n'}\delta_{m-m'+n-n'} + m'\alpha^\mu_{m-m'}\alpha_{n-n'\mu}\delta_{m'+n'}\big] \\ &= \frac{i}{2}\sum_{m'}\big[(m - m')\alpha_{m'\mu}\alpha^\mu_{m-m'+n} + m'\alpha^\mu_{m-m'}\alpha_{n+m'\mu}\big]. \end{aligned} \tag{2.123}$$

The above three steps are all valid at the quantum level. Next, we need to shift the variable $m' \longrightarrow m' + n$ in the second sum. This step is only valid as long as $m + n \neq 0$. For $m + n = 0$, the two infinite sums in the above equation suffer from normal ordering ambiguity at the quantum level. This ambiguity will involve at most a c-number. We must therefore have

$$[L_m, L_n] = (m - n)L_{m+n} + A(m)\delta_{m+n,\,0}. \tag{2.124}$$

This algebra is called the central extension of the Virasoro algebra, where the c-number term $A(m)$, is called the anomaly term of this algebra. Clearly, we must have

$$A(m) = -A(-m), \qquad A(0) = 0. \tag{2.125}$$

Thus we only need to determine $A(m)$ for $m{>}0$. We use the Jacobi identity

$$[L_k, [L_n, L_m]] + [L_n, [L_m, L_k]] + [L_m, [L_k, L_n]] = 0. \tag{2.126}$$

Equivalently

$$(-(m - n)A(k) + (m - k)A(n) + (k - n)A(m))\delta_{m+k+n,\,0} = 0. \tag{2.127}$$

Thus for $m + k + n = 0$, we must have

$$(n - m)A(k) + (m - k)A(n) + (k - n)A(m) = 0. \tag{2.128}$$

By setting $k = 1$ and $m = -n - 1$ we obtain

$$A(n + 1) = \frac{(n + 2)A(n) - (2n + 1)A(1)}{n - 1}. \tag{2.129}$$

This recursion relation will allow us to determine all $A(n)$ in terms of $A(1)$ and $A(2)$. Thus, the general form of $A(n)$ can be rewritten in terms of two unknown coefficients. In fact, we can check that the general solution of the above recursion relation is given by

$$A(n) = c_3 n^3 + c_1 n. \tag{2.130}$$

To determine c_1 and c_3 we calculate the expectation value of the commutator $[L_m, L_{-m}]$ in a suitable state. We choose the oscillator's ground state with $p^\mu = 0$. For $m = 1$ and $m = 2$ we find

$$\langle 0, 0|[L_1, L_{-1}]|0, 0\rangle = 0 \tag{2.131}$$

$$\langle 0, 0|[L_2, L_{-2}]|0, 0\rangle = \frac{D}{2}. \tag{2.132}$$

The proof goes as follows. We have

$$L_{\pm 2} = \frac{1}{2}\sum_n \alpha^\mu_{\pm 2 - n}(\alpha_n)_\mu. \tag{2.133}$$

Also

$$a_m^\mu|0, 0\rangle = 0, \quad m \geqslant 0 \tag{2.134}$$

$$\langle 0, 0|a_m^\mu = 0, \quad m \leqslant 0. \tag{2.135}$$

We compute immediately

$$\begin{aligned}
\langle 0, 0|[L_2, L_{-2}]|0, 0\rangle &= \frac{1}{4}\sum_{n<2}\sum_{n'<0}\langle 0, 0|\alpha_{2-n}^\mu[\alpha_{n\mu'}, \alpha_{-2-n'}^\nu]\alpha_{n'\nu}|0, 0\rangle \\
&+\frac{1}{4}\sum_{n<2}\sum_{n'>-2}\langle 0, 0|\alpha_{2-n}^\mu\alpha_{-2-n'}^\nu[\alpha_{n\mu'}, \alpha_{n'\nu}]|0, 0\rangle \\
&+\frac{1}{4}\sum_{n<0}\sum_{n'>0}\langle 0, 0|[\alpha_{2-n}^\mu, \alpha_{-2-n'}^\nu]\alpha_{n'\nu}\alpha_{n\mu}|0, 0\rangle \\
&+\frac{1}{4}\sum_{n\langle 0}\sum_{n'<-2}\langle 0, 0|\alpha_{-2-n'}^\nu[\alpha_{2-n}^\mu, \alpha_{n'\nu}]\alpha_{n\mu}|0, 0\rangle \\
&= -\frac{1}{2}\sum_{n<2} n(n-2)\eta_\mu^\nu\eta_\nu^\mu + \frac{1}{2}\sum_{n<0} n(n-2)\eta_\mu^\nu\eta_\nu^\mu \\
&= -\frac{1}{2}\sum_{n=0,1} n(n-2)\eta_\mu^\nu\eta_\nu^\mu \\
&= \frac{1}{2}\eta_\mu^\nu\eta_\nu^\mu \\
&= \frac{D}{2}.
\end{aligned} \tag{2.136}$$

This leads immediately to the two equations

$$2\langle 0, 0|L_0|0, 0\rangle + c_3 + c_1 = 0 \tag{2.137}$$

$$2\langle 0, 0|L_0|0, 0\rangle + 4c_3 + c_1 = \frac{D}{4}. \tag{2.138}$$

This leads to the solution

$$c_3 = -(2\langle 0, 0|L_0|0, 0\rangle + c_1) = \frac{D}{12}. \tag{2.139}$$

We note, from $[L_m, L_{-m}] = 2mL_0 + c_3m^3 + c_1m$ that the constant c_1 can be changed by shifting the definition of L_0 by a constant. On the other hand, by shifting L_0 by a constant, we also shift the constant a, and hence only the relation between c_1 and a has an invariant meaning. We shift L_0 by a constant such that

$$\langle 0, 0|L_0|0, 0\rangle = 0. \tag{2.140}$$

The quantum Virasoro algebra becomes

$$[L_m, L_n] = (m - n)L_{m+n} + \frac{D}{12}m(m^2 - 1)\delta_{m+n,\,0}. \tag{2.141}$$

The Virasoro generators L_1, L_0 and L_{-1} form an $SL(2, R) \sim SO(2, 1)$ subalgebra of the Virasoro algebra which is a non-compact form of $SU(2)$. Similarly, for closed strings, we have an $SO(2, 2)$ subalgebra of the Virasoro algebra formed by L_1, L_0, L_{-1}, $\tilde{L}_1$, $\tilde{L}_0$ and $\tilde{L}_{-1}$.

2.3 Spurious states and critical strings

2.3.1 Spurious states

Let us start by recalling that the physical states $|\phi\rangle$ must satisfy the Virasoro constraints

$$(L_m - a\delta_{m,\,0})|\phi\rangle = 0, \quad m \geqslant 0. \tag{2.142}$$

Next, we define spurious states $|\psi\rangle$ as the states which satisfy the mass-shell condition, and are orthogonal to all physical states, viz

$$(L_0 - a)|\psi\rangle = 0, \quad \langle\psi|\phi\rangle = 0. \tag{2.143}$$

It is obvious that the projection operator $O = |\psi\rangle\langle\psi|$ annihilates all physical states. Thus O can be rewritten as $O = \sum_{n>0} X_{-n}L_n$, where X_{-n} is some operator. Thus we can write the spurious state $|\psi\rangle$ as

$$|\psi\rangle = \sum_{n>0} L_{-n}|\chi_n\rangle. \tag{2.144}$$

In other words, $X_{-n} = |\psi\rangle\langle\chi_n|$. Since $(L_0 - a)|\psi\rangle = 0$, $|\chi_n\rangle$ must satisfy

$$(L_0 - a + n)|\chi_n\rangle = 0. \tag{2.145}$$

Furthermore, we can truncate the above infinite sum since L_{-n}, for $n \geqslant 3$, can always be rewritten in terms of L_{-1} and L_{-2}, by using the Virasoro algebra. For example, $L_{-3} = [L_{-1}, L_{-2}]$. Thus, we obtain

$$|\psi\rangle = L_{-1}|\chi_1\rangle + L_{-2}|\chi_2\rangle. \tag{2.146}$$

Again, since $(L_0 - a)|\psi\rangle = 0$, the $|\chi_1\rangle$ and $|\chi_2\rangle$ must satisfy (2.145). By construction, and as required by its definition, these states are orthogonal to all physical states.

Let us consider a state $|\psi\rangle$ which is physical and spurious. Then, this state must be orthogonal to itself, viz

$$\langle\psi|\psi\rangle = \sum_{n>0} \langle\chi_n|L_n|\psi\rangle = 0. \tag{2.147}$$

Thus, a physical and spurious state has zero norm. These states are also called null physical states. Hence, these states mark the boundary between the physical Hilbert subspace which is positive semi-definite, and the full Hilbert space which contains negative-norm states.

Let us consider the open string ground state of momentum k, viz $|0, k\rangle$. The mass-shell condition $L_0 = a$, where $L_0 = \alpha' p^\mu p_\mu + \sum_{n=1} n a_n^{\mu +} a_{n\mu}$, gives $\alpha' k^2 = a$. Next, we consider the first excited state $|\phi\rangle = \xi_\mu a_1^{\mu +}|0, k\rangle$ where ξ is some polarization vector. The mass-shell condition gives now $\alpha' k^2 = a - 1$. On the other hand, the constraint $L_1 = 0$ implies $L_1|\phi\rangle = l_s p^\mu a_{1\mu}|\phi\rangle = l_s k^\mu \xi_\mu|0, k\rangle = 0$, i.e. $k^\mu \xi_\mu = 0$. Therefore, this condition reduces the number of independent polarizations to $D-1$.

We can choose k to lie in the 0–1 plane. Thus, the $D-2$ polarizations normal to this plane are spacelike with positive norm. There are now three cases to consider:

- If $k^2 > 0$ ($a > 1$). In this case, k is spacelike, and therefore the last polarization is timelike with negative norm.
- If $k^2 < 0$ ($a < 1$). In this case, k is timelike, and therefore the last polarization is spacelike with positive norm.
- If $k^2 = 0$ ($a = 1$). In this case, k is lightlike, and therefore the last polarization is also lightlike, proportional to k, with zero norm.

We have no ghosts, i.e. no negative norm states, for

$$a \leqslant 1. \tag{2.148}$$

In the case $a = 1$, the ground state is a scalar tachyon, while the first excited state is a massless vector particle, with $D - 2$ transverse positive norm states, and one longitudinal zero norm state. The constraint $L_1 = 0$ is therefore the analogue of $\partial_\mu A^\mu = 0$ in electrodynamics.

This argument can be generalized as follows. We consider the spurious states of the form

$$|\psi\rangle = L_{-1}|\chi_1\rangle, \tag{2.149}$$

with $|\chi_1\rangle$ an arbitrary state satisfying

$$(L_0 - a + 1)|\chi_1\rangle = 0, \quad L_m|\chi_1\rangle = 0, \quad m > 0. \tag{2.150}$$

By demanding that $|\psi\rangle$ is physical, viz

$$L_m|\psi\rangle = (L_0 - a)|\psi\rangle = 0, \quad m = 1, 2, \ldots \tag{2.151}$$

All these conditions are satisfied apart from the condition $L_1|\psi\rangle = 0$. We use the Virasoro algebra

$$L_1 L_{-1} = 2L_0 + L_{-1}L_1. \tag{2.152}$$

We compute immediately

$$L_1|\psi\rangle = 2(a - 1)|\chi_1\rangle. \tag{2.153}$$

Thus, this condition is satisfied for $a = 1$. At this value, $|\psi\rangle$ is both spurious and physical and thus has zero norm.

The longitudinal component of the massless vector state considered above is an example of the state $|\psi\rangle$ with $|\chi_1\rangle = |0, k\rangle$.

2.3.2 Critical strings

Thus, the null physical states, the physical and spurious states, mark the boundary between positive-norm states and negative-norm states. The number of these zero-norm states increases if $a = 1$. As we will now show, this number increases also if $D = 26$.

To see this, we consider spurious states of the form (with $a = 1$)

$$|\psi\rangle = (L_{-2} + \gamma L_{-1}^2)|\tilde{\chi}\rangle, \tag{2.154}$$

with

$$\begin{aligned} L_m|\tilde{\chi}\rangle &= 0, \quad m > 0 \\ (L_0 + 1)|\tilde{\chi}\rangle &= 0. \end{aligned} \tag{2.155}$$

Clearly, $|\psi\rangle$ satisfies the mass-shell condition

$$(L_0 - 1)|\psi\rangle = 0. \tag{2.156}$$

$|\psi\rangle$ is by construction a spurious state. Thus, it will have zero norm if it is a physical state. We must then have $L_m|\psi\rangle = 0$. Obviously, this is true for all $m \geqslant 3$. Hence, it remains to check that

$$L_1|\psi\rangle = L_2|\psi\rangle = 0. \tag{2.157}$$

We compute

$$\begin{aligned} L_1|\psi\rangle &= [L_1, L_{-2} + \gamma L_{-1}^2]|\chi\rangle \\ &= (3L_{-1} + 2\gamma L_0 L_{-1} + 2\gamma L_{-1} L_0)|\tilde{\chi}\rangle \\ &= ((3 - 2\gamma)L_{-1} + 4\gamma L_0 L_{-1})|\tilde{\chi}\rangle \\ &= (3 - 2\gamma)L_{-1}|\tilde{\chi}\rangle. \end{aligned} \tag{2.158}$$

In the third line we have used the constraint $(L_0 + 1)|\tilde{\chi}\rangle = 0$. This vanishes if $\gamma = 3/2$.

Next, we compute

$$\begin{aligned} L_2|\psi\rangle &= [L_2, L_{-2} + \gamma L_{-1}^2]|\chi\rangle \\ &= (13L_0 + 9L_{-1}L_1 + \frac{D}{2})|\tilde{\chi}\rangle \\ &= (-13 + \frac{D}{2})|\tilde{\chi}\rangle. \end{aligned} \tag{2.159}$$

This vanishes for $D = 26$.

The physical Hilbert subspace is free of negative-norm states for $a = 1$ and $D =$ 26. The zero-norm spurious states, or the null physical states, are in fact unphysical, and decouple from physical processes. The resulting string theory is called critical string theory.

The physical Hilbert subspace is also free from negative-norm states for $a \leqslant 1$ and $D \leqslant 25$ corresponding to non-critical string theory.

2.4 Lightcone gauge quantization

2.4.1 Critical strings

In the covariant conformal gauge $h_{\alpha\beta} = \eta_{\alpha\beta}$, we have already found for open strings that the coordinates admit the mode expansion

$$X^\mu = x^\mu + l_s^2 p^\mu \tau + il_s \sum_{n\neq 0} \frac{1}{n}\alpha_n^\mu \exp(-in\tau)\cos n\sigma. \tag{2.160}$$

These coordinates must also satisfy the Virasoro conditions $T_{--} = T_{++} = 0$.

Now, we will use the residual gauge symmetry (2.115) to impose an additional non-covariant gauge known as the lightcone gauge. First, we introduce the spacetime lightcone coordinates X^+ and X^-, by analogy with the lightcone worldsheet coordinates (2.39), by the equations

$$X^+ = \frac{1}{\sqrt{2}}(X^0 + X^{D-1}), \quad X^- = \frac{1}{\sqrt{2}}(X^0 - X^{D-1}). \tag{2.161}$$

This is obviously an arbitrary choice. The metric in this system is $\eta_{ij} = \delta_{ij}$, $\eta_{+-} = \eta_{-+} = -1$. In this system, any vector will have components $V^\pm = (V^0 \pm V^{D-1})/\sqrt{2}$ and V^i, $i = 1, \ldots, D-2$. The inner product of any two vectors is $VW = V^iW^i - V^+W^- - V^-W^+$. We raise and lower indices by the rule

$$V^+ = -V_-, \quad V^- = -V_+, \quad V^i = V_i. \tag{2.162}$$

The above residual symmetry corresponds to the reparametrizations (2.119), i.e. to

$$\sigma^+ \longrightarrow \tilde{\sigma}^+(\sigma^+), \quad \sigma^- \longrightarrow \tilde{\sigma}^-(\sigma^-). \tag{2.163}$$

In other words, we have the transformations

$$\tilde{\tau} = \frac{1}{2}(\tilde{\sigma}^+ + \tilde{\sigma}^-), \quad \tilde{\sigma} = \frac{1}{2}(\tilde{\sigma}^+ - \tilde{\sigma}^-). \tag{2.164}$$

Thus, $\tilde{\tau}$ is a solution of the free massless wave function

$$\left(\frac{\partial^2}{\partial\sigma^2} - \frac{\partial^2}{\partial\tau^2}\right)\tilde{\tau} = 0. \tag{2.165}$$

The determination of $\tilde{\tau}$, will fix $\tilde{\sigma}$ up to a constant. This is the same equation obeyed by X^μ in conformal gauge. The residual symmetry allows us, therefore, to choose one of the X^μ equal to $\tilde{\tau}$. The lightcone gauge corresponds precisely to the choice

$$X^+(\sigma, \tau) = x^+ + l_s^2 p^+ \tau. \tag{2.166}$$

In other words,

$$\alpha_n^+ = 0, \quad n \neq 0. \tag{2.167}$$

For open strings, the Virasoro constraints $T_{--} = T_{++} = 0$ become $(\dot{X}^\mu \pm X'\mu)^2 = 0$. Equivalently

$$(\dot{X}^i \pm X'^i)^2 - 2(\dot{X}^+ \pm X'^+)(\dot{X}^- \pm X'^-) = 0 \Rightarrow \dot{X}^- \pm X'^- = \frac{1}{2l_s^2 p^+}(\dot{X}^i \pm X'^i)^2. \quad (2.168)$$

These two equations lead to

$$\dot{X}^- = \frac{1}{2l_s^2 p^+}(\dot{X}^{i2} + X'^{i2}). \quad (2.169)$$

This is a linear equation for X^-. The mode expansion of X^- is given by

$$X^- = x^- + l_s^2 p^- \tau + i l_s \sum_{n \neq 0} \frac{1}{n} \alpha_n^- \exp(-in\tau) \cos n\sigma. \quad (2.170)$$

Thus

$$\dot{X}^- = l_s \sum_{n=-\infty}^{+\infty} \alpha_n^- \exp(-in\tau) \cos n\sigma, \quad \alpha_0^- = l_s p^-. \quad (2.171)$$

Furthermore, we compute

$$\begin{aligned} \dot{X}^{i2} + X'^{i2} &= l_s^2 \sum_n \sum_m \alpha_n^i \alpha_m^i e^{-i(n+m)\tau} \cos(n+m)\sigma \\ &= l_s^2 \sum_{i=1}^{D-2} \sum_n \sum_m \alpha_{n-m}^i \alpha_m^i e^{-in\tau} \cos n\sigma, \quad \alpha_0^i = l_s p^i. \end{aligned} \quad (2.172)$$

By comparison we get

$$\sum_{n=-\infty}^{+\infty} \left(\alpha_n^- - \frac{1}{2l_s p^+} \sum_{i=1}^{D-2} \sum_{m=-\infty}^{+\infty} \alpha_{n-m}^i \alpha_m^i \right) \exp(-in\tau) \cos n\sigma = 0. \quad (2.173)$$

Since $[\alpha_n^\mu, \alpha_m^\nu] = n\delta_{n+m,0}\eta^{\mu\nu}$, we conclude immediately that

$$\alpha_n^- = \frac{1}{l_s p^+} \left(\frac{1}{2} \sum_{i=1}^{D-2} \sum_{m=-\infty}^{+\infty} :\alpha_{n-m}^i \alpha_m^i: - a\delta_{n,0} \right). \quad (2.174)$$

In summary, the lightcone gauge corresponds to the choice

$$X^+(\sigma, \tau) = x^+ + p^+ \tau. \quad (2.175)$$

The Virasoro constraints can then be solved for X^- (α_n^-) in terms of the transverse coordinates X^i (α_n^i) to give

$$\alpha_n^- = \frac{l_s}{p^+} \left(\frac{1}{2} \sum_{i=1}^{D-2} \sum_{m=-\infty}^{+\infty} :\alpha_{n-m}^i \alpha_m^i: - a\delta_{n,0} \right). \quad (2.176)$$

Thus, in the lightcone gauge the string has only transverse excitations in precisely the same way that a massless particle has only transverse polarizations. The mass-shell condition becomes using the above result given by

$$\begin{aligned} M^2 = -p_\mu p^\mu &= 2p^+p^- - p^i p^i \\ &= 2p^+ \frac{\alpha_0^-}{l_s} - \frac{1}{l_s^2}\alpha_0^i \alpha_0^i \\ &= \frac{2}{l_s^2}\left[\frac{1}{2}\sum_{i=1}^{D-2}\sum_{m=-\infty}^{+\infty} :\alpha_{n-m}^i \alpha_m^i: -a\right] - \frac{1}{l_s^2}\alpha_0^i\alpha_0^i \\ &= \frac{2}{l_s^2}(N-a). \end{aligned} \tag{2.177}$$

The transverse number operator is defined by

$$N = \sum_{i=1}^{D-2}\sum_{m=1}^{+\infty} :\alpha_{-m}^i \alpha_m^i. \tag{2.178}$$

The first excited state is given by $|\phi\rangle = \alpha_{-1}^i|0, p\rangle$. This corresponds to an irreducible representation of the transverse rotation group $SO(D-2)$ iff the corresponding particle is massless. Thus, this string theory is Lorentz invariant iff the vector state $|\phi\rangle = \alpha_{-1}^i|0, p\rangle$ is massless. In other words, we must have

$$M^2|\phi\rangle = \frac{2}{l_s^2}(1-a)|\phi\rangle = 0 \Rightarrow a = 1. \tag{2.179}$$

The normal ordering constant a arises from the expression

$$\frac{1}{2}\sum_{i=1}^{D-2}\sum_{m=-\infty}^{+\infty} \alpha_{-m}^i\alpha_m^i = \frac{1}{2}\sum_{i=1}^{D-2}\sum_{m=-\infty}^{+\infty} :\alpha_{-m}^i\alpha_m^i: + \frac{D-2}{2}\sum_{m=1}^{\infty} m, \tag{2.180}$$

since

$$\alpha_{-m}^i\alpha_m^i =: \alpha_{-m}^i\alpha_m^i: , \quad m > 0 \tag{2.181}$$

$$\alpha_{-m}^i\alpha_m^i =: \alpha_m^i\alpha_{-m}^i :+ [\alpha_{-m}^i, \alpha_m^i], \quad m < 0. \tag{2.182}$$

The second term in (2.180) is divergent and can be regularized using the Riemann zeta function

$$\zeta(s) = \sum_{m=1}^{\infty} m^{-s}, \quad \text{Re } s > 1. \tag{2.183}$$

There exists a unique analytical continuation of the Riemann zeta function to $s = -1$ where we get

$$\zeta(-1) = -\frac{1}{12}. \tag{2.184}$$

Thus we get

$$\frac{1}{2}\sum_{i=1}^{D-2}\sum_{m=-\infty}^{+\infty}\alpha_{-m}^{i}\alpha_{m}^{i}=\frac{1}{2}\sum_{i=1}^{D-2}\sum_{m=-\infty}^{+\infty}:\alpha_{-m}^{i}\alpha_{m}^{i}:-\frac{D-2}{24}. \tag{2.185}$$

By comparing with (2.176) we obtain

$$a=\frac{D-2}{24}\Rightarrow D=26. \tag{2.186}$$

2.4.2 Rigorous proof

The Lorentz generators are:

$$J^{\mu\nu}=l^{\mu\nu}+E^{\mu\nu} \tag{2.187}$$

$$l^{\mu\nu}=x^{\mu}p^{\nu}-x^{\nu}p^{\mu} \tag{2.188}$$

$$E^{\mu\nu}=-i\sum_{n=1}^{\infty}\frac{1}{n}(\alpha_{-n}^{\mu}\alpha_{n}^{\nu}-\alpha_{-n}^{\nu}\alpha_{n}^{\mu}). \tag{2.189}$$

The Lorentz algebra is

$$[J^{\mu\nu},J^{\rho\lambda}]=-i\eta^{\nu\rho}J^{\mu\lambda}+i\eta^{\mu\rho}J^{\nu\lambda}+i\eta^{\nu\lambda}J^{\mu\rho}-i\eta^{\mu\lambda}J^{\nu\rho}. \tag{2.190}$$

The Lorentz generators associated with the transverse space generate an $SO(D-2)$ subgroup which is a manifest symmetry of the lightcone gauge formalism. The transformations which affect X^{+}, and as a consequence affect the gauge condition, are generated by J^{+-} and J^{i-}. The corresponding commutators may then involve a quantum anomaly. Indeed, we find that the commutator $[J^{i-},J^{j-}]$ is non-zero for generic values of a and D. Explicitly, we obtain

$$[J^{i-},J^{j-}]=-\frac{1}{(p^{+})^{2}}\sum_{m=1}^{\infty}\Delta_{m}(\alpha_{-m}^{i}\alpha_{m}^{j}-\alpha_{-m}^{j}\alpha_{m}^{i}). \tag{2.191}$$

The coefficients Δ_m are c-numbers. A non-trivial computation gives

$$\Delta_{m}=m\frac{26-D}{12}+\frac{1}{m}\left(\frac{D-26}{12}+2(1-a)\right). \tag{2.192}$$

The above commutator should vanish if Lorentz invariance is maintained in the quantum theory. We get immediately $a=1$ and $D=26$.

2.4.3 Spectrum

We will follow the very brief and concise discussion of the spectrum found in [2].

- Open strings: for open string theory, the ground state $|0,k\rangle$ is a tachyon of negative mass squared since, from (2.177), we get $\alpha' M^2=-a$, by substituting $N=0$. The next state is a massless gauge boson given by $\alpha_{-1}^{i}|0,k\rangle$

corresponding to $N = 1$. This provides a vector representation of $SO(d)$, $d = 24$.

For $N = 2$ we get a positive mass squared given by $\alpha' M^2 = 2 - a$. The corresponding states are explicitly given by

$$\alpha_{-2}^{i}|0, k\rangle, \quad \alpha_{-1}^{i}\alpha_{-1}^{j}|0, k\rangle. \tag{2.193}$$

We have obviously $324 = d + d(d + 1)/2 = 24 + 24.25/2$ of these states. This is the dimension of the symmetric traceless second rank tensor representation of $SO(25)$ since $324 = 25.26/2 - 1$. This corresponds therefore to a massive spin 2 particle.

- Closed strings: The mass-shell condition for closed strings is

$$\alpha' M^2 = 4(N - a) = 4(\tilde{N} - a). \tag{2.194}$$

The level-matching condition is given by

$$N = \tilde{N}. \tag{2.195}$$

The spectrum of closed string is obtained by taking the tensor product of two copies of the open string states. These two copies correspond to the left-movers and the right-movers. The ground state is again a tachyon with mass squared $\alpha' M^2 = -4a$ corresponding to $N = \tilde{N} = 0$.

The first excited level corresponds to $N = \tilde{N} = 1$ with mass squared $\alpha' M^2 = 4(1 - a)$ given explicitly by the states

$$|\Omega^{ij}\rangle = \alpha_{-1}^{i}\tilde{\alpha}_{-1}^{j}|0, k\rangle. \tag{2.196}$$

These states correspond to the tensor product of two massless vectors, one of them is left-moving and the other one is right-moving, with a total number of $24.24 = 576$ states.

The part of $|\Omega^{ij}\rangle$ which is symmetric and traceless in i and j transforms as a massless spin 2 particle under $SO(24)$. This is precisely the graviton. The antisymmetric part $|\Omega^{ij}\rangle - |\Omega^{ji}\rangle$ transforms as a massless antiysmmetric second rank tensor under $SO(24)$. The trace part $\delta_{ij}|\Omega^{ij}\rangle$ transforms as a massless scalar under $SO(24)$ called the dilaton.

2.5 Exercises

Exercise 1: Derive the reparametrization invariance of the point particle action and show that the constraint is exactly equivalent to the mass-shell condition.

Solution 1: The action of a massive point particle is proportional to the invariant length of its worldline, viz

$$S = -m\int ds, \quad ds^2 = -g_{\mu\nu}(x)dx^{\mu}dx^{\nu}. \tag{2.197}$$

Let τ be an arbitrary parameter which labels the points along the worldline, i.e $x^\mu \equiv x^\mu(\tau)$. Then

$$S = -m \int d\tau \sqrt{-\dot{x}^2}, \quad \dot{x}^2 = g_{\mu\nu}(x) \frac{dx^\mu}{d\tau} \frac{dx^\nu}{d\tau}. \tag{2.198}$$

In flat space $g_{\mu\nu}(x) = \eta_{\mu\nu} = (-1, +1, +1, \ldots, +1)$. We compute in this case $\sqrt{-\dot{x}^2} = \sqrt{1 - \vec{v}^2}$ where $\vec{v}$ is the velocity of the particle and where we have made the identification $x^0 = t \equiv \tau$. It is then clear that for $v \ll 1$ we obtain

$$S = \int dt \left(\frac{1}{2} m \vec{v}^2 - m \right). \tag{2.199}$$

This is indeed the action of a nonrelativistic point particle of mass m where the potential energy is given by the rest mass m.

The action S is invariant under the reparametrization transformations

$$\tau \longrightarrow \tilde{\tau}(\tau). \tag{2.200}$$

Indeed

$$\sqrt{-\dot{x}^2} \longrightarrow \frac{d\tilde{\tau}}{d\tau} \sqrt{-\dot{x}^2}. \tag{2.201}$$

Let us introduce another field on the world-line, the metric $\gamma_{\tau\tau}(\tau)$. We introduce the tetrad $\eta(\tau) = \sqrt{-\gamma_{\tau\tau}(\tau)}$ and we write the action

$$S' = \frac{1}{2} \int d\tau \left(\frac{1}{\eta} \dot{x}^2 - \eta m^2 \right). \tag{2.202}$$

The equation of motion of η is given by

$$\eta^2 = -\frac{\dot{x}^2}{m^2}. \tag{2.203}$$

Using this in S' we can check that S' reduces to S. The reparametrization invariance of S' can be described as follows. From the term $\int d\tau \eta$ we conclude that we must have

$$d\tau \eta(\tau) = d\tilde{\tau} \tilde{\eta}(\tilde{\tau}). \tag{2.204}$$

Infinitesimally we have $\tilde{\tau} = \tau + \xi(\tau)$ where $\xi(\tau)$ is an infinitesimal parameter with arbitrary dependence on τ. Thus $d\tilde{\tau}/d\tau = 1 + d\xi/d\tau$. Also $\tilde{\eta} = \tilde{\eta}(\tilde{\tau}) = \tilde{\eta}(\tau) + \xi d\tilde{\eta}/d\tau$. Hence

$$\delta\eta = \eta(\tau) - \tilde{\eta}(\tau) = \frac{d}{d\tau}(\xi \tilde{\eta}). \tag{2.205}$$

From the invariance of the first term in S' we conclude that we must have

$$\left(\frac{d\tau}{d\tilde{\tau}}\right)^2 \dot{x}^2 = \dot{\tilde{x}}^2. \quad (2.206)$$

In the above $x^\mu = x^\mu(\tau)$ and $\tilde{x}^\mu = \tilde{x}^\mu(\tilde{\tau}) \equiv x^\mu(\tau)$. Hence we can compute the infinitesimal transformation

$$\delta x^\mu = x^\mu(\tau) - \tilde{x}^\mu(\tau) = \xi \dot{x}^\mu. \quad (2.207)$$

Using this invariance we can make the gauge choice

$$\eta = \frac{1}{m}. \quad (2.208)$$

The constraint (2.203) becomes

$$-\dot{x}^2 = 1. \quad (2.209)$$

This is the mass-shell condition.

Exercise 2: Derive the Nambu–Goto action from first principles.

Solution 2: The string is a 1-dimensional object. It sweeps out a 2-dimensional worldsheet in spacetime (the target space). The worldsheet of the string can be described in terms of two parameters σ and τ. The coordinate σ runs from 0 to π and determines the locations of points on the string, whereas the timelike coordinate τ is the time coordinate for an observer sitting at the position σ along the string. The worldsheet is described by specifying the positions $X^\mu \equiv X^\mu(\sigma, \tau)$ where $\mu = 0, 1, .., d$ of the points σ on the string for all times τ. The coordinates $X^\mu(\sigma, \tau)$ define a map from the parameter space to the target space. Locally this map is one-to-one so to each point (σ, τ) of the parameter space we assign a point on the worldsheet surface.

The area of a small element of the parameter space surface is given by $d\sigma d\tau$. This will correspond to a small element of the target space surface (the worldsheet) of area dA which can be computed as follows. The tangent vector to the worldsheet at the point (σ, τ) along the σ direction is given by

$$dv_1^\mu = \frac{\partial X^\mu}{\partial \sigma} d\sigma. \quad (2.210)$$

Similarly the tangent vector to the worldsheet at the point (σ, τ) along the τ direction is given by

$$dv_2^\mu = \frac{\partial X^\mu}{\partial \tau} d\tau. \quad (2.211)$$

Let us assume for simplicity that the metric is Euclidean. Using the formula for the area of a parallelogram we find that the area dA of a small element of the target space surface is given by

$$\begin{aligned} dA &= |dv_1||dv_2||\sin\theta| \\ &= |dv_1||dv_2|\sqrt{1-\cos^2\theta} \\ &= \sqrt{dv_1^2 dv_2^2 - dv_1^2 dv_2^2\cos^2\theta} \\ &= \sqrt{(dv_1^\mu dv_{1\mu})(dv_2^\mu dv_{2\mu}) - (dv_1^\mu dv_{2\mu})^2} \\ &= d\sigma d\tau\sqrt{\left(\frac{\partial X^\mu}{\partial\sigma}\frac{\partial X_\mu}{\partial\sigma}\right)\left(\frac{\partial X^\mu}{\partial\tau}\frac{\partial X_\mu}{\partial\tau}\right) - \left(\frac{\partial X^\mu}{\partial\sigma}\frac{\partial X_\mu}{\partial\tau}\right)^2}. \end{aligned} \tag{2.212}$$

For Minkowskian metric we have

$$\left(\frac{\partial X^\mu}{\partial\sigma}\frac{\partial X_\mu}{\partial\sigma}\right)\left(\frac{\partial X^\mu}{\partial\tau}\frac{\partial X_\mu}{\partial\tau}\right) - \left(\frac{\partial X^\mu}{\partial\sigma}\frac{\partial X_\mu}{\partial\tau}\right)^2 \leqslant 0. \tag{2.213}$$

The area dA of a small element of the target space surface is given in this case by

$$dA = d\sigma d\tau\sqrt{\left(\frac{\partial X^\mu}{\partial\sigma}\frac{\partial X_\mu}{\partial\tau}\right)^2 - \left(\frac{\partial X^\mu}{\partial\sigma}\frac{\partial X_\mu}{\partial\sigma}\right)\left(\frac{\partial X^\mu}{\partial\tau}\frac{\partial X_\mu}{\partial\tau}\right)}. \tag{2.214}$$

Let us introduce the induced metric

$$G_{ab} = \partial_a X^\mu \partial_b X_\mu. \tag{2.215}$$

The indices a, b run over the values σ and τ. Hence

$$G = -\det G = \left(\frac{\partial X^\mu}{\partial\sigma}\frac{\partial X_\mu}{\partial\tau}\right)^2 - \left(\frac{\partial X^\mu}{\partial\sigma}\frac{\partial X_\mu}{\partial\sigma}\right)\left(\frac{\partial X^\mu}{\partial\tau}\frac{\partial X_\mu}{\partial\tau}\right). \tag{2.216}$$

In other words

$$dA = d\sigma d\tau\sqrt{G}. \tag{2.217}$$

The basic string action (the so-called Nambu–Goto action) is taken to be proportional to the area of the worldsheet, viz

$$S_{NG}[X] = -T\int_\Sigma dA = -T\int_\Sigma d\sigma d\tau\sqrt{G}. \tag{2.218}$$

Σ denotes the worldsheet. T is the string tension, it is related to the Regge slope by

$$T = \frac{1}{2\pi\alpha'}. \tag{2.219}$$

The action S_{NG} is the analogue of the point particle action S.

Exercise 3: Give a complete derivation of the energy–momentum, angular momentum and Virasoro operators.

Solution 3: The variation of the action was computed to be given by

$$\delta_X S_P = T\int d^2\sigma \delta X^\mu \left[\partial^a\partial_a X_\mu\right]. \tag{2.220}$$

The D-dimensional Poincaré invariance is a global invariance with respect to the worldsheet theory and hence it must be associated with conserved Noether currents. First let us consider an infinitesimal translation $\delta X^\mu = \epsilon^\mu$. The variation of the action reads

$$\delta_X S_P = -T\int d^2\sigma \partial\epsilon^\mu \partial_a X_\mu. \tag{2.221}$$

The corresponding Noether current is

$$P_a^\mu = T\partial_a X^\mu. \tag{2.222}$$

Lorentz transformations are given by $X'^\mu = \Lambda^\mu_\nu X^\nu$ where $\Lambda^\mu_\alpha g_{\mu\nu}\Lambda^\nu_\beta = g_{\alpha\beta}$. The infinitesimal Lorentz transformation is $\Lambda^\mu_\nu = g_\nu^{\ \mu} + \omega_\nu^{\ \mu}$ where $\omega_{\mu\nu} = -\omega_{\nu\mu}$. Hence the variation is $\delta X^\mu = \omega_\nu^{\ \mu} X^\nu$ and the corresponding variation of the action is

$$\delta_X S_P = \frac{T}{2}\int d^2\sigma \partial_a\omega_{\mu\nu}(X^\mu\partial^a X^\nu - X^\nu\partial^a X^\mu). \tag{2.223}$$

The corresponding current is

$$J_a^{\mu\nu} = T(X^\mu\partial^a X^\nu - X^\nu\partial^a X^\mu). \tag{2.224}$$

By construction $\partial^a P_a^\mu = 0$ and $\partial^a J_a^{\mu\nu} = 0$ where we must use the equations of motion.

The constant T is the string tension, i.e. the energy per unit length. To see this consider P_0^0 in the time gauge $X^0 = \tau$. In this case we see that $P_0^0 = T$. The total conserved energy–momentum and the total conserved angular momentum of the string are obtained by integrating the currents P_a^μ and $J_a^{\mu\nu}$ over σ at $\tau = 0$. Current conservation guarantees that we will get the same result by integrating over any spacelike (i.e. fixed τ) that cuts the worldsheet once. The total energy–momentum vector and the angular momentum tensor are given by

$$P^\mu = \int d\sigma P_\tau^\mu = T\int_0^\pi d\sigma\frac{dX^\mu}{d\tau} \tag{2.225}$$

$$J^{\mu\nu} = \int d\sigma J_\tau^{\mu\nu} = T\int_0^\pi d\sigma\left(X^\mu\frac{dX^\nu}{d\tau} - X^\nu\frac{dX^\mu}{d\tau}\right). \tag{2.226}$$

We have

$$\begin{aligned} \frac{dX^\mu}{d\tau}|_{\tau=0} &= l^2 p^\mu + \frac{l}{2}\sum_{n\neq 0}(\alpha_n^\mu + \alpha_{-n}^\mu)\cos n\sigma, \quad \text{open string}\\ \frac{dX^\mu}{d\tau}|_{\tau=0} &= l^2 p^\mu + \frac{l}{2}\sum_{n\neq 0}((\alpha_n^\mu + \tilde{\alpha}_{-n}^\mu)e^{2in\sigma} + (\tilde{\alpha}_n^\mu + \alpha_{-n}^\mu)e^{-2in\sigma}), \quad \text{closed string.} \end{aligned} \tag{2.227}$$

Obviously

$$P^\mu = p^\mu. \tag{2.228}$$

We use the results

$$\begin{aligned}\int_0^\pi d\sigma \cos m\sigma \cos n\sigma &= \frac{\pi}{2}\delta_{mn}\\ \int_0^\pi d\sigma \sin m\sigma \sin n\sigma &= \frac{\pi}{2}\delta_{mn}\\ \int_0^\pi d\sigma \cos m\sigma \sin n\sigma &= 0\\ \int_0^\pi d\sigma e^{2in\sigma}e^{-2im\sigma} &= \pi\delta_{n,\,m}\\ \int_{-\pi}^\pi d\sigma e^{in\sigma}e^{-im\sigma} &= 2\pi\delta_{n,\,m}.\end{aligned} \tag{2.229}$$

Then we compute

$$\begin{aligned}&J^{\mu\nu} = L^{\mu\nu} + S^{\mu\nu} + \tilde{S}^{\mu\nu}, \quad \text{closed string}\\ &L^{\mu\nu} = x^\mu p^\nu - x^\nu p^\mu\\ &S^{\mu\nu} = -i\sum_{n=1}^{\infty}\frac{1}{n}(\alpha_{-n}^\mu\alpha_n^\nu - \alpha_{-n}^\nu\alpha_n^\mu), \qquad \tilde{S}^{\mu\nu} = -i\sum_{n=1}^{\infty}\frac{1}{n}(\tilde{\alpha}_{-n}^\mu\tilde{\alpha}_n^\nu - \tilde{\alpha}_{-n}^\nu\tilde{\alpha}_n^\mu)\end{aligned} \tag{2.230}$$

$$\begin{aligned}&J^{\mu\nu} = L^{\mu\nu} + S^{\mu\nu}, \quad \text{open string}\\ &L^{\mu\nu} = x^\mu p^\nu - x^\nu p^\mu, \qquad S^{\mu\nu} = -i\sum_{n=1}^{\infty}\frac{1}{n}(\alpha_{-n}^\mu\alpha_n^\nu - \alpha_{-n}^\nu\alpha_n^\mu).\end{aligned} \tag{2.231}$$

The Lagrangian of the string is

$$L = \frac{T}{2}\int_0^\pi d\sigma(\partial_\tau X^\mu \partial_\tau X_\mu - \partial_\sigma X^\mu \partial_\sigma X_\mu). \tag{2.232}$$

The conjugate momentum of X^μ is

$$P_\tau^\mu = \frac{\partial L}{\partial_\tau X_\mu} = T\partial_\tau X^\mu. \tag{2.233}$$

The Hamiltonian of the string is

$$\begin{aligned}H &= \int_0^\pi d\sigma(\partial_\tau X_\mu P_\tau^\mu) - L\\ &= \frac{T}{2}\int_0^\pi d\sigma(\partial_\tau X^\mu \partial_\tau X_\mu + \partial_\sigma X^\mu \partial_\sigma X_\mu).\end{aligned} \tag{2.234}$$

We find

$$\begin{aligned}\frac{T}{2}\int_0^\pi d\sigma\partial_\tau X^\mu(\partial_\tau X_\mu)^* &= \frac{T}{2}\int_0^\pi d\sigma\partial_\sigma X^\mu(\partial_\sigma X_\mu)^*\\ &= \frac{1}{4}\sum_n \alpha_n^\mu(\alpha_{-n})_\mu, \quad \text{open string.}\end{aligned} \tag{2.235}$$

Remark that $\alpha_0^\mu = lp^\mu$. We get

$$H = \frac{1}{2}\sum_n \alpha_n^\mu(\alpha_{-n})_\mu, \quad \text{open string.} \tag{2.236}$$

Also we compute

$$\begin{aligned}\frac{T}{2}\int_0^\pi d\sigma\partial_\tau X^\mu(\partial_\tau X_\mu)^* = \frac{1}{2}\sum_n\Big(&\alpha_n^\mu(\alpha_{-n})_\mu + \tilde{\alpha}_n^\mu(\tilde{\alpha}_{-n})_\mu \\ &+ 2\alpha_n^\mu(\tilde{\alpha}_n)_\mu^{-4in\tau}\Big), \quad \text{closed string}\end{aligned} \tag{2.237}$$

$$\begin{aligned}\frac{T}{2}\int_0^\pi d\sigma\partial_\sigma X^\mu(\partial_\sigma X_\mu)^* = \frac{1}{2}\sum_n\Big(&\alpha_n^\mu(\alpha_{-n})_\mu + \tilde{\alpha}_n^\mu(\tilde{\alpha}_{-n})_\mu \\ &- 2\alpha_n^\mu(\tilde{\alpha}_n)_\mu^{-4in\tau}\Big), \quad \text{closed string.}\end{aligned} \tag{2.238}$$

Hence[1]

$$H = \sum_n\Big(\alpha_n^\mu(\alpha_{-n})_\mu + \tilde{\alpha}_n^\mu(\tilde{\alpha}_{-n})_\mu\Big), \quad \text{closed string.} \tag{2.239}$$

In the above $lp^\mu = \alpha_0^\mu + \tilde{\alpha}_0^\mu$ and $\alpha_0^\mu = \tilde{\alpha}_0^\mu$.

Next we introduce the Virasoro operators which are the Fourier modes of the energy–momentum tensor. We compute for the closed string the combinations

$$\begin{aligned}\partial_+ X_L^\mu &= l\sum_n \tilde{\alpha}_n^\mu e^{-2in\tau}e^{-2in\sigma}, \\ \partial_- X_R^\mu &= l\sum_n \alpha_n^\mu e^{-2in\tau}e^{2in\sigma}, \quad \text{closed string.}\end{aligned} \tag{2.240}$$

The Fourier components of the constraints $T_{++} = 0$ and $T_{--} = 0$ for the closed string are given by

$$\begin{aligned}\tilde{L}_m(\tau) &= \frac{T}{2}\int_0^\pi d\sigma e^{2im\sigma}T_{++} \\ &= \frac{T}{2}\int_0^\pi d\sigma e^{2im\sigma}\partial_+ X_L^\mu\partial_+(X_L)_\mu \\ &= \frac{1}{2}\sum_n \tilde{\alpha}_n^\mu(\tilde{\alpha}_{m-n})_\mu e^{-2im\tau} \\ &= 0, \quad \text{closed string}\end{aligned} \tag{2.241}$$

[1] In [1] there is an extra 1/2 in front of this expression.

$$
\begin{aligned}
L_m(\tau) &= \frac{T}{2}\int_0^{\pi} d\sigma e^{-2im\sigma} T_{--} \\
&= \frac{T}{2}\int_0^{\pi} d\sigma e^{-2im\sigma} \partial_- X_R^{\mu} \partial_- (X_R)_{\mu} \\
&= \frac{1}{2}\sum_n \alpha_n^{\mu} (\alpha_{m-n})_{\mu} e^{-2im\tau} \\
&= 0, \quad \text{closed string.}
\end{aligned}
\tag{2.242}
$$

We can immediately see that (with $L_m(0) \equiv L_m$ and $\tilde{L}_m(0) \equiv \tilde{L}_m$)

$$H = 2(L_0 + \tilde{L}_0), \quad \text{closed string.} \tag{2.243}$$

For the open string we proceed as follows. First we compute the combinations

$$
\begin{aligned}
\partial_+ X_L^{\mu} &= \frac{l}{2}\sum_n \alpha_n^{\mu} e^{-in\tau} e^{-in\sigma}, \\
\partial_- X_R^{\mu} &= \frac{l}{2}\sum_n \alpha_n^{\mu} e^{-in\tau} e^{in\sigma}, \quad \text{open string.}
\end{aligned}
\tag{2.244}
$$

We remark that since $T_{++} = \partial_+ X_L^{\mu} \partial_+ X_{L\mu}$ and $T_{--} = \partial_- X_R^{\mu} \partial_- X_{R\mu}$ and $\partial_- X_R^{\mu}(-\sigma) = \partial_+ X_L^{\mu}(\sigma)$ we have

$$T_{--}(-\sigma) = T_{++}(\sigma), \quad \text{open string.} \tag{2.245}$$

Thus the constraints $T_{++} = T_{--} = 0$ over the interval $0 \leqslant \sigma \leqslant \pi$ are equivalent to the constraints $T_{++} = 0$ over the interval $-\pi \leqslant \sigma \leqslant \pi$. The Fourier components of the constraints $T_{++} = 0$ for the open string are given by

$$
\begin{aligned}
L_m(\tau) &= T\int_{-\pi}^{\pi} d\sigma e^{im\sigma} T_{++} \\
&= T\int_{-\pi}^{\pi} e^{im\sigma} \partial_+ X_L^{\mu} \partial_+ X_{L\mu} \\
&= \frac{1}{2}\sum_n \alpha_n^{\mu} (\alpha_{m-n})_{\mu} e^{-im\tau} \\
&= 0, \quad \text{open string.}
\end{aligned}
\tag{2.246}
$$

Now we see that (again with $L_m(0) \equiv L_m$)

$$H = L_0, \quad \text{open string.} \tag{2.247}$$

By using the fact that $p^{\mu} = \alpha_0^{\mu}/l$ for the open string and the constraint $L_0 = 0$ we compute the mass squared $M^2 = -p_{\mu}p^{\mu}$ of the open string to be given by

$$M^2 = \frac{1}{\alpha'}\sum_{n=1} \alpha_n^{\mu} (\alpha_{-n})_{\mu}, \quad \text{open string.} \tag{2.248}$$

Similarly for the closed string the fact $p^\mu = 2\alpha_0^\mu/l = 2\tilde{\alpha}_0^\mu/l$ and the constraints $L_0 = \tilde{L}_0 = 0$ lead to the mass squared

$$\begin{aligned} M^2 &= \frac{4}{\alpha'}\sum_{n=1}\alpha_n^\mu(\alpha_{-n})_\mu \\ &= \frac{4}{\alpha'}\sum_{n=1}\tilde{\alpha}_n^\mu(\tilde{\alpha}_{-n})_\mu \\ &= \frac{2}{\alpha'}\sum_{n=1}(\alpha_n^\mu(\alpha_{-n})_\mu + \tilde{\alpha}_n^\mu(\tilde{\alpha}_{-n})_\mu), \quad \text{closed string.} \end{aligned} \tag{2.249}$$

Exercise 4: Show by studying null states that the spectrum of the quantum theory is free from ghosts (negative norm states) only when (1) $a = 1$, $D = 26$ or (2) $a \leqslant 1$, $D \leqslant 25$.

Solution 4: We will only consider for simplicity the open strings. Generalization to the closed strings is straightforward. The ground state of the open string with center-of-mass momentum k^μ is $|0; k\rangle$. The constraint $L_0|0; k\rangle = a|0; k\rangle$ leads to the result

$$L_0|0; k\rangle = \frac{l^2}{2}p^\mu p_\mu|0; k\rangle = a|0; k\rangle \quad \Leftrightarrow \alpha' k^\mu k_\mu = a. \tag{2.250}$$

This ground state satisfies the other constraints by construction. The first excited level contains the states

$$|1; k\rangle \equiv \xi_\mu(k)\alpha_{-1}^\mu|0; k\rangle. \tag{2.251}$$

The $\xi^\mu(k)$ is a polarization D-vector. The constraint $L_0|1; k\rangle = a|1; k\rangle$ leads in this case to the result

$$\begin{aligned} L_0|1; k\rangle &= \frac{l^2}{2}p^\mu p_\mu|1; k\rangle + \xi_\nu\alpha_{-1}^\mu[(\alpha_1)_\mu, \alpha_{-1}^\nu]|0; k\rangle \\ &= a|1; k\rangle \quad \Leftrightarrow \alpha' k^\mu k_\mu = a - 1. \end{aligned} \tag{2.252}$$

For these excited states the constraint $L_1|1; k\rangle = 0$ leads also to a non-trivial result given by

$$L_1|1; k\rangle = lp_\mu\xi_\nu[\alpha_1^\mu, \alpha_{-1}^\nu]|0; k\rangle = 0 \quad \Leftrightarrow lk^\mu\xi_\mu = 0. \tag{2.253}$$

The norm of the state $|1; k\rangle$ is

$$\langle 1; k|1; k\rangle = \xi^\mu\xi_\mu. \tag{2.254}$$

The condition $k^\mu\xi_\mu = 0$ means that only $D - 1$ components of ξ^μ are independent. In other words there can only be $D - 1$ independent polarizations. We also remark that we can always choose the momentum k such that $k^\mu = (k^0, k^1, 0, \ldots, 0)$. The condition $k^\mu\xi_\mu = 0$ reads then $k^0\xi^0 = k^1\xi^1$. Clearly the $D - 2$ spacelike polarization vectors which are normal to the plane (0,1) have positive norm since they have $\xi^0 = 0$. They also have $\xi^1 = 0$. It remains to determine the sign of the norm of the one polarization vector which lies in the plane (0, 1). As it turns out this depends on the value of a since we must always have $-(k^0)^2 + (k^1)^2 = (a-1)/\alpha'$. If $a - 1 < 0$ then we can choose $k^1 = 0$ and hence from $k^0\xi^0 = k^1\xi^1$ we deduce $\xi^0 = 0$. In other words $\xi^\mu\xi_\mu > 0$. If $a - 1 > 0$ then we can choose $k^0 = 0$ and hence from $k^0\xi^0 = k^1\xi^1$ we deduce $\xi^1 = 0$. In other words $\xi^\mu\xi_\mu < 0$. Thus we get a negative norm state (a ghost) for $a > 1$. For $a - 1 = 0$ we see that we must have $k^0 = \pm k^1$ and hence $\xi^0 = \pm\xi^1$. In this case we can check that $k^\mu k_\mu = \xi^\mu\xi_\mu = 0$ and ξ^μ is proportional to k^μ. Thus for $a = 1$ the ground state is a scalar tachyon, the first excited state is a massless vector particle with $D - 2$ positive norm transverse polarizations and one zero norm longitudinal polarization state ($\xi^\mu \propto k^\mu$). For $a < 1$ this longitudinal component becomes of positive norm whereas for $a > 1$ it becomes of negative norm, i.e. a ghost. The condition for the absence of ghosts at the first excited level is therefore given by

$$a \leqslant 1. \tag{2.255}$$

As in QED the zero norm state (for $a = 1$) decouples from the S matrix.

The longitudinal polarization state $\xi^\mu \propto k^\mu$ is only the first null state we get for $a = 1$. In fact there is an infinite number of null states for $a = 1$ which can be constructed as follows. We define physical states $|\phi\rangle$ and spurious states $|\psi\rangle$ by the requirements

$$L_0|\phi\rangle = a|\phi\rangle, \quad L_m|\phi\rangle = 0, \quad m > 0, \quad \text{physical state} \tag{2.256}$$

$$L_0|\psi\rangle = a|\psi\rangle, \quad \langle\phi|\psi\rangle = 0, \quad \text{spurious state.} \tag{2.257}$$

Let P be the projector on the spurious state $|\psi\rangle$, viz $P = |\psi\rangle\langle\psi|$. Then by construction $P|\phi\rangle = 0$. Hence we must have

$$P = |\psi\rangle\langle\psi| = \sum_{m>0} X_{-m}L_m. \tag{2.258}$$

The X_{-m} are some operators. Clearly we have $\langle\psi| = \sum_{m>0}\langle\psi|X_{-m}L_m$ or equivalently

$$|\psi\rangle = \sum_{m>0} L_m^+ X_{-m}^+|\psi\rangle = \sum_{m>0} L_{-m}|\chi_m\rangle. \tag{2.259}$$

From the condition $L_0|\psi\rangle = a|\psi\rangle$ we deduce that

$$L_0|\chi_m\rangle = (a - m)|\chi_m\rangle. \tag{2.260}$$

A null state is a state of zero norm. A state which is both physical and spurious is a null physical state. Indeed given a spurious state $|\psi\rangle$ which is also physical, viz

$$\langle\phi|\psi\rangle = 0, \quad L_0|\psi\rangle = a|\psi\rangle, \quad L_m|\psi\rangle = 0, \quad m > 0, \tag{2.261}$$

where ϕ is an arbitrary physical state we can show that

$$\langle\psi|\psi\rangle = \sum_{m>0} \langle\psi|X_{-m}L_m|\psi\rangle = 0. \tag{2.262}$$

Null states for $a = 1$ can be constructed as follows. Let $|\tilde{\chi}\rangle$ be the state defined by the equation

$$L_0|\tilde{\chi}\rangle = (a-1)|\tilde{\chi}\rangle, \quad L_m|\tilde{\chi}\rangle = 0, \quad m > 0. \tag{2.263}$$

Then the state

$$|\psi\rangle = L_{-1}|\tilde{\chi}\rangle \tag{2.264}$$

is a null physical state when $a = 1$. Indeed we can verify that $|\psi\rangle$ is a spurious state, viz

$$\langle\phi|\psi\rangle = \langle\tilde{\chi}|L_1|\phi\rangle^* = 0, \quad L_0|\psi\rangle = L_0L_{-1}|\tilde{\chi}\rangle = a|\psi\rangle. \tag{2.265}$$

We also compute

$$L_1|\psi\rangle = [L_1, L_{-1}]|\tilde{\chi}\rangle = 2L_0|\tilde{\chi}\rangle = 2(a-1)|\tilde{\chi}\rangle. \tag{2.266}$$

$$L_m|\psi\rangle = [L_m, L_{-1}]|\tilde{\chi}\rangle = (m+1)L_{m-1}|\tilde{\chi}\rangle = 0, \quad m > 1. \tag{2.267}$$

Clearly for $a = 1$ we have $L_m|\psi\rangle = 0$, $m > 0$ and hence the spurious state $|\psi\rangle$ is also a physical state and as a consequence it is of zero norm. The first example of the null states $|\psi\rangle$ is obtained by taking $|\tilde{\chi}\rangle = |0; k\rangle$. In this case we get $|\psi\rangle = L_{-1}|\tilde{\chi}\rangle = lk_\mu\alpha_{-1}^\mu|0; k\rangle$ which is the longitudinal null state of the massless vector particle discussed earlier. Note that $L_0|0; k\rangle = 0$ since $k^\mu k_\mu = 0$ and $L_m|0; k\rangle = 0$ for $m > 0$.

There are many more null states for $a = 1$ when $D = 26$. Indeed introduce the state $|\tilde{\chi}\rangle$ defined by the equation

$$L_0|\tilde{\chi}\rangle = -|\tilde{\chi}\rangle, \quad L_m|\tilde{\chi}\rangle = 0, \quad m > 0. \tag{2.268}$$

Next consider the state

$$|\psi\rangle = L_{-2}|\tilde{\chi}\rangle + \gamma L_{-1}^2|\tilde{\chi}\rangle. \tag{2.269}$$

We can immediately compute

$$L_0|\psi\rangle = |\psi\rangle. \tag{2.270}$$

In other words $a = 1$. Also given any physical state $|\phi\rangle$ we compute

$$\langle\phi|\psi\rangle = \langle\tilde{\chi}|L_2|\phi\rangle^* + \gamma\langle\tilde{\chi}|L_1^2|\phi\rangle^* = 0. \tag{2.271}$$

Hence $|\psi\rangle$ is a spurious state. Also we compute

$$\begin{aligned} L_1|\psi\rangle &= (3-2\gamma)L_{-1}|\tilde{\chi}\rangle \\ L_2|\psi\rangle &= \left(\frac{D}{2} - 4 - 6\gamma\right)|\tilde{\chi}\rangle \\ L_m|\psi\rangle &= 0, \quad m > 2. \end{aligned} \tag{2.272}$$

In other words $|\psi\rangle$ is a physical state if and only if $3 - 2\gamma = 0$ and $\frac{D}{2} - 4 - 6\gamma = 0$ or equivalently $\gamma = 3/2$ and $D = 26$. Thus for $a = 1$ all the states $|\psi\rangle$ with $\gamma = 3/2$ are null physical states when $D = 26$.

Physical states of negative norm, i.e. ghosts can only be constructed when $D > 26$. It can be shown in general that the spectrum of the theory is free from ghosts (negative norm states) only when (1) $a = 1, D = 26$ or (2) $a \leqslant 1, D \leqslant 25$. The first case corresponds to critical strings which can only have transverse excitations. The physical Hilbert space contains in the critical dimension many more null states and the corresponding spectrum is generated by $D - 2 = 24$ oscillators.

Exercise 5: Construct explicitly the light-cone gauge.

Solution 5: The starting point of the covariant approach is the invariance of the action under the diffeomorphism (or reparametrization) transformations of the worldsheet and under the Weyl scalings given, respectively, by

$$\begin{aligned} X'^{\mu}(\sigma', \tau') &= X^{\mu}(\sigma, \tau) \\ h'^{cd}(\sigma', \tau') &= h^{ab}(\sigma, \tau)\frac{\partial\sigma'^{c}}{\partial\sigma^{a}}\frac{\partial\sigma'^{d}}{\partial\sigma^{b}} \end{aligned} \tag{2.273}$$

$$\begin{aligned} X'^{\mu}(\sigma, \tau) &= X^{\mu}(\sigma, \tau) \\ h'_{ab}(\sigma, \tau) &= e^{2\omega(\sigma, \tau)}h_{ab}(\sigma, \tau). \end{aligned} \tag{2.274}$$

We use these symmetries to choose the metric h such that $h_{ab} = \eta_{ab} = (-1,1)$. As it turns out this gauge choice is preserved under particular combinations of reparametrization transformations and Weyl scalings. These residual symmetries can be constructed as follows. Infinitesimal reparametrization transformations and Weyl scalings are obtained by setting $\sigma_c' = \sigma_c + \xi_c$ and $2\omega = -\Lambda$. We get

$$\begin{aligned} \delta X^{\mu} &= X^{\mu}(\sigma, \tau) - X'^{\mu}(\sigma, \tau) = \xi^{c}\partial_{c}X^{\mu}(\sigma, \tau) \\ \delta h^{ab} &= h^{ab}(\sigma, \tau) - h'^{ab}(\sigma, \tau) \\ &= \xi^{c}\partial_{c}h^{ab}(\sigma, \tau) - h^{ac}(\sigma, \tau)\partial_{c}\xi^{b} - h^{bc}(\sigma, \tau)\partial_{c}\xi^{a} \end{aligned} \tag{2.275}$$

$$\delta h^{ab} = h^{ab}(\sigma, \tau) - h'^{ab}(\sigma, \tau) = \Lambda h^{ab}. \tag{2.276}$$

We can see that if we start from $h_{ab} = \eta_{ab}$ then under an infinitesimal reparametrization transformation we get the metric

$$h^{ab}(\sigma, \tau) = \eta^{ab} + \partial^{a}\xi^{b} + \partial^{b}\xi^{a}. \tag{2.277}$$

Next we perform a Weyl scaling such that

$$\eta_{ab} = e^{-\Lambda(\sigma, \tau)}h_{ab}(\sigma, \tau). \tag{2.278}$$

In other words the metric can be brought back to the gauge choice $h_{ab} = \eta_{ab}$ if and only if we choose the Weyl scaling such that

$$\partial^a \xi^b + \partial^b \xi^a = \Lambda \eta^{ab}. \tag{2.279}$$

This last equation is equivalent to (with $\xi^{\pm} = \xi^0 \pm \xi^1$)

$$\partial_- \xi^+ = \partial_+ \xi^- = 0, \quad \partial_+ \xi^+ + \partial_- \xi^- = \Lambda. \tag{2.280}$$

In other words ξ^+ is a function of σ_+ alone and $\xi^-$is a function of σ_- alone. Hence the residual symmetries of the gauge-fixed theory consist of transforming the world-sheet coordinates $\sigma_\pm$ as

$$\sigma_+ \longrightarrow \sigma'_+ = \sigma'_+(\sigma_+), \quad \sigma_- \longrightarrow \sigma'_- = \sigma'_-(\sigma_-). \tag{2.281}$$

For closed strings these transformations are independent reparametrizations of σ_+ and σ_- while for open strings we must take into account the boundary conditions. The last equation above can be put in the form

$$\begin{aligned} \tau &\longrightarrow \tau' = \frac{1}{2}(\sigma'_+(\sigma_+) + \sigma'_-(\sigma_-)) \\ \sigma &\longrightarrow \sigma' = \frac{1}{2}(\sigma'_+(\sigma_+) - \sigma'_-(\sigma_-)). \end{aligned} \tag{2.282}$$

We can immediately verify that τ' must satisfy the equation of motion

$$(-\partial_\tau^2 + \partial_\sigma^2)\tau' = -4\partial_+\partial_-\tau' = 0. \tag{2.283}$$

Let us recall that in the conformal gauge $h_{ab} = \text{diag}(-1, 1)$ the equations of motion of the space–time coordinates X^μ take the form

$$(-\partial_\tau^2 + \partial_\sigma^2)X_\mu = -4\partial_+\partial_- X_\mu = 0. \tag{2.284}$$

Hence we can use the residual symmetries of the gauge-fixed string action to choose τ' to be equal to any one of the space–time coordinates X^μ. The light-cone gauge corresponds to the choice

$$X^+(\sigma, \tau) = x^+ + l^2 p^+ \tau. \tag{2.285}$$

Thus we are setting the oscillators α_n^+, $n > 0$, to 0. The X^+ is the time coordinate as seen in a frame in which the string moves at infinite momentum. This time is the same for all points on the string since X^+ is independent of σ.

Exercise 6: Give a complete proof of Lorentz invariance.

Solution 6. For open strings the quantum Lorentz generators are given by the classical expressions since there is no ordering ambiguity, viz

$$J^{\mu\nu} = L^{\mu\nu} + S^{\mu\nu}, \quad \text{open string}$$
$$L^{\mu\nu} = x^\mu p^\nu - x^\nu p^\mu, \quad S^{\mu\nu} = -i\sum_{n=1}^{\infty}\frac{1}{n}(\alpha_{-n}^{\mu}\alpha_n^{\nu} - \alpha_{-n}^{\nu}\alpha_n^{\mu}). \tag{2.286}$$

Let us recall that $\alpha_0^\mu = lp^\mu$, $[x^\mu, x^\nu] = 0$, $[x^\mu, p^\nu] = i\eta^{\mu\nu}$ and

$$[p^\mu, p^\nu] = 0. \tag{2.287}$$

By using also the commutation relations $[p^\mu, \alpha_n^\mu] = 0$ we compute

$$[p^\mu, J^{\nu\rho}] = [p^\mu, L^{\nu\rho}] = -i\eta^{\mu\nu}p^\rho + i\eta^{\mu\rho}p^\nu. \tag{2.288}$$

We note the results

$$[L^{\mu\nu}, x^\rho] = -ix^\mu\eta^{\nu\rho} + ix^\nu\eta^{\mu\rho} \tag{2.289}$$

$$[L^{\mu\nu}, p^\lambda] = -ip^\mu\eta^{\nu\lambda} + ip^\nu\eta^{\mu\lambda} \tag{2.290}$$

$$[L^{\mu\nu}, x^\rho p^\lambda] = -ix^\mu p^\lambda\eta^{\nu\rho} + ix^\nu p^\lambda\eta^{\mu\rho} + ix^\rho p^\nu\eta^{\mu\lambda} - ix^\rho p^\mu\eta^{\nu\lambda}. \tag{2.291}$$

Hence we obtain

$$[L^{\mu\nu}, L^{\rho\lambda}] = iL^{\mu\rho}\eta^{\nu\lambda} - iL^{\mu\lambda}\eta^{\nu\rho} - iL^{\nu\rho}\eta^{\mu\lambda} + iL^{\nu\lambda}\eta^{\mu\rho}. \tag{2.292}$$

Similarly we compute

$$[S^{\mu\nu}, \alpha_m^\rho] = -i\alpha_m^\mu\eta^{\nu\rho} + i\alpha_m^\nu\eta^{\mu\rho} \tag{2.293}$$

$$\begin{aligned}[S^{\mu\nu}, \alpha_{-m}^\rho\alpha_m^\lambda] = &- i\alpha_{-m}^\mu\alpha_m^\lambda\eta^{\nu\rho} + i\alpha_{-m}^\nu\alpha_m^\lambda\eta^{\mu\rho} \\ &- i\alpha_{-m}^\rho\alpha_m^\mu\eta^{\nu\lambda} + i\alpha_{-m}^\rho\alpha_m^\nu\eta^{\mu\lambda}.\end{aligned} \tag{2.294}$$

Thus we obtain

$$[S^{\mu\nu}, S^{\rho\lambda}] = iS^{\mu\rho}\eta^{\nu\lambda} - iS^{\mu\lambda}\eta^{\nu\rho} - iS^{\nu\rho}\eta^{\mu\lambda} + iS^{\nu\lambda}\eta^{\mu\rho}. \tag{2.295}$$

Also by using the commutation relations $[x^\mu, \alpha_n^\nu] = 0$ we compute

$$[L^{\mu\nu}, S^{\rho\lambda}] = 0. \tag{2.296}$$

Putting the above equations together we obtain

$$[J^{\mu\nu}, J^{\rho\lambda}] = iJ^{\mu\rho}\eta^{\nu\lambda} - iJ^{\mu\lambda}\eta^{\nu\rho} - iJ^{\nu\rho}\eta^{\mu\lambda} + iJ^{\nu\lambda}\eta^{\mu\rho}. \tag{2.297}$$

From equations (2.287), (2.288) and (2.297) we conclude that the covariant quantum operators $J^{\mu\nu}$ form a Poincaré algebra without any anomaly. This is expected since the covariant quantization approach is manifestly Lorentz invariant.

The situation in the light-cone formalism is much more involved. For example, Lorentz transformations which affect the coordinate X^+ must be compensated by reparametrization transformations of the worldsheet which are compatible with the gauge choice $h_{ab} = \eta_{ab}$ in order to preserve the form of the light-cone gauge condition. Thus, to prove Lorentz invariance in the light-cone gauge will be a complicated exercise. The quantum Lorentz generators in the light-cone gauge are given by

$$\begin{aligned} J^{ij} &= L^{ij} + S^{ij}, \quad \text{open string} \\ L^{ij} &= x^i p^j - x^j p^i, \\ S^{ij} &= -i \sum_{n=1}^{\infty} \frac{1}{n}(\alpha_{-n}^i \alpha_n^j - \alpha_{-n}^j \alpha_n^i) \end{aligned} \tag{2.298}$$

$$\begin{aligned} J^{-+} &= L^{-+} = x^- p^+ - x^+ p^- \\ J^{i+} &= L^{i+} = x^i p^+ - x^+ p^i \end{aligned} \tag{2.299}$$

$$\begin{aligned} J^{i-} &= L^{i-} + S^{i-} \\ L^{i-} &= x^i p^- - x^- p^i, \\ S^{i-} &= -i \sum_{n=1}^{\infty} \frac{1}{n}(\alpha_{-n}^i \alpha_n^- - \alpha_{-n}^- \alpha_n^i). \end{aligned} \tag{2.300}$$

In the second equation we have used the fact that $S^{\mu+} = 0$ since in the light-cone gauge we have $\alpha_n^+ = 0$. In the third equation we must on the other hand use the fact that α_n^-, $n \neq 0$ is given by

$$\begin{aligned} \alpha_n^- &= \frac{1}{2lp^+} \sum_m \alpha_{n-m}^i \alpha_m^i = \frac{p^i}{2p^+}\alpha_n^i + \frac{1}{2lp^+} \sum_{m\neq 0} \alpha_{n-m}^i \alpha_m^i = \frac{p^i}{p^+}\alpha_n^i \\ &+ \frac{1}{2lp^+} \sum_{m\neq 0,n} \alpha_{n-m}^i \alpha_m^i. \end{aligned} \tag{2.301}$$

In other words S^{i-} (and hence J^{i-}) is cubic in the transverse oscillators. This is the main potential obstruction to exact Lorentz invariance. We can verify that all commutators of the light-cone Lorentz generators are in fact consistent with the Poincaré algebra with the exception of $[J^{i-}, J^{j-}]$ which we will now compute in detail.

From the identities $[x^j, \alpha_n^i] = [x^j, \alpha_{-n}^i] = 0$ and

$$[x^j, \alpha_n^-] = \frac{i}{p^+}\alpha_n^j, \quad [x^j, \alpha_{-n}^-] = \frac{i}{p^+}\alpha_{-n}^j. \tag{2.302}$$

We compute

$$[x^j, S^{i-}] = -\frac{i}{p^+}S^{ji}. \tag{2.303}$$

Next we derive the identity

$$\begin{aligned} \left[x^-, \frac{1}{p^+}\right] &= \left[x^-, \frac{1}{(p^+)^2}\right]p^+ - \frac{i}{(p^+)^2} \\ &= \left[x^-, \frac{1}{p^+}\right] + \frac{1}{p^+}\left[x^-, \frac{1}{p^+}\right]p^+ - \frac{i}{(p^+)^2}. \end{aligned} \tag{2.304}$$

From here we get the result

$$\frac{1}{p^+}\left[x^-, \frac{1}{p^+}\right]p^+ - \frac{i}{(p^+)^2} = 0 \Leftrightarrow \left[x^-, \frac{1}{p^+}\right] = \frac{i}{(p^+)^2}. \tag{2.305}$$

Using this result we compute the commutator

$$\begin{aligned}[x^-, S^{i-}] &= \frac{i}{(p^+)^2}\hat{S}^{i-} + \frac{1}{p^+}[x^-, \hat{S}^{i-}] \\ &= \frac{i}{p^+}S^{i-}.\end{aligned} \tag{2.306}$$

In the above $S^{i-} = \hat{S}^{i-}/p^+$ and we have used the result $[x^-, \hat{S}^{i-}] = 0$ since $\hat{S}^{i-}$ does not contain either p^0 or p^{D-1}. Furthermore since the momenta p^i and $p^{\pm}$ commute we get immediately the commutators

$$[p^j, S^{i-}] = [p^{\pm}, S^{i-}] = 0. \tag{2.307}$$

Then we compute

$$[L^{i-}, S^{j-}] = -\frac{i}{p^+}S^{ij}p^- - \frac{i}{p^+}S^{j-}p^i \tag{2.308}$$

$$[L^{i-}, L^{j-}] = 0. \tag{2.309}$$

Thus

$$\begin{aligned}[J^{i-}, J^{j-}] = &- \frac{2i}{p^+}S^{ij}p^- - \frac{i}{p^+}S^{j-}p^i \\ &+ \frac{i}{p^+}S^{i-}p^j + [S^{i-}, S^{j-}].\end{aligned} \tag{2.310}$$

Since we want the commutator $[S^{i-}, S^{j-}]$ we need to evaluate the commutators $[S^{i-}, \alpha_n^j]$ and $[S^{i-}, \alpha_n^-]$. For n, $m \neq 0$ we compute using the identity $[\alpha_n^i, \alpha_m^j] = n\delta_{n+m,0}\eta^{ij}$ the basic commutator

$$[\alpha_n^-, \alpha_m^j] = -\frac{m}{lp^+}\alpha_{n+m}^j. \tag{2.311}$$

This result holds also for $n = 0$. Immediately we obtain

$$[S^{i-}, \alpha_n^j] = 2i\eta^{ij}\alpha_n^- + \frac{in}{lp^+}\sum_{m\neq 0}\frac{1}{m}\alpha_{-m}^i\alpha_{m+n}^j. \tag{2.312}$$

Next by using the commutator

$$\begin{aligned}[\alpha_n^-, \alpha_m^-] &= \frac{1}{lp^+}(n - m)\alpha_{n+m}^- + \frac{R(n)}{l^2(p^+)^2}\delta_{n+m,0}, \\ R(n) &= \frac{D-2}{12}(m^3 - m) + 2am.\end{aligned} \tag{2.313}$$

We compute for $n \neq 0$ the commutator

$$
\begin{aligned}
[S^{i-}, \alpha_n^-] = &- \frac{i}{l^2(p^+)^2}\frac{R(n)}{n}\alpha_n^i + \frac{i}{lp^+}\sum_{m=1}\frac{n}{m}(\alpha_{-m}^i\alpha_{m+n}^- - \alpha_{n-m}^-\alpha_m^i) \\
&+ \frac{i}{lp^+}\sum_{m=1}(\alpha_{n-m}^i\alpha_m^- + \alpha_{-m}^-\alpha_{n+m}^i) \\
&- \frac{i}{lp^+}\sum_{m=1}(\alpha_{-m}^i\alpha_{m+n}^- + \alpha_{-m+n}^-\alpha_m^i).
\end{aligned}
\tag{2.314}
$$

For $n > 0$ we can bring this to the form

$$
\begin{aligned}
[S^{i-}, \alpha_n^-] = &- \frac{i}{l^2(p^+)^2}\frac{R(n)}{n}\alpha_n^i + \frac{i}{lp^+}\sum_{m=1}\frac{n}{m}(\alpha_{-m}^i\alpha_{m+n}^- - \alpha_{n-m}^-\alpha_m^i) \\
&+ \frac{i}{lp^+}\sum_{m=1}^{n}\alpha_{n-m}^i\alpha_m^- - \frac{i}{lp^+}\sum_{m=0}^{n-1}\alpha_m^-\alpha_{n-m}^i \\
= &- \frac{i}{l^2(p^+)^2}\frac{R(n)}{n}\alpha_n^i + \frac{i}{lp^+}\sum_{m=1}\frac{n}{m}(\alpha_{-m}^i\alpha_{m+n}^- - \alpha_{n-m}^-\alpha_m^i) \\
&+ \frac{i}{p^+}(p^i\alpha_n^- - p^-\alpha_n^i) + \frac{i}{l^2(p^+)^2}\frac{n(n-1)}{2}\alpha_n^i.
\end{aligned}
\tag{2.315}
$$

For $-n < 0$ we obtain

$$
\begin{aligned}
[S^{i-}, \alpha_{-n}^-] = &- \frac{i}{l^2(p^+)^2}\frac{R(n)}{n}\alpha_{-n}^i - \frac{i}{lp^+}\sum_{m=1}\frac{n}{m}(\alpha_{-m}^i\alpha_{m-n}^- - \alpha_{-n-m}^-\alpha_m^i) \\
&- \frac{i}{lp^+}\sum_{m=0}^{n-1}\alpha_{m-n}^i\alpha_{-m}^- + \frac{i}{lp^+}\sum_{m=1}^{n}\alpha_{-m}^-\alpha_{m-n}^i \\
= &- \frac{i}{l^2(p^+)^2}\frac{R(n)}{n}\alpha_{-n}^i - \frac{i}{lp^+}\sum_{m=1}\frac{n}{m}(\alpha_{-m}^i\alpha_{m-n}^- - \alpha_{-n-m}^-\alpha_m^i) \\
&+ \frac{i}{p^+}(p^i\alpha_{-n}^- - p^-\alpha_{-n}^i) + \frac{i}{l^2(p^+)^2}\frac{n(n-1)}{2}\alpha_{-n}^i.
\end{aligned}
\tag{2.316}
$$

We are now ready to compute the commutator $[S^{i-}, S^{j-}]$. We start from

$$
\begin{aligned}
[S^{i-}, S^{j-}] = &- i\sum_{n=1}\frac{1}{n}([S^{i-}, \alpha_{-n}^j]\alpha_n^- - \alpha_{-n}^-[S^{i-}, \alpha_n^j]) \\
&- i\sum_{n=1}\frac{1}{n}(\alpha_{-n}^j[S^{i-}, \alpha_n^-] - [S^{i-}, \alpha_{-n}^-]\alpha_n^j).
\end{aligned}
\tag{2.317}
$$

The first line is equal to

$$
\begin{aligned}
\text{first line} = &- \frac{1}{lp^+}\sum_{n=1}\sum_{m=1}\frac{1}{m}(\alpha_{-m}^i\alpha_{m-n}^j\alpha_n^- + \alpha_{-n}^-\alpha_{-m}^i\alpha_{m+n}^j \\
&- \alpha_m^i\alpha_{-m-n}^j\alpha_n^- - \alpha_{-n}^-\alpha_m^i\alpha_{-m+n}^j).
\end{aligned}
\tag{2.318}
$$

Th second line is equal to

$$\text{second line} = \frac{i}{p^+}(S^{ij}p^- + S^{j-}p^i) + \frac{1}{l^2(p^+)^2}\sum_{n=1}\left(\frac{n-1}{2} - \frac{R(n)}{n^2}\right)(\alpha_{-n}^j\alpha_n^i - \alpha_{-n}^i\alpha_n^j) + \text{second line}_1 \tag{2.319}$$

$$\text{second line}_1 = \frac{1}{lp^+}\sum_{n=1}\sum_{m=1}\frac{1}{m}(\alpha_{-m}^i\alpha_{-n}^j\alpha_{m+n}^- + \alpha_{-m}^i\alpha_{m-n}^-\alpha_n^j - \alpha_{-n}^j\alpha_{n-m}^-\alpha_m^i - \alpha_{-m-n}^-\alpha_m^i\alpha_n^j). \tag{2.320}$$

We have

$$\begin{aligned}
\text{first line} + \text{second line}_1 &= -\frac{1}{lp^+}\sum_{m=1}\frac{1}{m}\sum_{n=1}^{m}(\alpha_{-m}^i\alpha_{m-n}^j\alpha_n^- - \alpha_{-n}^-\alpha_m^i\alpha_{n-m}^j)\\
&+ \frac{1}{lp^+}\sum_{m=1}\frac{1}{m}\sum_{n=0}^{m-1}(-\alpha_m^i\alpha_{n-m}^j\alpha_{-n}^- + \alpha_n^-\alpha_{-m}^i\alpha_{-n+m}^j)\\
&+ \frac{1}{l^2(p^+)^2}\sum_{n=1}\sum_{m=1}(\alpha_{-n}^j\alpha_n^i - \alpha_{-n}^i\alpha_n^j) + \frac{p^-}{p^+}\eta^{ij}\sum_{m=1}1\\
&= -\frac{1}{lp^+}\sum_{m=1}\frac{1}{m}\sum_{n=1}^{m}(\alpha_{-m}^i\alpha_{m-n}^j\alpha_n^- - \alpha_{-n}^-\alpha_m^i\alpha_{n-m}^j)\\
&+ \frac{1}{lp^+}\sum_{m=1}\frac{1}{m}\sum_{n=1}^{m}(-\alpha_m^i\alpha_{n-m}^j\alpha_{-n}^- + \alpha_n^-\alpha_{-m}^i\alpha_{-n+m}^j)\\
&+ i\frac{p^-}{p^+}S^{ij} - \frac{p^-}{p^+}\eta^{ij}\sum_{m=1}1 - i\frac{p^j}{p^+}S^{i-}\\
&+ \frac{1}{l^2(p^+)^2}\sum_{n=1}\sum_{m=1}(\alpha_{-n}^j\alpha_n^i - \alpha_{-n}^i\alpha_n^j) + \frac{p^-}{p^+}\eta^{ij}\sum_{m=1}1\\
&= \frac{1}{lp^+}\sum_{m=1}\frac{1}{m}\sum_{n=1}^{m}([\alpha_{-n}^-, \alpha_m^i\alpha_{n-m}^j] + [\alpha_n^-, \alpha_{-m}^i\alpha_{m-n}^j])\\
&+ i\frac{p^-}{p^+}S^{ij} - i\frac{p^j}{p^+}S^{i-}\\
&+ \frac{1}{l^2(p^+)^2}\sum_{n=1}\sum_{m=1}(\alpha_{-n}^j\alpha_n^i - \alpha_{-n}^i\alpha_n^j)\\
&= -\frac{1}{l^2(p^+)^2}\sum_{m=1}\sum_{n=1}^{m}(\alpha_{n-m}^j\alpha_{m-n}^i - \alpha_{n-m}^i\alpha_{m-n}^j)\\
&- \frac{1}{l^2(p^+)^2}\sum_{m=1}\frac{1-m}{2}(\alpha_{-m}^j\alpha_m^i - \alpha_{-m}^i\alpha_m^j)\\
&+ i\frac{p^-}{p^+}S^{ij} - i\frac{p^j}{p^+}S^{i-}\\
&+ \frac{1}{l^2(p^+)^2}\sum_{n=1}\sum_{m=1}(\alpha_{-n}^j\alpha_n^i - \alpha_{-n}^i\alpha_n^j).
\end{aligned} \tag{2.321}$$

Define $V_{m-n} = \alpha^j_{-(m-n)}\alpha^i_{m-n} - \alpha^i_{-(m-n)}\alpha^j_{m-n}$. We use the crucial identity

$$\begin{aligned}\sum_{m=1}\sum_{n=1}^{m} V_{m-n} &= \sum_{m=0}\sum_{n=0}^{m} V_{m-n} - \sum_{m=1} V_m - V_0 \\ &= \sum_{n=1}\sum_{m=1} V_m - \sum_{m=1}(m+1)V_m + \left(\sum_{n=1} 1\right)V_0 - V_0.\end{aligned} \tag{2.322}$$

We then use the result $V_0 = 0$ in this last equation. We then obtain

$$\begin{aligned}\text{first line + second line}_1 &= \frac{1}{l^2(p^+)^2}\sum_{m=1}\frac{1+3m}{2}(\alpha^j_{-m}\alpha^i_m - \alpha^i_{-m}\alpha^j_m) \\ &\quad + i\frac{p^-}{p^+}S^{ij} - i\frac{p^j}{p^+}S^{i-}.\end{aligned} \tag{2.323}$$

The commutator of the spin generators S^{i-} is therefore given by

$$\begin{aligned}[S^{i-}, S^{j-}] &= \frac{i}{p^+}(2S^{ij}p^- + S^{j-}p^i - S^{i-}p^j) \\ &\quad + \frac{1}{l^2(p^+)^2}\sum_{n=1}\left(2n - \frac{R(n)}{n^2}\right)(\alpha^j_{-n}\alpha^i_n - \alpha^i_{-n}\alpha^j_n).\end{aligned} \tag{2.324}$$

As a consequence the commutator of the total angular momentum generators J^{i-} will be given by

$$\begin{aligned}[J^{i-}, J^{j-}] &= \frac{1}{l^2(p^+)^2}\sum_{n=1}\left(2n - \frac{R(n)}{n^2}\right)(\alpha^j_{-n}\alpha^i_n - \alpha^i_{-n}\alpha^j_n) \\ &= \frac{1}{l^2(p^+)^2}\sum_{n=1}\left(\frac{26-D}{12}n + \left[\frac{D-26}{12} + 2(1-a)\right]\frac{1}{n}\right) \\ &\quad (\alpha^j_{-n}\alpha^i_n - \alpha^i_{-n}\alpha^j_n).\end{aligned} \tag{2.325}$$

If the light-cone formalism is Lorentz invariant then we must have

$$[J^{i-}, J^{j-}] = 0. \tag{2.326}$$

We conclude immediately that we must have

$$D = 26, \quad a = 1. \tag{2.327}$$

References

[1] Green M B, Schwarz J H and Witten E 1987 *Superstring Theory. Vol. 1: Introduction* (Cambridge Monographs On Mathematical Physics) (Cambridge: Cambridge University Press)

[2] Becker K, Becker M and Schwarz J H 2006 *String theory and M-theory: A modern introduction* (Cambridge: Cambridge University Press)

IOP Publishing

Matrix Models of String Theory

Badis Ydri

Chapter 3

Polyakov path integral

The quantization of the bosonic string via the Polyakov path integral is crucial for a proper and profound understanding of the spectrum and symmetries of the open and closed strings. In this chapter we will mainly follow [1, 2].

3.1 Gauge fixing and Fadeev–Popov ghosts

The starting point is the path integral

$$Z = \int Dh(\sigma)DX(\sigma)\, e^{iS[h,\, X]}$$
$$S[h,\, X] = -\frac{1}{2\pi}\int_{\Sigma} d^2\sigma\sqrt{h}\;\; h^{ab}\partial_a X^{\mu}\partial_b X_{\mu}. \tag{3.1}$$

We have set $T = 1/\pi$ or equivalently $\alpha' = 1/2$. We use the two reparametrization invariances and the Weyl symmetry to impose the gauge conditions

$$h_{ab} = e^{\phi}\eta_{ab}, \quad h^{ab} = e^{-\phi}\eta^{ab}. \tag{3.2}$$

Thus we have the metric $ds^2 = e^{\phi}(-d\tau^2 + d\sigma^2) = -e^{\phi}d\sigma^+ d\sigma^- = h_{ab}d\sigma^a d\sigma^b$ where $\sigma^{\pm} = \tau \pm \sigma$. The gauge conditions read therefore

$$h_{++} = h_{--} = 0, \quad h_{+-} = h_{-+} = -\frac{1}{2}e^{\phi}. \tag{3.3}$$

The inverse metric is defined by $ds^2 = -e^{-\phi}d\sigma_+ d\sigma_- = h^{ab}d\sigma_a d\sigma_b$ and thus we can also write the gauge conditions in the form

$$h^{++} = h^{--} = 0, \quad h^{+-} = h^{-+} = -2e^{-\phi}. \tag{3.4}$$

We lower and raise indices by the rules

$$t^+ = -2e^{-\phi}t_-, \quad t^- = -2e^{-\phi}t_+$$
$$t_+ = -\frac{1}{2}e^{\phi}t^-, \quad t_- = -\frac{1}{2}e^{\phi}t^+. \tag{3.5}$$

Under reparametrization transformations of the worldsheet coordinates we have

$$\begin{aligned}\delta h^{ab} = h^{ab}(\sigma, \tau) - h'^{ab}(\sigma, \tau) &= \xi^c\partial_c h^{ab}(\sigma, \tau) - h^{ac}(\sigma, \tau)\partial_c\xi^b - h^{bc}(\sigma, \tau)\partial_c\xi^a \\ &= -\xi^c\partial_c\phi\ h^{ab} - \partial^a\xi^b - \partial^b\xi^a.\end{aligned} \tag{3.6}$$

Thus, the gauge conditions $h^{++} = h^{--} = 0$ transform as

$$\delta h^{++} = -2\partial^+\xi^+, \quad \delta h^{--} = -2\partial^-\xi^-. \tag{3.7}$$

The first equation $\delta h^{++} = -2\partial^+\xi^+$ can be put in the equivalent form

$$\begin{aligned}(h^{+-})^2\delta h_{--} &= -2h^{+-}\partial_-\xi^+ \\ &= -2(h^{+-})^2\partial_-\xi_- - 2h^{+-}\partial_-h^{+-}\xi_-.\end{aligned} \tag{3.8}$$

This leads to

$$\delta h_{--} = -2(\partial_-\xi_- - \partial_-\phi\xi_-). \tag{3.9}$$

Similarly, we compute

$$\delta h_{++} = -2(\partial_+\xi_+ - \partial_+\phi\xi_+). \tag{3.10}$$

These two last equations can be rewritten in the form

$$\delta h_{--} = -2\nabla_-\xi_-, \quad \delta h_{++} = -2\nabla_+\xi_+. \tag{3.11}$$

Indeed, the covariant derivative is defined by

$$\nabla_a\xi_b = \partial_a\xi_b - \Gamma^c_{ab}\xi_c. \tag{3.12}$$

The Christoffel connection is defined as usual by

$$\Gamma^c_{ab} = \frac{1}{2}h^{cd}(\partial_a h_{bd} + \partial_b h_{ad} - \partial_d h_{ab}). \tag{3.13}$$

For the present conformally flat metric we compute

$$\Gamma^+_{++} = \partial_+\phi, \quad \Gamma^-_{--} = \partial_-\phi. \tag{3.14}$$

The rest are zero.

In gauge theory when we impose a gauge-fixing condition $G(A) = 0$ (say the Lorentz gauge $G(A) = \partial_\mu A^\mu$) we do so by inserting in the path integral an identity in the form

$$1 = \int dG\delta(G) = \int dG^g\delta(G^g) = \int Dg \det\left(\frac{\delta G^g}{\delta g}\right)\delta(G^g) \tag{3.15}$$

where $G \equiv G(A)$ and $G^g \equiv G(A^g)$, and A^g is the gauge-transformed field. The Dg is the measure over the group manifold.

In our case the gauge group is replaced by the group of reparametrizations of the string worldsheet Σ and Dg is the integration over this group manifold. The field is the metric h and we denote the transformed metric by h^g. The gauge-fixing conditions we wish to impose reads here $h_{++} = h_{--} = 0$. The analogue of the above identity is then

$$1 = \int Dg \det\left(\frac{\delta h^g_{++}}{\delta g}\right)\det\left(\frac{\delta h^g_{--}}{\delta g}\right)\delta(h^g_{++})\delta(h^g_{--}). \tag{3.16}$$

The path integral becomes

$$\begin{aligned} Z &= \int Dg \int DhDX \det\left(\frac{\delta h^g_{++}}{\delta g}\right)\det\left(\frac{\delta h^g_{--}}{\delta g}\right)\delta(h^g_{++})\delta(h^g_{--})e^{iS[h,\,X]} \\ &= \int Dg \int Dh^g DX \det\left(\frac{\delta h^g_{++}}{\delta g}\right)\det\left(\frac{\delta h^g_{--}}{\delta g}\right)\delta(h^g_{++})\delta(h^g_{--})e^{iS[h^g,\,X]}. \end{aligned} \tag{3.17}$$

In the above equation (3.17) we have used the reparametrization invariance of the action and the measure. Then we make the change of variable $h^g \longrightarrow h'$ and we observe that the integral over g decouples. We thus get

$$Z = \int Dh'DX \det\left(\frac{\delta h'_{++}}{\delta g}\right)\det\left(\frac{\delta h'_{--}}{\delta g}\right)\delta(h'_{++})\delta(h'_{--})e^{iS[h',\,X]}. \tag{3.18}$$

We compute

$$\frac{\delta h'_{++}(\sigma)}{\delta g(\rho)} = \frac{\delta h'_{++}(\sigma)}{\delta \xi_+(\rho)} = -2\nabla_+\delta(\sigma - \rho) \tag{3.19}$$

$$\frac{\delta h'_{--}(\sigma)}{\delta g(\rho)} = \frac{\delta h'_{--}(\sigma)}{\delta \xi_-(\rho)} = -2\nabla_-\delta(\sigma - \rho). \tag{3.20}$$

Hence, we can represent the determinants by the path integrals

$$\det\left(\frac{\delta h'_{++}}{\delta g}\right) = N_{++}\int Dc^- Db_{--} e^{-\frac{1}{\pi}\int d^2\sigma c^- \nabla_+ b_{--}} \tag{3.21}$$

$$\det\left(\frac{\delta h'_{--}}{\delta g}\right) = N_{--}\int Dc^+ Db_{++} e^{-\frac{1}{\pi}\int d^2\sigma c^+ \nabla_- b_{++}}. \tag{3.22}$$

The precise values of the normalizations $\mathcal{N}_{++}$ and $\mathcal{N}_{--}$ are not needed. The anticommuting numbers b_{--} and b_{++} are the ghosts and the anticommuting numbers c^- and c^+ are the antighosts.

Similarly, we compute

$$\int Dh' \delta(h'_{++})\delta(h'_{--}) e^{iS[h', X]} = \int D\phi e^{iS[X]},$$
$$S[X] = -\frac{1}{2\pi}\int_\Sigma d^2\sigma \partial_a X^\mu \partial^a X_\mu. \tag{3.23}$$

We then get the path integral

$$Z = \int D\phi \int DXDcDb\; e^{iS[X, b, c]}. \tag{3.24}$$

The action $S[X, b, c]$ is given by

$$S[X, b, c] = S[X] - \frac{1}{i\pi}\int d^2\sigma c^- \nabla_+ b_{--} - \frac{1}{i\pi}\int d^2\sigma\; c^+ \nabla_- b_{++}. \tag{3.25}$$

The integral over ϕ formally decouples from the rest because only Γ^+_{++} and Γ^-_{--} are non-zero. However, due to regularization this property truly holds only in $D = 26$ where the conformal anomaly happens to vanish. We will thus assume that ϕ decouples and study the path integral

$$Z = \int DXDcDb\; e^{iS[X, b, c]}. \tag{3.26}$$

The action is given by

$$S[X, b, c] = S[X] - \frac{1}{i\pi}\int d^2\sigma\; c^- \partial_+ b_{--} - \frac{1}{i\pi}\int d^2\sigma\; c^+ \partial_- b_{++}. \tag{3.27}$$

The ghost fields c^+ and c^- are the components of a vector ghost field c^a whereas the antighosts b_{++} and b_{--} are the components of a traceless symmetric antighost tensor field b_{ab}. The ghost action can be put in the form

$$iS_g = -\frac{1}{\pi}\int d^2\sigma\; c^- \nabla_+ b_{--} - \frac{1}{\pi}\int d^2\sigma\; c^+ \nabla_- b_{++}$$
$$= \frac{1}{2\pi}\int d^2\sigma \sqrt{h}\;\; h^{ab} c^c\;\; \nabla_a b_{bc}. \tag{3.28}$$

In the above we have reinserted the scalar field ϕ. We have also used $\sqrt{h} = e^\phi$, $h^{+-} = h^{-+} = -2e^{-\phi}$, $h^{++} = h^{--} = 0$ and $b_{-+} = b_{+-} = 0$ since b_{ab} is a traceless symmetric tensor.

3.2 The energy–momentum tensor

We can write the above action in the form

$$iS_g = -\frac{1}{2\pi}\int d^2\sigma \sqrt{h}\;\; h^{ab} \nabla_a c^c\;\; b_{bc}. \tag{3.29}$$

We have used the fact that the metric is covariantly constant viz $\nabla_c h^{ab} = 0$ and the property $\int d^2\sigma \sqrt{h}\, \nabla_a X^a =$ boundary terms. Since b_{bc} is traceless we consider instead the action

$$iS_g = -\frac{1}{2\pi}\int d^2\sigma\sqrt{h}\ \left(h^{ab}\nabla_a c^c\ b_{bc} + h^{ab}cb_{ab}\right). \tag{3.30}$$

The auxiliary field c is a Lagrange multiplier. We want to compute the contribution of the ghosts to the worldsheet energy–momentum tensor defined by

$$T^g_{ab} = -\frac{2\pi}{\sqrt{h}}\frac{\delta S_g}{\delta h^{ab}}. \tag{3.31}$$

We use the definitions

$$h^{ab}h_{bc} = \eta^a_c \tag{3.32}$$

$$\begin{aligned}\nabla_a X_b &= \partial_a X_b - \Gamma^c_{ab}X_c\\ \nabla_a X^b &= \partial_a X^b + \Gamma^b_{ac}X^c\\ \nabla_c X_{ab} &= \partial_c X_{ab} - \Gamma^d_{ca}X_{db} - \Gamma^d_{cb}X_{ad}.\end{aligned} \tag{3.33}$$

We compute immediately

$$\begin{aligned}\delta(\nabla_a c^c) &= \delta\Gamma^c_{ab}c^b\\ \delta\Gamma^c_{ab} &= \frac{1}{2}h^{cd}(\nabla_a\delta h_{db} + \nabla_b\delta h_{da} - \nabla_d\delta h_{ab}).\end{aligned} \tag{3.34}$$

We recall the result

$$\delta(\sqrt{h}\,h^{ab}) = \sqrt{h}\,\delta h^{cd}\left(h^a_c\ h^b_d - \frac{1}{2}h_{cd}h^{ab}\right). \tag{3.35}$$

The variation of the action S_g when we vary the metric h_{ab} is given by

$$\begin{aligned}-i\delta_h S_g = &\frac{1}{2\pi}\int d^2\sigma\sqrt{h}\,\delta h^{ab}(\nabla_a c^c b_{bc} + cb_{ab}\\ &-\frac{1}{2}h_{ab}(\nabla^c c^d b_{cd} + ch^{cd}b_{cd}))\\ &+\frac{1}{2\pi}\int d^2\sigma\sqrt{h}\,h^{ad}\delta\Gamma^c_{ab}c^b b_{dc}.\end{aligned} \tag{3.36}$$

We also compute

$$\begin{aligned}h^{ad}\delta\Gamma^c_{ab}c^b b_{dc} &= \frac{1}{2}h^{ad}h^{ce}(\nabla_a\delta h_{eb} + \nabla_b\delta h_{ea} - \nabla_e\delta h_{ab})c^b b_{dc}\\ &= \frac{1}{2}h^{ad}h^{ce}(\nabla_b\delta h_{ea})c^b b_{dc}\\ &= \frac{1}{2}\nabla_b(h^{ad}h^{ce}\delta h_{ea})c^b b_{dc}\\ &= -\frac{1}{2}\nabla_b\delta h^{cd}c^b b_{dc}.\end{aligned} \tag{3.37}$$

In the second line we have used the fact that $b_{ab} = b_{ba}$. By using partial integration we get finally

$$i\delta_h S_g = -\frac{1}{2\pi}\int d^2\sigma\sqrt{h}\,\delta h^{ab}(\nabla_a c^c b_{bc} + cb_{ab} - \frac{1}{2}h_{ab}(\nabla^c c^d b_{cd} + ch^{cd}b_{cd}) + \frac{1}{2}\nabla_c(c^c b_{ab})). \tag{3.38}$$

The equations of motion are found as follows. The variation of S_g with respect to b_{ab} is

$$\begin{aligned} i\delta_b S_g &= -\frac{1}{2\pi}\int d^2\sigma\sqrt{h}\left(\frac{1}{2}\nabla^b c^c + \frac{1}{2}\nabla^c c^b + ch^{bc}\right)\delta b_{bc} \\ &\Longrightarrow \frac{1}{2}\nabla^b c^c + \frac{1}{2}\nabla^c c^b + ch^{bc} = 0. \end{aligned} \tag{3.39}$$

From this equation we deduce the relation

$$c = -\frac{1}{2}\nabla_a c^a. \tag{3.40}$$

The variation of S_g with respect to c^a is

$$i\delta_c S_g = -\frac{1}{2\pi}\int d^2\sigma\sqrt{h}(-\nabla^b b_{bc})\delta c^c \Longrightarrow \nabla^b b_{bc} = 0. \tag{3.41}$$

The variation of S_g with respect to c gives the obvious equation of motion

$$h^{ab}b_{ab} = 0. \tag{3.42}$$

By using (3.40), (3.42) and $b_{ab} = b_{ba}$ we can put the variation of the action with respect to the metric h given by $\delta_h S_g$ in the form

$$i\delta_h S_g = -\frac{1}{2\pi}\int d^2\sigma\sqrt{h}\,\delta h^{ab}\left(\frac{1}{2}\nabla_a c^c b_{bc} + \frac{1}{2}\nabla_b c^c b_{ac} + \frac{1}{2}c^c\nabla_c b_{ab} - \frac{1}{2}h_{ab}\nabla^c c^d b_{cd}\right). \tag{3.43}$$

The contribution of the ghosts to the energy–momentum tensor is given by

$$iT^g_{ab} = \frac{1}{2}\nabla_a c^c b_{bc} + \frac{1}{2}\nabla_b c^c b_{ac} + \frac{1}{2}c^c\nabla_c b_{ab} - \frac{1}{2}h_{ab}\nabla^c c^d b_{cd}. \tag{3.44}$$

We verify that this tensor is traceless as it should be, viz

$$h^{ab}T^g_{ab} = 0. \tag{3.45}$$

The b_{ab} is a traceless symmetric tensor which means that $b_{+-} = b_{-+} = 0$. The equations of motion $\nabla^b b_{bc} = 0$ are equivalent to $\nabla^+ b_{++} = 0$ and $\nabla^- b_{--} = 0$ or $\nabla_- b_{++} = 0$ and $\nabla_+ b_{--} = 0$. We compute

$$iT^g_{++} = \nabla_+ c^+ b_{++} + \frac{1}{2} c^+ \nabla_+ b_{++} \tag{3.46}$$

$$iT^g_{--} = \nabla_- c^- b_{--} + \frac{1}{2} c^- \nabla_- b_{--}. \tag{3.47}$$

3.3 Quantization of the ghosts

Now we quantize the action. We have

$$S_g = -\frac{1}{i\pi}\int d^2\sigma\ c^-\partial_+ b_{--} - \frac{1}{i\pi}\int d^2\sigma\ c^+\partial_- b_{++}. \tag{3.48}$$

This form of the action means that c^+ and b_{++} are conjugate variables and similarly $c^-$and b_{--} are also conjugate variables since $\partial_\pm \longrightarrow \frac{1}{2}\partial_\tau + \cdots$. We have then the equal-time anticommutation relations

$$\{c^+(\sigma, \tau), b_{++}(\sigma', \tau)\} = 2\pi\delta(\sigma - \sigma') \tag{3.49}$$

$$\{c^-(\sigma, \tau), b_{--}(\sigma', \tau)\} = 2\pi\delta(\sigma - \sigma'). \tag{3.50}$$

The equations of motion are obviously given by

$$\begin{aligned} \partial_+ b_{--} = \partial_- b_{++} = 0 \\ \partial_+ c^- = \partial_- c^+ = 0. \end{aligned} \tag{3.51}$$

We can immediately conclude that c^+ and b_{++} are functions of $\sigma^+ = \tau + \sigma$ only and similarly c^- and b_{--} are functions of $\sigma^- = \tau - \sigma$ only.

3.3.1 The open string

For the open string we impose the boundary conditions

$$c^+|_{\sigma=0,\,\pi} = c^-|_{\sigma=0,\,\pi},\ b_{++}|_{\sigma=0,\,\pi} = b_{--}|_{\sigma=0,\,\pi}. \tag{3.52}$$

We have the general expansions

$$\begin{aligned} c^+ &= \sum_{-\infty}^{\infty} c_n^+ e^{-in(\tau+\sigma)}, \quad c^- = \sum_{-\infty}^{\infty} c_n^- e^{-in(\tau-\sigma)} \\ b_{++} &= \sum_{-\infty}^{\infty} (b_n)_{++} e^{-in(\tau+\sigma)}, \quad b_{--} = \sum_{-\infty}^{\infty} (b_n)_{--} e^{-in(\tau-\sigma)}. \end{aligned} \tag{3.53}$$

The boundary conditions imply that $c_n^+ = c_n^- = c_n$ and $(b_n)_{++} = (b_n)_{--} = b_n$. Hence

$$\begin{aligned} c^+ &= \sum_{-\infty}^{\infty} c_n e^{-in(\tau+\sigma)}, \quad c^- = \sum_{-\infty}^{\infty} c_n e^{-in(\tau-\sigma)} \\ b_{++} &= \sum_{-\infty}^{\infty} b_n e^{-in(\tau+\sigma)}, \quad b_{--} = \sum_{-\infty}^{\infty} b_n e^{-in(\tau-\sigma)}. \end{aligned} \tag{3.54}$$

We have

$$\delta(\sigma - \sigma') = \frac{1}{2\pi} \sum_n e^{-in(\sigma-\sigma')}. \tag{3.55}$$

Indeed, we can check that

$$\text{if } \sigma - \sigma' \neq 0 \quad \text{then } \frac{1}{2\pi} \sum_n e^{-in(\sigma-\sigma')} = 0 \tag{3.56}$$

by using the integral

$$\int_{-\pi}^{\pi} d\sigma e^{-in\sigma} = 2\pi\delta_{n,\,0}. \tag{3.57}$$

We check the identity

$$\frac{1}{2\pi} \int_{-\infty}^{\infty} \sum_n e^{-in(\sigma-\sigma')} = 1. \tag{3.58}$$

We then verify the anticommutation relations

$$\{c_n,\, b_m\} = \delta_{n+m,\,0}, \quad \{c_n,\, c_m\} = \{b_n,\, b_m\} = 0. \tag{3.59}$$

3.3.2 The closed string

For the closed string we impose the periodic boundary conditions

$$\begin{aligned} c^+|_{\sigma+\pi} &= c^+|_{\sigma}, \quad c^-|_{\sigma+\pi} = c^-|_{\sigma}, \\ b_{++}|_{\sigma+\pi} &= b_{++}|_{\sigma}, \quad b_{--}|_{\sigma+\pi} = b_{--}|_{\sigma}. \end{aligned} \tag{3.60}$$

We obtain the expansions

$$\begin{aligned} c^+ &= \sqrt{2} \sum_{-\infty}^{\infty} c_n e^{-2in(\tau+\sigma)}, \quad c^- = \sqrt{2} \sum_{-\infty}^{\infty} \tilde{c}_n e^{-2in(\tau-\sigma)} \\ b_{++} &= \sqrt{2} \sum_{-\infty}^{\infty} b_n e^{-2in(\tau+\sigma)}, \quad b_{--} = \sqrt{2} \sum_{-\infty}^{\infty} \tilde{b}_n e^{-2in(\tau-\sigma)}. \end{aligned} \tag{3.61}$$

Now we use

$$\delta(\sigma - \sigma') = \frac{1}{\pi} \sum_n e^{-2in(\sigma-\sigma')} \tag{3.62}$$

since

$$\text{if } \sigma - \sigma' \neq 0 \text{ then } \frac{1}{\pi}\sum_n e^{-2in(\sigma-\sigma')} = 0. \tag{3.63}$$

By using the integral

$$\int_0^{\pi} d\sigma e^{-2in\sigma} = \pi\delta_{n,\,0} \tag{3.64}$$

we check the identity

$$\frac{1}{\pi}\sum_n e^{-2in(\sigma-\sigma')} = 1. \tag{3.65}$$

We then verify the anticommutation relations

$$\begin{aligned}
&\{c_n, b_m\} = \delta_{n+m,\,0}, \quad \{c_n, c_m\} = \{b_n, b_m\} = 0\\
&\{\tilde{c}_n, \tilde{b}_m\} = \delta_{n+m,\,0}, \quad \{\tilde{c}_n, \tilde{c}_m\} = \{\tilde{b}_n, \tilde{b}_m\} = 0\\
&\{c_n, \tilde{c}_m\} = \{c_n, \tilde{b}_m\} = \{b_n, \tilde{b}_m\} = 0.
\end{aligned} \tag{3.66}$$

3.3.3 Virasoro generators

Let us recall the definition of the Virasoro generators. For the open strings it is clear that $T^g_{--}(\sigma) = T^g_{++}(-\sigma)$ so we can consider only T^g_{++}. The Virasoro generators L^g_m are the Fourier modes of T^g_{++}, viz

$$\begin{aligned}
L^g_m(\tau) &= \frac{1}{\pi}\int_{-\pi}^{\pi} d\sigma e^{im\sigma} T^g_{++}\\
&= \frac{1}{i\pi}\int_{-\pi}^{\pi} d\sigma e^{im\sigma}\left(\frac{1}{2}c^+\partial_+ b_{++} + \partial_+ c^+ b_{++}\right).
\end{aligned} \tag{3.67}$$

We have $\partial_+ = \frac{1}{2}(\partial_\tau + \partial_\sigma)$ and $\partial_+ e^{-in(\tau+\sigma)} = -ine^{-in(\tau+\sigma)}$. Hence we compute

$$\begin{aligned}
L^g_m(\tau) &= -\sum_n (m+n) c_n b_{m-n} e^{-im\tau}\\
&= \sum_n (m-n) b_{m+n} c_{-n} e^{-im\tau}.
\end{aligned} \tag{3.68}$$

Thus

$$L^g_m \equiv L^g_m(0) = \sum_n (m-n) b_{m+n} c_{-n}. \tag{3.69}$$

Since $\{b_{m+n}, c_{-n}\} = \delta_{m,\,0}$ we have normal ordering problems only for L^g_0. Thus in the quantum theory the L^g_m, $m \neq 0$ will be given by the same expressions whereas L^g_0 will be defined by the equation

$$L^g_0 = 2\sum_{n<0}(-n) b_n c_{-n}. \tag{3.70}$$

In the commutation of the different commutators below we can assume that L_m^g for all m are given by (3.69) since the normal ordering constant in L_0^g will drop. Using this formula (3.69) we can immediately compute the identities

$$\begin{aligned}[L_m^g, b_n] &= (m-n)b_{m+n}\\ [L_m^g, c_n] &= -(2m+n)c_{m+n}.\end{aligned} \tag{3.71}$$

Next we compute the Virasoro algebra

$$\begin{aligned}[L_m^g, L_n^g] = &\sum_k (n-k)(m-n-k)b_{m+n+k}c_{-k}\\ &-\sum_k (n-k)(2m-k)b_{n+k}c_{m-k}.\end{aligned} \tag{3.72}$$

There is no normal ordering ambiguity in the commutators $[L_m^g, L_n^g]$ only when $n+m \neq 0$. In this case b_{m+n+k} and c_{-k} anticommute and b_{n+k} and $c_{m-\mathrm{k}}$ anticommute. Thus for $n+m \neq 0$ we can shift the summation index k and rewrite the above commutators in the form

$$\begin{aligned}[L_m^g, L_n^g] = &\sum_k (n-k)(m-n-k)b_{m+n+k}c_{-k}\\ &-\sum_k (m-k)(n-m-k)b_{n+m+k}c_{-k}.\end{aligned} \tag{3.73}$$

This yields the result

$$[L_m^g, L_n^g] = (m-n)L_{m+n}^g, \quad m+n \neq 0. \tag{3.74}$$

For $m+n=0$ we obtain instead from equation (3.72)

$$\begin{aligned}[L_m^g, L_{-m}^g] = &\sum_k (-m-k)(2m-k)b_k c_{-k}\\ &-\sum_k (-m-k)(2m-k)b_{-m+k}c_{m-k}\\ = &-\sum_{k=-\infty}^{\infty} (m+k)(2m-k)b_k c_{-k}\\ &+\sum_{k=-\infty-m}^{\infty-m} (m-k)(2m+k)b_k c_{-k}.\end{aligned} \tag{3.75}$$

We get the result

$$[L_m^g, L_{-m}^g] = 2mL_0^g + A^g(m). \tag{3.76}$$

Thus, in general we should get the commutators

$$[L_m^g, L_n^g] = (m-n)L_{m+n}^g + A^g(m)\delta_{m+n,0}. \tag{3.77}$$

In order to determine the anomaly $A^g(m)$ we follow the same method used for the Virasoro generators L_m. We get immediately that we must have

$$A^g(m) = c_3 m^3 + c_1 m. \tag{3.78}$$

We determine the constants c_1 and c_3 from the identity

$$A^g(m) = \langle 0;\, 0|[L^g_m,\, L^g_{-m}]|0;\, 0\rangle - 2\, m\langle 0;\, 0|L^g_0|0;\, 0\rangle. \tag{3.79}$$

Reality requires that $c_n^+ = c_{-n}$ and $b_n^+ = b_{-n}$ and hence $(L^g_m)^+ = L^g_{-m}$. Furthermore, since $A^g(-m) = -A^g(m)$ and $A^g(0) = 0$ we only consider $m > 0$. The operators c_n with $n \geqslant 0$ are annihilation operators where the corresponding creation operators are given by b_{-n}. Similarly the operators c_n with $n < 0$ are creation operators where the corresponding annihilation operators are given by b_{-n}. In the quantum theory the operator L_0 is defined by the normal ordered expression (3.70). We can immediately compute that $L^g_0|0;\, 0\rangle = 0$. The operators L^g_m, $m \neq 0$ are defined in the quantum theory by the same formula as the classical operators. Using these expressions we compute $L^g_m|0;\, 0\rangle = 0$ for $m > 0$. Hence for $m > 0$ equation (3.79) reduces to

$$A^g(m) = \langle 0;\, 0|L^g_m L^g_{-m}|0;\, 0\rangle. \tag{3.80}$$

Next, we compute

$$L^g_{-m}|0;\, 0\rangle = \sum_{n=1}^{m}(m+n)c_{-n}b_{n-m}|0;\, 0\rangle. \tag{3.81}$$

$$\langle 0;\, 0|L^g_m = \sum_{n=1}^{m}(m+n)\langle 0;\, 0|b_{m-n}c_n. \tag{3.82}$$

Hence

$$A^g(m) = -\sum_{n=1}^{m}(m+n)(2m-n) = -\frac{13}{6}m^3 + \frac{m}{6}. \tag{3.83}$$

Hence, we get the Virasoro algebra

$$[L^g_m,\, L^g_n] = (m-n)L^g_{m+n} + \left(-\frac{13}{6}m^3 + \frac{m}{6}\right)\delta_{m+n,\,0}. \tag{3.84}$$

We recall the Virasoro algebra

$$[L_m,\, L_n] = (m-n)L_{m+n} + \frac{D}{12}(m^3 - m)\delta_{m+n,\,0}. \tag{3.85}$$

The total Virasoro generators are defined by

$$L_m^{TOT} = L_m + L_m^g - a\delta_{m,0}. \tag{3.86}$$

The constraints read now

$$L_m^{TOT} = 0. \tag{3.87}$$

The Virasoro algebra reads now

$$\begin{aligned}[L_m^{TOT}, L_n^{TOT}] &= (m-n)L_{m+n}^{TOT} + \left(2am + \frac{D}{12}(m^3 - m) - \frac{13}{6}m^3 + \frac{m}{6}\right)\delta_{m+n,0} \\ &= (m-n)L_{m+n}^{TOT} + \left(\frac{D-26}{12}m^3 + \frac{2+24a-D}{12}m\right)\delta_{m+n,0}.\end{aligned} \tag{3.88}$$

The conformal anomaly vanishes iff

$$D = 26, \quad a = 1. \tag{3.89}$$

This means in particular that only for these values that the theory is truly conformally invariant.

3.4 BRST symmetry

3.4.1 Gauge theory

The BRST formalism is more general than the Faddeev–Popov gauge fixing. The Batalin–Vilkovisky formalism is a complicated generalization of the BRST formalism. The BRST symmetry is a supersymmetry since it has an anticommuting continuous parameter. More precisely, the BRST symmetry in the case of a pure non-abelian gauge theory is a global symmetry of the gauge-fixed Lagrangian for any value of the gauge parameter ξ. Indeed, the gauge-fixed Lagrangian of a non-abelian gauge theory is given by

$$\mathcal{L} = -\frac{1}{2}\operatorname{Tr} F_{\mu\nu}F^{\mu\nu} + \xi \operatorname{Tr}(\partial^\mu A_\mu)^2 + 2\operatorname{Tr} \partial^\mu b D_\mu c. \tag{3.90}$$

This can be put in the form

$$\mathcal{L} = -\frac{1}{2}\operatorname{Tr} F_{\mu\nu}F^{\mu\nu} - \frac{1}{\xi}\operatorname{Tr} B^2 + 2\operatorname{Tr}(\partial^\mu A_\mu)B + 2\operatorname{Tr} \partial^\mu b D_\mu c. \tag{3.91}$$

The BRST transformations of the fields are given by

$$\begin{aligned}\delta A_\mu^a &= \lambda (D_\mu c)^a \\ \delta c^a &= -\frac{1}{2} g\lambda f^{abc} c^b c^c \\ \delta b^a &= \lambda B^a \\ \delta B^a &= 0.\end{aligned} \tag{3.92}$$

In the above D_μ is defined by $(D_\mu\phi)^a = \partial_\mu\phi^a + gf^a_{bc}A^b_\mu\phi^c$ where ϕ is a field transforming in the adjoint representation of the gauge group. Using the above transformations we can immediately compute the following variation of the Lagrangian density

$$\delta\mathcal{L} = -\frac{1}{2}\,\mathrm{Tr}\,\delta(F_{\mu\nu}F^{\mu\nu}) + 2\,\mathrm{Tr}\,\partial^\mu b\delta(D_\mu c). \tag{3.93}$$

Let us recall that infinitesimal gauge transformations take the form $\delta A^a_\mu = (D_\mu\Lambda)^a/g$. The first term of the Lagrangian density $\mathcal{L}$ is by construction invariant under these transformations. Hence, this term must be invariant under $\delta A^a_\mu = \lambda(D_\mu c)^a$ where we have to make the identification $\Lambda = g\lambda c$. We are left with

$$\delta\mathcal{L} = 2\,\mathrm{Tr}\,\partial^\mu b\delta(D_\mu c). \tag{3.94}$$

We compute

$$\begin{aligned}
\delta(D_\mu c)^a &= (D_\mu\delta c)^a + gf^a_{bc}\delta A^b_\mu c^c \\
&= \partial_\mu\delta c^a + gf^a_{bc}A^b_\mu\delta c^c + gf^a_{bc}\delta A^b_\mu c^c \\
&= -\frac{1}{2}g\lambda\partial_\mu\left(f^a_{bc}c^bc^c\right) - \frac{1}{2}g^2\lambda f^a_{bc}f^c_{pq}A^b_\mu c^pc^q \\
&\quad + g\lambda f^a_{bc}\partial_\mu c^bc^c + g^2\lambda f^a_{bc}f^b_{pq}A^p_\mu c^qc^c \\
&= -\frac{g^2\lambda}{2}\left(f^a_{bc}f^c_{pq} + f^a_{qc}f^c_{bp} + f^a_{pc}f^c_{qb}\right)A^b_\mu c^pc^q \\
&= 0.
\end{aligned} \tag{3.95}$$

We have clearly used the Jacobi identity $f^a_{bc}f^c_{pq} + f^a_{qc}f^c_{bp} + f^a_{pc}f^c_{qb} = 0$. This establishes the fact that $\mathcal{L}$ is BRST invariant. Let us note also that the above identity can be put in the form

$$\delta^2 A^a_\mu = 0. \tag{3.96}$$

In fact we can show that for all fields ϕ we must have $\delta^2\phi = 0$. Let Q be the conserved charge, i.e. the time component of the conserved current, of the BRST symmetry then we must have the operator identity

$$Q^2 = 0. \tag{3.97}$$

In other words, Q is nilpotent. The action of Q on the fields ϕ is defined by

$$Q\phi = [\lambda Q, \phi]. \tag{3.98}$$

3.4.2 General case

Let us assume now that we have symmetry operators which form a closed Lie algebra G, viz

$$[K_i, K_j] = f^k_{ij}K_k. \tag{3.99}$$

In BRST quantization we introduce antighosts b_i which transform in the adjoint representation and ghosts c^i which transform in the dual of the adjoint representation. In the case of compact Lie algebras the dual of the adjoint is the same as the adjoint. We assume that the ghosts and antighosts obey canonical anticommutation relations, viz

$$\{c^i, b_j\} = \delta^i_j. \tag{3.100}$$

The BRST operator is defined by

$$Q = c^i K_i - \frac{1}{2} f_{ij}^k c^i c^j b_k. \tag{3.101}$$

By using (8.112) and (3.100) we compute

$$\begin{aligned} Q^2 &= c^i c^j K_i K_j - \frac{1}{2} f_{ij}^k c^i c^j \{b_k, c^r\} K_r + \frac{1}{4} f_{ij}^k f_{i'j'}^{k'} c^i c^j b_k c^{i'} c^{j'} b_{k'} \\ &= \frac{1}{4} f_{ij}^k f_{i'j'}^{k'} c^i c^j b_k c^{i'} c^{j'} b_{k'} \\ &= \frac{1}{2} f_{ij}^k f_{kl}^r c^i c^j c^l b_r \\ &= \frac{1}{6} \left(f_{ij}^k f_{kl}^r + f_{li}^k f_{kj}^r + f_{jl}^k f_{ki}^r \right) c^i c^j c^l b_r. \end{aligned} \tag{3.102}$$

Next, we note the Jacobi identity

$$f_{ij}^k f_{kl}^r + f_{li}^k f_{kj}^r + f_{jl}^k f_{ki}^r = 0. \tag{3.103}$$

Hence, we must have

$$Q^2 = 0. \tag{3.97}$$

In other words, Q is nilpotent.

The ghost numbers U is defined by

$$U = \sum_i c^i b_i. \tag{3.105}$$

Since the antighosts b_i transform in the adjoint representation and the ghosts c^i transform in the dual of the adjoint we have n (where n is the dimension of the Lie algebra G) antighosts b_i and n ghosts c^i. We will think of the antighosts b_i as ghost annihilation operators and the ghosts c^i as the corresponding ghost creation operators. Thus, each term $c^i b_i$ in U will have the eigenvalues 0 and 1 and as a consequence the eigenvalues of U are between 0 and n.

Let χ be a state of ghost number k. This state is said to be BRST invariant if $Q\chi = 0$. Given any state λ of ghost number $k-1$ the state $Q\lambda$ has ghost number k (since Q increases the ghost number by 1) and furthermore this state is BRST invariant due to the nilpotency of Q. The most important BRST invariant states are those states which are not of the form $Q\lambda$. Two solutions χ_1 and χ_2 of $Q\chi = 0$ are equivalent if their difference $\chi_1 - \chi_2$ is of the form $\chi_1 - \chi_2 = Q\lambda$. We can define the equivalence classes

$$[\chi] = \{\chi' : \chi' - \chi = Q\lambda\}. \tag{3.106}$$

For example

$$[0] = \{\chi' : \chi' = Q\lambda\}. \tag{3.107}$$

The equivalence classes $[\chi]$ of ghost number k (called also the cohomology classes) form the kth cohomology group $H^n(G; R)$ of the Lie algebra G with values in the representation R defined by the matrices K_i. In this sense the BRST operator Q computes the cohomology of the Lie algebra G with values in the representation R.

Let χ be a BRST invariant state of ghost number 0. In other words, we must have $U\chi = \sum_i c^i b_i \chi = 0$. Since the antighosts b_i are annihilation operators we see that we must have $b_i\chi = 0$ for any state of ghost number 0. Therefore we will have

$$Q\chi = \sum_i c^i K_i \chi = 0. \tag{3.108}$$

Since the state χ is annihilated by the b_i it cannot be annihilated by the ghosts creation operators c^i and hence

$$K_i \chi = 0. \tag{3.109}$$

In conclusion a state χ of ghost number 0 is BRST invariant if and only if it is G invariant. Elements of $H^0(G; R)$ are the equivalence classes $[\chi]$. Clearly $[\chi] = \{\chi\}$ since states of ghost number 0 (or their differences) cannot be rewritten as $Q\lambda$ because there are no states of ghost numbers -1. Therefore, all BRST states of ghost number 0 must satisfy $K_i\chi = 0$ (so they are G invariant) and they are identical to the cohomolgy equivalence classes of ghost number 0. The cohomolgy group $H^0(G; R)$ is thus identical to the space of G invariant states of ghost number 0. These states clearly do not contain ghosts since they are annihilated by b_i.

3.4.3 String theory

In this case G is an infinite dimensional Lie algebra which is the Virasoro algebra.

We start by recalling the contributions of the ghosts to the energy–momentum tensor given by the non-zero components

$$\begin{aligned} iT^g_{++} &= \partial_+ c^+ b_{++} + \frac{1}{2} c^+ \partial_+ b_{++} \\ iT^g_{--} &= \partial_- c^- b_{--} + \frac{1}{2} c^- \partial_- b_{--}. \end{aligned} \tag{3.110}$$

From the equations of motion $\partial_- c^+ = \partial_+ c^- = 0$ and $\partial_- b_{++} = \partial_+ b_{--} = 0$ we conclude that $\partial_- T^g_{++} = \partial_+ T^g_{--} = 0$.

We also recall the non-zero components of the energy–momentum tensor

$$\begin{aligned} T_{++} &= \partial_+ X^\mu \partial_+ X_\mu \\ T_{--} &= \partial_- X^\mu \partial_- X_\mu . \end{aligned} \tag{3.111}$$

Again by the equations of motion $\partial_+\partial_- X^\mu = 0$ we have $\partial_+ T_{--} = \partial_- T_{++} = 0$.

We define the BRST currents J_+^B and J_-^B and the ghost-number currents J_+ and J_- by

$$\begin{aligned} J_+^B &= 2c^+\left(T_{++} + \frac{1}{2}T_{++}^g\right) \\ J_-^B &= 2c^-\left(T_{--} + \frac{1}{2}T_{--}^g\right) \\ J_+ &= c^+ b_{++} \\ J_- &= c^- b_{--}. \end{aligned} \tag{3.112}$$

The corresponding conserved BRST charge and conserved ghost-number charge are given by

$$Q = \frac{1}{2\pi}\int_0^\pi d\sigma\left(J_+^B + J_-^B\right). \tag{3.113}$$

$$U = \frac{1}{2\pi}\int_0^\pi d\sigma(J_+ + J_-). \tag{3.114}$$

We consider open strings. We recall

$$\begin{aligned} T_{++}(\tau, \sigma) &= \frac{1}{2}\sum_m e^{-im\sigma}L_m(\tau) \\ T_{++}^g(\tau, \sigma) &= \frac{1}{2}\sum_m e^{-im\sigma}L_m^g(\tau) \\ c^+ &= \sum_n c_n e^{-in(\tau+\sigma)} \\ b_{++} &= \sum_n b_n e^{-in(\tau+\sigma)}. \end{aligned} \tag{3.115}$$

We compute

$$\frac{1}{2\pi}\int_{-\pi}^{\pi} d\sigma J_+^B = \sum_m c_m\left(L_{-m}(\tau) + \frac{1}{2}L_{-m}^g(\tau)\right)e^{-im\tau}. \tag{3.116}$$

Similarly,

$$\begin{aligned} T_{--}(\tau, \sigma) &= \frac{1}{2}\sum_m e^{-im\sigma}L_{-m}(\tau) \\ T^{g}_{--}(\tau, \sigma) &= \frac{1}{2}\sum_m e^{-im\sigma}L^{g}_{-m}(\tau) \\ c^{-} &= \sum_n c_n e^{-in(\tau-\sigma)} \\ b_{--} &= \sum_n b_n e^{-in(\tau-\sigma)}. \end{aligned} \tag{3.117}$$

And

$$\frac{1}{2\pi}\int_{-\pi}^{\pi} d\sigma J^{B}_{-} = \sum_m c_m\left(L_{-m}(\tau) + \frac{1}{2}L^{g}_{-m}(\tau)\right)e^{-im\tau}. \tag{3.118}$$

Hence at $\tau = 0$ we have

$$\frac{1}{2\pi}\int_{-\pi}^{\pi} d\sigma\left(J^{B}_{+} + J^{B}_{-}\right) = 2\sum_m c_m\left(L_{-m} + \frac{1}{2}L^{g}_{-m}\right). \tag{3.119}$$

Since $c^{+}(\tau = 0, -\sigma) = c^{-}(\tau = 0, \sigma)$, $b_{++}(\tau = 0, -\sigma) = b_{--}(\tau = 0, \sigma)$, $T_{++}(\tau, -\sigma) = T_{--}(\tau, \sigma)$, $T^{g}_{++}(\tau, -\sigma) = T^{g}_{--}(\tau, \sigma)$ we obtain

$$Q = \sum_m c_m\left(L_{-m} + \frac{1}{2}L^{g}_{-m}\right). \tag{3.120}$$

Using now

$$L^{g}_{m} = \sum_n (m - n)b_{m+n}c_{-n}. \tag{3.121}$$

Then

$$\begin{aligned} Q = \sum_m L_{-m}c_m &- \frac{1}{2}\sum_{m,n}(m - n)c_{-m}c_{-n}b_{m+n} \\ &+ \frac{1}{2}\sum_{m,n}(m - n)c_{-m}\{c_{-n}, b_{m+n}\}. \end{aligned} \tag{3.122}$$

In the quantum theory normal ordering is required. We will have $\{c_n, b_m\} = \delta_{m+n,0}$. The c_n, $n \geqslant 0$, are annihilation operators with b_{-n} the corresponding creation operators, whereas, c_n, $n < 0$, are creation operators with b_{-n} the corresponding annihilation operators. Thus we write

$$Q = \sum_m L_{-m}c_m - \frac{1}{2}\sum_{m,n}(m - n){:}\, c_{-m}c_{-n}b_{m+n}{:} - ac_0. \tag{3.123}$$

Similarly, we compute

$$\frac{1}{2\pi}\int_{-\pi}^{\pi} d\sigma J_{+} = \frac{1}{2\pi}\int_{-\pi}^{\pi} d\sigma J_{-} = \sum_{n} c_{-n}b_{n}. \tag{3.124}$$

Then we obtain

$$U = \sum_{n} c_{-n}b_{n}. \tag{3.125}$$

Again in the quantum theory we require normal ordering, viz

$$U = \sum_{n} : c_{-n}b_{n}: . \tag{3.126}$$

This is the ghost number operator.

A direct calculation leads to the crucial results

$$-\frac{1}{2}\sum_{n,k}(n-k): c_{-n}c_{-k}b_{n+k} := -\frac{1}{2}\sum_{n,k}(n-k)c_{-n}c_{-k}b_{n+k} - \sum_{n<0} nc_{0}. \tag{3.127}$$

$$-\frac{1}{2}\sum_{n,k}(n-k)\{c_{-n}c_{-k}b_{n+k}, b_{m}\} = \sum_{n}(m-n)b_{m+n}c_{-n} + \sum_{n} n\delta_{m,0}. \tag{3.128}$$

$$\begin{aligned} -\frac{1}{2}\sum_{n,k}(n-k)\{:c_{-n}c_{-k}b_{n+k}: , b_{m}\} &= \sum_{n}(m-n)b_{m+n}c_{-n} + \sum_{n>0} n\delta_{m,0} \\ &= \sum_{n}(m-n): b_{m+n}c_{-n}: \\ &= L_{m}^{g}. \end{aligned} \tag{3.129}$$

Hence

$$\begin{aligned} \{Q, b_{m}\} &= L_{m} - \frac{1}{2}\sum_{n,k}(n-k)\{:c_{-n}c_{-k}b_{n+k}: , b_{m}\} - a\delta_{m,0} \\ &= L_{m} + L_{m}^{g} - a\delta_{m,0} \\ &= L_{m}^{TOT}. \end{aligned} \tag{3.130}$$

Next, we write

$$\begin{aligned} &Q = \sum_{m} K_{-m}c_{m} - \frac{1}{2}\sum_{m,n}(m-n)c_{-m}c_{-n}b_{m+n}, \ K_{-m} = L_{-m} - a'\delta_{m,0}, \\ &a' = a + \sum_{k<0} k. \end{aligned} \tag{3.131}$$

Immediately, we compute

$$Q^2 = \frac{1}{2}\sum_{m,n}([L_m, L_n] - (m-n)L_{m+n} + a'(m-n)\delta_{m+n,\,0})c_{-m}c_{-n} + \frac{1}{4}\left(\sum_{m,n}(m-n)c_{-m}c_{-n}b_{m+n}\right)^2. \tag{3.132}$$

$$\begin{aligned}\frac{1}{4}\left(\sum_{m,n}(m-n)c_{-m}c_{-n}b_{m+n}\right)^2 &= \frac{1}{2}\sum_{m,n,k}(m-n)(m+n-k)c_{-m}c_{-n}c_{-k}b_{m+n+k}\\ &= \frac{1}{6}\sum_{m,n,k}[(m-n)(m+n-k)\\ &\quad + (k-m)(k+m-n)\\ &\quad + (n-k)(n+k-m)]c_{-m}c_{-n}c_{-k}b_{m+n+k}\\ &= 0.\end{aligned} \tag{3.133}$$

On the other, hand we have

$$\begin{aligned}\frac{1}{4}\left(\sum_{m,n}(m-n)c_{-m}c_{-n}b_{m+n}\right)^2 &= \frac{1}{2}\sum_{m,n,k}(m-n)(m+n-k)c_{-m}c_{-n}c_{-k}b_{m+n+k}\\ &= -\frac{1}{2}\sum_{k<0}k\sum_{m,n}(m-n)c_{-m}c_{-n}\delta_{m+n,\,0}\\ &\quad - \frac{1}{2}\sum_{m,n}(m-n)c_{-m}c_{-n}L^{g}_{m+n}\\ &= \frac{1}{4}\sum_{m,n}c_{-m}c_{-n}[L^{g}_{m}, L^{g}_{n}].\end{aligned} \tag{3.134}$$

In other words,

$$\frac{1}{4}\sum_{m,n}c_{-m}c_{-n}[L^{g}_{m}, L^{g}_{n}] = 0. \tag{3.135}$$

Hence

$$\begin{aligned}Q^2 &= \frac{1}{2}\sum_{m,n}c_{-m}c_{-n}([L_m, L_n] - (m-n)L^{TOT}_{m+n})\\ &= \frac{1}{2}\sum_{m,n}c_{-m}c_{-n}([L^{TOT}_{m}, L^{TOT}_{n}] - (m-n)L^{TOT}_{m+n}) - \frac{1}{2}\sum_{m,n}c_{-m}c_{-n}[L^{g}_{m}, L^{g}_{n}]\\ &= \frac{1}{2}\sum_{m,n}c_{-m}c_{-n}([L^{TOT}_{m}, L^{TOT}_{n}] - (m-n)L^{TOT}_{m+n}).\end{aligned} \tag{3.136}$$

Therefore we have $Q^2 = 0$ because of the vanishing of the anomaly $[L_m^{TOT}, L_n^{TOT}] -(m-n)L_{m+n}^{TOT} = 0$. We also note

$$[L_m, Q] = [\{Q, b_m\}, Q] = [b_m, Q^2] = 0. \tag{3.137}$$

The c_n, $n > 0$, are annihilation operators with b_{-n} the corresponding creation operators, whereas, c_n, $n < 0$, are creation operators with b_{-n} the corresponding annihilation operators. The ghost number operator is thus given by

$$U =: c_0 b_0 :+ \sum_{n=1}^{\infty}(c_{-n}b_n - b_{-n}c_n). \tag{3.138}$$

Clearly there is no a natural way to decide whether the ghost (antighost) zero modes c_0 (b_0) are creation or annihilation operators. Thus we choose the symmetrical normal ordering prescription

$$\begin{aligned} U &= U_0 + \sum_{n=1}^{\infty}(c_{-n}b_n - b_{-n}c_n) \\ &= \frac{1}{2}(c_0 b_0 - b_0 c_0) + \sum_{n=1}^{\infty}(c_{-n}b_n - b_{-n}c_n). \end{aligned} \tag{3.139}$$

This choice makes gauge-invariant string field theory as simple as possible. The operators c_0 and b_0 satisfy the algebra $c_0^2 = b_0^2 = 0, \{c_0, b_0\} = 1$. The corresponding irreducible representation is given in terms of two states $|+\rangle$, $|-\rangle$ defined by the equations

$$c_0|-\rangle = |+\rangle, \quad b_0|+\rangle = |-\rangle, \quad c_0|+\rangle = b_0|-\rangle = 0. \tag{3.140}$$

Hence these states have the following ghost numbers

$$U_0|+\rangle = \frac{1}{2}|+\rangle, \quad (U_0)_+ = \frac{1}{2}. \tag{3.141}$$

$$U_0|-\rangle = -\frac{1}{2}|-\rangle, \quad (U_0)_- = -\frac{1}{2}. \tag{3.142}$$

From here we conclude that all eigenvalues of the ghost-number operator U are half integral since the eigenvalues of the operator $U - U_0$ are obviously integral (every term $c_{-n}b_n$ or $b_{-n}c_n$ has eigenvalues 0 and 1) while the eigenvalues of the operator U_0 are half integral as we have just shown.

Physical states $|\psi\rangle$ must contain no ghost excitations. Hence, it is possible to bring any physical state $|\psi\rangle$, which is not expected to contain ghosts, using a transformation $|\psi\rangle \longrightarrow |\psi\rangle + Q|\lambda\rangle$, to a form where the ghost wave function is proportional to one of the two ghost ground states $|+\rangle$ and $|-\rangle$. Therefore, physical states are BRST cohomology classes of ghost number $\pm\frac{1}{2}$. The choice between these two values does matter since the theory is not symmetric between ghosts and antighosts (the ghost c_n has conformal dimension -1 and the antighost b_n has conformal dimension 2). We now determine the correct ghost number of physical states $|\psi\rangle$. Since physical states must contain no ghost excitations they must satisfy

$$b_n|\psi\rangle = c_n|\psi\rangle = 0, \quad n > 0. \tag{3.143}$$

We can immediately compute

$$-\frac{1}{2}\sum_{n,k}(n-k)\colon c_{-n}c_{-k}b_{n+k}\colon |\psi\rangle = -\sum_n nc_{-n}c_n b_0|\psi\rangle. \tag{3.144}$$

Thus

$$Q|\psi\rangle = c_0(L_0 - a)|\psi\rangle + \sum_{m>0} c_{-m}L_m|\psi\rangle - \sum_n nc_{-n}c_n b_0|\psi\rangle. \tag{3.145}$$

Let us assume now that $|\psi\rangle$ has ghost number equal to $-\frac{1}{2}$, viz

$$U|\psi\rangle = -\frac{1}{2}|\psi\rangle. \tag{3.146}$$

Multiplying both sides by b_0 we conclude that

$$b_0|\psi\rangle = 0. \tag{3.147}$$

Thus we get

$$Q|\psi\rangle = c_0(L_0 - a)|\psi\rangle + \sum_{m>0} c_{-m}L_m|\psi\rangle. \tag{3.148}$$

Therefore, the condition that the physical state $|\psi\rangle$ is BRST invariant is exactly equivalent, if $|\psi\rangle$ has ghost number equal to $-\frac{1}{2}$, to the physical state conditions of the covariant quantization, i.e.

$$Q|\psi\rangle = 0 \Leftrightarrow (L_0 - a)|\psi\rangle = 0, \quad L_m|\psi\rangle = 0, \quad m > 0. \tag{3.149}$$

If instead $|\psi\rangle$ had ghost number equal to $+\frac{1}{2}$ then we will get instead of (4.168) the condition $c_0|\psi\rangle = 0$ and therefore

$$Q|\psi\rangle = \sum_{m>0} c_{-m}L_m|\psi\rangle - \sum_n nc_{-n}c_n b_0|\psi\rangle. \tag{3.150}$$

Thus the condition of BRST invariance $Q|\psi\rangle = 0$ does not lead in this case to the correct physical state conditions of the covariant quantization. The physical states must therefore have ghost number equal to $-\frac{1}{2}$.

In summary physical states $|\psi\rangle$ must satisfy equations (3.143) and (4.168) and hence they have ghost number equal to $-\frac{1}{2}$ and they must be BRST invariant, i.e. $Q|\psi\rangle = 0$, which is equivalent to the physical state conditions $L_0|\psi\rangle = a|\psi\rangle$ and $L_m|\psi\rangle = 0$, $m>0$ of the covariant quantization.

Let $|\phi\rangle$ be a physical state which is trivial as a cohomology class so that $|\phi\rangle = Q|\lambda\rangle$ for some state $|\lambda\rangle$. Clearly $\langle\phi|\psi\rangle = \langle\lambda|Q|\psi\rangle = 0$ for any physical state $|\psi\rangle$ since $Q|\psi\rangle = 0$. Since general physical states are annihilated by L_m for $m > 0$ we conclude that we must have $\langle\phi| = \sum_{m>0}\langle\lambda_m|L_m$ for some states $|\lambda_m\rangle$ or equivalently

$$|\phi\rangle = \sum_{m>0} L_{-m}|\lambda_m\rangle. \tag{3.151}$$

These are exactly the null states of the covariant quantization.

We have thus shown that physical states must be BRST cohomology classes of ghost number equal to $-\frac{1}{2}$. This is because $|\psi\rangle + Q|\lambda\rangle$ for all $|\lambda\rangle$ represent the same physical state $|\psi\rangle$ which is BRST invariant, i.e. $Q|\psi\rangle = 0$. BRST invariance of these states yields the physical state conditions of the covariant quantization. The physical states which are trivial as a cohomology class, i.e. those states which can be written as $|\phi\rangle = Q|\lambda\rangle$ are the null states of the covariant quantization.

3.5 Exercises

Exercise 1: Compute the contribution of the ghosts to the worldsheet energy–momentum tensor.

Exercise 2: Compute the total Virasoro generators and show that the conformal anomaly vanishes if and only if $D = 26$ and $a = 1$.

Exercise 3: Show that the BRST charge Q is nilpotent if and only if the conformal anomaly vanishes.

References

[1] Green M B, Schwarz J H and Witten E 1987 *Superstring Theory. Vol. 1: Introduction* (Cambridge Monographs On Mathematical Physics) (Cambridge: Cambridge University Press)

[2] Polchinski J 2005 *String Theory. Vol. 1: An Introduction to the Bosonic String* (Cambridge: Cambridge University Press)

IOP Publishing

Chapter 4

Introduction to conformal field theory

Conformal field theory is a topic of fundamental importance to string theory. In this chapter we introduce the conformal group, representation theory of the Virasoro algebra, the operator product expansion, and the role of conformal symmetry in the Polyakov action and BRST quantization. We will follow closely the presentations of [1, 2, 3]. See also [4, 5–10].

4.1 The conformal groups $SO(p+1, q+1)$

We will assume the signature $(-1, \ldots, +1, \ldots)$ with q minus signs and p plus signs. We have then $q = 0$ and $p = D$ for Euclidean, and $q = 1$ and $p = D - 1$ for Lorentzian. We will also assume a flat space and a particular field theory in it.

We will consider both the case of a pseudo-Riemannian manifold with Lorentzian signature, and the case of a Riemannian manifold with Euclidean signature. We start with the group of diffeomorphisms, i.e. the group of general coordinate transformations. The conformal group is a subgroup of the diffeomorphism group which preserves the conformal flatness of the metric. The corresponding transformations preserve the angles but change the lengths.

A D-dimensional manifold, with $D = q + p$, is called conformally flat if the metric takes the following form

$$ds^2 = \exp(\omega(x))dx^\mu dx_\mu. \tag{4.1}$$

We are interested in studying flat space quantum field theory which are invariant under the conformal transformations

$$x^\mu \longrightarrow x'^\mu(x), \quad dx'^\mu dx'_\mu = \Omega^2(x)dx_\mu dx^\mu. \tag{4.2}$$

Thus, if the flat space is conformally flat, it will remain so under conformal transformations. These conformal transformations are generalizations of the scale transformation

$$x^{\mu} \longrightarrow x'^{\mu} = \alpha x^{\mu}, \quad dx'^{\mu}dx'_{\mu} = \alpha^2 dx_{\mu}dx^{\mu}. \tag{4.3}$$

The conformal group contains: (1) Lorentz transformations $SO(p, q)$: rotations and boosts, (2) translations, (3) the scale (dilatation) transformation $x^{\mu} \longrightarrow \lambda x^{\mu}$ where λ is a constant, and (4) special conformal transformations. These latter can be derived as follows. First, we note that the conformal group contains the inversion

$$x^{\mu} \longrightarrow \frac{x^{\mu}}{x^2} \Rightarrow dx^{\mu} \longrightarrow \frac{1}{x^2}\left(\eta^{\mu}_{\nu} - \frac{2x^{\mu}x_{\nu}}{x^2}\right)dx^{\nu}. \tag{4.4}$$

The inversion is a global transformation which preserves the angles but does not have an infinitesimal form. The metric transforms then as

$$dx^{\mu}dx_{\mu} \longrightarrow \frac{dx^{\mu}dx_{\mu}}{(x^2)^2}. \tag{4.5}$$

Thus, the above inversion leaves the metric conformally flat. We consider now a transformation consisting of an inversion, then a translation, then an inversion as follows

$$x^{\mu} \longrightarrow \frac{x^{\mu}}{x^2} \longrightarrow \frac{x^{\mu}}{x^2} + b^{\mu} \longrightarrow \frac{x^{\mu} + b^{\mu}x^2}{1 + 2b^{\mu}x_{\mu} + b^2x^2}. \tag{4.6}$$

This is the finite version of special conformal transformations. By taking the translation b^{μ}, which now plays the role of the parameter of the special conformal transformations, to be infinitesimal we get

$$\delta x^{\mu} = b^{\mu}x^2 - 2x^{\mu}bx. \tag{4.7}$$

In summary, the conformal group is given by the following infinitesimal transformations

$$\delta x^{\mu} = a^{\mu} + \omega^{\mu}{}_{\nu}x^{\nu} + \lambda x^{\mu} + b^{\mu}x^2 - 2x^{\mu}bx. \tag{4.8}$$

We can arrive at this important result by a different route as follows.

We consider infinitesimal conformal transformations $x'_{\mu} = x_{\mu} + v_{\mu}$, $\Omega = 1 + \omega/2$. We get immediately from (4.2) the condition

$$\partial_{\nu}v^{\mu} + \partial^{\mu}v_{\nu} = \omega\eta^{\mu}_{\nu}. \tag{4.9}$$

By taking the trace of both sides we get $\omega = 2\partial_{\mu}v^{\mu}/D$. Thus the condition on infinitesimal conformal transformations becomes

$$\partial_{\nu}v^{\mu} + \partial^{\mu}v_{\nu} - \frac{2}{D}\partial_{\rho}v^{\rho}\eta^{\mu}_{\nu} = 0. \tag{4.10}$$

In $D = 2$ this equation admits an infinite number of solutions (given in Euclidean signature by all holomorphic functions on the plane as we will show below), and thus, the conformal group in two dimensions is infinite dimensional.

In $D \neq 2$ there is a finite number of solutions given precisely by $v_\mu = \delta x_\mu$ where δx_μ is given by equation (4.8).

The elementary proof of these results can be found below equation (1.16) of [4].

We have then (1) D translations, (2) $D(D-1)/2$ Lorentz transformations since $\omega_{\mu\nu} = -\omega_{\nu\mu}$, (3) one scale transformation and (4) D special conformal transformations. Altogether we have then $(D+1)(D+2)/2$ conformal transformations. This is exactly the number of generators of the rotation group $SO(D+2)$. However, the conformal group must be a non-compact group. It is therefore given by $SO(p+1, q+1)$. In Lorentzian signature the conformal group is $SO(D,2)$, whereas in Euclidean signature the conformal group is $SO(D+1,1)$.

For $D > 2$ the conformal group is actually given by $O(p+1, q+1)$ and consists of two disconnected components since the inversion element is not infinitesimally generated. The generators of $SO(p+1, q+1)$ are the translations P_μ, the Lorentz generators $M_{\mu\nu}$, the scale transformation D and the special conformal transformations K_μ. The Lorentz generators satisfy the $SO(p, q)$ algebra, viz

$$[M_{\mu\nu}, M_{\rho\sigma}] = i(\eta_{\nu\rho}M_{\mu\sigma} + \eta_{\mu\sigma}M_{\nu\rho} - \eta_{\mu\rho}M_{\nu\sigma} - \eta_{\nu\sigma}M_{\mu\rho}). \tag{4.11}$$

For $D = 2$ the conformal algebra $so(3,1)$ or $o(3,1)$ is a subalgebra of a much larger algebra (in fact infinite dimensional algebra) called the Virasoro algebra.

The scale or dilatation transformations play a major role. A field theory which is invariant under scale transformations will also, under mild conditions, be invariant under all conformal transformations. The conserved current J_μ associated with the conformal transformation δx^μ is constructed from the stress–energy tensor $T_{\mu\nu}$ as follows

$$J_\mu = T_{\mu\nu}\delta x^\nu. \tag{4.12}$$

The conservation of the current associated with translations yields the conservation of the stress–energy tensor, viz $\partial^\mu T_{\mu\nu} = 0$, whereas the conservation of the current associated with Lorentz transformations yields $T_{\mu\nu} = T_{\nu\mu}$, i.e. the stress–energy tensor is symmetric. On the other hand, the conservation of the current $J_\mu = T_{\mu\nu}x^\nu$ associated with dilatation leads to the condition that the stress–energy tensor is traceless, i.e. $T^\mu_\mu = 0$.

Thus in a Poincaré and scale invariant theory with a symmetric traceless conserved stress–energy tensor all conformal currents are conserved, almost trivially, since

$$\partial^\mu J_\mu = \partial^\mu(T_{\mu\nu}v^\nu) = T_{\mu\nu}\partial^\mu v^\nu = \frac{1}{2}T_{\mu\nu}(\partial^\mu v^\nu + \partial^\nu v^\mu) = \frac{D}{2}\partial_\alpha v^\alpha T^\mu_\mu = 0. \tag{4.13}$$

4.2 The conformal group in two dimensions

The gauge-fixed Polyakov action in Euclidean signature is given by

$$S = \frac{1}{4\pi\alpha'}\int d^2\sigma(\partial_1X^\mu\partial_1X_\mu + \partial_2X^\mu\partial_2X_\mu). \tag{4.14}$$

Obviously, we have performed the Wick rotation $\sigma^2 = i\sigma^0 = i\tau$, and we have as before $\sigma^1 = \sigma$. We define the complex coordinates

$$z = \sigma^1 + i\sigma^2, \quad \bar{z} = \sigma^1 - i\sigma^2. \tag{4.15}$$

In other words, $\sigma^- = -z$, $\sigma^+ = \bar{z}$. The action can be put into the form

$$S = \frac{1}{2\pi\alpha'}\int d^2z\partial X^\mu\bar{\partial}X_\mu. \tag{4.16}$$

The equations of motion are (with $\partial = \partial_z$ and $\bar{\partial} = \partial_{\bar{z}}$)

$$\partial\bar{\partial}X_\mu = 0. \tag{4.17}$$

Thus, ∂X_μ is a holomorphic, i.e. an analytic function of z referred to as 'right-moving'. Similarly, $\bar{\partial}X_\mu$ is an antiholomorphic, i.e. an analytic function of $\bar{z}$ referred to as 'left-moving'. This terminology is the same as that found in [6] but different from the one used in [2, 3].

We recall that reparametrization gauge invariance is given by the finite transformations $\sigma^\alpha \longrightarrow \sigma^{'\alpha} = f^\alpha(\sigma)$ and infinitesimally by $\sigma^\alpha \longrightarrow \sigma^{\alpha'} = \sigma^\alpha\xi^\alpha$. The residual reparametrization gauge symmetries are given by $\xi^\pm = \xi^\pm(\sigma^\pm)$ or equivalently by $\sigma^\pm \longrightarrow \sigma^{\pm'} = f^\pm(\sigma^\pm)$.

Thus, the residual symmetries of the above Euclidean action in the conformal gauge are given by the transformations

$$z \longrightarrow f(z), \quad \bar{z} \longrightarrow \bar{z} = \bar{f}(\bar{z}). \tag{4.18}$$

These are conformal mappings. They are angle-preserving transformations when f and its inverse are both holomorphic, i.e. f is biholomorphic. For example, $z \longrightarrow z + a$ is a translation, $z \longrightarrow \zeta z$ where $|\zeta| = 1$ is a rotation, and $z \longrightarrow \zeta z$ where ζ is real not equal to 1 is a scale transformation also called dilatation.

We can then work, for example, with the complex coordinates

$$f(z) = e^{-2iz} = e^{2(\sigma^2 - i\sigma^1)}, \quad \bar{f}(\bar{z}) = e^{-2i\bar{z}} = e^{2(\sigma^2 + i\sigma^1)}. \tag{4.19}$$

The factor of 2 in the exponent is included for consistency with the periodicity of the closed string given by $\sigma^1 \longrightarrow \sigma^1 + \pi$. The worldsheet which is topologically an $\mathbf{R}^2$ can now be regarded as a Riemann surface, i.e. as a deformation of the complex plane. The Euclidean time σ^2 on the worldsheet corresponds to the radial distance $r = \exp(2\sigma^2)$ on the complex plane, with the infinite past $\sigma^2 = -\infty$ at $r = 0$, and the infinite future $\sigma^2 = +\infty$ is a circle at $r = \infty$. For example, an infinitely long cylinder is mapped to the complex plane minus the origin.

We will write for simplicity

$$z = e^{2(\sigma^2 - i\sigma^1)}, \quad \bar{z} = e^{2(\sigma^2 + i\sigma^1)}. \tag{4.20}$$

The conformal mappings are $z \longrightarrow f(z)$ and $\bar{z} \longrightarrow \bar{f}(\bar{z})$. The corresponding infinitesimal transformations are given by $z \longrightarrow z + a(z)$ and $\bar{z} \longrightarrow \bar{z} + \bar{a}(\bar{z})$. The generators are given by the transformations

$$z \longrightarrow z' = z - \epsilon_n z^{n+1}, \quad \bar{z} \longrightarrow \bar{z}' = \bar{z} - \bar{\epsilon}_n \bar{z}^{n+1}, \quad n \in Z. \tag{4.21}$$

The generators are immediately given by

$$l_n = -z^{n+1}\partial, \quad \bar{l}_n = -\bar{z}^{n+1}\bar{\partial}, \quad n \in Z. \tag{4.22}$$

The generators with $n < -1$ are defined on the punctured complex plane, whereas the generators with $n > 1$ are defined on the complex plane with the point at infinity removed. The generators l_{-1}, l_0, l_1 are defined on the whole Riemann sphere, i.e. the complex plane + the point at infinity.

The generators l_n are clearly non-singular at $z = 0$ for all $n \geqslant -1$. In order to study the behavior at $z = \infty$ we perform the following conformal transformation $z = 1/w$. Under this transformation, the generator l_n becomes

$$l_n = -w^{1-n}\partial_w. \tag{4.23}$$

This is non-singular at $w = 0$ for all $n \leqslant 1$. The remaining range is therefore $-1 \leqslant n \leqslant 1$ for which the generators l_n are defined on the whole Riemann sphere.

It is not difficult to show that the above generators l_n and $\bar{l}_n$ satisfy the classical Virasoro algebra

$$[l_m, l_n] = (m - n)l_{m+n}, \quad [\bar{l}_m, \bar{l}_n] = (m - n)\bar{l}_{m+n}, \quad [l_m, \bar{l}_n] = 0. \tag{4.24}$$

It can also be easily seen that the Virasoro algebra is the same as the algebra of infinitesimal diffeomorphisms of the circle $\mathbf{S}^1$.

An arbitrary conformal transformation on functions $f(z, \bar{z})$ is then given by the linear combination $\sum_n (l_n + \bar{l}_n)$ with $\bar{\epsilon}_n = \epsilon_n^+$ where the parameters ϵ_n and $\bar{\epsilon}_n$ are inside l_n and $\bar{l}_n$ respectively. The two commuting algebra generated by l_n and $\bar{l}_n$ can be treated independently if we drop the reality condition $\bar{\epsilon}_n = \epsilon_n^+$. In this case the conformal transformation is a well defined map on $\mathbf{C}^2$ and it does not simply take a point $(\sigma^1, \sigma^2) \in \mathbf{R}^2$ into another point in $\mathbf{R}^2$. In other words, z and $\bar{z}$ can be treated as independent, and the field theory is defined on a complexified worldsheet $\mathbf{C}^2$.

The group $SO(3,1)$ is called the restricted conformal group. The full conformal group in two dimensions is infinite dimensional. The finite dimensional subgroup $SO(3,1)$ is generated by l_0, $l_{\pm 1}$, $\bar{l}_0$, $\bar{l}_{\pm 1}$. These are given explicitly by

$$\begin{aligned} l_{-1}&: z \longrightarrow z - \epsilon \\ l_0&: z \longrightarrow z - \epsilon z \\ l_1&: z \longrightarrow z - \epsilon z^2 \end{aligned} \tag{4.25}$$

$$\begin{aligned} \bar{l}_{-1}&: \bar{z} \longrightarrow \bar{z} - \bar{\epsilon} \\ \bar{l}_0&: \bar{z} \longrightarrow \bar{z} - \bar{\epsilon}\bar{z} \\ \bar{l}_1&: \bar{z} \longrightarrow \bar{z} - \bar{\epsilon}\bar{z}^2. \end{aligned} \tag{4.26}$$

We have the following interpretation:

$$\begin{aligned} &l_{-1},\ \bar{l}_{-1}\ \text{translations}\\ &i(l_0 - \bar{l}_0)\ \ \text{rotations}\\ &l_0 + \bar{l}_0\ \ \text{scalings}\\ &l_1,\quad \bar{l}_1\ \ \text{special conformal transformations.} \end{aligned} \tag{4.27}$$

The finite or global form of these transformations are:

$$\begin{aligned} l_{-1}&:\ z \longrightarrow z + \alpha\\ l_0&:\ z \longrightarrow \lambda z\\ l_1&:\ z \longrightarrow \frac{z}{1 - \beta z}. \end{aligned} \tag{4.28}$$

By combining these transformations we obtain

$$z \longrightarrow \frac{az + b}{cz + d}, \qquad ad - bc = 1. \tag{4.29}$$

For infinitesimal z we obtain

$$z \longrightarrow \frac{b}{d} + \frac{1}{d^2}z - \frac{c}{d^3}z^2 \tag{4.30}$$

Only three parameters are independent as it should be. We obtain a linear combination of the transformations (4.25).

The group given by the relation (4.29) is $SO(3, 1) = SL(2, C)/\mathbf{Z}_2$. The division by $\mathbf{Z}_2$ is to take into account the property that the above transformations remain unchanged if $a, b, c, d \longrightarrow -a, -b, -c, -d$. The Lorentzian analogue is the group $SO(2, 2) = SL(2, R) \times SL(2, R)$ where one factor of $SL(2, R)$ stands for left-movers and the other factor stands for right-movers.

4.3 The energy–momentum tensor

We recall that Weyl symmetry of the Polyakov action yields the tracelessness of the energy–momentum tensor $T^{\alpha}{}_{\alpha} = 0$ which, in light cone coordinates, is equivalent to the vanishing of the components T_{+-} and T_{-+} of the energy–momentum tensor, i.e. $T_{+-} = T_{-+} = 0$. The invariance under diffeomorphisms yields the conservation of the energy–momentum tensor $\partial^{\alpha} T_{\alpha\beta} = 0$ which, in light cone coordinates, reads $\partial_- T_{++} = 0$ and $\partial_+ T_{--} = 0$.

After Wick rotation $\sigma^- \longrightarrow -z$ and $\sigma^+ \longrightarrow \bar{z}$, Weyl symmetry becomes given by $T_{\bar{z}z} = T_{z\bar{z}} = 0$, while reparametrization invariance becomes

$$\partial T_{\bar{z}\bar{z}} = 0, \qquad \bar{\partial} T_{zz} = 0. \tag{4.31}$$

Thus T_{zz} is holomorphic while $T_{\bar{z}\bar{z}}$ is anti-holomorphic. We write

$$T_{\bar{z}\bar{z}} = \tilde{T}(\bar{z}), \qquad \partial T_{zz} = T(z). \tag{4.32}$$

We recall also the formula

$$T_{++} = \partial_+ X^\mu \partial_+ X_\mu = 2l_s^2 \sum_m \tilde{L}_m e^{-2im(\tau+\sigma)}. \tag{4.33}$$

After Wick rotation we obtain

$$T_{\bar{z}\bar{z}} = \bar{\partial} X^\mu \bar{\partial} X_\mu = 2l_s^2 \sum_m \frac{\tilde{L}_m}{\bar{f}(\bar{z})^m}. \tag{4.34}$$

Equivalently,

$$T_{\bar{z}\bar{z}} = -4\bar{f}^2(\bar{z}) \frac{\partial}{\partial \bar{f}} X^\mu \frac{\partial}{\partial \bar{f}} X_\mu = 2l_s^2 \sum_m \frac{\tilde{L}_m}{\bar{f}(\bar{z})^m}. \tag{4.35}$$

Thus

$$T_{\bar{f}\bar{f}} = \frac{T_{\bar{z}\bar{z}}}{-4\bar{f}^2(\bar{z})} = \frac{\partial}{\partial \bar{f}} X^\mu \frac{\partial}{\partial \bar{f}} X_\mu = -\frac{l_s^2}{2} \sum_m \frac{\tilde{L}_m}{\bar{f}(\bar{z})^{m+2}}. \tag{4.36}$$

We get then (with $l_s = 1$ and by setting as before $\bar{f} = \bar{z}$ for simplicity)

$$T_{\bar{z}\bar{z}} = \bar{\partial} X^\mu \bar{\partial} X_\mu = -\frac{1}{2} \sum_m \frac{\tilde{L}_m}{\bar{z}^{m+2}}. \tag{4.37}$$

Similarly, we obtain

$$T_{zz} = \partial X^\mu \partial X_\mu = -\frac{1}{2} \sum_m \frac{L_m}{z^{m+2}}. \tag{4.38}$$

We will write these two equations as

$$\tilde{T}(\bar{z}) = \tilde{T}_X(\bar{z}) = \bar{\partial} X^\mu \bar{\partial} X_\mu = -\frac{1}{2} \sum_m \frac{\tilde{L}_m}{\bar{z}^{m+2}} \tag{4.39}$$

$$T(z) = T_X(z) = \partial X^\mu \partial X_\mu = -\frac{1}{2} \sum_m \frac{L_m}{z^{m+2}}. \tag{4.40}$$

Also we recall that

$$X_R^\mu = \frac{1}{2} x^\mu + \frac{1}{2} l_s^2 p^\mu (\tau - \sigma) + \frac{i}{2} l_s \sum_{n \neq 0} \frac{\alpha_n^\mu}{n} e^{-2in(\tau-\sigma)}. \tag{4.41}$$

After Wick rotation, we get

$$X_R^\mu(z) = \frac{1}{2} x^\mu - \frac{i}{4} p^\mu \ln z + \frac{i}{2} \sum_{n \neq 0} \frac{\alpha_n^\mu}{n z^n}. \tag{4.42}$$

Similarly, we obtain

$$X_L^{\mu}(\bar{z}) = \frac{1}{2}x^{\mu} - \frac{i}{4}p^{\mu}\ln\bar{z} + \frac{i}{2}\sum_{n\neq 0}\frac{\tilde{\alpha}_n^{\mu}}{n\bar{z}^n}. \tag{4.43}$$

Thus

$$\partial X^{\mu}(z, \bar{z}) = -\frac{i}{2}\sum_{n\neq 0}\frac{\alpha_n^{\mu}}{z^{n+1}} \tag{4.44}$$

$$\bar{\partial} X^{\mu}(z, \bar{z}) = -\frac{i}{2}\sum_{n\neq 0}\frac{\tilde{\alpha}_n^{\mu}}{\bar{z}^{n+1}}, \tag{4.45}$$

where

$$X^{\mu}(z, \bar{z}) = X_R^{\mu}(z) + X_L^{\mu}(\bar{z}). \tag{4.46}$$

The conformal algebra in two dimensions is thus infinite dimensional. There exists therefore an infinite number of conserved charges which are given essentially by the Virasoro generators, which are the modes of the energy–momentum tensor, viz L_n and $\tilde{L}_n$.

4.4 The operator product expansion

The Euclidean Polyakov action gauge-fixed in the conformal gauge is given by

$$S = \frac{1}{4\pi\alpha'}\int d^2\sigma(\partial_1 X^{\mu}\partial_1 X_{\mu} + \partial_2 X^{\mu}\partial_2 X_{\mu}). \tag{4.47}$$

The conformal gauge is related to Weyl scale invariance which is part of conformal invariance. The Polyakov action before imposing the conformal gauge is conformally invariant. We define the complex coordinates

$$z = \sigma^1 + i\sigma^2, \quad \bar{z} = \sigma^1 - i\sigma^2. \tag{4.48}$$

The action can be put into the form

$$S = \frac{1}{2\pi\alpha'}\int d^2z\partial X^{\mu}\bar{\partial}X_{\mu}. \tag{4.49}$$

The equations of motion are (with $\partial = \partial_z$ and $\bar{\partial} = \partial_{\bar{z}}$)

$$\partial\bar{\partial}X_{\mu} = 0. \tag{4.50}$$

The general solution is of the form

$$X^{\mu} = X_R^{\mu}(z) + X_L^{\mu}(\bar{z}). \tag{4.51}$$

After Wick rotation $\sigma^2 = i\sigma^0 = i\tau$, we see that the holomorphic functions $X_R^{\mu}(z)$ become functions of $z = -(\tau - \sigma)$, i.e. right-moving, whereas the antiholomorphic functions X_L^{μ} become functions of $\bar{z} = \tau + \sigma$, i.e. left-moving.

The conformal group in two dimensions is defined by the Virasoro algebra which in terms of the Virasoro generators L_m reads

$$[L_m, L_n] = (m - n)L_{m+n} + \frac{c}{12}(m^3 - m)\delta_{m+n,0}. \tag{4.52}$$

The $c = 0$ is the classical result while the term proportional to c is the central charge due to quantum effects. The conformal group in two dimensions is infinite dimensional and the corresponding algebra does not close in the usual sense because of the central charge.

The Virasoro generators L_m and their conjugate $\tilde{L}_m$, or equivalently the components T_{zz} and $\tilde{T}_{\bar{z}\bar{z}}$ of the stress–energy tensor, generate conformal symmetry operators with corresponding conserved charges.

The global subgroup is $SO(3, 1) = SL(2, C)/\mathbf{Z}_2$ which is generated by the algebra L_0, $L_{\pm 1}$ and their conjugates $\bar{L}_0$, $\bar{L}_{\pm}$. The translations P_i, the rotation M, the scaling D and the special conformal transformations are given in terms of these Virasoro generators by equation (1.43) of [4].

We will work now with the Euclidean action

$$\begin{aligned} S &= \frac{1}{4\pi\alpha'}\int d^2x \partial_\alpha\Phi(x)\partial^\alpha\Phi(x) \\ &= \frac{1}{2\pi\alpha'}\int dz d\bar{z} \partial\Phi(z, \bar{z})\bar{\partial}\Phi(z, \bar{z}). \end{aligned} \tag{4.53}$$

Recall that $T = 1/2\pi\alpha'$, $\alpha' = l_s^2/2$ and we choose $l_s = 1$. The complex and real coordinates are related by

$$z = x^1 - ix^2, \quad \bar{z} = x^1 + ix^2. \tag{4.54}$$

A conformal field of conformal dimension (p, q), also called a primary field, is a tensor field of rank $n = p + q$ under conformal transformations with components $\Phi_{z\ldots z,\, \bar{z}\ldots\bar{z}}(z, \bar{z})$, which transforms under the conformal transformations $z \longrightarrow w(z)$ as follows

$$\Phi(z, \bar{z}) \longrightarrow \left(\frac{\partial w}{\partial z}\right)^p \left(\frac{\partial \bar{w}}{\partial \bar{z}}\right)^q \Phi(w, \bar{w}). \tag{4.55}$$

Thus the differential

$$\Phi(z, \bar{z})(dz)^p(d\bar{z})^q \tag{4.56}$$

is invariant under conformal transformations.

The stress–energy tensor arises from the conformal symmetries $z \longrightarrow z' = z + \epsilon(z)$, $\bar{z} \longrightarrow \bar{z}' = \bar{z} + \tilde{\epsilon}(\bar{z})$. The action is invariant under the more general transformations $z \longrightarrow z' = z + \epsilon(z, \bar{z})$, $\bar{z} \longrightarrow \bar{z}' = \bar{z} + \tilde{\epsilon}z, (\bar{z})$ if we also change the (flat) metric as

$$\eta_{\alpha\beta} \longrightarrow h_{\alpha\beta} = \eta_{\alpha\beta} + \delta h_{\alpha\beta}, \quad \delta h_{\alpha\beta} = -\partial_\alpha\xi_\beta - \partial_\beta\xi_\alpha. \tag{4.57}$$

This means that the change in the action under the transformation of the coordinates is the opposite of what we get if we just transform the metric, viz

$$\delta S = -\int d^2x \frac{\partial S}{\partial h_{\alpha\beta}} \delta h^{\alpha\beta}. \tag{4.58}$$

The stress–energy tensor is therefore given by

$$\begin{aligned} T_{\alpha\beta} &= -\frac{4\pi\alpha'}{\sqrt{-h}} \frac{\delta S}{\delta h^{\alpha\beta}} \\ &= \partial_\alpha \Phi \partial_\beta \Phi - \frac{1}{2} h_{\alpha\beta} \partial_\gamma \Phi \partial^\gamma \Phi. \end{aligned} \tag{4.59}$$

We compute also (using $\partial = \partial_z = (\partial_1 + i\partial_2)/2$, $\bar{\partial} = \partial_{\bar{z}} = (\partial_1 - i\partial_2)/2$ and the Weyl symmetry $T_{11} = -T_{22}$)

$$\begin{aligned} T_{zz} &= \frac{1}{2}(T_{11} + iT_{12}) \\ &= \frac{1}{2}\left(\frac{1}{2}(\partial_1\Phi)^2 - \frac{1}{2}(\partial_2\Phi)^2 + i(\partial_1\Phi)(\partial_2\Phi)\right) \\ &= (\partial_z\Phi)^2 \end{aligned} \tag{4.60}$$

$$\begin{aligned} T_{\bar{z}\bar{z}} &= \frac{1}{2}(T_{11} - iT_{12}) \\ &= \frac{1}{2}\left(\frac{1}{2}(\partial_1\Phi)^2 - \frac{1}{2}(\partial_2\Phi)^2 - i(\partial_1\Phi)(\partial_2\Phi)\right) \\ &= (\partial_{\bar{z}}\Phi)^2 \end{aligned} \tag{4.61}$$

$$T_{\bar{z}z} = T_{z\bar{z}} = 0. \tag{4.62}$$

The conservation of the stress–energy tensor now gives the equations

$$\bar{\partial} T_{zz} = \partial T_{\bar{z}\bar{z}} = 0. \tag{4.63}$$

In Minkowski signature the conserved charge, which generates the symmetry transformation on the quantum Hilbert space, is defined in terms of the conserved current J^μ by the 1-dimensional integral (in two dimensions)

$$Q = \int dx^1 J^0(x^1, x^0). \tag{4.64}$$

We impose a boundary condition in the x^1 direction. We can choose for convenience $x^1 \in [0, 2\pi]$ without any loss of generality because of scale invariance. We then Wick rotate as $x^0 \longrightarrow x^2 = ix^0$, $J^0(x^1, x^0) \longrightarrow J^2(x^1, x^2) = iJ^0(x^1, x^0)$ to obtain

$$Q = \frac{1}{2\pi}\int_0^{2\pi} dx^1(-iJ^2(x^1, x^2)). \tag{4.65}$$

We have also included a factor of 2π for convenience. The Euclidean coordinates are therefore coordinates on a cylinder. By passing to complex coordinates we have

$$J_z = \frac{\partial x^1}{\partial z}J_1 + \frac{\partial x^2}{\partial z}J_2 = \frac{1}{2}(J_1 + iJ_2) \tag{4.66}$$

$$J_{\bar{z}} = \frac{\partial x^1}{\partial \bar{z}}J_1 + \frac{\partial x^2}{\partial \bar{z}}J_2 = \frac{1}{2}(J_1 - iJ_2). \tag{4.67}$$

We rewrite the current as

$$Q = -\frac{1}{2\pi}(\oint dzJ_z - \oint d\bar{z}J_{\bar{z}}). \tag{4.68}$$

In the second term we have changed the integration variable from z to $\bar{z}$ since $\mathrm{Re}\, z = \mathrm{Re}\, \bar{z} = x^1$. The integration is now over a closed contour which encircles the cylinder and it is such that $\oint dz = \oint d\bar{z} = 2\pi$.

We perform now the conformal transformation

$$w = \exp(iz) = \exp(x^2 + ix^1). \tag{4.69}$$

The circle at the Euclidean time $x^2 = -\infty$ is mapped to the origin $w = 0$ of the w complex plane, whereas the circle at the Euclidean time $x^2 = +\infty$ is mapped to the circle at $|w| = \infty$, i.e. to the circle at infinity on the complex plane. We obtain thus the complex plane as our domain of integration. The integration over z becomes therefore a contour integration around the origin. The direction of the contour is fixed by the Cauchy integrals

$$\oint dz = \oint \frac{dw}{iw} = 2\pi, \quad \oint d\bar{z} = \oint \frac{d\bar{w}}{i\bar{w}} = 2\pi. \tag{4.70}$$

The conserved charge becomes

$$Q = -\frac{1}{2\pi}\left(\oint \frac{dw}{iw} J_z^{\mathrm{cyl}}(z(w), \bar{z}(\bar{w})) - \oint \frac{d\bar{w}}{i\bar{w}} J_{\bar{z}}^{\mathrm{cyl}}(z(w), \bar{z}(\bar{w}))\right). \tag{4.71}$$

The currents are labeled cyl since they are defined on the cylinder. The plane currents are obtained from the cylinder currents by conformal transformations. In general, the plane and cylinder fields are related by

$$\Phi^{\mathrm{pla}}(w, \bar{w}) = \left(\frac{\partial z}{\partial w}\right)^h \left(\frac{\partial \bar{z}}{\partial \bar{w}}\right)^{\bar{h}} \Phi^{\mathrm{cyl}}(z(w), \bar{z}(\bar{w})), \tag{4.72}$$

where Φ is a primary field of conformal dimension $(h, \bar{h})$. In our case, we get

$$\Phi^{\mathrm{cyl}}(z(w), \bar{z}(\bar{w})) = (iw)^h(-i\bar{w})^{\bar{h}}\Phi^{\mathrm{pla}}(w, \bar{w}). \tag{4.73}$$

The currents J_z and $J_{\bar{z}}$ are primary fields with conformal dimensions $(h, \bar{h}) = (1, 0)$, $(h, \bar{h}) = (0, 1)$ respectively. We then get the conserved charge

$$Q = -\frac{1}{2\pi}\left(\oint dw(iw)^{h-1}J_w^{\text{pla}}(w, \bar{w}) - \oint d\bar{w}\bar{w}^{\bar{h}-1}J_{\bar{w}}^{\text{pla}}(w, \bar{w})\right). \tag{4.74}$$

The current J_z is holomorphic, viz $J_z(z, \bar{z}) = J(z)$, while the current $J_{\bar{z}}$ is antiholomorphic, viz $J_{\bar{z}}(z, \bar{z}) = \bar{J}(\bar{z})$. Thus the conservation law $\partial^\mu J_\mu = 0$ is automatically satisfied since

$$\partial^\mu J_\mu = 2(\partial_z J_{\bar{z}} + \partial_{\bar{z}} J_z). \tag{4.75}$$

We compute now (with $\xi_1 = (\bar{\epsilon} + \epsilon)/2$, $\xi_2 = (\bar{\epsilon} - \epsilon)/2i$)

$$\begin{aligned} \delta S &= -\frac{1}{2\pi\alpha'}\int d^2x T_{\alpha\beta}\partial^\alpha\xi^\beta \\ &= -\frac{1}{2\pi\alpha'}\int dzd\bar{z}(T_{zz}\bar{\partial}\epsilon + T_{\bar{z}\bar{z}}\partial\bar{\epsilon}). \end{aligned} \tag{4.76}$$

Since T_{zz} is holomorphic and $T_{\bar{z}\bar{z}}$ antiholomorphic the conserved currents in terms of the stress–energy tensor are given by (with $T(z) = T_{zz}$, $\bar{T}(\bar{z}) = T_{\bar{z}\bar{z}}$)

$$J_z = iT(z)\epsilon(z), \quad J_{\bar{z}} = i\bar{T}(\bar{z})\bar{\epsilon}(\bar{z}). \tag{4.77}$$

The conserved charge Q_ϵ on the complex plane, associated with the infinitesimal conformal transformation $T(z)\epsilon(z)$, is then defined by

$$Q_\epsilon = \frac{1}{2\pi i}\oint dwT(w)\epsilon(w). \tag{4.78}$$

This generates the conformal transformation

$$\begin{aligned} \Phi(w, \bar{w}) \longrightarrow \Phi'(w, \bar{w}) = \left(\frac{\partial f(w)}{\partial w}\right)^h \Phi(f(w), \bar{w}), \quad f(w) \\ = \quad w + \epsilon(w). \end{aligned} \tag{4.79}$$

The corresponding infinitesimal transformation reads

$$\delta\Phi(w, \bar{w}) = \epsilon(w)\partial_w\Phi + h\Phi\partial_w\epsilon(w). \tag{4.80}$$

The quantum analogue of this equation is

$$\begin{aligned} \delta\Phi(w, \bar{w}) &= [Q_\epsilon, \Phi(w, \bar{w})] \\ &= \frac{1}{2\pi i}\oint dz\epsilon(z)[T(z), \Phi(w, \bar{w})] \\ &= \frac{1}{2\pi i}\oint dz\epsilon(z)(T(z)\Phi(w, \bar{w}) - \Phi(w, \bar{w})T(z)). \end{aligned} \tag{4.81}$$

We need to specify the integration contour. The main result is that the operator products $T(z)\Phi(w, \bar{w})$ and $\Phi(w, \bar{w})T(z)$ are only convergent for radially ordered operators which is the analogue of time ordering for field theory on the cylinder [4]. More precisely, the first term has a convergent series of expansions for $|z| > |w|$, while the second term has a convergent series of expansion for $|z| < |w|$.

On the cylinder Q_ϵ is a conserved charge and thus evaluating it at different times gives the same answer. This means that on the plane, where the integration over z is a contour integration around the origin, the charge Q_ϵ must be independent of the chosen contour at least classically. We use this classical freedom to rewrite the two terms in the above equation in such a way that they are well-behaved at the quantum level. We write then

$$\begin{aligned}\delta\Phi(w, \bar{w}) &= \frac{1}{2\pi i}\oint_{|z|>|w|} dz\epsilon(z)T(z)\Phi(w, \bar{w}) \\ &- \frac{1}{2\pi i}\oint_{|z|<|w|} dz\Phi(w, \bar{w})T(z) \\ &= \frac{1}{2\pi i}[\oint_{|z|>|w|} - \oint_{|z|<|w|}]dz\epsilon(z)R(T(z)\Phi(w, \bar{w})),\end{aligned} \tag{4.82}$$

where R is the radial ordering operator defined in an obvious way. The first z-contour encircles the origin and the point w, whereas the second contour does not encircle w. Their difference is a contour that encircles w. See figure (3.2) of [1]. We then write simply

$$\delta\Phi(w, \bar{w}) = \frac{1}{2\pi i}\oint dz\epsilon(z)R(T(z)\Phi(w, \bar{w})). \tag{4.83}$$

The integration contour over z is understood now to encircle the point w.

The radially ordered product must be analytic in the neighborhood of the point w in order for the integral to make sense. Thus one must have the Laurent expansion

$$R(T(z)\Phi(w, \bar{w})) = \sum_n (z - w)^n O_n(w, \bar{w}), \tag{4.84}$$

for some operators $O_n(w, \bar{w})$. We reproduce the infinitesimal transformations (4.80) if and only if the radially ordered product is given by

$$\begin{aligned}R(T(z)\Phi(w, \bar{w})) = &\frac{h}{(z - w)^2}\Phi(w, \bar{w}) + \frac{1}{z - w}\partial_w\Phi(w, \bar{w}) \\ &+ \text{regular power series in } (z - w).\end{aligned} \tag{4.85}$$

Indeed,

$$\begin{aligned}\frac{1}{2\pi i}\oint dz\epsilon(z)\frac{h}{(z - w)^2}\Phi(w, \bar{w}) &= \frac{1}{2\pi i}\oint dz[\epsilon(w) \\ &\quad + (z - w)\partial_w\epsilon(w) + \cdots]\frac{h}{(z - w)^2}\Phi(w, \bar{w}) \\ &= h\partial_w\epsilon(w).\ \Phi(w, \bar{w})\end{aligned} \tag{4.86}$$

$$\frac{1}{2\pi i}\oint dz\epsilon(z)\frac{1}{z - w}\partial_w\Phi(w, \bar{w}) = \epsilon(w)\partial_w\Phi(w, \bar{w}). \tag{4.87}$$

The conjugate analogue of equation (4.85) is

$$R(\bar{T}(\bar{z})\Phi(w, \bar{w})) = \frac{\bar{h}}{(\bar{z}-\bar{w})^2}\Phi(w, \bar{w}) + \frac{1}{\bar{z}-\bar{w}}\partial_{\bar{w}}\Phi(w, \bar{w}) + \text{regular power series in } (\bar{z}-\bar{w}). \tag{4.88}$$

We will generalize the action (4.53) to

$$\begin{aligned} S &= \frac{1}{4\pi\alpha'}\int d^2x\partial_a\Phi^i(x)\partial^a\Phi^i(x) \\ &= \frac{1}{2\pi\alpha'}\int dzd\bar{z}\partial\Phi^i(z, \bar{z})\bar{\partial}\Phi^i(z, \bar{z}). \end{aligned} \tag{4.89}$$

The fields are defined on the cylinder, viz $\Phi^i(x^0, 0) = \Phi^i(x^0, 2\pi)$. A field satisfying this boundary condition and solving the classical equation of motion $(\partial_0^2 - \partial_1^2)\Phi = 0$ can be put into the form (with $\alpha_n^i = (\alpha_{-n}^i)^*$, $\tilde{\alpha}_n^i = (\tilde{\alpha}_{-n}^i)^*$)

$$\Phi^i = q^i + \frac{p^i x^0}{2} + \frac{i}{2}\sum_{n\neq 0}\frac{1}{n}(\alpha_n^i e^{-in(x^0+x^1)} + \tilde{\alpha}_n^i e^{-in(x^0-x^1)}). \tag{4.90}$$

The Fourier coefficients α_n^i and $\tilde{\alpha}_n^i$ are found to satisfy quantum mechanically the commutation relations

$$[\alpha_n^i, \alpha_m^j] = [\tilde{\alpha}_n^i, \tilde{\alpha}_m^j] = n\delta^{ij}\delta_{n+m,0}, \quad [\alpha_n^i, \tilde{\alpha}_m^j] = 0, \quad [q^i, p^j] = i\delta^{ij}. \tag{4.91}$$

The vacuum state satisfies

$$\alpha_n^i|0\rangle = 0, \quad n > 0, \quad p^i|0\rangle = 0. \tag{4.92}$$

By transforming to the complex plane, viz $w = \exp(i(x^0 + x^1))$ and $\bar{w} = \exp(i(x^0 - x^1))$, we obtain

$$2\Phi^i = 2q^i - i\frac{p^i}{2}(\ln w + \ln \bar{w}) + i\sum_{n\neq 0}\frac{1}{n}(\alpha_n^i w^{-n} + \tilde{\alpha}_n^i \bar{w}^{-n}). \tag{4.93}$$

This is not holomorphic since the function $\ln w$ is not holomorphic. In the following we will write Φ in place of 2Φ and q in place of 2q and p in place of $p/2$ for simplicity.

The operator $\Phi^i(z, \bar{z})\Phi^j(w, \bar{w})$ is replaced quantum mechanically by the radially ordered product $R(\Phi^i(z, \bar{z})\Phi^j(w, \bar{w}))$ with $|z| > |w|$. We have

$$\begin{aligned} R(\Phi^i(z, \bar{z})\Phi^j(w, \bar{w})) = {} & q^i q^j - iq^i p^j(\ln w + \ln \bar{w}) - ip^i q^j(\ln z + \ln \bar{z}) \\ & - p^i p^j(\ln z + \ln \bar{z})(\ln w + \ln \bar{w}) + \cdots \end{aligned} \tag{4.94}$$

The normal ordering of p^i and q^i is such that we always place p^i to the right of q^i because $p^i|0\rangle = 0$. Thus, in the above operator product, all terms with the exception of the third are already normal ordered, and as a consequence we obtain

$$R(\Phi^i(z,\bar{z})\Phi^j(w,\bar{w})) = q^iq^j - iq^ip^j(\ln w + \ln\bar{w}) - iq^ip^j(\ln z + \ln\bar{z})$$
$$- p^ip^j(\ln z + \ln\bar{z})(\ln w + \ln\bar{w}) + \cdots$$
$$- i[p^i, q^j](\ln z + \ln\bar{z})$$
$$= \ :\Phi^i(z,\bar{z})\Phi^j(w,\bar{w}): -\delta^{ij}(\ln z + \ln\bar{z}). \tag{4.95}$$

This generalizes in such a way that only commutators between operators which do not commute contribute. The normal ordering of α_n and $\alpha_{-\mathrm{n}}$ for $n > 0$ is such that we always have α_n to the right of $\alpha_{-\mathrm{n}}$ because $\alpha_n|0\rangle = 0$, and similarly for $\tilde{\alpha}_n$ and $\tilde{\alpha}_{-n}$ for $n > 0$. We get then

$$R(\Phi^i(z,\bar{z})\Phi^j(w,\bar{w})) = \ :\Phi^i(z,\bar{z})\Phi^j(w,\bar{w}): -i[p^i, q^j](\ln z + \ln\bar{z})$$
$$+ \left[i\sum_{n>0}\frac{1}{n}\alpha_n^i z^{-n}, i\sum_{n<0}\frac{1}{n}\alpha_n^j w^{-n}\right]$$
$$+ \left[i\sum_{n>0}\frac{1}{n}\tilde{\alpha}_n^i \bar{z}^{-n}, i\sum_{n<0}\frac{1}{n}\tilde{\alpha}_n^j \bar{w}^{-n}\right] \tag{4.96}$$
$$= \ :\Phi^i(z,\bar{z})\Phi^j(w,\bar{w}):$$
$$+ \delta^{ij}\Big(-\ln z\bar{z} + \sum_{n>0}\frac{1}{n}\left(\frac{w}{z}\right)^n + \sum_{n>0}\frac{1}{n}\left(\frac{\bar{w}}{\bar{z}}\right)^n\Big).$$

Since the operators are radially ordered then $|z| > |w|$ and hence the sums converge. We obtain

$$R(\Phi^i(z,\bar{z})\Phi^j(w,\bar{w})) = \ :\Phi^i(z,\bar{z})\Phi^j(w,\bar{w}):$$
$$- \delta^{ij}(\ln(z - w) + \ln(\bar{z} - \bar{w})). \tag{4.97}$$

This is not meromorphic. However, the operator $\partial\Phi$ which is of conformal dimension (1,0) has meromorphic operator product expansions with itself and with Φ. We compute immediately from the above expression

$$R(\partial\Phi^i(z,\bar{z}).\ \Phi^j(w,\bar{w})) = \ :\partial\Phi^i(z,\bar{z}).\ \Phi^j(w,\bar{w}): -\delta^{ij}\left(\frac{1}{z-w}\right) \tag{4.98}$$

$$R(\partial\Phi^i(z,\bar{z})\partial\Phi^j(w,\bar{w})) = \ :\partial\Phi^i(z,\bar{z})\partial\Phi^j(w,\bar{w}): -\delta^{ij}\left(\frac{1}{(z-w)^2}\right). \tag{4.99}$$

The normal ordered objects behave as classical quantities and thus they are finite in the limit $z \longrightarrow w$. We rewrite the above result as

$$\partial\Phi^i(z)\partial\Phi^j(w) = -\delta^{ij}\left(\frac{1}{(z-w)^2}\right). \tag{4.100}$$

The classical stress–energy tensor $T(z) = \partial\Phi^i(z)\partial\Phi^i(z)$ is defined quantum mechanically by the normal ordered expression

$$T(z) = \ :\partial\Phi^i(z)\partial\Phi^i(z): \\ = \lim_{z\longrightarrow w}\left(\partial\Phi^i(z)\partial\Phi^i(w) + \frac{\delta^{ii}}{(z-w)^2}\right). \tag{4.101}$$

The generalization of (4.97) to three operators is straightforwardly given by

$$R(\Phi^i(z,\bar{z})\Phi^j(w,\bar{w})\Phi^l(y,\bar{y})) = \ :\Phi^i(z,\bar{z})\Phi^j(w,\bar{w})\Phi^l(y,\bar{y}): \\ + \sum\text{contractions}. \tag{4.102}$$

The sum runs over all ways of choosing one pair of fields (or two or more pairs in the case we have an arbitrary product of fields) from the product and replacing each term with the contraction in (4.97), viz

$$R(\Phi^i(z,\bar{z})\Phi^j(w,\bar{w})\Phi^l(y,\bar{y})) = \ :\Phi^i(z,\bar{z})\Phi^j(w,\bar{w})\Phi^l(y,\bar{y}): \\ -\delta^{ij}(\ln(z-w)+\ln(\bar{z}-\bar{w}))\Phi^l(y,\bar{y}) \\ -\delta^{il}(\ln(z-y)+\ln(\bar{z}-\bar{y}))\Phi^j(w,\bar{w}) \\ -\delta^{jl}(\ln(w-y)+\ln(\bar{w}-\bar{y}))\Phi^i(z,\bar{z}). \tag{4.103}$$

By differentiation we then obtain

$$R(\partial\Phi^i(z)\partial\Phi^j(w)\partial\Phi^l(y)) = \ :\partial\Phi^i(z)\partial\Phi^j(w)\partial\Phi^l(y): \\ -\delta^{ij}\frac{1}{(z-w)^2}\partial\Phi^l(y) \\ -\delta^{il}\frac{1}{(z-y)^2}\partial\Phi^j(w) \\ -\delta^{jl}\frac{1}{(w-y)^2}\partial\Phi^i(z). \tag{4.104}$$

From this result we obtain immediately

$$R(T(z)\partial\Phi^l(y)) = \ :\partial\Phi^i(z)\partial\Phi^i(w)\partial\Phi^l(y): -\frac{2}{(z-y)^2}\partial\Phi^l(z). \tag{4.105}$$

By substituting $\partial\Phi^l(z,\bar{z}) = \partial\Phi^l(y,\bar{z}) + (z-y)\partial^2\Phi^i(y) + \cdots$ we obtain

$$R(T(z)\partial\Phi^l(y)) = \ :\partial\Phi^i(z)\partial\Phi^i(w)\partial\Phi^l(y): \\ -\frac{2}{(z-y)^2}\partial\Phi^l(y) - \frac{2}{z-y}\partial^2\Phi^l(y) + \cdots \tag{4.106}$$

We rewrite this result as

$$T(z)\partial\Phi^l(y) = =-\frac{2}{(z-y)^2}\partial\Phi^l(y) - \frac{2}{z-y}\partial^2\Phi^l(y). \tag{4.107}$$

This means in particular that $\partial\,\Phi^i$ is a conformal field of weight $h = 1$.

Following the same trick, equation (4.102), we find

$$\begin{aligned}R(\partial\Phi^i(z)\partial\Phi^j(w)\partial\Phi^l(y)\partial\Phi^k(x)) = &: \partial\Phi^i(z)\partial\Phi^j(w)\partial\Phi^l(y)\partial\Phi^k(x) :\\ &+ \delta^{ij}\delta^{lk}\frac{1}{(z-w)^2}\frac{1}{(y-x)^2}\\ &- \delta^{ij}\frac{1}{(z-w)^2} : \partial\Phi^l(y)\partial\Phi^k(x) :\\ &+ \delta^{il}\delta^{jk}\frac{1}{(z-y)^2}\frac{1}{(w-x)^2}\\ &- \delta^{il}\frac{1}{(z-y)^2} : \partial\Phi^j(w)\partial\Phi^k(x) :\\ &+ \delta^{ik}\delta^{jl}\frac{1}{(z-x)^2}\frac{1}{(w-y)^2}\\ &- \delta^{ik}\frac{1}{(z-x)^2} : \partial\Phi^j(w)\partial\Phi^l(y) :\\ &- \delta^{jl}\frac{1}{(w-y)^2} : \partial\Phi^i(z)\partial\Phi^k(x) :\\ &- \delta^{jk}\frac{1}{(w-x)^2} : \partial\Phi^i(z)\partial\Phi^l(y) :\\ &- \delta^{lk}\frac{1}{(y-x)^2} : \partial\Phi^i(z)\partial\Phi^j(w) : .\end{aligned} \tag{4.108}$$

We compute, by setting $i = j$ and $l = k$ and summing over i and l and also by taking the limit $w \longrightarrow z$ and $x \longrightarrow y$, first

$$\begin{aligned}&-\frac{1}{(w-y)^2} : \partial\Phi^i(z)\partial\Phi^i(x) : -\frac{1}{(w-x)^2} : \partial\Phi^i(z)\partial\Phi^i(y) :\\ &= -\frac{2}{(z-y)^2}T(z) + \frac{2}{z-y} : \partial\Phi^i(z)\partial^2\Phi^i(w) : .\end{aligned} \tag{4.109}$$

$$\begin{aligned}&-\frac{1}{(z-y)^2} : \partial\Phi^i(w)\partial\Phi^i(x) : -\frac{1}{(z-x)^2} : \partial\Phi^i(w)\partial\Phi^i(y) :\\ &= -\frac{2}{(z-y)^2}T(z) + \frac{2}{z-y} : \partial^2\Phi^i(z)\partial\Phi^i(w) : .\end{aligned} \tag{4.110}$$

From this we get immediately the result

$$\begin{aligned}R(\partial\Phi^i(z)\partial\Phi^i(w)\partial\Phi^l(y)\partial\Phi^l(x)) = &: \partial\Phi^i(z)\partial\Phi^i(w)\partial\Phi^l(y)\partial\Phi^l(x) :\\ &+ \delta^{ii}\delta^{ll}\frac{1}{(z-w)^2}\frac{1}{(y-x)^2}\\ &- \delta^{ii}\frac{1}{(z-w)^2}T(y)\\ &+ 2\delta^{ii}\frac{1}{(z-y)^4} - \frac{4}{(z-y)^2}T(z)\\ &+ \frac{2}{z-y}\partial T(z) - \delta^{ll}\frac{1}{(y-x)^2}T(z).\end{aligned} \tag{4.111}$$

On the other hand, we have

$$R(T(z)T(y)) = R(R(\partial\Phi^i(z)\partial\Phi^i(w))R(\partial\Phi^l(y)\partial\Phi^l(x)))$$
$$+ \delta^{ii}\delta^{ll}\frac{1}{(z-w)^2}\frac{1}{(y-x)^2}$$
$$+ \delta^{ii}\frac{1}{(z-w)^2}\left(T(y) - \frac{\delta^{ll}}{(y-x)^2}\right) \tag{4.112}$$
$$+ \delta^{ll}\frac{1}{(y-x)^2}\left(T(z) - \frac{\delta^{ii}}{(z-w)^2}\right).$$

Hence, we obtain

$$R(T(z)T(y)) = \ :\partial\Phi^i(z)\partial\Phi^i(w)\partial\Phi^l(y)\partial\Phi^l(x):$$
$$+ 2\delta^{ii}\frac{1}{(z-y)^4} - \frac{4}{(z-y)^2}T(z) + \frac{2}{z-y}\partial T(z). \tag{4.113}$$

We rewrite this as

$$T(z)T(y) = 2\delta^{ii}\frac{1}{(z-y)^4} - \frac{4}{(z-y)^2}T(z) + \frac{2}{z-y}\partial T(z). \tag{4.114}$$

Or equivalently

$$T(z)T(y) = 2\delta^{ii}\frac{1}{(z-y)^4} - \frac{4}{(z-y)^2}T(y) - \frac{2}{z-y}\partial T(y). \tag{4.115}$$

By comparing with (4.85) we can also see that $-T(z)/2$ is a conformal field of weight $h = 2$, which is the classical value, if $\delta^{ii} = 0$. The first term in the above equation is therefore a conformal anomaly due to quantum effects.

4.5 Conformal field theory and BRST quantization

The gauge-fixed action in the BRST quantization of the string is given in Euclidean worldsheet by

$$S_E = \frac{1}{2\pi}\int d^2z(2\partial X^\mu\bar{\partial}X_\mu - ib_{--}\bar{\partial}c^- + ib_{++}\partial c^+). \tag{4.116}$$

We have used $i\int d^2\sigma \longrightarrow \int d^2z/2$, and $\tau + \sigma = \bar{z} = -i\sigma^2 + \sigma^1$, $\tau - \sigma = -z = -i\sigma^2 - \sigma^1$. Recall also that the ghost fields c^+ and $c^-$are the components of a vector ghost field c^a, whereas the antighosts b_{++} and b_{--} are the components of a traceless symmetric antighost tensor field b_{ab}. We scale the ghost fields as $b \longrightarrow b$ and $c^- \longrightarrow ic^-$ and $c^+ \longrightarrow -ic^+$ to obtain

$$S_E = \frac{1}{2\pi}\int d^2z(2\partial X^\mu\bar{\partial}X_\mu + b_{--}\bar{\partial}c^- + b_{++}\partial c^+). \tag{4.117}$$

The energy–momentum tensor was found to be given by (including also Wick rotation)

$$-2T^g_{++} = -2b_{++}\bar{\partial}c^{+}+c^{+}\bar{\partial}b_{++} \tag{4.118}$$

$$-2T^g_{--} = -2b_{--}\partial c^{-}+c^{-}\partial b_{--}. \tag{4.119}$$

The basic anticommutation relations become

$$\{c^{+}(\sigma,\tau), b_{++}(\sigma',\tau)\} = 2\pi i\delta(\sigma-\sigma'), \quad \{c^{-}(\sigma,\tau), b_{--}(\sigma',\tau)\} = -2\pi i\delta(\sigma-\sigma'). \tag{4.120}$$

The above theory involves a free fermionic conformal field theory termed *bc* CFT given in terms of anticommuting fields b and c by

$$S_E = \frac{1}{2\pi}\int d^2z b\bar{\partial}c. \tag{4.121}$$

This is conformally invariant for all b and c transforming under conformal transformations as tensors of weights $(\lambda, 0)$ and $(1-\lambda, 0)$. In the BRST quantization of the string action we have found that $\lambda = 2$. Thus, the ghost field c has conformal dimension -1, whereas the antighost field b has conformal dimension 2. In other words, b transforms as the energy–momentum tensor while c transforms as the gauge transformations (diffeomorphism) parameter.

We compute

$$\begin{aligned} 0 &= \int [db][dc]\frac{\delta}{\delta c(z,\bar{z})}[\exp(-S_E)c(0)] \\ &\Rightarrow \langle\bar{\partial}b(z)\cdot c(0)\rangle = 2\pi\delta^2(z,\bar{z}). \end{aligned} \tag{4.122}$$

Using the same method we obtain the quantum equations of motion

$$\langle\bar{\partial}c(z)\rangle = \langle\bar{\partial}b(z)\rangle = 0. \tag{4.123}$$

We compute the radial ordering in terms of the normal ordering and a contraction (propagator) as follows

$$R(b(z_1)c(z_2)) =: b(z_1)c(z_2) : +\langle b(z_1)c(z_2)\rangle. \tag{4.124}$$

The propagator is given by

$$\langle b(z_1)c(z_2)\rangle = \frac{1}{z_1 - z_2}. \tag{4.125}$$

Let us redo this for the bosonic action

$$S = \frac{1}{2\pi\alpha'}\int d^2z\partial\phi\bar{\partial}\phi. \tag{4.126}$$

We compute then

$$0 = \int [d\phi] \frac{\delta}{\delta\phi(z, \bar{z})} [\exp(-S)\phi(0)]$$
$$\Rightarrow \frac{1}{\pi\alpha'} \langle \bar{\partial}\partial\phi(z) \cdot \phi(0) \rangle = -\delta^2(z, \bar{z}). \tag{4.127}$$

We get the propagator

$$\frac{1}{\pi\alpha'} \bar{\partial}\partial \langle \phi(z) \cdot \phi(0) \rangle = -\delta^2(z, \bar{z}). \tag{4.128}$$

However, we obtain from (4.97) (after we insert the coefficient of α')

$$\langle \phi(z) \cdot \phi(0) \rangle = \langle R(\phi(z)\phi(0)) \rangle = -\frac{\alpha'}{2} \ln|z|^2. \tag{4.129}$$

Thus

$$\bar{\partial}\partial \ln|z|^2 = 2\pi\delta^2(z, \bar{z}). \tag{4.130}$$

Equivalently

$$\bar{\partial}\frac{1}{z} = 2\pi\delta^2(z, \bar{z}). \tag{4.131}$$

This should be checked directly by integrating by part over a region containing the origin. By using (4.122) and (4.131) we can see that the fermion propagator is indeed given by (4.125). We then get the operator product expansions

$$b(z_1)c(z_2) \sim \frac{1}{z_1 - z_2}, \quad c(z_1)b(z_2) \sim \frac{\epsilon}{z_1 - z_2} \tag{4.132}$$

$$b(z_1)b(z_1) \sim O(z_1 - z_2), \quad c(z_1)c(z_2) = O(z_1 - z_2). \tag{4.133}$$

The ϵ in (4.132) is equal to +1 in our case. But in the case where b and c satisfy Bose statistics we must set $\epsilon = -1$. The generalization of the first equation of (4.118) is given by (including also normal ordering to be precise)

$$2T_{bc} = \, : -\lambda b(z)\partial c(z) + \epsilon(\lambda - 1)c(z)\partial b(z) :$$
$$= \lim_{z \longrightarrow w}\Big(-\lambda b(z)\partial c(w) + \epsilon(\lambda - 1)c(z)\partial b(w) + \frac{1}{(z - w)^2}\Big). \tag{4.134}$$

The proof goes as follows. We compute the following radial products

$$R(b(z)c(w)b(y)) =: \ldots :+ \frac{b(y)}{z - w} - \frac{b(z)}{y - w} \tag{4.135}$$

$$-\lambda R(b(z)\partial c(w).\, b(y)) =: \ldots :-\lambda \frac{b(y)}{(z - w)^2} + \lambda \frac{b(z)}{(y - w)^2} \tag{4.136}$$

$$(\lambda - 1)R(c(w)\partial b(z).\ b(y)) =: \ldots : (\lambda - 1)\frac{b(y)}{(z-w)^2} + (\lambda - 1)\frac{\partial b(z)}{y-w}. \quad (4.137)$$

Thus

$$R(T(z)b(y)) =: \ldots :+\frac{\lambda}{(z-y)^2}b(y) + \frac{1}{z-y}\partial b(y). \quad (4.138)$$

This shows explicitly that b is of conformal dimension λ as it should be and that $T(z)$ is the correct form of the energy–momentum tensor. Similarly we compute

$$R(T(z)c(y)) =: \ldots :+\frac{1-\lambda}{(z-y)^2}c(y) + \frac{1}{z-y}\partial c(y). \quad (4.139)$$

Thus c is of conformal weight equal to $1 - \lambda$ and $T(z)$ is the correct energy–momentum tensor.

The above formulas generalize to the Ward identities

$$R(T(z)\mathcal{O}(y)) =: \ldots :+\frac{h}{(z-w)^2}\mathcal{O}(y) + \frac{1}{z-y}\partial\mathcal{O}(y), \quad (4.140)$$

for the general conformal transformations $z \longrightarrow z' = f(z)$ under which the primary field $\mathcal{O}$ with weights $(h, \bar{h})$ transforms as

$$\mathcal{O}'(z', \bar{z}') = (\partial_z z')^{-h}(\partial_{\bar{z}}\bar{z}')^{-\bar{h}}\mathcal{O}(z, \bar{z}). \quad (4.141)$$

The sum $h + \bar{h}$ is called the dimension of $\mathcal{O}$ and it determines its behavior under scalings while $h - \bar{h}$ is the spin of $\mathcal{O}$ and it determines its behavior under rotations.

For example, X^μ is of weights (0,0), ∂X^μ is of weight (1,0), $\bar{\partial}X^\mu$ is weight (0,1), $\partial^2 X^\mu$ is of weight (2,0), and :exp(ikX): is of weights ($\alpha' k^2/4$, $\alpha' k^2/4$). More importantly, the operator product expansion of the energy–momentum tensor with itself is given by

$$T(z)T(y) = \frac{c}{2(z-y)^4} + \frac{2}{(z-y)^2}T(y) + \frac{1}{z-y}\partial T(y). \quad (4.142)$$

The constant c is called the central charge. The energy–momentum is not conformal unless $c = 0$. In this case it is a primary field of weights (2,0). The central charge is therefore the conformal anomaly. The above equation gives the Ward identities for the conformal transformation law of the energy–momentum tensor given by (where $z' = z + \epsilon v$)

$$-\delta T(z) = -\frac{c}{12}\partial^3 v(z) - 2\partial v(z).\ T(z) - v(z)\partial T(z) \quad (4.143)$$

using the method which has led to equation (4.85).

An example of (6.58) is (4.115) rewritten as (with the scaling $T(z) \longrightarrow -T(z)/2$)

$$T(z)T(y) = \frac{D}{2(z-y)^4} + \frac{2}{(z-y)^2}T(y) + \frac{1}{z-y}\partial T(y). \quad (4.144)$$

Thus, the central charge is equal to one for a single scalar field.
What is the analogue result for the bc CFT above?
First, we compute

$$\begin{aligned} R(b(z)c(w)b(y)c(x)) =&: \ldots : + \frac{1}{z-w} : b(y)c(x) : \\ &+ \frac{1}{z-x} : c(w)b(y) : + \frac{1}{w-y} : b(z)c(x) : \\ &+ \frac{1}{y-x} : b(z)c(w) : + \frac{1}{z-w} . \frac{1}{y-x} \\ &+ \frac{1}{z-x} . \frac{1}{w-y} . \end{aligned} \tag{4.145}$$

Then we compute (by taking also the limit $x \longrightarrow y$)

$$\begin{aligned} R(b(z)c(w)T(y)) =&: \ldots : + \frac{1}{z-w} T(y) - \frac{\lambda}{w-y} : b(z)\partial c(y) : \\ &- \frac{\lambda - 1}{z-y} : c(w)\partial b(y) : \\ &- \frac{\lambda}{(z-y)^2} R(c(w)b(y)) - \frac{(\lambda - 1)}{(w-y)^2} R(b(z)c(y)). \end{aligned} \tag{4.146}$$

Further

$$\begin{aligned} R(b(z)\partial c(w)T(y)) =&: \ldots : + \frac{1}{(z-w)^2} T(y) + \frac{\lambda}{(w-y)^2} : b(z)\partial c(y) : \\ &- \frac{\lambda - 1}{z-y} : \partial c(w)\partial b(y) : \\ &- \frac{\lambda}{(z-y)^2} : \partial c(w)b(y) : + \frac{2(\lambda - 1)}{(w-y)^3} : b(z)c(y) : \\ &+ \frac{3\lambda - 2}{(z-y)^4} \end{aligned} \tag{4.147}$$

$$\begin{aligned} - R(c(z)\partial b(w)T(y)) =&: \ldots : - \frac{1}{(z-w)^2} T(y) - \frac{\lambda}{z-y} : \partial b(w)\partial c(y) : \\ &+ \frac{\lambda - 1}{(w-y)^2} : c(z)\partial b(y) : \\ &+ \frac{2\lambda}{(w-y)^3} : c(z)b(y) : - \frac{(\lambda - 1)}{(z-y)^2} : \partial b(w)c(y) : \\ &+ \frac{3\lambda - 1}{(z-y)^4} . \end{aligned} \tag{4.148}$$

The first and last terms in the above two equations yield immediately

$$R(T(z)T(y)) = :\dots: + \frac{c}{2(z-y)^4} + \cdots,$$
$$c = -2\epsilon(6\lambda^2 - 6\lambda + 1). \tag{4.149}$$

This is the conformal anomaly. Now we compute the other terms. The terms proportional to $1/z^3$ vanish. Indeed, we have (by expanding y around z or w and expanding z around w)

$$\begin{aligned}
O\left(\frac{1}{z^3}\right) &= -\frac{2\lambda(\lambda-1)}{(w-y)^3} : b(z)c(y) : \\
&\quad -\frac{2\lambda(\lambda-1)}{(w-y)^3} : c(z)b(y) : \\
&= \frac{2\lambda(\lambda-1)}{(y-z)^2}(:b(z)\partial c(w): + : c(w)\partial b(z) :) \\
&\quad + \frac{\lambda(\lambda-1)}{y-z}(:b(z)\partial^2 c(w): + : c(w)\partial^2 b(z) :).
\end{aligned} \tag{4.150}$$

By adding the first line of this equation to the terms proportional to $1/z^2$ (and also by expanding y around w, etc) we obtain

$$\begin{aligned}
O\left(\frac{1}{z^2}\right) + O\left(\frac{1}{z^3}\right)_{\text{first line}} &= 2\frac{T(z)}{(z-y)^2} - \frac{\lambda^2}{y-w} : b(z)\partial^2 c(w) : \\
&\quad - \frac{(\lambda-1)^2}{y-w} : c(z)\partial^2 b(w) : \\
&\quad - \frac{\lambda^2}{y-z} : \partial b(y)\partial c(w) : - \frac{(\lambda-1)^2}{y-z} : \partial c(z)\partial b(w) : .
\end{aligned} \tag{4.151}$$

Thus

$$\begin{aligned}
O\left(\frac{1}{z^2}\right) + O\left(\frac{1}{z^3}\right) &= 2\frac{T(z)}{(z-y)^2} - \frac{\lambda}{y-w} : b(z)\partial^2 c(w) : \\
&\quad + \frac{\lambda-1}{y-w} : c(z)\partial^2 b(w) : \\
&\quad - \frac{\lambda^2}{y-z} : \partial b(y)\partial c(w) : - \frac{(\lambda-1)^2}{y-z} : \partial c(z)\partial b(w) : .
\end{aligned} \tag{4.152}$$

We add now the terms proportional to $O(1/z)$ to find

$$\begin{aligned}
O\left(\frac{1}{z}\right) + O\left(\frac{1}{z^2}\right) + O\left(\frac{1}{z^3}\right) &= 2\frac{T(z)}{(z-y)^2} + \frac{\lambda}{z-y} : b(z)\partial^2 c(w) : \\
&\quad - \frac{\lambda-1}{z-y} : c(z)\partial^2 b(w) : \\
&\quad + \frac{\lambda}{z-y} : \partial b(z)\partial c(w) : - \frac{\lambda-1}{z-y} : \partial c(z)\partial b(w) : .
\end{aligned} \tag{4.153}$$

We use

$$\partial T(z) = -\lambda : b(z)\partial^2 c(w) : +(\lambda - 1) : c(z)\partial^2 b(w) : \\ -\lambda : \partial b(z)\partial c(w) : +(\lambda - 1) : \partial c(z)\partial b(w) : . \tag{4.154}$$

Hence

$$O\left(\frac{1}{z}\right) + O\left(\frac{1}{z^2}\right) + O\left(\frac{1}{z^3}\right) = 2\frac{T(z)}{(z-y)^2} - \frac{\partial T(z)}{z-y} \\ = 2\frac{T(y)}{(z-y)^2} + \frac{\partial T(y)}{z-y}. \tag{4.155}$$

We then get the operator product expansion

$$R(T(z)T(y)) = :...: + \frac{c}{2(z-y)^4} + 2\frac{T(y)}{(z-y)^2} + \frac{\partial T(y)}{z-y}, \\ c = -2\epsilon(6\lambda^2 - 6\lambda + 1). \tag{4.156}$$

The Faddeev–Popov ghosts of the gauge-fixed Polyakov action have weights corresponding to $\lambda = 2$, viz $(2, 0)$ for the antighost b and $(-1, 0)$ for the ghost c, with conformal anomaly $c = -26$. Another important case is given by fermions $b = \psi$ and $c = \bar{\psi}$ with equal weights $\lambda = 1 - \lambda = 1/2$ and with conformal anomaly $c = 1$.

The conformal anomaly $c(\lambda = 2, \epsilon = 1) = -26$ can be canceled by 26 spacetime coordinates X^μ since the weight of every X^μ is 1.

We enhance conformal symmetry to the superconformal case. The gauge fixing in this case using the so-called superconformal gauge introduces, besides the two fermion ghosts b and c with $\lambda = 2$, two boson ghosts with $\lambda = 3/2$. Thus the total conformal anomaly coming from the ghosts in this case is $c(2, 1) + c(3/2, -1) = -26 + 11 = 15$. This can be canceled by the contribution of the D coordinates X^μ and their D superpartners ψ^μ with $\lambda = 1/2$. Indeed, the contribution of the dynamical fields to the conformal anomaly given by $c = D(1 + 1/2) = 3D/2$ cancels exactly for $3D/2 = 15$, i.e. $D = 10$.

The total energy–momentum tensor corresponding to the Euclidean action (4.117) (with $\alpha' = 1/2$ and $b_{++} = \tilde{b}$, $c^+ = \tilde{c}$, $b_{--} = b$, $c^- = c$)

$$S_E = \frac{1}{2\pi}\int d^2z(2\partial X^\mu \bar{\partial} X_\mu + b\bar{\partial}c + \tilde{b}\partial\tilde{c}) \tag{4.157}$$

is given by

$$T(z) = T_X(z) + T_{bc}(z) \\ = -2 : \partial X^\mu \partial X_\mu : + : -\lambda b\partial c + (\lambda - 1)c\partial b : . \tag{4.158}$$

This action is invariant under the BRST symmetry

$$\delta X^\mu = \eta c\partial X^\mu, \quad \delta c = \eta c\partial c, \quad \delta b = \eta T. \tag{4.159}$$

The BRST charge (see the contribution coming from the BRST current J_{-}^{B} in equation (3.113)) is given by

$$Q_B = -\frac{1}{2\pi i}\oint (cT_X + :bc\partial c:\,)dz. \tag{4.160}$$

Using this formula we compute (using $\{c(z), b(z')\} = -2\pi i\delta(z - z')$) for example

$$\begin{aligned}\{Q_B, b(z)\} &= -\frac{1}{2\pi i}\oint \{c(z')T_X(z') + b(z')c(z')\partial c(z'), b(z)\}\\ &= T_X(z) - \partial(b(z)c(z)) - b(z)\partial c(z)\\ &= T(z).\end{aligned} \tag{4.161}$$

Thus

$$\delta b(z) = [\eta Q_B, b(z)] = \eta T(z). \tag{4.162}$$

A generalization is

$$\delta Y = [\eta Q_B, Y]. \tag{4.163}$$

In particular, the calculation of the transformation law for X^μ requires the use of the formal commutation relation

$$[\partial X^\mu(z), X^\nu(z')] = i\frac{\pi}{2}\eta^{\mu\nu}\delta(z - z'). \tag{4.164}$$

Recall that $\partial = -\partial_{-} \longrightarrow -\partial_\tau/2$, and $T = 1/\pi$, $\alpha' = 1/2$, $l_s = 1$, and $P^\mu = T\dot{X}^\mu$.

The conserved ghost-number charge is obviously given by the number operator (see the contribution coming from the ghost-number current J_{-} in equation (3.114))

$$U = -\frac{1}{2\pi i}\oint dzcb. \tag{4.165}$$

Clearly, c is the raising operator (ghost number +1) and b is the lowering operator (ghost number −1).

The BRST charge defined by (4.160) is nilpotent, viz $Q_B^2 = 0$. We have already given the proof in (3.136). Thus, Q_B is like the exterior derivative d of differential geometry which satisfies $d^2 = 0$.

Every differentiable manifold $\mathcal{M}$ comes with an exterior derivative d which is nilpotent, i.e. $d^2 = 0$. The differential forms ω, generalization of functions, defined on the manifold $\mathcal{M}$ can be either closed satisfying $d\omega = 0$ or exact satisfying $\omega = d\phi$ where ϕ is a differential form of lower degree. The closed forms, which are not exact, encode the topological properties of the underlying manifold. Any two closed forms are said to be equivalent if and only if their difference is exact. We obtain thus equivalence classes of closed forms. The vector space of equivalence classes of closed n-forms is called the n de Rham cohomology group $H^n(\mathcal{M})$.

By analogy, the physical string states must be BRST cohomology classes. Indeed, a physical string state is a BRST closed state which must be annihilated by the BRST charge Q_B. Two such states are equivalent if their difference is a BRST exact state.

In other words, a physical string state is an equivalence class of BRST closed states under the effect of the BRST charge Q_B.

The ghost number operator is given explicitly by equation (3.139), viz

$$U = \frac{1}{2}(c_0 b_0 - b_0 c_0) + \sum_{n=1}^{\infty}(c_{-n}b_n - b_{-n}c_n). \tag{4.166}$$

The c_n, $n > 0$, are annihilation operators with $b_{-\mathrm{n}}$ the corresponding creation operators, whereas, c_n, $n < 0$, are creation operators with $b_{-\mathrm{n}}$ the corresponding annihilation operators.

The situation with the ghost zero modes c_0 and b_0 is more involved. The corresponding irreducible representation is given in terms of two states $|+\rangle$, $|-\rangle$ defined by (3.140), viz

$$c_0|-\rangle = |+\rangle, \quad b_0|+\rangle = |-\rangle, \quad c_0|+\rangle = b_0|-\rangle = 0. \tag{4.167}$$

The physical states $|\psi\rangle$ must contain no ghost excitations, i.e. they must satisfy equation (3.143). Thus, the ghost wave function is proportional to one of the two ghost ground states $|+\rangle$ and $|-\rangle$. These ghost states $|+\rangle$ and $|-\rangle$ have ghost numbers $+1/2$ and $-1/2$ respectively. This shows that physical states are *a priori* doubly degenerate or more precisely they are BRST cohomology classes with ghost numbers equal to $\pm 1/2$.

As pointed out before, the choice between the values $\pm 1/2$ does matter since the theory is not symmetric between the ghosts c and the antighosts b which have different conformal dimensions. As it turns out, the tachyon ground state and other physical states correspond to the ghost number $-1/2$.

Indeed, by assuming first that $|\psi\rangle$ has ghost number equal to $-1/2$ we conclude that

$$b_0|\psi\rangle = 0. \tag{4.168}$$

Since the physical states $|\psi\rangle$ must contain no b-oscillator and c-oscillator excitations we must also have

$$b_n|\psi\rangle = c_n|\psi\rangle = 0, \quad n > 0. \tag{4.169}$$

Then in this case the condition that the physical state $|\psi\rangle$ is BRST invariant is exactly equivalent to the physical state conditions of the covariant quantization, i.e.

$$Q|\psi\rangle = 0 \Leftrightarrow (L_0 - a)|\psi\rangle = 0, \quad L_m|\psi\rangle = 0, \quad m > 0. \tag{4.170}$$

We have thus shown that physical states $|\psi\rangle$ must be BRST cohomology classes of ghost number equal to $-1/2$ which contain no b-oscillator and c-oscillator excitations. This is because $|\psi\rangle + Q|\lambda\rangle$ for all $|\lambda\rangle$ represent the same physical state $|\psi\rangle$ which is BRST invariant, i.e. $Q|\psi\rangle = 0$. The BRST invariance of these states yields precisely the physical state conditions of the covariant quantization. The state $|\psi\rangle$ so constructed provides thus a unique representative of the BRST cohomology class.

4.6 Representation theory of the Virasoro algebra

4.6.1 Virasoro algebra revisited

We have found that the operator product expansion for the energy–momentum tensor is given by equation (4.115) which we rewrite as (with $c = \delta^{ii}$ the number of bosons)

$$T(z)T(y) = \frac{c}{2(z-y)^4} + \frac{2}{(z-y)^2}T(y) + \frac{1}{z-y}\partial T(y). \tag{4.171}$$

The first term is the anomaly which is absent if the energy–momentum tensor were really of conformal weight 2. The conserved conformal currents are defined by

$$J_z(z) = iT(z)\epsilon(z). \tag{4.172}$$

We expand the holomorphic function $\epsilon(z)$ in modes. We expect that $J_z(z)$ generates the transformation $z \longrightarrow z' = z - z^{n+1}$ if we choose $\epsilon(z) = z^{n+1}$. This corresponds to an expansion on the Riemann sphere where everything is continuous on contours around the origin. This is what we have done implicitly previously in writing the Virasoro generators L_n as the differential operators $l_n = -z^{n+1}\partial$. In terms of $T(z)$ the correctly normalized Virasoro generators L_n are given by the formula

$$L_n = \frac{1}{2\pi i}\oint dz z^{n+1}T(z). \tag{4.173}$$

We recall also that

$$T(z) = \sum_n \frac{L_n}{z^{n+2}}. \tag{4.174}$$

The proof goes as follows. We have

$$\begin{aligned} L_nL_m &= \oint \frac{dz}{2\pi i}\frac{dw}{2\pi i} z^{n+1}T(z)w^{m+1}T(w) \\ &= \oint_{|z|>||w|} \frac{dz}{2\pi i}\frac{dw}{2\pi i} z^{n+1}T(z)w^{m+1}T(w) \\ &= \oint_{|z|>|w|} \frac{dz}{2\pi i}\frac{dw}{2\pi i} R(z^{n+1}T(z)w^{m+1}T(w)). \end{aligned} \tag{4.175}$$

We will do the z integral first. We have assumed that the z integrand is radially ordered since the operator product $z^{n+1}T(z)w^{m+1}T(w)$ is only convergent for $|z| > |w|$. Similarly,

$$\begin{aligned} L_mL_n &= \oint \frac{dw}{2\pi i}\frac{dz}{2\pi i} w^{m+1}T(w)z^{n+1}T(z) \\ &= \oint_{|w|>|z|} \frac{dw}{2\pi i}\frac{dz}{2\pi i} w^{m+1}T(w)z^{n+1}T(z) \\ &= \oint_{|w|>|z|} \frac{dw}{2\pi i}\frac{dz}{2\pi i} R(w^{m+1}T(w)z^{n+1}T(z)). \end{aligned} \tag{4.176}$$

Thus

$$[L_n, L_m] = \Big(\oint_{|z|>|w|} \frac{dz}{2\pi i}\frac{dw}{2\pi i} - \oint_{|w|>|z|} \frac{dw}{2\pi i}\frac{dz}{2\pi i}\Big) R(z^{n+1}T(z)w^{m+1}T(w)). \quad (4.177)$$

The first contour encircles the origin and w, whereas the second contour encircles only the origin. Thus we get a contour C which encircles w only. See figure 3.2 of [1]. We get then

$$[L_n, L_m] = \oint_C \frac{dw}{2\pi i}\frac{dz}{2\pi i} R(z^{n+1}T(z)w^{m+1}T(w)). \quad (4.178)$$

The contribution comes only from the limit $w \longrightarrow z$. We substitute then the operator product expansion (4.171) to obtain

$$[L_n, L_m] = \oint_C \frac{dw}{2\pi i}\frac{dz}{2\pi i} z^{n+1}w^{m+1}\Bigg(\frac{c}{2(z-w)^4} + \frac{2}{(z-w)^2}T(w) + \frac{1}{z-w}\partial T(w)\Bigg). \quad (4.179)$$

We now use the Cauchy theorem to find

$$\begin{aligned}
[L_n, L_m] &= \oint_C \frac{dw}{2\pi i}\Bigg(\frac{1}{6}\frac{c}{2}(n+1)n(n-1)w^{n-2}w^{m+1} \\
&\quad + 2(n+1)w^n w^{m+1}T(w) + w^{n+1}w^{m+1}\partial T(w)\Big) \\
&= \oint_C \frac{dw}{2\pi i}\Bigg(\frac{1}{6}\frac{c}{2}(n+1)n(n-1)w^{n-2}w^{m+1} \\
&\quad + (n+1)w^n w^{m+1}T(w) - (m+1)w^{n+1}w^m T(w)\Big). \\
&= \frac{c}{12}(n^3 - n)\oint_C \frac{dw}{2\pi i} w^{n+m-1} + (n+1)L_{n+m} - (m+1)L_{n+m} \\
&= \frac{c}{12}(n^3 - n)\delta_{n,-m} + (n-m)L_{n+m}.
\end{aligned} \quad (4.180)$$

The first term should be thought of as an operator which commutes with all elements of the algebra called the central charge. As a consequence of this term, the classical symmetry is not preserved quantum mechanically and we cannot impose the condition $L_n|0\rangle = 0$ for all n. The $SL(2, \mathbf{C})$ subalgebra generated by the operators L_+, L_- and L_0 is not affected by the anomaly term. This is why the energy–momentum tensor can still be said to have a conformal weight in a consistent way.

4.6.2 Overview of representation theory

This is a rather complex issue and as such it will suffice for the time being to simply follow the excellent pedagogical presentation [1, 4, 5] and the seminal papers [11, 12].

We choose as maximal the set of commuting operators, the central charge c and the Hamiltonian L_0. These are the analogue of J^2 and J_3 here. We are interested in unitary representations which are representations where the Virasoro generators L_n are realized as operators on a Hilbert space satisfying the conditions $L_n^+ = L_{-n}$ (corresponding to reality of the energy–momentum tensor).

More precisely, we are interested in unitary highest weight representations which are representations containing a state with a smallest value of L_0. If $|\psi\rangle$ is an eigenvector of L_0 with eigenvalue h, viz

$$L_0|\psi\rangle = h|\psi\rangle, \tag{4.181}$$

then $L_n|\psi\rangle$ is also an eigenvector of L_0 with eigenvalue $h - n$, i.e.

$$L_0(L_n|\psi\rangle) = (h - n)(L_n|\psi\rangle). \tag{4.182}$$

In other words, L_n decreases the eigenvalues of L_0 by n. Thus, if $|h\rangle$ is a highest weight state of L_0 defined by

$$L_0|h\rangle = h|h\rangle, \tag{4.183}$$

then

$$L_n|h\rangle = 0, \quad n > 0. \tag{4.184}$$

The generators L_n, $n < 0$, are used to generate other states in the representation called descendant states. A general descendant state is of the form

$$L_{-n_1} L_{-n_2} \ldots L_{-n_k}|h\rangle. \tag{4.185}$$

This infinite collection of descendant states defines a representation of the Virasoro algebra known as Verma module.

The conformal vacuum $|0\rangle$ is defined by the requirement that it must respect the maximum number of symmetries. We will assume that the theory contains a unique state with this property. Explicitly, the conformal vacuum is given by the conditions

$$L_n|0\rangle = 0, \quad n > -2, L_{-2}|0\rangle = |h\rangle. \tag{4.186}$$

This can be shown as follows.

First we define the following state-operator correspondence. The highest weight $|h\rangle$ is associated with a conformal primary field $\Phi(z)$ of dimension h by the relation

$$|h\rangle = \lim_{z \longrightarrow 0}\Phi(z)\,|0\rangle. \tag{4.187}$$

The point $z = 0$ is the infinite Euclidean past. The field can be decomposed in modes as follows

$$\Phi(z) = \sum_{n=-\infty}^{+\infty} \frac{\Phi_n}{z^{n+h}}. \tag{4.188}$$

Immediately we get

$$\Phi_n|0\rangle = 0, \quad n > -h, \Phi_{-h}|0\rangle = |h\rangle. \tag{4.189}$$

An irreducible unitary highest weight representation is completely specified by the two real numbers c and h. Thus two representations with the same c and h are actually equivalent. The c takes a constant value on the representation since the central charge commutes with all elements of the algebra. On the other hand, h does not take a constant value on the representation but it is defined uniquely as the eigenvalue of L_0 on the highest weight state, i.e. L_0 is diagonal in the representation and not proportional to the identity. By requiring unitarity, i.e. the absence of negative norm states in the Hilbert space we obtain the restriction on the possible values of c and h given by the equation

$$c \geqslant 0, \quad h \geqslant 0. \tag{4.190}$$

A unitary conformal field theory with $c = 0$ is trivial in the sense that the vacuum representation contains only the state $|0\rangle$ since we can show that $L_{-n}|0\rangle = 0$ for all $n > 0$ and thus $h = 0$.

Let us give an example. The physical states of the open string satisfy

$$L_0|\phi\rangle = a|\phi\rangle, \quad L_n|\phi\rangle = 0, \quad n \geqslant 1. \tag{4.191}$$

There are no negative norm states for $a \leqslant 1$, $c = D \leqslant 25$ or $a = 1$ and $c = D = 26$. They are highest weight states with $h = a$. Recall that $|\phi\rangle$ is also characterized by the momentum k^μ of the center of mass since it is constructed from the string ground state $|0, k\rangle$. Thus, the vacuum state $|0\rangle = |0, 0\rangle$ corresponds to $k^\mu = 0$.

The highest weight states are the ground states of the Virasoro representations which can always be generated by acting with conformal primary fields or operators on the conformal vacuum. Let $\Phi(z, \bar{z})$ be a conformal field with weights h and $\bar{h}$. Then we can associate to this operator a highest weight state $|h, \bar{h}\rangle$ given by the formula

$$|h, \bar{h}\rangle = \Phi(0, 0)\, |0\rangle. \tag{4.192}$$

Indeed, we can check that

$$L_0|h, \bar{h}\rangle = h|h, \bar{h}\rangle, \quad L_n|h, \bar{h}\rangle = 0, \quad n > 0. \tag{4.193}$$

In fact, we can also generate descendant states, i.e. states obtained from the highest weight state by the action of L_{-k} for $k > 0$, by acting with conformal fields on the vacuum. We consider the operator product

$$T(z)\Phi(w, \bar{w}) = \sum_{k \geqslant 0}(z - w)^{k-2}\Phi^{(-k)}(w, \bar{w}). \tag{4.194}$$

Immediately we have

$$\Phi^{(-k)}(w, \bar{w}) = \oint \frac{dz}{2\pi i} \frac{1}{(z - w)^{k-1}} T(z)\Phi(w, \bar{w}). \tag{4.195}$$

Then we compute that

$$\begin{aligned}\Phi^{(-k)}(0, 0)|0\rangle &= \oint \frac{dz}{2\pi i} \frac{1}{(z)^{k-1}} T(z)\Phi(0, 0)|0\rangle \\ &= L_{-k}|h, \bar{h}\rangle.\end{aligned} \tag{4.196}$$

In other words, $\Phi^{(-k)}(0, 0)$ generates the $L_{-\text{K}}$ descendant of $|h, \bar{h}\rangle$.

The Verma module can be defined more carefully by defining the descendant states at excitation level N by the formula (exploiting the fact that we can order the generators by using commutation relations)

$$\begin{aligned} |\Psi_i\rangle &= L_{-i}^{n_i} \ldots L_{-2}^{n_2} L_{-1}^{n_1} |h\rangle \\ L_0 |\Psi_i\rangle &= (h + N)|\Psi_i\rangle, \quad N = n_1 + n_2 + \cdots + i n_i. \end{aligned} \tag{4.197}$$

The non-negative integer N is called the level of the state $|\Psi_i\rangle$. At level 1 we have the state $L_{-1}|h\rangle$. At level 2 we have the two states $L_{-2}|h\rangle$, $L_{-1}^2|h\rangle$. At level 3 we have the three states $L_{-3}|h\rangle$, $L_{-2}L_{-1}|h\rangle$, $L_{-1}^3|h\rangle$, etc. States on different levels have different eigenvalues of L_0 and thus they are orthogonal. We consider, therefore, only the states in a given level N. In order for the inner product to be positive semidefinite, and hence there are no negative norm states or ghosts in the Hilbert space, it is necessary that the Hermitian matrix

$$M_N(c, h) = \left[\langle \Psi_i, \Psi_j \rangle\right]_{1 \leqslant i, j \leqslant \pi(N)} \tag{4.198}$$

is positive semidefinite, i.e. it should only have positive eigenvalues. In this equation $\pi(N)$ is the number of states at level N. Indeed, it is not sufficient to check that the states have positive norms separately because their linear combinations may still have zero or negative norms.

The determinant of M_N is called the Kac determinant. Clearly, the positivity of this determinant is a necessary condition for the absence of negative norm states but not a sufficient one. So if det $M_N < 0$ then certainly ghosts exists there. But if det $M_N \geqslant 0$ there could still be ghosts corresponding to an even number of negative eigenvalues.

The values of c and h for which an irreducible unitary highest weight representation can exist are given by the theorem of Fiedan, Qiu and Schenker which is stated as follows in the next section.

4.7 Theorem

In order for an irreducible unitary highest weight representation, corresponding to given values of c and h, to exist it is necessary that either

-
$$c \geqslant 1, \quad h \geqslant 0, \tag{4.199}$$

or

-
$$\begin{aligned} c &= 1 - \frac{6}{(m+2)(m+3)} < 1 \\ h = h_{(p,q)}(c) &= \frac{[(m+3)p - (m+2)q]^2 - 1}{4(m+2)(m+3)} \\ m = 0, 1, 2, \ldots, \quad & p = 1, 2, \ldots, m+1, \quad q = 1, 2, \ldots, p. \end{aligned} \tag{4.200}$$

The second case corresponds to the so-called minimal models and it includes the trivial representation $c = h = 0$. The proof relies on properties of the Kac determinant.

The above theorem provides the necessary conditions for unitary representations. As it turns out, the first necessary conditions in the theorem $c \geqslant 1$ and $h \geqslant 0$ are also sufficient conditions.

Similarly, the second set of conditions in the theorem are also sufficient. Indeed, we can explicitly construct the unitary representations by exploiting the fact that to each affine so-called Kac–Moody algebra we can associate a Virasoro algebra via something called the Sugawara construction.

4.8 Exercises

Exercise 1: Show that the group given by the relation (4.29) is $SO(3, 1)= SL(2, C)/\mathbf{Z}_2$.

Solution: Let us discuss this in some more detail. The group $SL(2, C)$ is the set of 2×2 complex matrices with determinant 1 given by

$$\begin{pmatrix} a & b \\ c & d \end{pmatrix}, \quad ad - bc = 1. \tag{4.201}$$

We consider complex 2-dimensional vectors (z_1, z_2) with vectors related by an overall complex scale identified. Thus the vector (z_1, z_2) is identified with the vector $(z, 1)$ where $z = z_1/z_2$. We observe now

$$\begin{pmatrix} a & b \\ c & d \end{pmatrix}\begin{pmatrix} z_1 \\ z_2 \end{pmatrix} = \begin{pmatrix} az_1 + bz_2 \\ cz_1 + dz_2 \end{pmatrix}. \tag{4.202}$$

Thus the action of the above $SL(2, C)$ transformation is given by

$$z \longrightarrow \frac{az + b}{cz + d}. \tag{4.203}$$

This transformation is unchanged if the matrix is multiplied by an overall complex factor which can be fixed by the determinant condition. There remains obviously a freedom with regard to the sign. The correct group is therefore $SL(2, C)/\mathbf{Z}_2$ instead of $SL(2, C)$.

This for the holomorphic sector. We get from the non-holomorphic sector another group $SL(2, C)$. The total group is $SL(2, C) \times SL(2, C)$ which is twice as large as $SO(3,1)$. By imposing the reality condition $\bar{z} = z^+$ we reduce the number of real generators of $SL(2, C) \times SL(2, C)$ to that of $SO(3, 1)$.

Exercise 2: Derive the conserved charges which generate conformal symmetries.

Solution: First, we write conformal transformations as

$$z' = z + \epsilon(z), \quad \bar{z}' = \bar{z} + \tilde{\epsilon}(\bar{z}). \tag{4.204}$$

At the end, we need to impose the reality condition

$$\tilde{\epsilon} = \bar{\epsilon}. \tag{4.205}$$

For example, constants ϵ and $\tilde{\epsilon}$ represent translation while $\epsilon \sim z$ and $\tilde{\epsilon} \sim \bar{z}$ represent rotation and dilatation. In order to derive conserved Noether's currents, we promote ϵ and $\tilde{\epsilon}$ to general fields on the worldsheet, viz

$$\epsilon(z) \longrightarrow \epsilon(z, \bar{z}), \quad \tilde{\epsilon}(z) \longrightarrow \tilde{\epsilon}(z, \bar{z}). \tag{4.206}$$

The real case $\tilde{\epsilon} = \bar{\epsilon}$, and in the Lorentzian signature, corresponds to the conformal transformations $\sigma \longrightarrow \sigma' = \sigma + \xi$. These are symmetries of the action if we also transform the metric $\eta_{\alpha\beta}$ as

$$\eta_{\alpha\beta} \longrightarrow h_{\alpha\beta} = \eta_{\alpha\beta} + \delta h_{\alpha\beta}, \quad \delta h_{\alpha\beta} = -\partial_\alpha \xi_\beta - \partial_\beta \xi_\alpha. \tag{4.207}$$

This means that the change in the action under the transformation of the coordinates is opposite of what we get if we just transform the metric, viz

$$\begin{aligned}
\delta S &= - \int d^2\sigma \frac{\partial S}{\partial h_{\alpha\beta}} \delta h^{\alpha\beta} \\
&= \frac{1}{4\pi\alpha'} \int d^2\sigma T_{\alpha\beta} \delta h^{\alpha\beta} \\
&= - \frac{1}{2\pi\alpha'} \int d^2\sigma T_{\alpha\beta} \partial^\alpha \xi^\beta \\
&= - \frac{1}{2\pi\alpha'} \int d^2\sigma (T_{++}\partial^+\xi^+ + T_{--}\partial^-\xi^-) \\
&= \frac{1}{\pi\alpha'} \int d^2\sigma (T_{++}\partial_-\xi^+ + T_{--}\partial_+\xi^-),
\end{aligned} \tag{4.208}$$

where T is the stress–energy tensor. In the fourth line we have used Weyl symmetry, viz $T_{+-} = T_{-+} = 0$. We perform Wick rotation: $\sigma^- \longrightarrow -z$, $\sigma^+ \longrightarrow \bar{z}$ and thus $d^2\sigma \longrightarrow -id^2z/2$. We have $T_{++} = T_{\bar{z}\bar{z}}$, $T_{--} = T_{zz}$. We will also have by analogy $\xi^- \longrightarrow -\epsilon$, $\xi^+ \longrightarrow \bar{\epsilon}$. We get then

$$\delta S = \frac{i}{2\pi\alpha'} \int d^2z (T_{zz}\bar{\partial}\epsilon + T_{\bar{z}\bar{z}}\partial\bar{\epsilon}). \tag{4.209}$$

This is zero if ϵ is holomorphic and $\bar{\epsilon}$ is anti-holomorphic. This is conformal invariance. If we want the holomorphic and anti-holomorphic sectors to rotate independently, we only need to replace $\bar{\epsilon}$ in the above equation by $\tilde{\epsilon}$. We get

$$\delta S = \frac{i}{2\pi\alpha'} \int d^2z (T_{zz}\bar{\partial}\epsilon + \tilde{T}_{\bar{z}\bar{z}}\partial\tilde{\epsilon}). \tag{4.210}$$

Let us find the conserved currents and conserved charges. First, we consider the transformation

$$\epsilon(z, \bar{z}) = \epsilon(z)\bar{f}(\bar{z}), \quad \tilde{\epsilon}(z, \bar{z}) = 0. \tag{4.211}$$

We get the variation

$$\delta S = \frac{i}{2\pi\alpha'}\int d^2z T_{zz}(z)\epsilon(z)\bar{\partial}\bar{f}(\bar{z}). \tag{4.212}$$

This is of the form

$$\delta S = \int d^2z J^{\alpha}\partial_{\alpha}\bar{f}\,. \tag{4.213}$$

We get the conserved current with components in complex coordinates

$$J^z = 0, \quad J^{\bar{z}} = \frac{i}{2\pi\alpha'}T_{zz}(z)\epsilon(z). \tag{4.214}$$

This is obviously conserved

$$\partial_{\alpha}J^{\alpha} = \partial_{\bar{z}}J^{\bar{z}} = 0. \tag{4.215}$$

The conserved charge is then automatically given by

$$Q_{\epsilon} = \frac{i}{2\pi\alpha'}\oint T_{zz}(z)\epsilon(z)dz. \tag{4.216}$$

Similarly, we consider the transformation

$$\epsilon(z, \bar{z}) = 0, \quad \tilde{\epsilon}(z, \bar{z}) = \tilde{\epsilon}(\bar{z})f(z). \tag{4.217}$$

We get the variation

$$\delta S = \frac{i}{2\pi\alpha'}\int d^2z \tilde{T}_{\bar{z}\bar{z}}(\bar{z})\tilde{\epsilon}(\bar{z})\partial f(z). \tag{4.218}$$

We now get the conserved current with components in complex coordinates

$$J^z = \frac{i}{2\pi\alpha'}\tilde{T}_{\bar{z}\bar{z}}(\bar{z})\tilde{\epsilon}(\bar{z}), \quad J^{\bar{z}} = 0. \tag{4.219}$$

This is obviously conserved

$$\partial_{\alpha}J^{\alpha} = \partial_z J^z = 0. \tag{4.220}$$

The conserved charge is then automatically given by

$$Q_{\tilde{\epsilon}} = \frac{i}{2\pi\alpha'}\oint \tilde{T}_{\bar{z}\bar{z}}(\bar{z})\tilde{\epsilon}(\bar{z})d\bar{z}. \tag{4.221}$$

The conserved charge which generates the complete transformation $\delta z = \epsilon(z)$, $\delta\bar{z} = \tilde{\epsilon}(\bar{z})$ is given by the sum

$$\begin{aligned} Q &= Q_{\epsilon} + Q_{\tilde{\epsilon}} \\ &= \frac{i}{2\pi\alpha'}\oint [T_{zz}(z)\epsilon(z)dz + \tilde{T}_{\bar{z}\bar{z}}(\bar{z})\tilde{\epsilon}(\bar{z})d\bar{z}]. \end{aligned} \tag{4.222}$$

We perform the integral over a circle of a fixed radius.

References

[1] Becker K, Becker M and Schwarz J H 2006 *String Theory and M-theory: A Modern Introduction* (Cambridge: Cambridge University Press)

[2] Polchinski J 2005 *String Theory. Vol. 1: An Introduction to the Bosonic String* (Cambridge: Cambridge University Press)

[3] Schellekens A N Conformal field theory. 2016 (Downloaded from author website)

[4] Polchinski J 2005 *String Theory. Vol. 2: Superstring Theory and Beyond* (Cambridge: Cambridge University Press)

[5] Lundholm D 2005 The Virasoro algebra and its representations in physics *Report for the course "Lie algebras and quantum groups" at KTH.*

[6] Green M B, Schwarz J H and Witten E 1987 *Superstring Theory. Vol. 1: Introduction* (Cambridge Monographs On Mathematical Physics) (Cambridge: Cambridge University Press)

[7] Green M B, Schwarz J H and Witten E 1987 *Superstring Theory. Vol. 2: Loop Amplitudes, Anomalies And Phenomenology* (Cambridge Monographs On Mathematical Physics) (Cambridge: Cambridge University Press)

[8] Johnson C V 2003 *D-branes* (Cambridge: Cambridge University Press)

[9] Zwiebach B 2009 *A First Course in String Theory* (Cambridge: Cambridge University Press)

[10] Szabo R J 2004 *An Introduction to String Theory and D-Brane Dynamics* (London: Imperial College Press)

[11] Goddard P, Kent A and Olive D I 1986 Unitary representations of the Virasoro and supervirasoro algebras *Commun. Math. Phys.* **103** 105

[12] Friedan D, Shenker S H and Qiu Z A 1986 Details of the nonunitarity proof for highest weight representations of the Virasoro algebra *Commun. Math. Phys.* **107** 535

IOP Publishing

Matrix Models of String Theory

Badis Ydri

Chapter 5

Superstring theory essentials

In this chapter we introduce the Green–Schwarz and the Ramond–Neveu–Schwarz superstrings and discuss canonical quantization (super-Virasoro algebra) as well as light cone/path integral quantization (GSO projection). We will follow closely the presentation of [1]. See also [2–8].

5.1 The superparticle

5.1.1 Bosonic particle

We consider a relativistic particle moving in a D-dimensional curved spacetime manifold with metric $g_{\mu\nu}$ of Minkowski signature, viz

$$g_{\mu\nu} = (-1, +1, \ldots, +1).$$

The coordinates of this particle in spacetime are denoted by X^μ. The infinitesimal line element is then defined by

$$ds^2 = -g_{\mu\nu}(X)dX^\mu dX^\nu.$$

The worldlines, i.e. trajectories in spacetime, of this particle are timelike geodesics of the spacetime manifold. Let τ be an affine parameter on the worldline which can be identified with the proper time of the particle. Since the worldline is a timelike geodesic and since timelike geodesics by construction are the curves with maximum proper time, i.e. minimum spacetime length, we can immediately conclude that the action of the particle must be proportional to the line element s. The constant of proportionality can be easily shown to be equal to the mass of the particle. We have then

$$S = -m\int ds.$$

doi:10.1088/978-0-7503-1726-9ch5

We can rewrite this action as

$$S_0 = -m \int d\tau \sqrt{-g_{\mu\nu}(X)\dot{X}^\mu \dot{X}^\nu}, \quad \dot{X}^\mu = \frac{dX^\mu}{d\tau}.$$

We are immediately faced with two problems: (1) quantization of this action is very difficult because of the square root, and (2) in this form this action cannot be used to describe massless particles. These two problems are solved at once by introducing the metric e on the worldline as an independent variable as follows

$$\tilde{S}_0 = \frac{1}{2} \int d\tau (e^{-1} g_{\mu\nu}(X)\dot{X}^\mu \dot{X}^\nu - m^2 e).$$

By substituting with the equation of motion of e in $\tilde{S}_0$ we obtain the action S_0 which shows explicitly that these two actions are classically equivalent.

The action $\tilde{S}_0$ or S_0 enjoys, almost obviously, by construction, the following symmetries:

- Poincaré invariance in D dimensions.
- Diffeomorphism invariance which is given by the reparametrizations

$$\tau \longrightarrow \tau' = f(\tau).$$

5.1.2 The superparticle

The relativistic particle which appears as a non-perturbative topological configuration in type IIA string theory is a superparticle and as such it is invariant under supersymmetric transformations in addition to its invariance under the above-mentioned symmetries (Poincaré and diffeomorphisms). This supersymmetric particle or superparticle is also called a D_0-brane. It enjoys $\mathcal{N} = 2$ supersymmetry in $D = 10$ dimensions which is expressed in terms of two Majorana–Weyl spinors with opposite chiralities. The Majorana condition means that the spinor is real (neutral) while the Weyl condition means that the spinor has a definite chirality (either left-handed or right-handed). These two conditions can be simultaneously imposed only in 10 dimensions.

Let us then consider in what follows a superparticle with $\mathcal{N} = 2$ supersymmetry in $D = 10$ dimensions corresponding to two Majorana–Weyl spinors Θ^{aA} where the index A denotes the number of supersymmetries, i.e. $A = 1, \ldots, \mathcal{N}$, while the index a denotes the number of components of the spinor in spacetime, i.e. $a = 1, \ldots, 2^{D/2}$.

The bosonic particle is described by the bosonic coordinates X^μ, whereas the superparticle is described in addition to these by the fermionic coordinates Θ^{aA}. The bosonic coordinates are real numbers, whereas the fermionic coordinates are Grassmannian numbers. The supersymmetric transformations, which are a generalization of Poincaré (translation and Lorentz) transformations, view the bosonic and fermionic coordinates X^μ and Θ^{aA} collectively as the coordinates of a superspace. These transformations mix between the bosonic and fermionic coordinates as follows. The supersymmetric transformation appears as a translation in the fermionic direction:

$$\Theta^{aA} \longrightarrow \Theta^{aA} + \delta\Theta^{aA}, \quad \delta\Theta^{aA} = \epsilon^{aA}.$$

The constant Majorana–Weyl spinor ϵ^{aA} is the supersymmetry parameter. The effect of the supersymmetric transformation on X^μ must be linear in ϵ^{aA}, it must be a vector, and it must be linear. There is one solution which meets these three requirements given by

$$X^\mu \longrightarrow X^\mu + \delta X^\mu, \quad \delta X^\mu = \bar{\epsilon}^A \Gamma^\mu \Theta^A.$$

The Γ^μ are the Dirac matrices in $D = 10$, i.e. $\mu = 0, 1, \ldots, 9$. They are 32-dimensional Hermitian matrices satisfying

$$\Gamma^\mu \Gamma^\nu + \Gamma^\nu \Gamma^\mu = 2g^{\mu\nu}.$$

The chirality matrix satisfies

$$\Gamma_{11} = \Gamma_0 \Gamma_1 \ldots \Gamma_9.$$

It satisfies

$$\Gamma_{11}^2 = 1, \quad \{\Gamma_\mu, \Gamma_{11}\} = 0.$$

The set of Poincaré transformations form a group called the Poincaré group. The set of Poincaré plus supersymmetric transformations form the super-Poincaré group. All these symmetries are global symmetries of the action since they do not depend on the affine parameter τ of the worldline.

5.1.3 Action

The next question is to determine the action of the superparticle. First, we note that the action S_0 depends on the velocities $\dot{X}^\mu$. The variation of these velocities under supersymmetry is

$$\begin{aligned} \delta \dot{X}^\mu &= \bar{\epsilon}^A \Gamma^\mu \dot{\Theta}^A \\ &= \delta \bar{\Theta}^A \cdot \Gamma^\mu \dot{\Theta}^A \\ &= \delta(\bar{\Theta}^A \Gamma^\mu \dot{\Theta}^A). \end{aligned}$$

We conclude immediately that the expression

$$\Pi_0^\mu = \dot{X}^\mu - \bar{\Theta}^A \Gamma^\mu \dot{\Theta}^A$$

is invariant under supersymmetric transformations. We get then the supersymmetric action by making in S_0 the replacement $\dot{X}^\mu \longrightarrow \Pi_0^\mu$. We get the action

$$S_1 = -m \int d\tau \sqrt{-g_{\mu\nu}(X)\Pi_0^\mu \Pi_0^\nu}.$$

In terms of the metric e on the worldline of the superparticle we obtain the action

$$\tilde{S}_1 = \frac{1}{2} \int d\tau (e^{-1} g_{\mu\nu}(X)\Pi_0^\mu \Pi_0^\nu - m^2 e).$$

We define a new Majorana spinor Θ in terms of the Majorana–Weyl spinors Θ^1 and Θ^2 by the relation

$$\Theta = \Theta^1 + \Theta^2.$$

Thus Θ^1 is the positive chirality component of Θ and Θ^2 is the negative chirality component. We can express this using the chirality matrix Γ_{11} as follows

$$\Theta^1 = \frac{1 + \Gamma_{11}}{2}\Theta, \quad \Theta^2 = \frac{1 - \Gamma_{11}}{2}\Theta.$$

We express Π_0^μ in terms of the spinor Θ as follows

$$\Pi_0^\mu = \dot{X}^\mu - \bar{\Theta}\Gamma^\mu\dot{\Theta}.$$

We compute now the equations of motion. The conjugate momentum corresponding to X^μ is

$$P^\mu = \frac{\delta S_1}{\delta \dot{X}_\mu} = \frac{m}{\sqrt{-g_{\mu\nu}\Pi_0^\mu\Pi_0^\nu}}\Pi_0^\mu.$$

Thus not all the components of P^μ are linearly independent since they satisfy the mass-shell condition

$$g_{\mu\nu}P^\mu P^\nu = -m^2.$$

The equations of motion of X^μ are given by

$$\dot{P}^\mu = -0.$$

The equations of motion of Θ^a are given by

$$P_\mu\Gamma^\mu\Theta = 0. \tag{5.1}$$

We remark that $(P_\mu\Gamma^\mu)^2 = P_\mu P^\mu = -m^2$. Thus for $m = 0$ the matrix $P_\mu\Gamma^\mu$ is not invertible since half of its eigenvalues vanish. Hence the number of linearly independent equations is only half the number of components of the Majorana spinor Θ and as a consequence the above equation (5.1) constrains only half of the components of Θ by making them constant while leaving the other half free variables.

This remark means in particular that the action S_1 for $m = 0$ enjoys in addition to the symmetries mentioned above extra local fermionic symmetries which cause half of the degrees of freedom of the spinor Θ to be gauge degrees of freedom and thus they can be removed from the theory. These new fermionic local symmetries are known as κ-symmetries.

Before we write these κ-symmetries explicitely we note that the case $m \neq 0$ must also be κ-symmetric since there is no intrinsic physical difference between the massless case and the massive case. But in the massive case the action must be modified in such a way that the last equation above (5.1) changes so that half of the components of Θ become gauge degrees of freedom with no physical significance. The question is then simply what is the matrix in the case $m \neq 0$ that should replace $P_\mu\Gamma^\mu$, i.e. what is the matrix that squares to zero in this case? The answer is very simply given by

$P_\mu\Gamma^\mu + m\Gamma_{11}$ where Γ_{11} is the chirality matrix. Indeed $(P_\mu\Gamma^\mu + m\Gamma_{11})^2 = 0$. The last equation above (5.1) should then change to

$$(P_\mu\Gamma^\mu + m\Gamma_{11})\dot{\Theta} = 0.$$

This equation of motion can be derived from the action

$$S_2 = -m\int d\tau\sqrt{-g_{\mu\nu}(X)\Pi_0^\mu\Pi_0^\nu} - m\int d\tau\bar{\Theta}\Gamma_{11}\dot{\Theta}.$$

5.1.4 The κ-symmetry

We derive now the exact form of κ-symmetry which leaves S_2 invariant. This is a fermionic symmetry like supersymmetry but local. X^μ transforms under supersymmetry as $\delta X^\mu = -\bar{\Theta}\Gamma^\mu\delta\Theta$. Under the new fermionic symmetry we will then have

$$\delta X^\mu = \bar{\Theta}\Gamma^\mu\delta\Theta.$$

In supersymmetric transformations $\delta\Theta$ is a constant Majorana–Weyl spinor equal to ϵ but in κ-symmetry the variation $\delta\Theta$ is a Majorana–Weyl spinor which depends on the time τ which we will now determine.

From the variation δX^μ above we compute $\delta\Pi_0^\mu = -2\delta\bar{\Theta}\Gamma^\mu\dot{\Theta}$. Thus the variation of the action S_1 is given by

$$\begin{aligned}\delta S_1 &= m\int d\tau\frac{g_{\mu\nu}\Pi_0^\mu\delta\Pi_0^\nu}{\sqrt{-g_{\mu\nu}\Pi_0^\mu\Pi_0^\nu}}\\ &= -2m\int d\tau\frac{g_{\mu\nu}\Pi_0^\mu\delta\bar{\Theta}\Gamma^\nu\dot{\Theta}}{\sqrt{-g_{\mu\nu}\Pi_0^\mu\Pi_0^\nu}}\\ &= -2m\int d\tau\delta\bar{\Theta}\gamma\Gamma_{11}\dot{\Theta},\end{aligned}$$

where the matrix γ is given by

$$\gamma = \frac{g_{\mu\nu}\Gamma^\mu\Pi_0^\nu}{\sqrt{-g_{\mu\nu}\Pi_0^\mu\Pi_0^\nu}}\Gamma_{11}.$$

We remark that $\gamma^2 = 1$ and as a consequence we can define the projectors

$$P_\pm = \frac{1\pm\gamma}{2}.$$

We compute now the variation under κ-symmetry of the second term of the action S_2:

$$\delta\left(-m\int d\tau\bar{\Theta}\Gamma_{11}\dot{\Theta}\right) = -2m\int d\tau\delta\bar{\Theta}\Gamma_{11}\dot{\Theta}.$$

The variation under κ-symmetry of the action S_2 is then given by

$$\delta S_2 = -4m\int d\tau\delta\bar{\Theta}P_+\Gamma_{11}\dot{\Theta}.$$

This vanishes for the variations $\delta\bar{\Theta}$ given by

$$\delta\bar{\Theta} = \bar{\kappa}P_-,$$

where κ is an arbitrary Majorana spinor because the projectors P_+ and $P_-$are orthogonal.

5.2 The Green–Schwarz superstring

In this section we will follow mostly [1].

The Green–Schwarz string is a superstring which is supersymmetric in target space, i.e. in spacetime.

We recall the bosonic string action

$$S_1 = -\frac{1}{2}T\int d^2\sigma\sqrt{-h}\,h^{\alpha\beta}g_{\mu\nu}(X)\partial_\alpha X^\mu\partial_\beta X^\nu \tag{5.2}$$

and the superparticle action (in terms of the metric e)

$$S = \frac{1}{2}\int d\tau(e^{-1}g_{\mu\nu}(X)\Pi_0^\mu\Pi_0^\nu - m^2e) - m\int d\tau\bar{\Theta}\Gamma_{11}\dot{\Theta}, \tag{5.3}$$

where $\Pi_0^\mu = \dot{X}^\mu - \bar{\Theta}^A\Gamma^\mu\dot{\Theta}^A$. The supersymmetric analogue of the action S_1 is easy to determine. We need only to make the replacement

$$\partial_\alpha X^\mu \longrightarrow \Pi_\alpha^\mu = \partial_\alpha X^\mu - \bar{\Theta}^A\Gamma^\mu\partial_\alpha\Theta^A. \tag{5.4}$$

We get immediately

$$S_1 = -\frac{1}{2}T\int d^2\sigma\sqrt{-h}\,h^{\alpha\beta}g_{\mu\nu}(X)\Pi_\alpha^\mu\Pi_\beta^\nu. \tag{5.5}$$

This is the Polyakov type action. The Nambu–Goto type action is given on the other hand by (keeping the same symbol for simplicity)

$$S_1 = -T\int d^2\sigma\sqrt{-\det\,(g_{\mu\nu}\Pi_\alpha^\mu\Pi_\beta^\nu)}\,. \tag{5.6}$$

This has Lorentz invariance, local reparametrization invariance and $\mathcal{N}$ global supersymmetries since we are considering $\mathcal{N}$ Majorana–Weyl spinors Θ^A, $A = 1, \ldots, \mathcal{N}$. However, this action does not have local κ-symmetry and as a consequence the Θ^A contain twice as many degrees of freedom as required. We need thus to add to S_1 the analogue of $-m\int d\tau\bar{\Theta}\Gamma_{11}\dot{\Theta}$ for the superparticle which we will denote by S_2. This can only be done for $\mathcal{N} \leqslant 2$ though which is the most important case for type II string theories. Furthermore, it is not difficult to convince ourselves that S_2 is in fact a topological term of the Chern–Simons or Wess–Zumino type which is independent of $h_{\alpha\beta}$ and thus does not contribute to the energy–momentum tensor.

We will follow the construction of [1]. Since S_2 does not depend on the worldsheet metric $h_{\alpha\beta}$ it must be a topological term of the Wess–Zumino type given by the integral of 2-form Ω_2, viz

$$S_2 = \int_M \Omega_2 = \frac{1}{2}\int d^2\sigma \epsilon^{\alpha\beta}\Omega_{\alpha\beta}. \tag{5.7}$$

We imagine that the string worldsheet is a boundary of a 3-dimensional region D, i.e. $M = \partial D$. By Stokes' theorem we can rewrite S_2 as

$$S_2 = \int_D \Omega_3. \tag{5.8}$$

Ω_3 is a characteristic class invariant under the symmetries of the problem, whereas Ω_2 is a Chern–Simons form where the variation is given by a total derivative.

The only 1-forms which are invariant under supersymmetry are $d\Theta^\alpha$ and $\Pi^\mu = dX^\mu - \bar{\Theta}^A\Gamma\mu d\Theta^A$. The 3-form Ω_3 is a supersymmetric invariant object which must therefore be constructed out of these 1-forms. Ω_3 must also be a Lorentz invariant object. The correct expression is given by

$$\Omega_3 = c(d\bar{\Theta}^1 \wedge \Gamma_\mu d\Theta^1 - d\bar{\Theta}^2 \wedge \Gamma_\mu d\Theta^2) \wedge \Pi^\mu. \tag{5.9}$$

This is a closed form as it should be, viz $d\Omega_3 = 0$. Let us now study the effect of κ-symmetry. This is defined on X^μ by

$$\delta X^\mu = \bar{\Theta}\Gamma^\mu\delta\Theta \Rightarrow \delta\Pi^\mu_\alpha = -2\delta\bar{\Theta}^A\Gamma^\mu\partial_\alpha\Theta^A. \tag{5.10}$$

We compute

$$\begin{aligned}\delta\Omega_3 = {} & 2c(d\delta\bar{\Theta}^1 \wedge \Gamma_\mu d\Theta^1 - d\delta\bar{\Theta}^2 \wedge \Gamma_\mu d\Theta^2) \wedge \Pi^\mu \\ & - 2c(d\bar{\Theta}^1 \wedge \Gamma_\mu d\Theta^1 - d\bar{\Theta}^2 \wedge \Gamma_\mu d\Theta^2) \wedge (\delta\bar{\Theta}^A\Gamma^\mu d\Theta^A).\end{aligned} \tag{5.11}$$

A Majorana–Weyl spinor in 10 dimensions satisfy the Fierz identity

$$\Gamma^\mu d\Theta \wedge (d\bar{\Theta} \wedge \Gamma_\mu d\Theta) = 0. \tag{5.12}$$

By using this identity in the second line of $\delta\Omega_3$ it is straightforward to show that

$$\begin{aligned}& - 2c(d\bar{\Theta}^1 \wedge \Gamma_\mu d\Theta^1 - d\bar{\Theta}^2 \wedge \Gamma_\mu d\Theta^2) \wedge (\delta\bar{\Theta}^A\Gamma^\mu d\Theta^A) \\ & = -2c(\delta\bar{\Theta}^1\Gamma_\mu d\Theta^1 - \delta\bar{\Theta}^2\Gamma_\mu d\Theta^2) \wedge d\Pi^\mu.\end{aligned} \tag{5.13}$$

Hence[1]

$$\delta\Omega_3 = d(2c(\delta\bar{\Theta}^1 \wedge \Gamma_\mu d\Theta^1 - \delta\bar{\Theta}^2 \wedge \Gamma_\mu d\Theta^2) \wedge \Pi^\mu). \tag{5.14}$$

In other words,

$$\delta\Omega_2 = 2c(\delta\bar{\Theta}^1 \wedge \Gamma_\mu d\Theta^1 - \delta\bar{\Theta}^2 \wedge \Gamma_\mu d\Theta^2) \wedge \Pi^\mu. \tag{5.15}$$

[1] Recall the Leibniz rule for forms: $d(\omega \wedge \rho) = d\omega \wedge \rho + (-1)^p\omega \wedge d\rho$ where $\omega \in \Omega^p(M)$ and $\rho \in \Omega^q(M)$.

The variation of the action S_2 under κ-symmetry is then given by

$$\begin{aligned}\delta S_2 &= \int \delta\Omega_2 \\ &= c\int d^2\sigma\epsilon^{\alpha\beta}(\delta\bar{\Theta}^1\Gamma_\mu\partial_\alpha\Theta^1 - \delta\bar{\Theta}^2\Gamma_\mu\partial_\alpha\Theta^2)\Pi^\mu_\beta.\end{aligned} \tag{5.16}$$

The variation of the action S_1 under κ-symmetry is trivial to compute given by (with $G_{\alpha\beta} = g_{\mu\nu}\Pi^\mu_\alpha\Pi^\nu_\beta$ and $\delta G_{\alpha\beta} = -4\Pi^\mu_\alpha\delta\bar{\Theta}^A\Gamma_\mu\partial_\beta\Theta^A$)

$$\begin{aligned}\delta S_1 &= -\frac{T}{2}\int d^2\sigma\sqrt{-\det G}\,G^{\alpha\beta}\delta G_{\alpha\beta} \\ &= 2T\int d^2\sigma\sqrt{-\det G}\,G^{\alpha\beta}\Pi^\mu_\alpha\delta\bar{\Theta}^A\Gamma_\mu\partial_\beta\Theta^A.\end{aligned} \tag{5.17}$$

The variation of the action $S_1 + S_2$ becomes (with $c = 6T$ and $P_\pm$ are projectors given by $P_\pm = (1 \pm \gamma)/2$)

$$\delta S = 2c\int d^2\sigma\epsilon^{\alpha\beta}(\delta\bar{\Theta}^1 P_+\Gamma_\mu\partial_\alpha\Theta^1 - \delta\bar{\Theta}^2 P_-\Gamma_\mu\partial_\alpha\Theta^2)\Pi^\mu_\beta. \tag{5.18}$$

The matrix γ is an idempotent defined by

$$\gamma = -\frac{1}{2\sqrt{-\det G}}\epsilon^{\alpha\beta}\Gamma_{\mu\nu}\Pi^\mu_\alpha\Pi^\nu_\beta. \tag{5.19}$$

This variation vanishes for

$$\delta\bar{\Theta}^1 = \bar{\kappa}^1 P_-, \quad \delta\bar{\Theta}^2 = \bar{\kappa}^2 P_+, \tag{5.20}$$

where $\kappa^{1,2}$ are arbitrary Majorana–Weyl spinors.

The total κ-symmetric action is finally given by

$$\begin{aligned}S &= S_1 + S_2 \\ &= -T\int d^2\sigma\sqrt{-\det(g_{\mu\nu}\Pi^\mu_\alpha\Pi^\nu_\beta)} + \int\Omega_2.\end{aligned} \tag{5.21}$$

The 2-form Ω_2 is defined by $\Omega_3 = d\Omega_2$ where Ω_3 is given by (5.9). We can immediately determine Ω_2 to be

$$\Omega_2 = c(\bar{\Theta}^1\Gamma_\mu d\Theta^1 - \bar{\Theta}^2\Gamma_\mu d\Theta^2)\wedge dX^\mu - c\bar{\Theta}^1\Gamma_\mu d\Theta^1 \wedge \bar{\Theta}^2\Gamma_\mu d\Theta^2. \tag{5.22}$$

So far we have only focused on $D = 10$. The Wess–Zumino type action S_2 is $N \leqslant 2$ supersymmetric only in $D = 3, 4, 6$ and $D = 10$. In $D = 3$ the spinor is Majorana, in $D = 4$ it is Majorana or Weyl, in $D = 6$ it is Weyl and in $D = 10$ it is Majorana–Weyl as we have seen. See [4] for the proof.

The Green–Schwarz string action (5.21) in the light-cone gauge is essentially equivalent to the Ramond–Neveu–Schwarz string action which will be introduced in the next section. The only difference is that the fermions in the Green–Schwarz formalism are in the spinor representation $\mathbf{8}_s$ of spin(8), whereas in the Ramond–Neveu–Schwarz formalism they are in the vector representation $\mathbf{8}_v$ of spin(8) while the X^μ in both cases belong to the vector representation $\mathbf{8}_v$. Putting it differently, in the Green–Schwarz formalism the spacetime fermions transform as worldsheet scalars before gauge fixing and as worldsheet spinors after gauge fixing, whereas in the Ramond–Neveu–Schwarz formalism the worldsheet fermions transform as spacetime vectors. As a consequence in the Ramond–Neveu–Schwarz formalism there are two sectors (R and NS) as we will see in the next section whereas in the Green–Schwarz formalism there is a single sector. See [4] for a detailed discussion.

5.3 The Ramond–Neveu–Schwarz superstring

In this section we will follow mostly [4].

5.3.1 Supersymmetric action on the worldsheet

We introduce D Majorana fermions $\psi_A^\mu(\sigma, \tau)$, $\mu = 0, 1, \dots, D-1$, where A denotes worldsheet spinor indices so it takes two values. Thus for every fixed value of A the $\psi_A^\mu(\sigma, \tau)$ is a vector under the Lorentz group $SO(D-1,1)$ and for every fixed value of μ the $\psi_A^\mu(\sigma, \tau)$ is a spinor with respect to the worldsheet. The Lorentz group $SO(D-1, 1)$ is an internal symmetry group from the worldsheet point of view. The $\psi_A^\mu(\sigma, \tau)$ are internal degrees of freedom that are free to propagate along the string. Consider the action

$$S = -\frac{1}{2\pi}\int d^2\sigma(\partial_\alpha X^\mu \partial^\alpha X_\mu - i\bar{\psi}^\mu \rho^\alpha \partial_\alpha \psi_\mu), \quad \bar{\psi} = \psi^+ \rho^0. \tag{5.23}$$

The 2-dimensional Dirac matrices are defined by

$$\rho^0 = \begin{pmatrix} 0 & -i \\ i & 0 \end{pmatrix}, \quad \rho^1 = \begin{pmatrix} 0 & i \\ i & 0 \end{pmatrix}. \tag{5.24}$$

We have

$$\{\rho^\alpha, \rho^\beta\} = -2h^{\alpha\beta}, \quad h^{\alpha\beta} = \mathrm{diag}(-1, 1). \tag{5.25}$$

The matrices $i\rho^\alpha$ are real and hence the Dirac operator $i\rho^\alpha\partial_\alpha$ is real. Thus we can take the components of the worldsheet spinors $\psi^\mu(\sigma, \tau)$ to be real. The real spinors $\psi^\mu(\sigma, \tau)$ are therefore Majorana spinors. They satisfy

$$\bar{\psi}\chi = \bar{\chi}\psi. \tag{5.26}$$

The equal τ commutation relations of the bosonic coordinates are

$$[X^\mu(\sigma), \partial_\tau X^\nu(\sigma')] = i\pi\eta^{\mu\nu}\delta(\sigma - \sigma'). \tag{5.27}$$

The Lorentz metric $\eta^{\mu\nu}$ is not positive definite since $\eta^{00} = -1$ and as a consequence the X^0 oscillators create modes of wrong metric which are the ghosts. On the other hand the action has an infinite-dimensional Virasoro symmetry algebra. As we have shown, we can use these symmetries to eliminate the ghosts in the critical dimension $D = 26$.

Going back to the action (5.23) we read the fermion term

$$S_F = \frac{i}{2\pi}\int d^2\sigma\bar{\psi}^\mu\rho^\alpha\partial_\alpha\psi_\mu\,. \tag{5.28}$$

We have

$$i\bar{\psi}\rho^\alpha\partial_\alpha\psi = i\psi_A\partial_\tau\psi_A + \ldots \tag{5.29}$$

The conjugate momentum is therefore given by

$$\pi_A = \frac{i}{2\pi}\psi_A\,. \tag{5.30}$$

We must therefore have the equal τ anticommutation relations

$$\left\{\psi^\mu_{A'}(\sigma),,\,,\psi^\nu_{B'}(\sigma')\right\} = 2\pi\delta_{AB}\eta^{\mu\nu}\delta(\sigma-\sigma'). \tag{5.31}$$

Again the Lorentz metric $\eta^{\mu\nu}$ which is not positive definite appears and as a consequence the ψ^0_A fermion oscillators will create modes of wrong metric. Therefore, we need new symmetries of the action which can do for fermions what the Virasoro symmetries did for the bosons. As it turns out these new symmetries are superconformal symmetries.

The supersymmetric transformations are defined by

$$\begin{aligned}\delta X^\mu &= \bar{\epsilon}\psi^\mu\\ \delta\psi^\mu &= -\,i\rho^\alpha\partial_\alpha X^\mu\epsilon\\ \delta\bar{\psi}^\mu &= i\partial_\alpha X^\mu\bar{\epsilon}\rho^\alpha\,.\end{aligned} \tag{5.32}$$

The ϵ is a constant anticommuting spinor. We can immediately check that

$$\begin{aligned}\delta S_F &= \frac{i}{\pi}\int d^2\sigma\bar{\psi}^\mu\rho^\alpha\partial_\alpha\delta\psi_\mu\\ &= \frac{1}{\pi}\int d^2\sigma\partial^\alpha X^\mu\partial_\alpha\delta X^\mu\\ &= -\,\delta S_B.\end{aligned} \tag{5.33}$$

This establishes the supersymmetric invariance of the action. We can also verify that the commutators of two supersymmetries gives a translation and thus the supersymmetry algebra closes.

5.3.2 Energy–momentum tensor and supercurrent

Supercurrent

Let us recall that the variation of the fermion action under supersymmetry is given by

$$\begin{aligned}\delta S_F &= \frac{i}{\pi}\int d^2\sigma\bar{\psi}^\mu\rho^\alpha\partial_\alpha\delta\psi_\mu\\ &= -\,\delta S_B + \frac{1}{\pi}\int d^2\sigma\bar{\psi}^\mu\rho^\alpha\rho^\beta\partial_\beta X_\mu\partial_\alpha\epsilon.\end{aligned} \tag{5.34}$$

In the above we assumed that ϵ is not constant. Thus we have

$$\begin{aligned}\delta S &= \frac{1}{\pi}\int d^2\sigma \partial_\alpha \bar{\epsilon}\rho^\beta \rho^\alpha \partial_\beta X^\mu \psi_\mu \\ &= \frac{2}{\pi}\int d^2\sigma \partial_\alpha \bar{\epsilon} J^\alpha .\end{aligned} \tag{5.35}$$

We derive the conserved supercurrent

$$J^\alpha = \frac{1}{2}\rho^\beta \rho^\alpha \partial_\beta X^\mu \psi_\mu . \tag{5.36}$$

Since $\rho^\alpha \rho_\alpha = -2$ we get $\rho_\alpha \rho^\beta \rho^\alpha = 0$ and hence $\rho_\alpha J^\alpha = 0$.

The fermion's equation of motion is

$$\rho^\alpha \partial_\alpha \psi^\mu = 0. \tag{5.37}$$

In terms of $\sigma^\pm = \tau \pm \sigma$, $\partial_\pm = \frac{1}{2}(\partial_\tau \pm \partial_\sigma)$ we can put the above equation of motion in the form

$$\partial_- \psi_+^\mu = \partial_+ \psi_-^\mu = 0. \tag{5.38}$$

The components $\psi_\pm$ are defined by

$$\psi = \begin{pmatrix} \psi_- \\ \psi_+ \end{pmatrix}. \tag{5.39}$$

The supersymmetry transformations relating fermions and bosons can be exhibited by contrasting (5.38) with the boson's equation of motion

$$\partial_-(\partial_+ X^\mu) = \partial_+(\partial_- X^\mu) = 0. \tag{5.40}$$

Explicitly we compute

$$J_0 = \begin{pmatrix} -\partial_- X_\mu \psi_-^\mu \\ -\partial_+ X_\mu \psi_+^\mu \end{pmatrix}, \quad J_1 = \begin{pmatrix} \partial_- X_\mu \psi_-^\mu \\ -\partial_+ X_\mu \psi_+^\mu \end{pmatrix}. \tag{5.41}$$

Hence

$$J_+ = -\frac{1}{2}(J_0 + J_1) = \begin{pmatrix} 0 \\ \partial_+ X_\mu \psi_+^\mu \end{pmatrix}, \quad J_- = -\frac{1}{2}(J_0 + J_1) = \begin{pmatrix} \partial_- X_\mu \psi_-^\mu \\ 0 \end{pmatrix}. \tag{5.42}$$

Clearly

$$\partial_+ J_- = \partial_- J_+ = 0. \tag{5.43}$$

The energy–momentum tensor

To derive the energy–momentum tensor we need more work. The covariant bosonic action is given by

$$S_B = -\frac{1}{2\pi}\int_\Sigma d^2\sigma \sqrt{-h}\, h^{ab}\partial_a X^\mu \partial_b X_\mu . \tag{5.44}$$

In the conformal gauge we have $h^{ab} = \mathrm{diag}(-1, 1)$. We introduce a vielbein field e_a^A with inverse vielbein e_A^a by the equations[2]

$$h_{ab} = \eta_{AB} e_a^A e_b^B, \quad h^{ab} = \eta_{AB} e^{Aa} e^{Bb}, \quad e^{Aa} e_b^A = \delta_b^a, \quad e_b^A e^{Bb} = \delta^{AB}. \tag{5.45}$$

$$h_{ab} = \eta_{AB} e_a^A e_b^B, \quad \eta_{AB} = h_{ab} e_A^a e_B^b, \quad e_a^A e_B^a = \delta_B^A, \quad e_A^a e_b^A = \delta_b^a. \tag{5.46}$$

h_{ab} or $h_{\alpha\beta}$ is the metric of the worldsheet Σ, whereas η_{AB} is the metric its tangent space not to be confused with the metric of the target spacetime. The set of e_a^A vielbein provides an orthonormal tangent vector. In some sense the vielbein e_a^A is the square root of the metrics h_{ab}.

We define the Dirac matrices ρ^A by

$$\rho^0 = \begin{pmatrix} 0 & -i \\ i & 0 \end{pmatrix}, \quad \rho^1 = \begin{pmatrix} 0 & i \\ i & 0 \end{pmatrix}. \tag{5.47}$$

We define the curved Dirac matrices ρ^a by

$$\rho^a = e_A^a \rho^A. \tag{5.48}$$

The covariant fermionic action is given by

$$S_F = \frac{i}{2\pi} \int_\Sigma d^2\sigma \sqrt{-h}\, \bar{\psi}^\mu e_A^a \rho^A \partial_a \psi_\mu. \tag{5.49}$$

For Majorana fermions in two dimensions we have $\bar{\psi}^\mu \rho^a \nabla_a \psi_\mu = \bar{\psi}^\mu \rho^a \partial_a \psi_\mu$ where ∇_a is the covariant derivative, i.e. the spin connection does not contribute due to the Fermi statistics. The full covariant action is thus

$$S = -\frac{1}{2\pi} \int_\Sigma d^2\sigma \sqrt{-h} \left(h^{ab} \partial_a X^\mu \partial_b X_\mu - i \bar{\psi}^\mu e_A^a \rho^A \partial_a \psi_\mu \right). \tag{5.50}$$

Now we have:

1. Invariance under reparametrizations of the worldsheet, i.e. invariance under two-dimensional general coordinates transformations.
2. Invariance under local two-dimensional Lorentz transformations.

Using this form of the action we can compute the energy–momentum tensor. The bosonic part gives

$$\delta_h S_B = -\frac{1}{2\pi} \int_\Sigma d^2\sigma \sqrt{-h}\, \delta h^{ab} \left(\partial_a X^\mu \partial_b X_\mu - \frac{1}{2} h_{ab} h^{cd} \partial_c X^\mu \partial_d X_\mu \right). \tag{5.51}$$

[2] The index a in e_a^A is the vector index tangent to the worldsheet manifold and thus it should be identified with α, whereas the index A takes the two values 0 an 1 which are the names of the vectors.

We can check that

$$\delta h^{ab} \cdot h_{ab} = 2e_a^A \cdot \delta e_A^a \tag{5.52}$$

Hence we can immediately guess that

$$\delta e_A^a = \frac{1}{2}\delta h^{ac} \cdot e_A^d h_{cd}. \tag{5.53}$$

Using this last equation we find that the fermionic part gives

$$\begin{aligned}\delta_h S_F =& -\frac{1}{2\pi}\int d^2\sigma\sqrt{-h}\,\delta h^{ab} \\ &\left(-\frac{i}{2}\bar{\psi}^\mu\rho_b\partial_a\psi_\mu + \frac{i}{2}h_{ab}\bar{\psi}^\mu\rho^c\partial_c\psi_\mu\right) \\ =& -\frac{1}{2\pi}\int d^2\sigma\sqrt{-h}\,\delta h^{ab} \\ &\left(-\frac{i}{4}\bar{\psi}^\mu\rho_a\partial_b\psi_\mu - \frac{i}{4}\bar{\psi}^\mu\rho_b\partial_a\psi_\mu + \frac{i}{2}h_{ab}\bar{\psi}^\mu\rho^c\partial_c\psi_\mu\right).\end{aligned} \tag{5.54}$$

We deduce the energy momentum tensor

$$\begin{aligned}T_{ab} &= T'_{ab} - \frac{1}{2}h_{ab}h^{cd}T'_{cd} + \frac{i}{4}h_{ab}\bar{\psi}^\mu\rho^a\partial_a\psi_\mu \\ T'_{ab} &= \partial_a X^\mu\partial_b X_\mu - \frac{i}{4}\bar{\psi}^\mu\rho_a\partial_b\psi_\mu - \frac{i}{4}\bar{\psi}^\mu\rho_b\partial_a\psi_\mu.\end{aligned} \tag{5.55}$$

By using the equation of motion $\rho^a\partial_a\psi_\mu = 0$ we get

$$T_{ab} = T'_{ab} - \frac{1}{2}h_{ab}h^{cd}T'_{cd}. \tag{5.56}$$

This is traceless.

We go back to the conformal gauge $h^{ab} = \text{diag}(-1, 1)$. We compute using the equation of motion $\rho^a\partial_a\psi^\mu = 0$ the following

$$\begin{aligned}T_{00} = T_{11} &= \frac{1}{2}(\partial_0 X^\mu\partial_0 X_\mu + \partial_1 X^\mu\partial_1 X_\mu) + \frac{i}{2}\bar{\psi}^\mu\rho^0\partial_0\psi_\mu \\ &= \frac{1}{2}(\partial_0 X^\mu\partial_0 X_\mu + \partial_1 X^\mu\partial_1 X_\mu) - \frac{i}{2}\bar{\psi}^\mu\rho^1\partial_1\psi_\mu\end{aligned} \tag{5.57}$$

$$T_{01} = T_{10} = \partial_0 X^\mu\partial_1 X_\mu + \frac{i}{4}\bar{\psi}^\mu(\rho^0\partial_1 - \rho^1\partial_0)\psi_\mu. \tag{5.58}$$

We compute

$$\begin{aligned}\frac{i}{2}\bar{\psi}^\mu\rho^0\partial_0\psi_\mu &= \frac{i}{2}(\psi_-^\mu\partial_0\psi_{-\mu} + \psi_+^\mu\partial_0\psi_{+\mu}) \\ &= \frac{i}{2}(\psi_-^\mu\partial_-\psi_{-\mu} + \psi_+^\mu\partial_+\psi_{+\mu})\end{aligned} \tag{5.59}$$

In the above we have used $\partial_0 = \partial_+ + \partial_-$ and the equations of motion $\partial_+\psi_-^\mu = \partial_-\psi_+^\mu = 0$. Also we compute

$$\frac{i}{4}\bar{\psi}^\mu(\rho^0\partial_1 - \rho^1\partial_0)\psi_\mu = \frac{i}{2}(-\psi_-^\mu\partial_-\psi_{-\mu} + \psi_+^\mu\partial_+\psi_{+\mu}). \tag{5.60}$$

Hence

$$\begin{aligned} T_{++} &= \frac{1}{2}(T_{00} + T_{01}) = \partial_+X^\mu\partial_+X_\mu + \frac{i}{2}\psi_+^\mu\partial_+\psi_{+\mu} \\ T_{--} &= \frac{1}{2}(T_{00} - T_{01}) = \partial_-X^\mu\partial_-X_\mu + \frac{i}{2}\psi_-^\mu\partial_-\psi_{-\mu}. \end{aligned} \tag{5.61}$$

5.3.3 Super-Virasoro constraints

We found the covariant action

$$S_2 = -\frac{1}{2\pi}\int d^2\sigma\sqrt{-h}\left(h^{ab}\partial_aX^\mu\partial_bX_\mu - i\bar{\psi}^\mu e_A^a\rho^A\partial_a\psi_\mu\right). \tag{5.62}$$

As we said, it is invariant (a) under reparametrizations of the worldsheet, i.e. the two-dimensional general coordinates transformations, and (b) under local two-dimensional Lorentz transformations. This action also has the Weyl transformations

$$\delta X^\mu = 0, \quad \delta\psi^\mu = -\frac{1}{2}\Lambda\psi^\mu, \quad \delta e_a^A = \Lambda e_a^A. \tag{5.63}$$

Under these transformations we have

$$\delta h_{ab} = 2\Lambda h_{ab}, \quad \delta h^{ab} = -2\Lambda h^{ab}, \quad \delta\sqrt{-h} = 2\Lambda\sqrt{-h}. \tag{5.64}$$

Let us now consider the supersymmetric transformations

$$\begin{aligned} \delta X^\mu &= \bar{\epsilon}\psi^\mu \\ \delta\psi^\mu &= -\, i\rho^\alpha\partial_\alpha X^\mu\epsilon. \end{aligned} \tag{5.65}$$

We want to obtain a local supersymmetry, i.e. a supergravity theory. The variation of the action S_2 with ϵ being not a constant is given by

$$\delta S_2 = \frac{2}{\pi}\int d^2\sigma\sqrt{-h}\,\nabla_\alpha\bar{\epsilon}J^\alpha, \quad J^\alpha = \frac{1}{2}\rho^\beta\rho^\alpha\partial_\beta X^\mu\psi_\mu. \tag{5.66}$$

We introduce a supersymmetry gauge field χ_α (α is the vector index and χ_α for a given value of α is a spinor) with supersymmetry transformation

$$\delta\chi_\alpha = \nabla_\alpha\epsilon. \tag{5.67}$$

The variation δS_2 can be cancelled by a variation with respect to χ_α of the action

$$S_3 = -\frac{2}{\pi}\int d^2\sigma\sqrt{-h}\bar{\chi}_\alpha J^\alpha. \tag{5.68}$$

The χ_α is the gaugino (supersymmetry gauge) field. The variation of S_3 due to the variation of X_μ contains another term proportional to $\nabla_\alpha \epsilon$ which is given by

$$\begin{aligned}\delta_X S_3 &= -\frac{1}{\pi}\int d^2\sigma\sqrt{-h}\bar{\chi}_\alpha \rho^\beta \rho^\alpha \psi^\mu \cdot \bar{\psi}_\mu \nabla_\beta \epsilon + \ldots \\ &= \frac{1}{2\pi}\int d^2\sigma\sqrt{-h}\bar{\chi}_\alpha \rho^\beta \rho^\alpha \nabla_\beta \epsilon \cdot \bar{\psi}_\mu \psi^\mu + \ldots\end{aligned} \tag{5.69}$$

In the above '...' stands for terms coming from $\nabla_\beta \bar{\psi}_\mu$. Also we have used the results $\bar{\psi}^\mu \psi_\mu = 2\bar{\psi}^\mu_+ \psi_{\mu+} = 2\bar{\psi}^\mu_- \psi_{\mu-}$ and $\bar{\psi}^\mu_+ = -i\psi^\mu_-$, $\bar{\psi}^\mu_- = i\psi^\mu_+$. The above variation can be cancelled by a variation with respect to χ_β of the action

$$S_4 = -\frac{1}{4\pi}\int d^2\sigma\sqrt{-h}\bar{\chi}_\alpha \rho^\beta \rho^\alpha \chi_\beta \cdot \bar{\psi}_\mu \psi^\mu. \tag{5.70}$$

We get the total supergravity action

$$\begin{aligned}S = &-\frac{1}{2\pi}\int d^2\sigma\sqrt{-h}\left(h^{ab}\partial_a X^\mu \partial_b X_\mu - i\bar{\psi}^\mu e^a_A \rho^A \partial_a \psi_\mu\right) \\ &-\frac{1}{\pi}\int d^2\sigma\sqrt{-h}\bar{\chi}_\alpha \rho^\beta \rho^\alpha \partial_\beta X^\mu \psi_\mu - \frac{1}{4\pi}\int d^2\sigma\sqrt{-h}\bar{\chi}_\alpha \rho^\beta \rho^\alpha \chi_\beta \cdot \bar{\psi}_\mu \psi^\mu.\end{aligned} \tag{5.71}$$

There is no kinetic term for e^a_A since gravity in two dimensions is topological. Also, there is no kinetic term for χ_α since there is no third rank antisymmetric tensor $\rho^{\alpha\beta\gamma}$ in two dimensions. The full supergravity transformations are

$$\begin{aligned}&\delta X^\mu = \bar{\epsilon}\psi^\mu, \quad \delta\chi_\alpha = \nabla_\alpha \epsilon \\ &\delta\psi^\mu = -i\rho^\alpha \epsilon(\partial_\alpha X^\mu - \bar{\psi}^\mu \chi_\alpha), \\ &\delta e^a_\alpha = -2i\bar{\epsilon}\rho^a \chi_a.\end{aligned} \tag{5.72}$$

The above action also has the Weyl transformations

$$\delta X^\mu = 0, \quad \delta\psi^\mu = -\frac{1}{2}\Lambda\psi^\mu, \quad \delta e^A_a = \Lambda e^A_a, \quad \delta\chi_\alpha = \frac{1}{2}\Lambda\chi_\alpha. \tag{5.73}$$

Let us note that in the above equations the indices a and b denote the same thing as the indices α, β. As a consequence of the identity $\rho_\alpha \rho^\beta \rho^\alpha = 0$ we have also the following fermionic local symmetry

$$\delta\chi_\alpha = i\rho_\alpha \eta, \quad \delta X^\mu = \delta\psi^\mu = \delta e^A_a = 0. \tag{5.74}$$

The above action is therefore superconformal.

As for the bosonic string we have for the superstring the two local bosonic symmetries:

1. Two worldsheet reparametrizations $\xi_c = \delta\sigma_c$.
2. One Weyl scaling Λ.

In the case of the superstring we have an extra local bosonic symmetry, namely

1. One local Lorentz transformation (a boost along σ_1).

In total we have four local bosonic symmetries which can be used to set $e_a^A = \delta_a^A$.

Furthermore we have for the superstring four local fermionic symmetries:

1. Two supersymmetries ϵ.
2. Two superconformal symmetries η.

These can be used to set $\chi_\alpha = 0$. The total supergravity action reduces then to the original globally supersymmetric action (5.23).

Let us compute the variation of the action with respect to χ_α. We get

$$\begin{aligned}\delta_\chi S = &-\frac{1}{\pi}\int d^2\sigma\sqrt{h}\,\delta\bar{\chi}_\alpha\rho^\beta\rho^\alpha\partial_\beta X^\mu\psi_\mu\\ &-\frac{1}{2\pi}\int d^2\sigma\sqrt{h}\,\delta\bar{\chi}_\alpha\rho^\beta\rho^\alpha\chi_\beta\bar{\psi}^\mu\psi_\mu.\end{aligned} \tag{5.75}$$

The equation of motion of χ_α is

$$-\frac{\pi}{2}\frac{1}{\sqrt{h}}\frac{\delta_\chi S}{\delta\bar{\chi}_\alpha}|_{\chi=e=0} = \frac{1}{2}\rho^\beta\rho^\alpha\psi_\mu\partial_\beta X^\mu = 0. \tag{5.76}$$

This is equivalent to the statement that the supercurrent J^α must vanish, viz

$$J^\alpha = 0. \tag{5.77}$$

Similarly evaluating the variation of the action with respect to e_a^A at $\chi_\alpha = e_a^A = 0$ then setting it to 0 will yield the equations of motion of e_a^A which is equivalent to the statement that the energy–momentum tensor must vanish, viz

$$T_{ab} = 0. \tag{5.78}$$

The proof is obvious.

5.3.4 Boundary conditions (Ramond and Neveu–Schwarz)

Consider the action

$$\begin{aligned}S &= S_B + S_F\\ &= -\frac{1}{2\pi}\int d^2\sigma\partial_a X^\mu\partial^a X_\mu + \frac{i}{2\pi}\int d^2\sigma\bar{\psi}^\mu\rho^\alpha\partial_\alpha\psi_\mu.\end{aligned} \tag{5.79}$$

We compute the variation

$$\begin{aligned}\delta S &= \delta S_B + \delta S_F\\ \delta S_F &= \frac{i}{2\pi}\int d^2\sigma(2\delta\bar{\psi}^\mu\rho^\alpha\partial_\alpha\psi_\mu + \partial_\alpha(\bar{\psi}^\mu\rho^\alpha\delta\psi^\mu)).\end{aligned} \tag{5.80}$$

Setting the variation δS_B equal to 0 leads to the equation of motion of X^μ, viz $\partial^a\partial_a X^\mu = 0$ and to the boundary conditions for open or closed strings. Setting the variation $\delta\, S_F$ equal to 0 we get the equation of motion

$$\rho^\alpha \partial_\alpha \psi_\mu = 0 \Leftrightarrow \partial_- \psi_+^\mu = \partial_+ \psi_-^\mu = 0. \tag{5.81}$$

We also get the condition

$$\bar{\psi}^\mu \rho^1 \delta\psi_\mu|_{\sigma=0}^{\pi} = \psi_-^\mu \delta\psi_{-\mu}|_{\sigma=0}^{\pi} - \psi_+^\mu \delta\psi_{+\mu}|_{\sigma=0}^{\pi} = 0. \tag{5.82}$$

In the above we have assumed that $\delta\psi_\mu \longrightarrow 0$ when $\tau \longrightarrow \pm\infty$. We discuss separately the cases of open and closed strings.

Open strings
The requirement $\delta S_B = 0$ yields in this case the Neumann boundary condition

$$\partial_\sigma X_\mu|_{\sigma=0,\,\pi} = 0. \tag{5.83}$$

Next we must have $\delta S_F = 0$. The possibility of the Dirichlet boundary condition $\delta\psi_\pm^\mu|_{\sigma=0,\,\pi} = 0$ for the fermion field is too restrictive. Thus we must instead impose the boundary conditions $\psi_+^\mu|_{\sigma=0,\,\pi} = \pm\psi_-^\mu|_{\sigma=0,\,\pi}$. In other words, we can choose without any loss of generality one of two possibilities

$$\psi_+^\mu(\tau, 0) = \psi_-^\mu(\tau, 0), \quad \psi_+^\mu(\tau, \pi) = \psi_-^\mu(\tau, \pi), \quad \text{Ramond}. \tag{5.84}$$

$$\psi_+^\mu(\tau, 0) = \psi_-^\mu(\tau, 0), \psi_+^\mu(\tau, \pi) = -\psi_-^\mu(\tau, \pi), \quad \text{Neveu–Schwarz}. \tag{5.85}$$

Thus we obtain the mode expansions

$$\psi_-^\mu = \frac{1}{\sqrt{2}} \sum_n d_n^\mu e^{-in(\tau-\sigma)}, \psi_+^\mu = \frac{1}{\sqrt{2}} \sum_n d_n^\mu e^{-in(\tau+\sigma)}, \quad n = \text{integer}, \quad \text{R}. \tag{5.86}$$

$$\psi_-^\mu = \frac{1}{\sqrt{2}} \sum_r b_r^\mu e^{-ir(\tau-\sigma)}, \psi_+^\mu = \frac{1}{\sqrt{2}} \sum_r b_r^\mu e^{-ir(\tau+\sigma)}, \quad r + \frac{1}{2} = \text{integer}, \quad \text{NS}. \tag{5.87}$$

The reality conditions $\psi_\pm^\dagger = \psi_\pm$ yield $(d_n^\mu)^\dagger = d_{-n}^\mu$, $(b_r^\mu)^\dagger = b_{-r}^\mu$.

Closed strings
The requirement $\delta S_B = 0$ yields in this case the periodic boundary condition

$$X_\mu(\sigma + \pi, \tau) = X_\mu(\sigma, \tau). \tag{5.88}$$

Similarly, the requirement $\delta S_F = 0$ yields periodic and/or antiperiodic boundary conditions on the fermion fields ψ_-^μ and ψ_+^μ. Indeed, for the fermionic right-movers we can choose one of two conditions

$$\begin{aligned} \psi_-^\mu(\sigma + \pi, \tau) &= \psi_-^\mu(\sigma, \tau) \Leftrightarrow \psi_-^\mu(\sigma, \tau) \\ &= \frac{1}{\sqrt{2}} \sum_n d_n^\mu e^{-2in(\tau-\sigma)}, \quad n = \text{integer, R} \end{aligned} \tag{5.89}$$

$$\psi_-^\mu(\sigma+\pi,\tau) = -\psi_-^\mu(\sigma,\tau) \Leftrightarrow \psi_-^\mu(\sigma,\tau) = \frac{1}{\sqrt{2}}\sum_r b_r^\mu e^{-2ir(\tau-\sigma)}, \quad r+\frac{1}{2} = \text{integer, NS}. \tag{5.90}$$

For the fermionic left-movers we choose one of the two conditions

$$\psi_+^\mu(\sigma+\pi,\tau) = \psi_+^\mu(\sigma,\tau) \Leftrightarrow \psi_+^\mu(\sigma,\tau) = \frac{1}{\sqrt{2}}\sum_n \tilde{d}_n^\mu e^{-2in(\tau-\sigma)}, \quad n = \text{integer, R} \tag{5.91}$$

$$\psi_+^\mu(\sigma+\pi,\tau) = -\psi_+^\mu(\sigma,\tau) \Leftrightarrow \psi_+^\mu(\sigma,\tau) = \frac{1}{\sqrt{2}}\sum_r \tilde{b}_r^\mu e^{-2ir(\tau-\sigma)}, \quad r+\frac{1}{2} = \text{integer, NS}. \tag{5.92}$$

Thus, we have four sectors R–R, NS–NS, R–NS and NS–R.

5.4 Canonical quantization

In this section we will follow mostly [4].

5.4.1 Commutation relations

The quantization of the bosonic coordinates X^μ is found to be given by the commutation relations

$$[\partial_\tau X^\mu(\sigma), X^\nu(\sigma')] = -i\pi\delta(\sigma-\sigma')\eta^{\mu\nu}. \tag{5.93}$$

These lead to the commutation relations

$$[\alpha_n^\mu, \alpha_m^\nu] = n\delta_{n+m,0}\eta^{\mu\nu}, \quad \text{open string}. \tag{5.94}$$

$$[\alpha_n^\mu, \alpha_m^\nu] = [\tilde{\alpha}_n^\mu, \tilde{\alpha}_m^\nu] = n\delta_{n+m,0}\eta^{\mu\nu}, \quad [\alpha_n^\mu, \tilde{\alpha}_m^\nu] = 0, \quad \text{closed string}. \tag{5.95}$$

Similarly the quantization of the fermionic coordinates ψ_A^μ is found to be given by the anticommutation relations

$$\left\{\psi_A^\mu(\sigma), \psi_B^\nu(\sigma')\right\} = \pi\delta_{AB}\eta^{\mu\nu}\delta(\sigma-\sigma'). \tag{5.96}$$

For the open string there is a Ramond (R) fermionic sector and a Neveu–Schwarz (NS) bosonic sector, whereas for the closed string there are four fermionic sectors: R–R, R–NS, NS–R and NS–NS sectors.

For the open string we have the modes α_n^μ, d_n^μ in the R sector and the modes α_n^μ, b_r^μ in the NS sector. For the right-movers closed string we have the modes α_n^μ, d_n^μ in the R sector and the modes α_n^μ, b_r^μ in the NS sector while for the left-movers closed string we have the modes $\tilde{\alpha}_n^\mu$, $\tilde{d}_n^\mu$ in the R sector and the modes $\tilde{\alpha}_n^\mu$, $\tilde{b}_r^\mu$ in the NS sector.

For the open string we compute

$$
\begin{aligned}
d_n^\mu &= \frac{1}{\pi\sqrt{2}} \int_0^\pi d\sigma\Big(\psi_+^\mu(\tau,\sigma)e^{in(\tau+\sigma)} + \psi_-^\mu(\tau,\sigma)e^{in(\tau-\sigma)}\Big), \quad \mathrm{R} \\
b_r^\mu &= \frac{1}{\pi\sqrt{2}} \int_0^\pi d\sigma\Big(\psi_+^\mu(\tau,\sigma)e^{ir(\tau+\sigma)} + \psi_-^\mu(\tau,\sigma)e^{ir(\tau-\sigma)}\Big), \quad \mathrm{NS}.
\end{aligned}
\tag{5.97}
$$

For the right-movers closed string we compute

$$
\begin{aligned}
d_n^\mu &= \frac{2}{\pi\sqrt{2}} \int_0^\pi d\sigma\Big(\psi_-^\mu(\tau,\sigma)e^{2in(\tau-\sigma)}\Big), \quad \mathrm{R} \\
b_r^\mu &= \frac{2}{\pi\sqrt{2}} \int_0^\pi d\sigma\Big(\psi_-^\mu(\tau,\sigma)e^{2ir(\tau-\sigma)}\Big), \quad \mathrm{NS}.
\end{aligned}
\tag{5.98}
$$

For the left-movers closed string we compute

$$
\begin{aligned}
\tilde{d}_n^\mu &= \frac{2}{\pi\sqrt{2}} \int_0^\pi d\sigma\Big(\psi_+^\mu(\tau,\sigma)e^{2in(\tau-\sigma)}\Big), \quad \mathrm{R} \\
\tilde{b}_r^\mu &= \frac{2}{\pi\sqrt{2}} \int_0^\pi d\sigma\Big(\psi_+^\mu(\tau,\sigma)e^{2ir(\tau-\sigma)}\Big), \quad \mathrm{NS}.
\end{aligned}
\tag{5.99}
$$

We compute the fermionic anticommutation relations

$$
\left\{d_n^\mu, d_m^\nu\right\} = \eta^{\mu\nu}\delta_{n+m,0}, \quad \left\{b_r^\mu, b_s^\nu\right\} = \eta^{\mu\nu}\delta_{r+s,0}, \quad \text{open string.} \tag{5.100}
$$

$$
\left\{d_n^\mu, d_m^\nu\right\} = \eta^{\mu\nu}\delta_{n+m,0}, \quad \left\{b_r^\mu, b_s^\nu\right\} = \eta^{\mu\nu}\delta_{r+s,0}, \quad \text{right movers closed string.} \tag{5.101}
$$

$$
\left\{\tilde{d}_n^\mu, \tilde{d}_m^\nu\right\} = \eta^{\mu\nu}\delta_{n+m,0}, \quad \left\{\tilde{b}_r^\mu, \tilde{b}_s^\nu\right\} = \eta^{\mu\nu}\delta_{r+s,0}, \quad \text{left movers closed string.} \tag{5.102}
$$

The super-Virasoro generators are defined as the Fourier modes of $T_{\alpha\beta}$ and J_α with respect to σ. Since $T_{++}(\sigma) = T_{--}(-\sigma)$, $J_\pm(\sigma) = J_\mp(-\sigma)$ for the open string we have in this case

$$
\begin{aligned}
L_m(\tau) &= \frac{1}{\pi}\int_0^\pi d\sigma(e^{im\sigma}T_{++} + e^{-im\sigma}T_{--}) \\
&= \frac{1}{\pi}\int_{-\pi}^\pi d\sigma e^{im\sigma}T_{++}, \quad \text{open string}
\end{aligned}
\tag{5.103}
$$

$$
\begin{aligned}
F_m(\tau) &= \frac{\sqrt{2}}{\pi}\int_0^\pi d\sigma(e^{im\sigma}J_+ + e^{-im\sigma}J_-) \\
&= \frac{\sqrt{2}}{\pi}\int_{-\pi}^\pi d\sigma e^{im\sigma}J_+, \quad \mathrm{R}, \quad \text{open string}
\end{aligned}
\tag{5.104}
$$

$$G_r(\tau) = \frac{\sqrt{2}}{\pi}\int_0^{\pi} d\sigma(e^{ir\sigma}J_+ + e^{-ir\sigma}J_-)$$
$$= \frac{\sqrt{2}}{\pi}\int_{-\pi}^{\pi} d\sigma e^{ir\sigma}J_+, \quad \text{NS}, \quad \text{open string.} \tag{5.105}$$

For the closed string we have

$$L_m(\tau) = \frac{1}{2\pi}\int_0^{\pi} d\sigma e^{-2im\sigma}T_{--},$$
$$\tilde{L}_m(\tau) = \frac{1}{2\pi}\int_0^{\pi} d\sigma e^{2im\sigma}T_{++}, \quad \text{closed string} \tag{5.106}$$

$$F_m(\tau) = \frac{\sqrt{2}}{2\pi}\int_0^{\pi} d\sigma e^{-2im\sigma}J_-,$$
$$\tilde{F}_m(\tau) = \frac{\sqrt{2}}{2\pi}\int_0^{\pi} d\sigma e^{2im\sigma}J_+, \quad \text{R}, \quad \text{closed string} \tag{5.107}$$

$$G_r(\tau) = \frac{\sqrt{2}}{2\pi}\int_0^{\pi} d\sigma e^{-2ir\sigma}J_-,$$
$$\tilde{G}_r(\tau) = \frac{\sqrt{2}}{2\pi}\int_0^{\pi} d\sigma e^{2ir\sigma}J_+, \quad \text{NS}, \quad \text{closed string.} \tag{5.108}$$

5.4.2 Ramond (fermionic) and Neveu–Schwarz (bosonic) open string sectors

We consider in this section the open string.

The bosonic Virasoro generator is given by

$$L_m(\tau) = \frac{1}{\pi}\int_{-\pi}^{\pi} d\sigma e^{im\sigma}T_{++}$$
$$= \frac{1}{\pi}\int_{-\pi}^{\pi} d\sigma e^{im\sigma}\partial_+X^\mu\partial_+X_\mu + \frac{i}{2\pi}\int_{-\pi}^{\pi} d\sigma e^{im\sigma}\psi_+^\mu\partial_+\psi_{+\mu} \tag{5.109}$$
$$= L_m^\alpha(\tau) + \frac{i}{2\pi}\int_{-\pi}^{\pi} d\sigma e^{im\sigma}\psi_+^\mu\partial_+\psi_{+\mu}.$$

The first contribution $L_m^\alpha(\tau)$ was already computed. It is given by

$$L_m^\alpha(\tau) = \frac{1}{2}\sum_{n=-\infty}^{\infty} \alpha_{-n}^\mu(\alpha_{m+n})_\mu e^{-im\tau}. \tag{5.110}$$

The second contribution is given in the Ramond sector by (using the anticommutation of the fermionic degrees of freedom)

$$
\begin{aligned}
L_m^d(\tau) &= \frac{i}{2\pi}\int_{-\pi}^{\pi} d\sigma e^{im\sigma}\psi_+^\mu \partial_+ \psi_{+\mu} \\
&= \frac{1}{2}\sum_{n=-\infty}^{\infty}(n+m)d_{-n}^\mu(d_n+m)_\mu e^{-im\tau} \\
&= \frac{1}{2}\sum_{n=-\infty}^{\infty}(n+\frac{1}{2}m)d_{-n}^\mu(d_b+m)_\mu e^{-im\tau}, \quad \text{R.}
\end{aligned} \tag{5.111}
$$

Similarly, in the Neveu–Schwarz sector the second contribution is given by

$$
\begin{aligned}
L_m^b(\tau) &= \frac{i}{2\pi}\int_{-\pi}^{\pi} d\sigma e^{im\sigma}\psi_+^\mu \partial_+ \psi_{+\mu} \\
&= \frac{1}{2}\sum_{r=-\infty}^{\infty}(r+m)b_{-r}^\mu(d_n+m)_\mu e^{-im\tau} \\
&= \frac{1}{2}\sum_{r=-\infty}^{\infty}(r+\frac{1}{2}m)b_{-r}^\mu(d_n+m)_\mu e^{-im\tau}, \quad \text{NS}
\end{aligned} \tag{5.112}
$$

$r+\dfrac{1}{2}$ is an integer.

Recall that $T = 1/(2\pi\alpha') = 1/(\pi l^2)$, $p^\mu = \alpha_0^\mu/l$. We have chosen $T = 1/\pi$. Then

$$
\begin{aligned}
\alpha' M^2 &= -\alpha' p_\mu p^\mu \\
&= -L_0^\alpha(0) + \sum_{n=1}\alpha_{-n}^\mu \\
&= -L_0(0) + L_0^{d,b}(0) + \sum_{n=1}\alpha_{-n}^\mu(\alpha_n)_\mu \\
&= -L_0(0) + N
\end{aligned} \tag{5.113}
$$

$$
\begin{aligned}
&N = N^d + N^\alpha, \quad N^d = L_0^{(0)} = \sum_{n=1} n d_{-n}^\mu(d_n)_\mu, \\
&N^\alpha = \sum_{n=1}\alpha_{-n}^\mu(\alpha_n)_\mu, \quad \text{R}
\end{aligned} \tag{5.114}
$$

$$
\begin{aligned}
&N = N^b + N^\alpha, \quad N^b = L_0^b(0) = \sum_{r=1/2} r b_{-r}^\mu(b_r)_\mu, \\
&N^\alpha = \sum_{n=1}\alpha_{-n}^\mu(b_r)_\mu, \quad \text{NS.}
\end{aligned} \tag{5.115}
$$

Recall the classical constraints $T_{++} = T_{--} = J_+ = J_- = 0$. The quantum constraints are $L_m(\tau) = 0$, $F_m(\tau) = G_r(\tau) = 0$. Thus the mass-shell condition will reduce to

$$
\alpha' M^2 = N + \text{constant}. \tag{5.116}
$$

The constant is due to normal ordering.

The ground state is defined by

$$\alpha_n^\mu|0\rangle = d_n^\mu|0\rangle = 0, \quad n > 0, \quad \text{R} \tag{5.117}$$

$$\alpha_n^\mu|0\rangle = b_r^\mu|0\rangle = 0, \quad n, r > 0, \quad \text{NS}. \tag{5.118}$$

Let us recall that for $n > 0$ the operators α_n^μ are lowering operators while α_{-n}^μ are the raising operators. For $n < 0$ the operators α_n^μ are raising operators while α_{-n}^μ are the lowering operators. We can compute the identity

$$[\alpha_n^\mu, \alpha' M^2] = n\alpha_n^\mu, \quad n > 0. \tag{5.119}$$

Thus, if $|\phi\rangle$ is an eigenstate of $\alpha' M^2$ with eigenvalue $\alpha' m^2$ then we can show that $\alpha_n^\mu|\phi\rangle$ is an eigenstate of $\alpha' M^2$ with eigenvalue $\alpha' m^2 - n$. In other words, α_n^μ, $n > 0$, decreases the worldsheet energy by an amount n and hence α_n^μ, $n < 0$, will increase the worldsheet energy by an amount n.

Similarly, for $n > 0$ the operators $d_{\pm n}^\mu$ are lowering operators while $d_{\mp n}^\mu$ are the raising operators. We can also show that d_n^μ, $n > 0$, decreases the worldsheet energy by an amount n and d_n^μ, $n < 0$, increases the worldsheet energy by an amount n.

Furthermore, for $r > 0$ the operators $b_{\pm r}^\mu$ are lowering operators while $b_{\mp r}^\mu$ are the raising operators. We can show that b_r^μ, $r > 0$, decreases the worldsheet energy by an amount r and b_r^μ, $r < 0$, increases the worldsheet energy by an amount r.

Neveu–Schwarz sector (bosonic sector): The ground state is defined by $N^\alpha|0\rangle = N^b|0\rangle = 0$. This is a unique state which corresponds to a particle of spin 0. All excited modes are obtained by having the operators α_n^μ and/or b_r^μ, $n, r < 0$, act on the ground state. Since α_n^μ, b_r^μ transform as vectors we can only get integer spin states. Hence the ground state and all excited states in the Neveu–Schwarz sector are bosonic states.

Ramond sector (fermionic sector): The ground state is again defined by $N^\alpha|0\rangle = N^d|0\rangle = 0$. However in this case we also have the oscillators d_0^μ which satisfy

$$\{d_0^\mu, d_0^\nu\} = \eta^{\mu\nu}. \tag{5.120}$$

They commute with $\alpha' M^2$ since they do not raise or lower the worldsheet energy. These operators will be identified with the Dirac matrices as follows

$$d_0^\mu = \frac{1}{i\sqrt{2}}\Gamma^\mu. \tag{5.121}$$

The ground state is therefore an irreducible representation of $\{d_0^\mu, d_0^\nu\} = \eta^{\mu\nu}$ and since there are no other zero modes this irreducible representation corresponds to an $SO(1, d)$ spinor. All excited modes are obtained by having the operators α_n^μ and/or d_n^μ, $n < 0$ act on the ground state. Since α_n^μ, d_n^μ, transform as vectors we can only get half-integer spin states from a fermionic ground state. Hence the ground state and all excited states in the Ramond sector are fermionic states.

5.4.3 Super-Virasoro algebra

In the bosonic string we have found that the energy–momentum tensor $T_{\alpha\beta}$ is conserved because of spacetime translation symmetry. However, the requirement of conformal invariance led to the stronger condition that the energy–momentum tensor must vanish, viz $T_{++} = T_{--} = 0$.[3] These equations are also derived as the equations of motion for the worldsheet metric $h_{\alpha\beta}$. The Virasoro generators L_m are the Fourier modes of the energy–momentum tensor component T_{++}. They satisfy the quantum Virasoro algebra with a central extension, viz

$$[L_m, L_n] = (m - n)L_{m+n} + \frac{D}{12}m(m^2 - 1)\delta_{m+n,\,0}. \tag{5.122}$$

The constraints $T_{++} = T_{--} = 0$ become then $L_m = 0$, $m = 0, \pm 1, \ldots$, where $H = L_0$ is the Hamiltonian.

We generalize this to the RNS open superstring. We will have as before the energy–momentum tensor $T_{\alpha\beta}$ associated with global spacetime translations but now we will also have the supercurrent J_α associated with global supersymmetry. They are both conserved, viz $\partial_\pm T_{\mp\mp} = 0$, $\partial_\pm J_\mp = 0$. The requirement of superconformal invariance leads now to the vanishing of the energy–momentum tensor and the supercurrent, viz $T_{++} = T_{--} = J_+ = J_- = 0$. Alternatively, these equations can be derived as the equations of motion for the worldsheet metric $h_{\alpha\beta}$ and the gaugino field χ_α. The super-Virasoro generators consist of the bosonic Virasoro generators L_m, which are the Fourier modes of the energy–momentum tensor component T_{++}, and the fermionic generators F_m/G_r, which are the Fourier modes of the supercurrent J_+, in the Ramond/Neveu–Schwarz sectors. We have then:

- **The Ramond sector:** In this sector the spectrum in the Hilbert space of states will be fermionic in spacetime. The underlying degrees of freedom are the bosonic oscillators α_n^μ and the R fermionic oscillators d_n^μ where n is an integer. The bosonic Virasoro generators L_m will be given by the bosonic string contribution + a contribution coming from the R fermionic oscillators as follows (including now normal ordering explicitly)[4]

$$L_m = L_m^\alpha + L_m^d \tag{5.123}$$

$$L_m^\alpha = \frac{1}{2}\sum_{n=-\infty}^{\infty} :\alpha_{-n}^\mu(\alpha_{m+n})_\mu: \tag{5.124}$$

$$L_m^d = \frac{1}{2}\sum_{n=-\infty}^{\infty}\left(n + \frac{1}{2}m\right): d_{-n}^\mu(d_{n+m})_\mu: . \tag{5.125}$$

[3] The conditions $T_{+-} = T_{-+} = 0$ follow from Weyl invariance, i.e. from tracelessness of the energy–momentum tensor.

[4] $L_m = L_m(0)$ or L_m is the Fourier modes with respect to the light cone coordinate σ^+, etc.

The Fourier modes for the supercurrent J_+ in the Ramond sector are

$$F_m = \sum_{n=-\infty}^{\infty} \alpha_{-n}^{\mu}(d_{n+m})_{\mu}. \quad (5.126)$$

The generalization of the Virasoro algebra (5.122) is the super-Virasoro algebra which reads in the Ramond sector as follows

$$\begin{aligned}
[L_m, L_n] &= (m - n)L_{m+n} + \frac{D}{8}m^3\delta_{m+n,0} \\
[L_m, F_n] &= \left(\frac{m}{2} - n\right)F_{m+n} \\
\{F_m, F_n\} &= 2L_{m+n} + \frac{D}{2}m^2\delta_{m+n,0}.
\end{aligned} \quad (5.127)$$

The constraints $T_{++} = T_{--} = J_+ = J_- = 0$ become $L_m = F_m = 0$. However, as in the bosonic string, we only need to require that the positive modes of the super-Virasoro generators annihilate physical states, viz

$$\begin{aligned}
&(L_0 - a_{\mathrm{R}})|\phi\rangle = 0 \\
&L_m|\phi\rangle = 0, \quad m > 0 \\
&F_n|\phi\rangle = 0, \quad n \geqslant 0.
\end{aligned} \quad (5.128)$$

a_{R} is the normal ordering constant in the Ramond sector. Since $F_0^2 = L_0$ and since there is no normal ordering ambiguity in F_0 we conclude immediately that $a_{\mathrm{R}} = 0$. The F_0 equation is a wave equation called the Dirac–Ramond equation.

- **The Neveu–Schwarz sector:** In this sector the spectrum in the Hilbert space of states will be bosonic in spacetime. The underlying degrees of freedom are the bosonic oscillators α_n^{μ} and the NS fermionic oscillators d_r^{μ} where n is an integer and r is a half integer. Again, the bosonic Virasoro generators L_m will be given by the bosonic string contribution + a contribution coming from the NS fermionic oscillators as follows

$$L_m = L_m^{\alpha} + L_m^{b}. \quad (5.129)$$

L_m^{α} is as above and

$$L_m^{b} = \frac{1}{2}\sum_{r=-\infty}^{\infty}\left(r + \frac{1}{2}m\right): b_{-r}^{\mu}(b_{r+m})_{\mu}: . \quad (5.130)$$

The Fourier modes for the supercurrent J_+ in the Neveu–Schwarz sector are

$$G_r = \sum_{n=-\infty}^{\infty} \alpha_{-n}^{\mu}(b_{n+r})_{\mu}. \quad (5.131)$$

The generalization of the Virasoro algebra (5.122) in the Neveu–Schwarz sector is

$$[L_m, L_n] = (m - n)L_{m+n} + \frac{D}{8}(m^3 - m)\delta_{m+n,\,0}$$
$$[L_m, G_r] = \left(\frac{m}{2} - r\right)G_{m+r} \qquad (5.132)$$
$$\{G_r, G_s\} = 2L_{r+s} + \frac{D}{2}\left(r^2 - \frac{1}{4}\right)\delta_{r+s,\,0}.$$

The five generators L_1, L_0, L_-, $G_{1/2}$ and $G_{-1/2}$ form the $OSp(1|2)$ algebra which is the supersymmetric analogue of $SL(2,\mathrm{R})$.

The constraints $T_{++} = T_{--} = J_+ = J_- = 0$ become $L_m = G_r = 0$. More precisely

$$(L_0 - a_{\mathrm{NS}})|\phi\rangle = 0$$
$$L_m|\phi\rangle = 0, \quad m > 0 \qquad (5.133)$$
$$G_r|\phi\rangle = 0, \quad r > 0.$$

a_{NS} is the normal ordering constant in the Neveu–Schwarz sector. We can show that $a_{\mathrm{NS}} = 1/2$.

In both the R and the NS sectors the condition $(L_0 - a)|\phi\rangle = 0$ yields, as we have seen, the result that $\alpha' M^2 = N - a$ where M is the mass of the state $|\phi\rangle$ and N should be replaced by its eigenvalue on the state $|\phi\rangle$.

5.5 The light cone/path integral quantization

In this section we will follow the concise presentation of [1] but also [4].

5.5.1 The light cone quantization and critical dimension

The superconformal symmetry given by spacetime translations ($T_{\alpha\beta}$) and supersymmetry (J_α), which is expressed quantum mechanically by the super-Virasoro algebra, plays a crucial role where the vanishing of the super-Virasoro generators are precisely the Dirac constraints. The spectrum of the RNS open superstring contains unphysical negative-norm states which are called ghosts. These states exist for a certain range of D and a while for another range of D and a there are no ghosts. The boundary between the two regions is demarcated by the appearance, for special values of D and a, of extra zero-norm states which are physical states on the verge of turning into ghosts. These negative-norm states can be eliminated by using the superconformal symmetry only in the critical dimension $D = 10$.

Equivalently, the superconformal symmetry can be used to fix the light cone gauge where Lorentz invariance can be shown to hold quantum mechanically only in the critical dimension $D = 10$. To show this fundamental result we go essentially through the same steps as in the bosonic string case.

After imposing the conformal covariant gauge $h_{\alpha\beta} = \eta_{\alpha\beta}$ there remain a residual reparametrization invariance and a residual supersymmetry which preserve the conformal gauge but are also sufficient to gauge away the plus components X^+ and ψ^+ as follows

$$X^+(\sigma, \tau) = x^+ + p^+\tau, \ \psi^+=0. \tag{5.134}$$

Let us concentrate on the Neveu–Schwarz (bosonic) open superstring sector. In the above light cone gauge, the super-Virasoro constraints (vanishing of $T_{\alpha\beta}$ and J_α) can be solved for $X^-(\alpha_n^-)$ and $\psi^-(b_r^-)$ in terms of the transverse variables $X^i(\alpha_n^i)$ and $\psi^i(b_r^i)$.

The Lorentz generators for the RNS open superstring in the Neveu–Schwarz bosonic sector are:

$$J^{\mu\nu} = L^{\mu\nu} + K^{\mu\nu} = l^{\mu\nu} + E^{\mu\nu} + K^{\mu\nu}. \tag{5.135}$$

The $l^{\mu\nu}$ and $E^{\mu\nu}$ are as in the bosonic string, viz

$$l^{\mu\nu} = x^\mu p^\nu - x^\nu p^\mu \tag{5.136}$$

$$E^{\mu\nu} = -i\sum_{n=1}^{\infty}\frac{1}{n}(\alpha_{-n}^\mu\alpha_n^\nu - \alpha_{-n}^\nu\alpha_n^\mu). \tag{5.137}$$

The extra term $K^{\mu\nu}$ is due to the fermionic oscillators b_r^μ and it is given by

$$K^{\mu\nu} = -i\sum_{r=1/2}^{\infty}(b_{-r}^\mu b_r^\nu - b_{-r}^\nu b_r^\mu). \tag{5.138}$$

The Lorentz algebra is

$$[J^{\mu\nu}, J^{\rho\lambda}] = -i\eta^{\nu\rho}J^{\mu\lambda} + i\eta^{\mu\rho}J^{\nu\lambda} + i\eta^{\nu\lambda}J^{\mu\rho} - i\eta^{\mu\lambda}J^{\nu\rho}. \tag{5.139}$$

As in the bosonic open string, this algebra is satisfied except for the case $[J^{i-}, J^{j-}]$ which does not vanish generically as it should. We compute

$$[J^{i-}, J^{j-}] = -\frac{1}{(p^+)^2}\sum_{m=1}^{\infty}\Delta_m\left(\alpha_{-m}^i\alpha_m^j - \alpha_{-m}^j\alpha_m^i\right) \tag{5.140}$$

$$\Delta_m = m\frac{D-10}{8} + \frac{1}{m}\left(2a - \frac{D-2}{8}\right). \tag{5.141}$$

This commutator vanishes, and thus Lorentz invariance is maintained in the quantum theory, iff $a = 1/2$ and $D = 10$.

5.5.2 The GSO conditions and superstring spectrum

Spectrum of the open superstring

The spectrum of the RNS open superstring still contains a tachyon. This is true in both the Ramond (fermionic) and Neveu–Schwarz (bosonic) sectors with $a_{\mathrm{R}} = 0$ and $a_{\mathrm{NS}} = 1/2$ respectively. We have found the mass-shell conditions in both sectors in the canonical formalism. In the light cone gauge the same formulas hold except that the vector index μ of the various creation and annihilation operators will run over the transverse values only. We have then

$$\alpha' M^2 = \sum_{n=1} \alpha_{-n}^{\mu}(\alpha_n)_\mu + \sum_{r=1/2} r b_{-r}^{\mu}(b_r)_\mu - \frac{1}{2}, \quad \text{NS} \tag{5.142}$$

$$\alpha' M^2 = \sum_{n=1} \alpha_{-n}^{\mu}(\alpha_n)_\mu + \sum_{n=1} n d_{-n}^{\mu}(b_n)_\mu, \quad \text{R}. \tag{5.143}$$

The ground states are as follows:

1. The ground state $|0, k\rangle_{\rm NS}$ in the NS sector is a spacetime scalar tachyon since it satisfies (recall that $\alpha_0^\mu = \sqrt{2\alpha'}p^\mu$)

$$\begin{aligned} \alpha_n^i|0; k\rangle_{\rm NS} &= b_r^i|0; k\rangle_{\rm NS} = 0, \quad n, r > 0, \\ \alpha_0^i|0; k\rangle_{\rm NS} &= \sqrt{2\alpha'}k^\mu|0; k\rangle_{\rm NS}. \end{aligned} \tag{5.144}$$

   Then it has negative mass squared, viz $\alpha' M^2 = -1/2$. This is unphysical and it will be removed by means of the so-called GSO projection shortly.

2. After removing the tachyon the first excited state in the NS sector becomes the ground state. This is obtained by acting with the raising operator $b_{-1/2}^i$ on the original ground state $|0, k\rangle_{\rm NS}$. In the light cone gauge we get then the eight states

$$b_{-1/2}^i|0; k\rangle_{\rm NS}. \tag{5.145}$$

   These eight states are obviously massless since $b_{-1/2}^i$ raises energy by a 1/2 unit and $a_{\rm NS} = 1/2$ and hence they correspond to a gauge vector field in 10 dimensions.

3. The ground state $|0, k\rangle_{\rm R}$ in the R sector must satisfy

$$\alpha_n^i|0; k\rangle_{\rm R} = d_n^i|0; k\rangle_{\rm R} = 0, \quad n > 0. \tag{5.146}$$

   This state must also satisfy the Dirac–Ramond equation $F_0|\phi\rangle = 0$. This state is obviously massless and it is actually a spinor. Indeed, the oscillators d_0^μ which satisfy $\{d_0^\mu, d_0^\nu\} = \eta^{\mu\nu}$ commute with $\alpha' M^2$, since they do not raise or lower the worldsheet energy, and thus they can be identified with the 10-dimensional Dirac matrices. This ground state is therefore an irreducible representation of the 10-dimensional Clifford algebra, i.e. it is an $SO(1,9)$ spinor.

   Hence, the ground state in the R sector is a 32 component spinor. By construction this is a real spinor, i.e. we are really dealing with a 32 component Majorana spinor. By the GSO projection, which we will discuss next, only one chirality is allowed. In other words, this is in fact a 16 component Weyl spinor. This spinor is still required to solve the Dirac–Ramond equation which reduces to a massless equation for the ground state since excited oscillators do not contribute. This means that only half of the components of this spinor are physical, i.e. this is an $SO(8)$ spinor.

In summary, only eight fermionic degrees of freedom survive in the ground state of the R sector which match nicely with the eight bosonic degrees of freedom in the ground state of the NS sector, reflecting the underlying spacetime supersymmetry of the theory. Indeed, this is the field content of super Yang–Mills gauge theory in ten dimensions.

GSO projection

The GSO (Gliozzi, Scherk and Olive) projection is a very mysterious truncation of the spectrum of the RNS open superstring designed in such a way as to eliminate the tachyon and establish a supersymmetric theory in 10-dimensional spacetime.

First, we need to introduce the so-called G-parity defined in the NS bosonic sector by

$$G = (-1)^{F_{\mathrm{NS}}+1}, \quad F_{\mathrm{NS}} = \sum_{r=1/2}^{\infty} b_{-r}^{i} b_{r}^{i}, \quad \mathrm{NS}. \tag{5.147}$$

Since F is the number of b fermions it counts the number of fermion excitations on the worldsheet. Thus, G in the NS sector is equal to $+1$ if the number of fermion excitations is odd, and it is equal to -1 if the number of fermion excitations is even.

We remark that the tachyon in the NS sector has a negative G-parity, i.e. $G|0, k\rangle_{\mathrm{NS}} = -|0, k\rangle_{\mathrm{NS}}$. In order to eliminate this state we will demand that only states with a positive G-parity in the NS sector are kept, viz

$$(-1)^{F_{\mathrm{NS}}} = -1. \tag{5.148}$$

In other words, only states with an odd number of b fermions in the NS sector are allowed. After projecting out the tachyon, the ground state in the NS sector becomes the massless gauge vector field.

In the R fermionic sector we will define the G-parity by

$$G = \Gamma_{11}(-1)^{F_{\mathrm{R}}}, \quad F_{\mathrm{R}} = \sum_{n=1}^{\infty} d_{-n}^{i} d_{n}^{i}, \quad \mathrm{R}. \tag{5.149}$$

The Γ_{11} is the gamma five in 10 dimensions, i.e. $\Gamma_{11} = \Gamma_0\Gamma_1 \ldots \Gamma_9$. A Weyl spinor ψ satisfies $P_{\pm}\psi = \pm\psi$ where $P_{\pm} = (1 \pm \Gamma_{11})/2$ is the chirality projection operator.

As we have discussed, the ground state in the R sector is a massless Majorana–Weyl spinor. In other words, it has either a positive or a negative chirality. Hence, we will project in the R sector on states with positive or negative G-parity depending on the chirality of the ground state.

It can be proved that the GSO projection leaves an equal number of bosons and fermions at each mass level and thus it leaves a supersymmetric spectrum as required by spacetime supersymmetry in 10 dimensions. For a thorough discussion of the GSO condition see [4].

Spectrum of the closed superstring
For the closed superstring we have left-movers and right-movers. The left-movers can be in the R sector or the NS sector and similarly for the right-movers. In other words, we have four sectors: R–R, NS–NS, R–NS and NS–R. In the NS sector the GSO condition means that we project on states with positive G-parity in order to eliminate the tachyon, whereas in the R sector we project on states with positive or negative G-parity depending on the chirality of the ground state. Thus, we can obtain two different theories depending on whether the left- and right-movers in the R sectors have the same or opposite G-parity.

These two theories are:

1. **Type IIA**: In this case the R sector of the left-movers and the R sector of the right-movers are distinct, i.e. they have opposite chiralities. Again, the ground state of the left handed-movers in the R sector is a Majorana–Weyl spinor with positive chirality $|0, k, +\rangle_{\rm R}$ while in the NS sector the ground state is the vector field $b^{i}_{-1/2}|0, k\rangle_{\rm NS}$. Then, the ground states of the right-handed-movers are given by $|0, k, -\rangle_{\rm R}$ and $b^{i}_{-1/2}|0, k\rangle_{\rm NS}$. The ground states of the closed RNS superstring in its four sectors are then given by

$$|0, k, +\rangle_{\rm R} \otimes |0, k, -\rangle_{\rm R}, \quad {\rm R - R} \tag{5.150}$$

$$b^{i}_{-1/2}|0, k\rangle_{\rm NS} \otimes b^{i}_{-1/2}|0, k\rangle_{\rm NS}, \quad {\rm NS - NS} \tag{5.151}$$

$$b^{i}_{-1/2}|0, k\rangle_{\rm NS} \otimes |0, k, -\rangle_{\rm R}, \quad {\rm NS - R} \tag{5.152}$$

$$|0, k, +\rangle_{\rm R} \otimes b^{i}_{-1/2}|0, k\rangle_{\rm NS}, \quad {\rm R - NS.} \tag{5.153}$$

We write this tensor product as

$$\begin{aligned}(8_v \oplus 8_{s_+}) \otimes (8_v \oplus 8_{s_-}) = (1 \oplus 28 \oplus 35_v \oplus 8_v \oplus 56_v)_B \\ \oplus (8_{s+} \oplus 8_{s-} \oplus 56_{s+} \oplus 56_{s-})_F.\end{aligned} \tag{5.154}$$

Explicitly, we have

$$\begin{aligned}{\rm NS - NS} &= (1 \oplus 28 \oplus 35_v)_B \\ &= \phi(\text{scalar dilaton}) \\ &\quad \oplus A_{\mu\nu}(\text{antisymmetric 2-form gauge field}) \\ &\quad \oplus g_{\mu\nu}(\text{symmetric traceless rank-two tensor graviton})\end{aligned} \tag{5.155}$$

$$\begin{aligned}{\rm R - R} &= (8_v \oplus 56_v)_B \\ &= A_\mu(\text{gauge field}) \\ &\quad \oplus A_{\mu\nu\lambda}(\text{antisymmetric 3-form gauge field})\end{aligned} \tag{5.156}$$

$$\mathrm{NS}-\mathrm{R}=(8_{s+}\oplus 56_{s+})_F \\ =\psi_+(\text{spin half dilatino})\oplus\chi_+(\text{spin three half gravitino}) \tag{5.157}$$

$$\mathrm{R}-\mathrm{NS}=(8_{s-}\oplus 56_{s-})_F \\ =\psi_-(\text{spin half dilatino})\oplus\chi_-(\text{spin three half gravitino}). \tag{5.158}$$

This is the particle content of type IIA supergravity in 10 dimensions. And it is the particle content obtained from the dimensional reduction of 11-dimensional supergravity. Since we have two dilations and two gravitons we have $\mathcal{N}=2$ supersymmetry.

2. **Type IIB**: In this case the R sector of the left-movers and the R sector of the right-movers have the same chirality taken + for concreteness. The ground state of the left-handed-movers in the R sector is a Majorana–Weyl spinor with positive chirality $|0, k, +\rangle_{\mathrm{R}}$ while in the NS sector the ground state is the vector field $b^i_{-1/2}|0, k\rangle_{\mathrm{NS}}$. Similarly, the ground states of the right-handed-movers are $|0, k, +\rangle_{\mathrm{R}}$ and $b^i_{-1/2}|0, k\rangle_{\mathrm{NS}}$. The ground states of the closed RNS superstring in its four sectors are then given by

$$|0, k, +\rangle_{\mathrm{R}}\otimes|0, k, +\rangle_{\mathrm{R}},\quad \mathrm{R}-\mathrm{R} \tag{5.159}$$

$$b^i_{-1/2}|0, k\rangle_{\mathrm{NS}}\otimes b^i_{-1/2}|0, k\rangle_{\mathrm{NS}},\quad \mathrm{NS}-\mathrm{NS} \tag{5.160}$$

$$b^i_{-1/2}|0, k\rangle_{\mathrm{NS}}\otimes|0, k, +\rangle_{\mathrm{R}},\quad \mathrm{NS}-\mathrm{R} \tag{5.161}$$

$$|0, k, +\rangle_{\mathrm{R}}\otimes b^i_{-1/2}|0, k\rangle_{\mathrm{NS}},\quad \mathrm{R}-\mathrm{NS}. \tag{5.162}$$

We write this tensor product as

$$(8_v\oplus 8_{s_+})\otimes(8_v\oplus 8_{s_+})=(1\oplus 28\oplus 35_v\oplus 1\oplus 28\oplus 35_+)_B \\ \oplus(8_{s+}\oplus 8_{s+}\oplus 56_{s+}\oplus 56_{s+})_F. \tag{5.163}$$

Explicitly, we have

$$\mathrm{NS}-\mathrm{NS}=(1\oplus 28\oplus 35_v)_B \\ =\phi(\text{scalar dilaton})\oplus A_{\mu\nu}(\text{antisymmetric 2} \\ -\text{ form gauge field}) \\ \oplus g_{\mu\nu}(\text{symmetric traceless rank-two tensor graviton}) \tag{5.164}$$

$$\mathrm{R}-\mathrm{R}=(1\oplus 28\oplus 35_+)_B \\ =\phi(\text{scalar field})\oplus A_{\mu\nu}(\text{antisymmetric 2-form gauge field}) \\ \oplus A_{\mu\nu\alpha\lambda}(\text{antisymmetric 4-form gauge field with} \\ \text{self-dual field strength}) \tag{5.165}$$

$$\begin{aligned}\mathrm{NS}-\mathrm{R} &= (8_{s+}\oplus 56_{s+})_F\\ &= \psi_+(\text{spin half dilatino})\oplus\chi_+(\text{spin three half gravitino})\end{aligned} \tag{5.166}$$

$$\begin{aligned}\mathrm{R}-\mathrm{NS} &= (8_{s+}\oplus 56_{s+})_F\\ &= \psi_-(\text{spin half dilatino})\oplus\chi_-(\text{spin three half gravitino}).\end{aligned} \tag{5.167}$$

This is the particle content of type IIB supergravity in 10 dimensions which cannot be obtained by dimensional reduction of 11-dimensional supergravity.

5.6 Other very important topics

The two most important topics which remains in this part are:

- Heterotic superstrings.
- Green–Schwarz mechanism for anomaly cancellation.

5.7 Exercises

Exercise 1: Show that the threee form Ω_3 defined by equation (5.9) is closed, viz $d\Omega_3 = 0$.

Exercise 2: Show the Fierz identity (5.12).

Exercise 3: Show that γ defined by (5.19) is an idempotent.

Exercise 4: Compute the variation of the action $S_1 + S_2$ under κ-symmetry.

Solution 4:

The variation of the total action $S_1 + S_2$ is given by

$$\begin{aligned}\delta S &= 2T\int d^2\sigma\delta\bar{\Theta}^1\Gamma_\mu\partial_\alpha\Theta^1\left(\frac{c}{2T}\epsilon^{\alpha\beta}+\sqrt{-\det G}\,G^{\alpha\beta}\right)\Pi^\mu_\beta\\ &-2T\int d^2\sigma\delta\bar{\Theta}^2\Gamma_\mu\partial_\alpha\Theta^2\left(\frac{c}{2T}\epsilon^{\alpha\beta}-\sqrt{-\det G}\,G^{\alpha\beta}\right)\Pi^\mu_\beta.\end{aligned} \tag{5.168}$$

The Levi-Civita density in flat two dimensions satisfies the identity

$$\epsilon_{\alpha\beta}\epsilon^{\rho\sigma} = \delta^\rho_\alpha\delta^\sigma_\beta - \delta^\sigma_\alpha\delta^\rho_\beta. \tag{5.169}$$

In curved two dimensions the corresponding covariant tensor is given by

$$E^{\alpha\beta} = -\frac{\epsilon^{\alpha\beta}}{\sqrt{-\det G}}, \tag{5.170}$$

where

$$E_{\alpha\beta}E^{\rho\sigma} = G_\alpha^\rho G_\beta^\sigma - G_\alpha^\sigma G_\beta^\rho. \tag{5.171}$$

Hence

$$\epsilon_{\alpha\beta}\epsilon^{\rho\sigma} = -\det G(G_\alpha^\rho G_\beta^\sigma - G_\alpha^\sigma G_\beta^\rho). \tag{5.172}$$

From this we show

$$\epsilon^{\alpha\beta}\epsilon^{\rho\sigma} = \epsilon^{\alpha\rho}\epsilon^{\beta\sigma} - \epsilon^{\alpha\sigma}\epsilon^{\beta\rho} \tag{5.173}$$

$$\sqrt{-\det G}\,G^{\alpha\beta} = -\frac{1}{\sqrt{-\det G}}\epsilon^{\alpha\sigma}G_{\sigma\rho}\epsilon^{\rho\beta}. \tag{5.174}$$

Then

$$\begin{aligned}\sqrt{-\det G}\,G^{\alpha\beta}\Gamma_\mu\Pi^\mu_\beta &= -\frac{1}{2\sqrt{-\det G}}\epsilon^{\alpha\sigma}\epsilon^{\rho\beta}\Gamma_\nu\Gamma_\lambda\Gamma_\mu\Pi^\nu_\sigma\Pi^\lambda_\rho\Pi^\mu_\beta \\ &\quad -\frac{1}{2\sqrt{-\det G}}\epsilon^{\alpha\sigma}\epsilon^{\rho\beta}\Gamma_\lambda\Gamma_\nu\Gamma_\mu\Pi^\nu_\sigma\Pi^\lambda_\rho\Pi^\mu_\beta \\ &= \epsilon^{\alpha\beta}\Gamma_\mu\Pi^\mu_\beta\gamma \\ &\quad -\frac{1}{2\sqrt{-\det G}}(\epsilon^{\alpha\rho}\epsilon^{\sigma\beta} - \epsilon^{\alpha\beta}\epsilon^{\sigma\rho})\Gamma_\lambda\Gamma_\nu\Gamma_\mu\Pi^\nu_\sigma\Pi^\lambda_\rho\Pi^\mu_\beta \\ &= \epsilon^{\alpha\beta}\Gamma_\mu\Pi^\mu_\beta\gamma \\ &\quad + \epsilon^{\alpha\beta}\Gamma_\mu\Pi^\mu_\beta\gamma - \gamma\epsilon^{\alpha\beta}\Gamma_\mu\Pi^\mu_\beta.\end{aligned} \tag{5.175}$$

The matrices $\Gamma_{\mu\nu}$ and γ are defined by

$$\Gamma_{\mu\nu} = \frac{1}{2}[\Gamma_\mu, \Gamma_\nu] \tag{5.176}$$

$$\gamma = -\frac{1}{2\sqrt{-\det G}}\epsilon^{\alpha\beta}\Gamma_{\mu\nu}\Pi^\mu_\alpha\Pi^\nu_\beta = -\frac{2}{\sqrt{-\det G}}\epsilon^{\alpha\beta}V_\alpha V_\beta. \tag{5.177}$$

The gamma matrices V_α are defined by $V_\alpha = \Gamma_\mu\Pi^\mu_\alpha/2$ and they satisfy $G_{\rho\sigma}V^\rho V^\sigma = G^{\rho\sigma}G_{\rho\sigma}/4$. More importantly, we can verify that γ anticommutes with $\epsilon^{\rho\sigma}V_\sigma$ since

$$\begin{aligned}\{\gamma, \epsilon^{\rho\sigma}V_\sigma\} &= 2\sqrt{-\det G}[G_{\alpha\beta}V^\alpha V^\beta, V^\rho] \\ &= 0.\end{aligned} \tag{5.178}$$

Hence

$$\sqrt{-\det G}\,G^{\alpha\beta}\Gamma_\mu\Pi^\mu_\beta = -3\gamma\epsilon^{\alpha\beta}\Gamma_\mu\Pi^\mu_\beta. \tag{5.179}$$

Exercise 5: Derive equation (5.22).

Exercise 6: Derive the light-cone gauge GS action and show that it has the RNS form with the fermions being in the spinor representation of spin(8).

Exercise 7: Show that the commutator of two supersymmetry transformations (5.32) gives a translation. Use the Dirac equation.

Solution 7: We compute

$$\begin{aligned}[\delta_1, \delta_2]X^\mu &= \bar{\epsilon}_2\delta_1\psi^\mu - 1 \leftrightarrow 2 \\ &= 2i\bar{\epsilon}_1\rho^\alpha\epsilon_2\partial_\alpha X^\mu.\end{aligned} \quad (5.180)$$

In the above we have also used the identity $\bar{\epsilon}_1\rho^\alpha\epsilon_2 = -\bar{\epsilon}_2\rho^\alpha\epsilon_1$. Similarly we compute

$$[\delta_1, \delta_2]\psi_A^\mu = i\rho^\alpha_{AB}\partial_\alpha\psi_C^\mu(\bar{\epsilon}_{1C}\epsilon_{2B} - \bar{\epsilon}_{2C}\epsilon_{1B}). \quad (5.181)$$

We have

$$\psi = \begin{pmatrix}\psi_- \\ \psi_+\end{pmatrix}. \quad (5.182)$$

If we choose $A = +$ then $B = -$ and thus

$$[\delta_1, \delta_2]\psi_+^\mu = i\rho^\alpha_{+-}\partial_\alpha\psi_C^\mu(\bar{\epsilon}_{1C}\epsilon_{2-} - \bar{\epsilon}_{2C}\epsilon_{1-}). \quad (5.183)$$

By using the equation of motion $\rho^\alpha_{+-}\partial_\alpha\psi_-^\mu = 0$ we obtain

$$\begin{aligned}[\delta_1, \delta_2]\psi_+^\mu &= i\rho^\alpha_{+-}\partial_\alpha\psi_+^\mu(\bar{\epsilon}_{1+}\epsilon_{2-} - \bar{\epsilon}_{2+}\epsilon_{1-}) \\ &= 2i\rho^\alpha_{+-}\partial_\alpha\psi_+^\mu\bar{\epsilon}_{1+}\epsilon_{2-}.\end{aligned} \quad (5.184)$$

Now using the equation of motion $\rho^\alpha_{-+}\partial_\alpha\psi_+^\mu = 0$ we get

$$0 = 2i\rho^\alpha_{-+}\partial_\alpha\psi_+^\mu\bar{\epsilon}_{1-}\epsilon_{2+}. \quad (5.185)$$

Hence we must have

$$[\delta_1, \delta_2]\psi_+^\mu = 2i\bar{\epsilon}_1\rho^\alpha\epsilon_2\partial_\alpha\psi_+^\mu. \quad (5.186)$$

Exercise 8: Show that the supergravity action (5.71) is indeed invariant under the supergravity transformations (5.72), under the Weyl transformation (5.73), and under the fermionic local symmetry (5.74).

Exercise 9: Verify the anticommutation relations (5.100), (5.101) and (5.102).

Exercise 10:

- Derive the super-Virasoro algebras (5.127) and (5.132).
- Show that $(L_0 - a)|\phi\rangle = 0$ yields $\alpha' M^2 = N - a$.
- Show that $a_{\rm R} = 0$ and $a_{\rm NS} = 1/2$.
- Show that the critical dimension is $D = 10$ in the RNS open superstring by looking for extra zero-norm states for special values of a and D which demarcate the boundary between the region corresponding to the range of D and a where there are ghosts, i.e. negative-norm states, and the region corresponding to the range of D and a where there are no ghosts.

Exercise 11:

- Show that the Lorentz generators for the RNS open superstring in the NS sector takes the form (5.135) with $l^{\mu\nu}$, $E^{\mu\nu}$ and $K^{\mu\nu}$ given by (5.136), (5.137) and (5.138) respectively.
- Show that the commutators $[J^{i-}, J^{j-}]$ are given by (5.140).

References

[1] Becker K, Becker M and Schwarz J H 2006 *String Theory and M-theory: A Modern Introduction* (Cambridge: Cambridge University Press)
[2] Polchinski J 2005 *String Theory. Vol. 1: An Introduction to the Bosonic String* (Cambridge: Cambridge University Press)
[3] Polchinski J 2005 *String Theory Vol. 2: Superstring Theory and Beyond* (Cambridge: Cambridge University Press)
[4] Green M B, Schwarz J H and Witten E 1987 *Superstring Theory. Vol. 1: Introduction* (Cambridge Monographs On Mathematical Physics) (Cambridge: Cambridge University Press)
[5] Green M B, Schwarz J H and Witten E 1987 *Superstring Theory. Vol. 2: Loop Amplitudes, Anomalies And Phenomenology* (Cambridge Monographs On Mathematical Physics) (Cambridge: Cambridge University Press)
[6] Johnson C V 2003 *D-branes* (Cambridge: Cambridge University Press)
[7] Zwiebach B 2009 *A First Course in String Theory* (Cambridge: Cambridge University Press)
[8] Szabo R J 2004 *An Introduction to String Theory and D-Brane Dynamics* (London: Imperial College Press)

Part II

Matrix string theory

IOP Publishing

Matrix Models of String Theory

Badis Ydri

Chapter 6

A lightning introduction to superstring theory and some related topics

In this chapter we will start by reviewing very briefly quantum black holes and the information loss problem following the lectures [1], then we will introduce more results from string theory, conformal field theory, Dp-branes and T-duality following mostly [2], and finally we will discuss, again very briefly, aspects of quantum gravity in two dimensions and dynamical triangulations following the standard reviews [3, 4].

6.1 Quantum black holes

String theory provides one of the deepest insights into quantum gravity. Its single most central and profound result is the AdS/CFT correspondence or gauge/gravity duality [16]. See [17, 18] for a pedagogical introduction. As it turns out, this duality allows us to study in novel ways: (i) the physics of strongly coupled gauge theory (QCD in particular and the existence of Yang–Mills theories in four dimensions), as well as (ii) the physics of black holes (the information loss paradox and the problem of the reconciliation of general relativity and quantum mechanics). String theory reduces therefore for us to the study of the gauge/gravity duality and its most important example the AdS/CFT correspondence.

Indeed, the fundamental observation which drives the lectures in this chapter and [1] is that: 'BFSS matrix model [21] and the AdS/CFT duality [16, 19, 20] relates string theory in certain backgrounds to quantum mechanical systems and quantum field theories' which is a quotation taken from Polchinski [22]. The basic problem, which is of paramount interest to quantum gravity, is Hawking radiation of a black hole and the consequent evaporation of the hole and corresponding information loss [23, 24]. The BFSS and the AdS/CFT imply that there is no information loss paradox in the Hawking radiation of a black hole. This is the central question we would like to understand in great detail.

doi:10.1088/978-0-7503-1726-9ch6

Towards this end, we need to understand first quantum black holes, before we can even touch on the gauge/gravity duality and the AdS/CFT correspondence, which require in any case a great deal of conformal field theory and string theory as crucial ingredients. Thus, in this section we will only worry about black hole radiation, black hole thermodynamics and the information problem following [22, 25–30]. This section is a sort of a summary of the very detailed presentation [1] which contains an extensive list of references.

6.1.1 Schwarzschild black hole

We start by presenting the star of the show, the so-called Schwarzschild eternal black hole given by the metric

$$ds^2 = -\left(1 - \frac{2GM}{r}\right)dt^2 + \left(1 - \frac{2GM}{r}\right)^{-1} dr^2 + r^2 d\Omega^2. \tag{6.1}$$

Before we embark on the calculation of Hawking radiation it is very helpful to understand the physical origin behind this radiation in the most simple of terms.

The motion of a scalar particle of energy ν and angular momentum l in the background gravitational field of the Schwarzschild black hole is exactly equivalent to the motion of a quantum particle, i.e. a particle obeying the Schrodinger equation, with energy $E = \nu^2$ in a scattering potential given in the tortoise coordinate r^* with the expression

$$V(r_*) = \frac{r - r_s}{r}\left(\frac{r_s}{r^3} + \frac{l(l+1)}{r^2}\right), \quad r_* = r + r_s \ln\left(\frac{r}{r_s} - 1\right). \tag{6.2}$$

This potential vanishes at infinity and at the event horizon r_s and thus the particle is free at infinity and at the event horizon. See figure 6.1. This potential is characterized by a barrier at $r \sim 3r_s/2$ where the potential reaches its maximum and the height of this barrier is proportional to the square of the angular momentum, viz

$$V_{\max}(r_*) \sim \frac{l^2 + 1}{G^2M^2} \sim (l^2 + 1)T_{\rm H}, \tag{6.3}$$

where $T_{\rm H} = 1/(8\pi GM)$ is the Hawking temperature which we will compute shortly.

On the other hand, these particles are in thermal equilibrium at the Hawking temperature and thus the energy ν is proportional to $T_{\rm H}$. Thus, we can immediately see from this simple argument that only particles with no angular momentum, i.e. $l = 0$, can go through the potential barrier and escape from the black hole to infinity. These particles are precisely Hawking particles. The difference with the case of Rindler spacetime lies in the fact that in Rindler spacetime the potential barrier is infinite and thus no particles can go through and escape from the black hole to infinity. This is a very strong but simple physical description of Hawking radiation.

6.1.2 Hawking temperature

A systematic derivation of the Hawking radiation is given in three different ways.

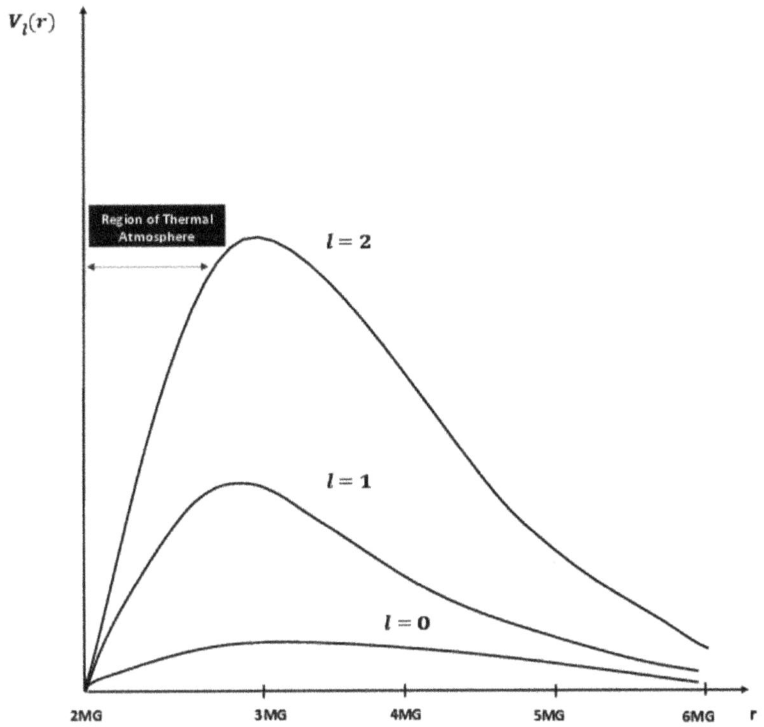

Figure 6.1. Schwarzschild potential.

The first is by employing the fact that the near-horizon geometry of Schwarzschild black hole is Rindler spacetime and then applying the Unruh effect in Rindler spacetime. Recall that Rindler spacetime is a uniformly accelerating observer with acceleration a related to the event horizon r_s by the relation

$$a = \frac{1}{2r_s}. \tag{6.4}$$

The Minkowski vacuum state is seen by the Rindler observer as a mixed thermal state at the temperature $T = a/2\pi$, which is precisely the Unruh effect. This is in one sense what lies at the basis of Hawking radiation.

The second way is by considering the eternal black hole geometry and studying the properties of the Kruskal vacuum state $|0_K\rangle$ with respect to the Schwarzschild observer. The Kruskal state for the Schwarzschild observer plays exactly the role of the Minkowski state for the Rindler observer the Schwarzschild vacuum plays the role of the Rindler vacuum. The Schwarzschild observer sees the Kruskal vacuum as a heat bath containing

$$\langle 0_k | N_\omega | 0_K \rangle = \frac{\delta(0)}{\exp(2\pi\omega/a) - 1} \tag{6.5}$$

particles. We can infer immediately from this result the correct value $T = a/2\pi$ of the Hawking temperature.

Thirdly, we computed the Hawking temperature by considering the more realistic situation in which a Schwarzschild black hole is formed by gravitational collapse of a thin mass shell as in the Penrose diagram shown in figure 6.2 (the thin shell is the red line). Then by deriving the actual incoming state known as the Unruh vacuum state. The Unruh state is a maximally entangled state describing a pair of particles with zero Killing energy. One of the pair $|n_{\rm R}\rangle$ goes outside the horizon and is seen as Hawking radiation, whereas the other pair $|n_L\rangle$ falls behind the horizon and goes into the singularity at the center and thus it corresponds to the information lost inside the black hole. This quantum state is given by the relation

$$|U\rangle \sim \sum_n \exp\left(-\frac{n\pi\omega}{a}\right)|n_{\rm R}\rangle|n_L\rangle. \tag{6.6}$$

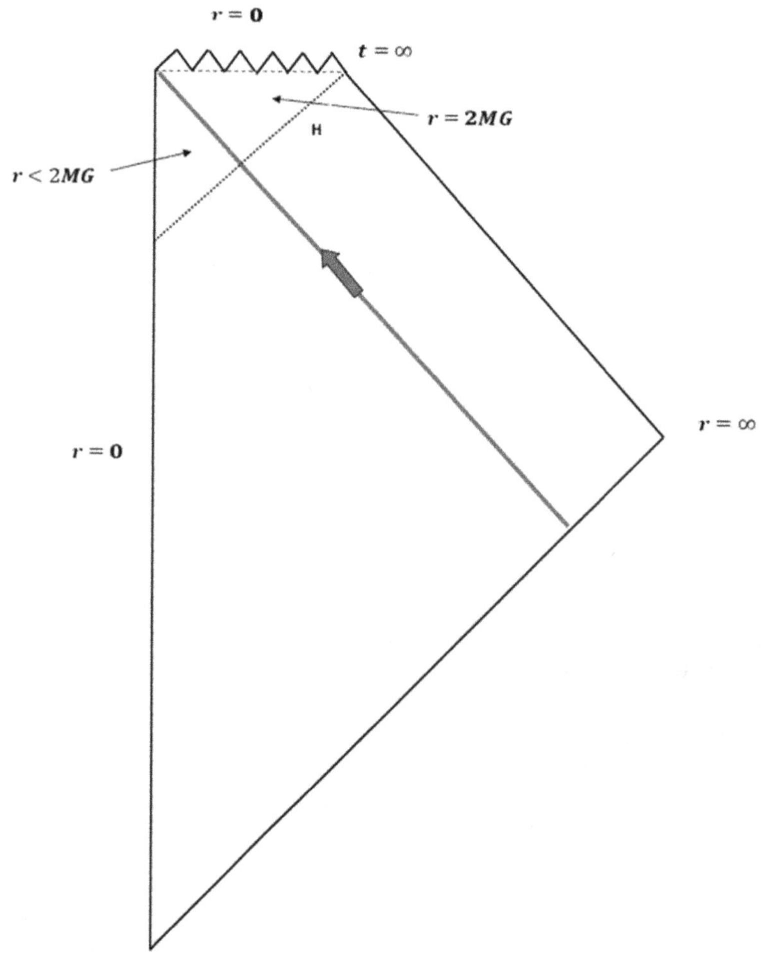

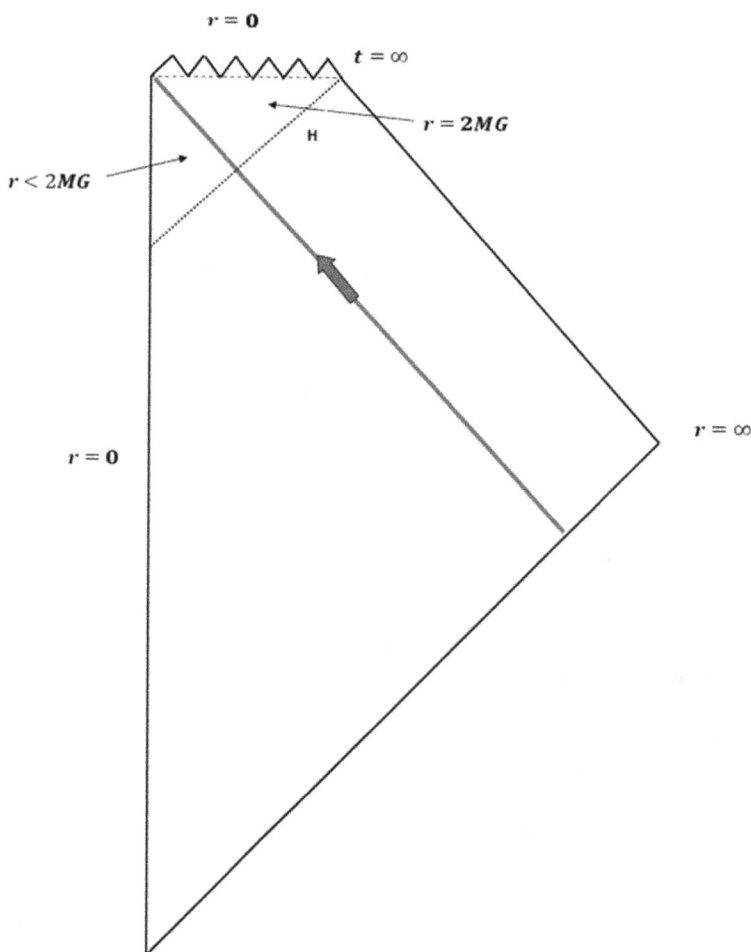

Figure 6.2. Penrose diagram of the formation of a black hole from gravitational collapse.

Although, the actual quantum state of the black hole is pure, the asymptotic Schwarzschild observer registers a thermal mixed state given by the density matrix

$$\rho_{\rm R} = {\rm Tr}_L|U\rangle\langle U| \sim \sum_n \exp\left(-\frac{2n\pi\omega}{a}\right)|n_{\rm R}\rangle\langle n_{\rm R}|. \tag{6.7}$$

Thus, the Schwarzschild observer registers a canonical ensemble with temperature

$$T = \frac{1}{8\pi GM}. \tag{6.8}$$

In summary, a correlated entangled pure state near the horizon gives rise to a thermal mixed state outside the horizon.

6.1.3 Page curve and unitarity

The information loss problem can then be summarized as follows (see [1] for an extensive discussion). The black hole starts in a pure state and after its complete evaporation the Hawking radiation is also in a pure state. This is the assumption of unitarity. The information is given by the difference between the thermal entropy of Boltzmann and the information entropy of Von Neumann, whereas the entanglement entropy is the Von Neumann entropy in the case where the system is described by a pure state. The entanglement entropy starts at zero value then it reaches a maximum value at the so-called Page time (also maybe called information retention time) then drops to zero again. The Page time is the time at which the black hole evaporates around one half of its mass and the information starts to get out with the radiation. Before the Page time only energy gets out with the radiation with little or no information, while only at the Page time does the information start to get out, and it gets out completely at the moment of evaporation (see figure 6.3). This is guaranteed to happen because of the second principle of thermodynamics and the assumption of unitarity. The behavior of the entanglement entropy with time is called the Page curve and a nice rough derivation of this curve can be outlined using the so-called Page theorem. The computation of the Page curve starting from first principles will provide, in some precise sense, the mathematical solution of the black hole information loss problem.

6.1.4 Information loss problem

Since this is a vital issue we state the information loss in different terms

We consider again a black hole formed by gravitational collapse as given by the above Penrose diagram. The Hilbert space $\mathbf{H}_{\rm in}$ of initial states $|\psi_{\rm in}\rangle$ is associated with null rays incoming from $\mathcal{J}^-$ at $r = \infty$, i.e. $\mathbf{H}_{\rm in} = \mathbf{H}_-$. The Hilbert space $\mathbf{H}_{\rm out}$ of final states $|\psi_{\rm out}\rangle$ is clearly a tensor product of the Hilbert space $\mathbf{H}_+$ of the scattered outgoing radiation which escapes to the infinity $\mathcal{J}^+$ and the Hilbert space $\mathbf{H}_S$ of the transmitted radiation which falls behind the horizon into the singularity. This is the assumption of locality. Indeed, the outgoing Hawking particle and the lost quantum

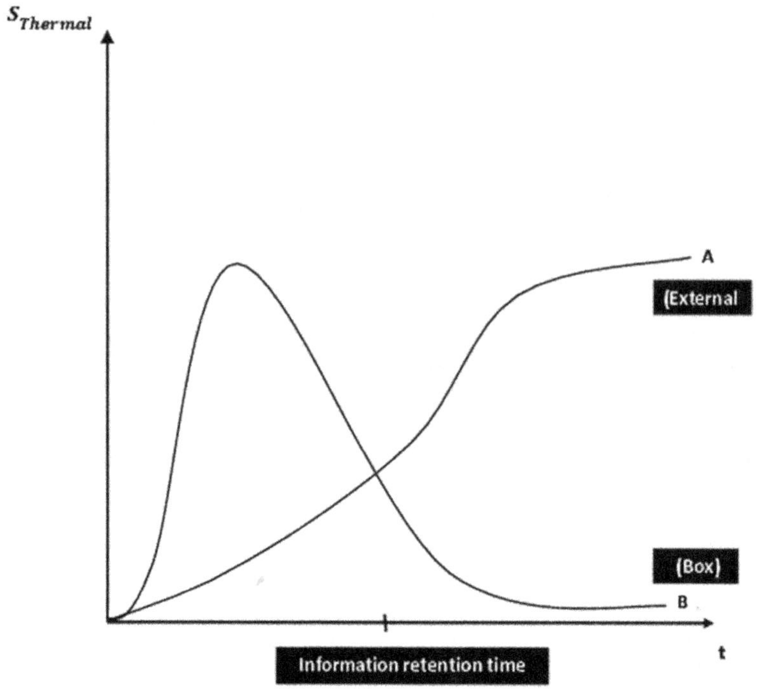

Figure 6.3. Page time, the entanglement entropy and information.

behind the horizon are maximally entangled, and thus they are space-like separated, and as a consequence localized operators on $\mathcal{J}^+$ and S must commute. We have then

$$H_{\text{in}} = H_-, \quad H_{\text{out}} = H_+ \otimes H_S. \tag{6.9}$$

From the perspective of observables at $\mathcal{J}^+$ (us), the outgoing Hawking particles can only be described by a reduced density matrix, even though the final state $|\psi_{\text{out}}\rangle$ is obtained from the initial state $|\psi_{\text{in}}\rangle$ by the action of a unitary S-matrix. This is the assumption of unitarity. This reduced density matrix is completely mixed despite the fact that the final state is a maximally entangled pure state. Eventually, the black hole will evaporate completely and it seems that we will end up only with the mixed state of the radiation. This the information paradox. There are six possibilities here:

1. Information is really lost which is Hawking's original stand.
2. Evaporation stops at a Planck-mass remnant which contains all the information with extremely large entropy.
3. Information is recovered only at the end of the evaporation when the singularity at $r = 0$ becomes a naked singularity. This contradicts the principle of information conservation with respect to the observation at $\mathcal{J}^+$ which states that by the time (Page or retention time) the black hole evaporates around one half of its mass the information must start coming out with the Hawking radiation.
4. Information is not lost during the entire process of formation and evaporation. This is the assumption of unitarity. But how?

5. Horizon is like a brick wall which cannot be penetrated. This contradicts the equivalence principle in an obvious way.
6. Horizon duplicates the information by sending one copy outside the horizon (as required by the principle of information conservation) while sending the other copy inside the horizon (as required by the equivalence principle). This is however forbidden by the linearity of quantum mechanics or the so-called quantum xerox principle.

6.1.5 Thermodynamics

The last point of primary importance concerns black hole thermodynamics. The thermal entropy is the maximum amount of information contained in the black hole. The entropy is mostly localized near the horizon, but quantum field theory (QFT) gives a divergent value, instead of the Bekenstein–Hawking value

$$S = A/4G, \tag{6.10}$$

where A is the surface area of the black hole. The number of accessible quantum microscopic states is determined by this entropy via the formula

$$n = \exp(S). \tag{6.11}$$

Since QFT gives a divergent entropy instead of the Bekenstein–Hawking value it must be replaced by quantum gravity (QG) near the horizon and this separation of the QFT and QG degrees of freedom can be implemented by the stretched horizon which is a timelike membrane, at a distance of one Planck length $l_P = \sqrt{G\hbar}$ from the actual horizon, and where the proper temperature gets very large and most of the black hole entropy accumulates.

6.2 Some string theory and conformal field theory

The standard text remains the classic book by Green, Schwarz and Witten [3, 4]. Of course, also a classic is the book by Polchinski [5, 6]. The modern text [2] is truly modern and as such it turns out to be extremely useful. Also the books [7–9] were used extensively. We also found the lectures [10, 11] on conformal field theory very illuminating and the seminal papers [12, 13] on the Virasoro algebra very helpful.

6.2.1 The conformal anomaly

The starting point is the path integral

$$\begin{aligned} Z &= \int Dh(\sigma)DX(\sigma)e^{iS[h,\,X]}, \\ S[h, X] &= -\frac{1}{2\pi}\int_\Sigma d^2\sigma\sqrt{h}\; h^{ab}\partial_a X^\mu \partial_b X_\mu. \end{aligned} \tag{6.12}$$

We use the two reparametrization invariances and the Weyl symmetry to impose the conformal gauge $h_{ab} = e^\phi \eta_{ab}$. The action $S[X, h]$ becomes

$$S[X] = -\frac{1}{2\pi}\int_\Sigma d^2\sigma \partial_a X^\mu \partial^a X_\mu. \tag{6.13}$$

The Fadeev–Popov gauge fixing procedure leads then to the ghost action

$$S[b, c] = \frac{i}{\pi}\int d^2\sigma\ c^- \nabla_+ b_{--} + \frac{i}{\pi}\int d^2\sigma\ c^+ \nabla_- b_{++}. \tag{6.14}$$

The ghost fields c^+ and $c^-$are the components of a vector ghost field c^a, whereas the antighosts b_{++} and b_{--} are the components of a traceless symmetric antighost tensor field b_{ab}. The total gauge-fixed action $S[X, b, c]$ is given by

$$\begin{aligned} S[X, b, c] = &-\frac{1}{2\pi}\int_\Sigma d^2\sigma \partial_a X^\mu \partial^a X_\mu + \frac{i}{\pi}\int d^2\sigma\ c^- \nabla_+ b_{--} \\ &+ \frac{i}{\pi}\int d^2\sigma\ c^+ \nabla_- b_{++}. \end{aligned} \tag{6.15}$$

We recall the non-zero components of the energy–momentum tensor given by

$$\begin{aligned} T_{++} &= \partial_+ X^\mu \partial_+ X_\mu \\ T_{--} &= \partial_- X^\mu \partial_- X_\mu. \end{aligned} \tag{6.16}$$

The contribution of the ghosts to the worldsheet energy–momentum tensor is given by

$$T^g_{++} = -i\nabla_+ c^+ b_{++} - \frac{i}{2} c^+ \nabla_+ b_{++} \tag{6.17}$$

$$T^g_{--} = -i\nabla_- c^- b_{--} - \frac{i}{2} c^- \nabla_- b_{--}. \tag{6.18}$$

The Virasoro generators L^g_m are the Fourier modes of T^g_{++} in the same way that L_m are the Fourier modes of T_{++}. They satisfy quantum mechanically the algebra

$$[L^g_m, L^g_n] = (m-n)L^g_{m+n} + \left(-\frac{13}{6}m^3 + \frac{m}{6}\right)\delta_{m+n,\,0}. \tag{6.19}$$

We recall the Virasoro algebra

$$[L_m, L_n] = (m-n)L_{m+n} + \frac{D}{12}(m^3 - m)\delta_{m+n,\,0}. \tag{6.20}$$

The total Virasoro generators are defined by

$$L^{\mathrm{TOT}}_m = L_m + L^g_m - a\delta_{m,\,0}. \tag{6.21}$$

The constraints read now

$$L^{\mathrm{TOT}}_m = 0. \tag{6.22}$$

The Virasoro algebra reads now

$$[L_m^{\mathrm{TOT}}, L_n^{\mathrm{TOT}}] = (m-n)L_{m+n}^{\mathrm{TOT}} + \left(2am + \frac{D}{12}(m^3 - m) - \frac{13}{6}m^3 + \frac{m}{6}\right)\delta_{m+n,\,0}$$
$$= (m-n)L_{m+n}^{\mathrm{TOT}} + \left(\frac{D-26}{12}m^3 + \frac{2+24a-D}{12}m\right)\delta_{m+n,\,0}. \tag{6.23}$$

The conformal anomaly vanishes iff

$$D = 26,\ a = 1. \tag{6.24}$$

This means, in particular, that only for these values is the theory truly conformally invariant.

6.2.2 The operator product expansion

We go now to Euclidean signature, i.e. $\sigma^2 = i\sigma^0 = i\tau$ and $\sigma^1 = \sigma$, and we define the complex coordinates $z = \sigma^1 + i\sigma^2$, $\bar{z} = \sigma^1 - i\sigma^2$. The action $S[X]$ becomes

$$S = \frac{1}{\pi}\int d^2z\,\partial X^\mu \bar{\partial} X_\mu. \tag{6.25}$$

The right-moving solution X_{R}^μ becomes a holomorphic function, i.e. an analytic function of z, while the left-moving solution X_L^μ becomes an antiholomorphic function, i.e. an analytic function of $\bar{z}$. The residual symmetries of this Euclidean action are given by the conformal mappings

$$z \longrightarrow f(z), \quad \bar{z} \longrightarrow \bar{z} = \bar{f}(\bar{z}). \tag{6.26}$$

These are angle-preserving transformations when f and its inverse are both holomorphic. For example, $z \longrightarrow z + a$ is a translation, $z \longrightarrow \zeta z$ where $|\zeta| = 1$ is a rotation, and $z \longrightarrow \zeta z$ where ζ is real not equal to one is a scale transformation called also dilatation.

We will work with the complex coordinates

$$w = e^{-2iz} = e^{2(\sigma^2 - i\sigma^1)}, \quad \bar{w} = e^{-2i\bar{z}} = e^{2(\sigma^2 + i\sigma^1)}. \tag{6.27}$$

The worldsheet is now regarded as a Riemann surface. The Euclidean time σ^2 corresponds to the radial distance $r = \exp(2\sigma^2)$ on the complex plane, with the infinite past $\sigma^2 = -\infty$ at $r = 0$, and the infinite future $\sigma^2 = +\infty$ is a circle at $r = \infty$. Thus, the time ordered product of operators on the cylinder becomes the radially ordered product of operators on the complex plane. We will rewrite in the following for simplicity

$$z = e^{2(\sigma^2 - i\sigma^1)}, \quad \bar{z} = e^{2(\sigma^2 + i\sigma^1)}. \tag{6.28}$$

We compute then

$$T_{\bar{z}\bar{z}} = -\,2\bar{\partial} X^\mu \bar{\partial} X^\mu = \sum_m \frac{\tilde{L}_m}{\bar{z}^{m+2}},$$
$$T_{zz} = -\,2\partial X^\mu \partial X^\mu = \sum_m \frac{L_m}{z^{m+2}}. \tag{6.29}$$

The classical stress–energy tensor $T(z) = T_{zz} = -2\partial X^\mu(z)\partial X^\mu(z)$ is defined quantum mechanically by the normal ordered expression

$$\begin{aligned} -2T(z) &= 4\colon \partial X^\mu(z)\partial X^\mu(z)\colon \\ &= \lim_{z\longrightarrow w}\left(4R(\partial X^\mu(z)\partial X^\mu(w)) + \frac{\delta^{\mu\mu}}{(z-w)^2}\right). \end{aligned} \tag{6.30}$$

The radially ordered product is related to the normal ordered product by the relation

$$\begin{aligned} 2^n R(X^\mu(z_1, \bar{z}_1) \cdots X^\nu(z_n, \bar{z}_n)) &= 2^n\colon X^\mu(z_1, \bar{z}_1) \cdots X^\nu(z_n, \bar{z}_n)\colon \\ &+ \sum \text{contractions}. \end{aligned} \tag{6.31}$$

The sum runs over all ways of choosing one pair of fields (or two or more pairs in the case where we have an arbitrary product of fields) from the product and replacing each term with the contraction

$$\begin{aligned} 4R(X^\mu(z, \bar{z})X^\nu(w, \bar{w})) &= 4\colon X^\mu(z, \bar{z})X^\nu(w, \bar{w})\colon \\ &- \delta^{\mu\nu}(\ln(z-w) + \ln(\bar{z}-\bar{w})). \end{aligned} \tag{6.32}$$

A primary field is a conformal field of conformal dimension $(h, \bar{h})$. In other words, it is a tensor field of rank $n = h + \bar{h}$ under conformal transformations with components $\Phi_{z\ldots z,\, \bar{z}\ldots\bar{z}}(z, \bar{z})$ transforming under the conformal transformations $z \longrightarrow w(z)$ and $\bar{z} \longrightarrow \bar{w}(\bar{z})$ as

$$\Phi(z, \bar{z}) \longrightarrow \left(\frac{\partial w}{\partial z}\right)^h \left(\frac{\partial \bar{w}}{\partial \bar{z}}\right)^{\bar{h}} \Phi(w, \bar{w}). \tag{6.33}$$

The rank $h + \bar{h}$ is called the dimension of Φ and it determines its behavior under scalings while $h - \bar{h}$ is the spin of Φ and it determines its behavior under rotations.

Since T_{zz} is holomorphic and $T_{\bar{z}\bar{z}}$ antiholomorphic the conserved currents in terms of the stress–energy tensor are given by (with $T(z) = T_{zz} = (\partial_z\Phi)^2$, $\bar{T}(\bar{z}) = T_{\bar{z}\bar{z}} = (\partial_{\bar{z}}\Phi)^2$)

$$J_z = iT(z)\epsilon(z), \quad J_{\bar{z}} = i\bar{T}(\bar{z})\bar{\epsilon}(\bar{z}). \tag{6.34}$$

We will only concentrate on the holomorphic part for simplicity. The conserved charge Q_ϵ on the complex plane, associated with the infinitesimal conformal transformation $T(z)\epsilon(z)$, is then defined by

$$Q_\epsilon = \frac{1}{2\pi i}\oint dw T(w)\epsilon(w). \tag{6.35}$$

This generates the infinitesimal conformal transformation

$$\delta\Phi(w, \bar{w}) = \epsilon(w)\partial_w\Phi + h\Phi\partial_w\epsilon(w). \tag{6.36}$$

The quantum analogue of this equation is

$$\begin{aligned}\delta\Phi(w,\bar{w}) &= [Q_{\epsilon}, \Phi(w,\bar{w})]\\ &= \frac{1}{2\pi i}\oint dz\epsilon(z)R(T(z)\Phi(w,\bar{w})).\end{aligned} \tag{6.37}$$

The integration contour over z is understood now to encircle the point w. The radially ordered product R must be analytic in the neighborhood of the point w in order for the integral to make sense. Thus, one must have a Laurent expansion. We reproduce the infinitesimal transformations (6.36) if and only if the radially ordered product is given by

$$\begin{aligned}R(T(z)\Phi(w,\bar{w})) &= \frac{h}{(z-w)^2}\Phi(w,\bar{w}) + \frac{1}{z-w}\partial_w\Phi(w,\bar{w})\\ &+ \text{regular power series in } (z-w).\end{aligned} \tag{6.38}$$

The conjugate analogue of this equation is

$$\begin{aligned}R(\bar{T}(\bar{z})\Phi(w,\bar{w})) &= \frac{\bar{h}}{(\bar{z}-\bar{w})^2}\Phi(w,\bar{w}) + \frac{1}{\bar{z}-\bar{w}}\partial_{\bar{w}}\Phi(w,\bar{w})\\ &+ \text{regular power series in } (\bar{z}-\bar{w}).\end{aligned} \tag{6.39}$$

As examples we compute

$$\begin{aligned}R(T(z)X^{\nu}(y)) &= -2\colon \partial X^{\mu}(z)\partial X^{\mu}(w)X^{\nu}(y)\colon\\ &+ \frac{1}{z-y}\partial X^{\nu}(y) + \cdots\end{aligned} \tag{6.40}$$

$$\begin{aligned}R(T(z)\partial X^{\nu}(y)) &= -2\colon \partial X^{\mu}(z)\partial X^{\mu}(w)\partial X^{\nu}(y)\colon\\ &+ \frac{1}{(z-y)^2}\partial X^{\nu}(y) + \frac{1}{z-y}\partial^2 X^{\nu}(y) + \cdots\end{aligned} \tag{6.41}$$

The normal ordered objects behave as classical quantities and thus they are finite in the limit $z \longrightarrow w$. We rewrite these results as

$$T(z)X^{\nu}(y) = \frac{1}{z-y}\partial X^{\nu}(y) + \cdots \tag{6.42}$$

$$T(z)\partial X^{\nu}(y) = \frac{1}{(z-y)^2}\partial X^{\nu}(y) + \frac{1}{z-y}\partial^2 X^{\nu}(y) + \cdots \tag{6.43}$$

This means in particular that X^{μ} and $\partial\, X^{\mu}$ are conformal fields of weights $h = 0$ and $h = 1$ respectively. These are examples of the operator product expansion.

The operator product expansion of the energy–momentum tensor with itself is found to be given by

$$\begin{aligned}R(T(z)T(y)) &= \colon \partial X^{\mu}(z)\partial X^{\mu}(w)\partial X^{\nu}(y)\partial X^{\nu}(x)\colon\\ &+ \delta^{\mu\mu}\frac{1}{2(z-y)^4} + \frac{2}{(z-y)^2}T(y) + \frac{1}{z-y}\partial T(y).\end{aligned} \tag{6.44}$$

We write this as

$$T(z)T(y) = \delta^{\mu\mu}\frac{1}{2(z-y)^4} + \frac{2}{(z-y)^2}T(y) + \frac{1}{z-y}\partial T(y). \tag{6.45}$$

We can immediately see that $T(z)$ is a conformal field of weight $h = 2$, which is the classical value, if $\delta^{\mu\mu} = 0$. The constant $c = \delta^{\mu\mu} = D$ is called the central charge. The energy–momentum tensor is not conformal unless $c = 0$. In this case it is a primary field of weight (2, 0). The central charge is therefore the conformal anomaly and it is due to quantum effects. Obviously, the central charge is equal to one for a single scalar field.

6.2.3 The bc CFT

The gauge-fixed action is given in Euclidean worldsheet by (with the scaling $b \longrightarrow b$ and $c^- \longrightarrow ic^-$ and $c^+ \longrightarrow -ic^+$)

$$S_E = \frac{1}{2\pi}\int d^2z(2\partial X^\mu\bar{\partial}X_\mu + b_{--}\bar{\partial}c^- + b_{++}\partial c^+). \tag{6.46}$$

The ghost energy–momentum tensor was found to be given by (including also Wick rotation)

$$\begin{aligned} -2T^g_{++} &= -2b_{++}\bar{\partial}c^+ + c^+\bar{\partial}b_{++}, \\ -2T^g_{--} &= -2b_{--}\partial c^- + c^-\partial b_{--}. \end{aligned} \tag{6.47}$$

The above theory involves a free fermionic conformal field theory termed bc CFT given in terms of anticommuting fields b and c by

$$S_E = \frac{1}{2\pi}\int d^2z b\bar{\partial}c. \tag{6.48}$$

This is conformally invariant for all b and c transforming under conformal transformations as tensors of weights $(\lambda, 0)$ and $(1-\lambda, 0)$. In the quantization of the string action we have found that $\lambda = 2$. Since the ghost fields c^+ and $c^-$are the components of a vector ghost field c^a, whereas the antighosts b_{++} and b_{--} are the components of a traceless symmetric antighost tensor field b_{ab}. Thus, the ghost field c has conformal dimension -1, whereas the antighost field b has conformal dimension 2. In other words, b transforms as the energy–momentum tensor while c transforms as the gauge transformations (diffeomorphism) parameter.

We compute the radial ordering in terms of the normal ordering and a contraction (propagator) as follows

$$R(b(z_1)c(z_2)) =: b(z_1)c(z_2) : + \langle b(z_1)c(z_2)\rangle. \tag{6.49}$$

The propagator is given by

$$\langle b(z_1)c(z_2)\rangle = \frac{1}{z_1 - z_2}. \tag{6.50}$$

We get then the operator product expansions

$$b(z_1)c(z_2) \sim \frac{1}{z_1 - z_2}, \quad c(z_1)b(z_2) \sim \frac{\epsilon}{z_1 - z_2} \tag{6.51}$$

$$b(z_1)b(z_1) \sim O(z_1 - z_2), \quad c(z_1)c(z_2) = O(z_1 - z_2). \tag{6.52}$$

The ϵ in (6.51) is equal to +1 in our case. But in the case where b and c satisfy Bose statistics we must set $\epsilon = -1$.

The generalization of the second equation of (6.47) is given by (including also normal ordering to be precise)

$$\begin{aligned} 2T_{bc} &=: -\lambda b(z)\partial c(z) + \epsilon(\lambda - 1)c(z)\partial b(z) : \\ &= \lim_{z \longrightarrow w}\left(-\lambda b(z)\partial c(w) + \epsilon(\lambda - 1)c(z)\partial b(w) + \frac{1}{(z - w)^2}\right). \end{aligned} \tag{6.53}$$

Indeed, we can check that

$$R(T_{bc}(z)b(y)) =: \ldots : +\frac{\lambda}{(z - y)^2}b(y) + \frac{1}{z - y}\partial b(y) \tag{6.54}$$

$$R(T_{bc}(z)c(y)) =: \ldots : +\frac{1 - \lambda}{(z - y)^2}c(y) + \frac{1}{z - y}\partial c(y). \tag{6.55}$$

This shows explicitly that b is of conformal dimension λ and c is of conformal weight equal to $1 - \lambda$, as it should be, and that $T(z)$ is then the correct form of the energy–momentum tensor.

The above formulas generalize, for a primary field $\mathcal{O}$ with weights $(h, \bar{h})$, to the Ward identities

$$R(T(z)\mathcal{O}(y, \bar{y})) =: \ldots : +\frac{h}{(z - w)^2}\mathcal{O}(y, \bar{y}) + \frac{1}{z - y}\partial\mathcal{O}(y, \bar{y}) \tag{6.56}$$

$$R(\bar{T}(\bar{z})\mathcal{O}(y, \bar{y})) =: \ldots : +\frac{\bar{h}}{(\bar{z} - \bar{w})^2}\mathcal{O}(y, \bar{y}) + \frac{1}{\bar{z} - \bar{y}}\partial\mathcal{O}(y, \bar{y}). \tag{6.57}$$

For example, X^μ is of weights (0,0), ∂X^μ is of weight (1,0), $\bar{\partial}X^\mu$ is weight (0,1), $\partial^2 X^\mu$ is of weight (2,0), and $:\exp(ikX):$ is of weights $(\alpha' k^2/4, \alpha' k^2/4)$. However, the operator product expansion of the energy–momentum tensor with itself is given in general by

$$T(z)T(y) = \frac{c}{2(z - y)^4} + \frac{2}{(z - y)^2}T(y) + \frac{1}{z - y}\partial T(y). \tag{6.58}$$

The constant c is called the central charge and the energy–momentum is therefore not a conformal field unless $c = 0$. We have found that $c = \delta^{\mu\mu} = D$ for D scalar fields.

What is the analogue result for the bc CFT above?

After some calculation we get the operator product expansion

$$T_{bc}(z)T_{bc}(y) = \frac{c}{2(z-y)^4} + 2\frac{T_{bc}(y)}{(z-y)^2} + \frac{\partial T_{bc}(y)}{z-y}$$
$$c = -2\epsilon(6\lambda^2 - 6\lambda + 1). \tag{6.59}$$

Th Faddeev–Popov ghosts of the gauge-fixed Polyakov action have weights corresponding to $\lambda = 2$, viz (2,0) for the antighost b and $(-1,0)$ for the ghost c, with conformal anomaly $c = -26$. Thus, the conformal anomaly $c(\lambda = 2, \epsilon = 1) = -26$ can be canceled by 26 spacetime coordinates X^μ since the weight of every X^μ is 1.

6.2.4 The superconformal field theory

We enhance conformal symmetry to the superconformal case. In this case we have D scalars X^μ with conformal dimension 1 and D Majorana–Weyl spinors corresponding to $b = \psi$ and $c = \bar{\psi}$ with equal conformal dimensions $\lambda = 1 - \lambda = 1/2$. The gauge fixing in this case using the so-called superconformal gauge introduces, besides the two fermion ghosts b and c with $\lambda = 2$, two boson ghosts with $\lambda = 3/2$. Thus the total conformal anomaly coming from the ghosts in this case is $c(2, 1) + c(3/2, -1) = -26 + 11 = 15$. This can be canceled by the contribution of the D coordinates X^μ and their D superpartners ψ^μ with $\lambda = 1/2$. Indeed, the contribution of the dynamical fields to the conformal anomaly given by $c = D(1 + 1/2) = 3D/2$ cancels exactly for $3D/2 = 15$, i.e. $D = 10$. As it turns out, $c = 3D/2$ is precisely the central charge of the energy–momentum tensor.

6.2.5 Vertex operators

In quantum field theory particles (or states) are created from the vacuum by quantum fields (operators). This provides a one-to-one map between states and operators. In closed string theory this map is given by vertex operators $V_\phi(z, \bar{z})$ which represent the absorption or emission of string states $|\phi\rangle$ from points z on the worldsheet. (Recall that the closed string cylinder is mapped to the complex plane under Wick rotation $\tau \longrightarrow i\tau$ and conformal mapping $z = \exp(\tau - i\sigma)$). The vertex operators $V_\phi(z, \bar{z})$ are clearly insertions of point-like operators at the points z on the complex plane. They are primary fields, whereas the string states are highest weight states. By summing over all insertion points we obtain $g_s \int d^2z V_\phi(z, \bar{z})$. The vertex operators for closed strings have conformal dimension (1,1).

The closed string ground state (tachyon) is the state with no oscillators excited but with momentum k, viz $|\phi\rangle = |0, k\rangle$. The corresponding vertex operator is $: \exp(ikX) :$ which has a conformal dimension equal to $(k^2/8, k^2/8) = (1, 1)$ (assuming $l_s = 1$). We should then take an average over the absorption or emission point on the worldsheet as $\int d^2z : \exp(ikX) :$ since the state is independent of the insertion point.

The vertex operators for excited states will contain additional factors, of conformal dimension (n, n) where n is a positive integer, which are induced by the creation operators α^μ_{-m}, $m > 0$. It is not difficult to convince ourselves that the desired rule to pass from the state to the operator is to replace α^μ_{-m} with $\partial^m X^\mu$.

The vertex operators for open strings are conformal fields of dimension 1. The tachyon state is again associated with the operator $: \exp(ikX) :$ which has a conformal dimension equal to $k^2/2 = 1$ in open string theory. Also, by summing over all insertion points on the open string worldsheet (upper complex half plane) we get an expression of the form $g_o \oint V_\phi(s)ds$ where g_o is the open string coupling, viz $g_o^2 = g_s$, and s labels the boundary.

The vertex operator for the vector gauge field (photon) state $|\phi\rangle = \xi_\mu(\alpha_{-1}^\mu)^\dagger|0, k\rangle$ (the first excited state in open string theory) is given by

$$\int ds : \xi_\mu \partial_t X^\mu \exp(ikX) : . \tag{6.60}$$

The insertions are located on the real axis, which is the boundary of the upper half complex plane, and ∂_t is the derivative along the boundary. The boundary of the upper half complex plane corresponds (if we undo the conformal mapping) to the boundaries $\sigma = 0$ and $\sigma = \pi$ of the open string worldsheet which is a strip in spacetime. Thus the photon is associated with the end points of the open string.

6.2.6 Background fields

The spectrum of closed strings at the first excited level consists of a graviton, an antisymmetric second rank tensor field and a scalar field. The most important background fields which can couple to the string are precisely those fields which are associated with these massless bosonic degrees of freedom in the spectrum. Namely, the metric $g_{\mu\nu}(X)$, the antisymmetric two-form gauge field $B_{\mu\nu}(X)$, and the dilaton field $\Phi(x)$. The metric $g_{\mu\nu}$ couples in the obvious way

$$S_1 = -\frac{1}{4\pi\alpha'}\int_M d^2\sigma\sqrt{-h}\,h^{\alpha\beta}g_{\mu\nu}(X)\partial_\alpha X^\mu\partial_\beta X^\nu. \tag{6.61}$$

The coupling of the two-form gauge field $B_{\mu\nu}$ is given by

$$S_2 = -\frac{1}{4\pi\alpha'}\int_M d^2\sigma\epsilon^{\alpha\beta}B_{\mu\nu}(X)\partial_\alpha X^\mu\partial_\beta X^\nu. \tag{6.62}$$

The epsilon symbol is defined such that $\epsilon^{01} = 1$ and it is a tensor density, i.e. $\epsilon^{\alpha\beta}/\sqrt{-h}$ transforms as a tensor. The above action changes by a total divergence under the gauge transformations

$$\delta B_{\mu\nu} = \partial_\mu\Lambda_\nu - \partial_\nu\Lambda_\mu. \tag{6.63}$$

Thus, this term which couples the two-form gauge field $B_{\mu\nu}$ to the worldsheet of the string is the analogue of the coupling of the one-form Maxwell field to the world line of a charged particle given by

$$S = q\int d\tau A_\mu \dot{x}^\mu. \tag{6.64}$$

Also, this term is only present for oriented strings, and it can be eliminated by a procedure called orientifold projection, i.e. a projection onto strings which are invariant under reversal of orientation. This term is also the source of much of the noncommutative geometry which appears from string theory.

The dilaton field is more interesting. It couples to the string via the scalar curvature $R^{(2)}(h)$ of the metric $h_{\mu\nu}$ on the 2-dimensional string worldsheet. The action reads

$$S_3 = \frac{1}{4\pi}\int_M d^2\sigma\sqrt{-h}\,\Phi(X)R^{(2)}(h). \qquad (6.65)$$

Because of the absence of explicit factors of X this action is one order higher than S_1 and S_2 in the α' expansion, i.e. S_3 should be thought of as an order α' correction compared to the first two actions S_1 and S_2. For $\Phi = 1$ we get The Hilbert–Einstein action in two dimensions. However, in two dimensions this action is exactly equal to the so-called Euler characteristic

$$\chi = \frac{1}{4\pi}\int_M d^2\sigma\sqrt{-h}\,R^{(2)}(h). \qquad (6.66)$$

This is a topological invariant which gives no dynamics to the 2-dimensional metric $h_{\mu\nu}$. The proof goes as follows. The variation of the Hilbert–Einstein action in any dimension is known to be given by

$$\delta\chi = \frac{1}{4\pi}\int d^2\sigma\sqrt{-h}\,\delta h^{\alpha\beta}\left(R_{\alpha\beta} - \frac{1}{2}h_{\alpha\beta}R\right). \qquad (6.67)$$

The Riemann tensor always satisfies $R_{\alpha\beta\gamma\delta} = -R_{\beta\alpha\gamma\delta} = -R_{\alpha\beta\delta\gamma}$. In two dimensions a second rank antisymmetric tensor can only be proportional to $\epsilon_{\alpha\beta}$. Thus, the Riemann tensor $R^{(2)}_{\alpha\beta\gamma\delta}$ must be proportional to the scalar curvature $R^{(2)}$. We find explicitly

$$\frac{1}{h}\epsilon_{\alpha\beta}\epsilon_{\gamma\delta} = h_{\alpha\gamma}h_{\beta\delta} - h_{\beta\gamma}h_{\alpha\delta}. \qquad (6.68)$$

Hence

$$R_{\alpha\beta\gamma\delta} = \frac{1}{2}(h_{\alpha\gamma}h_{\beta\gamma} - h_{\beta\gamma}h_{\alpha\delta})R^{(2)}. \qquad (6.69)$$

We deduce immediately that

$$R^{(2)}_{\alpha\beta} - \frac{1}{2}h_{\alpha\beta}R^{(2)} = 0 \Rightarrow \delta\chi = 0. \qquad (6.70)$$

The Hilbert–Einstein action in two dimensions is therefore invariant under any continuous change in the metric. This does not mean that the Hilbert–Einstein action vanishes in two dimensions but it means that it depends only on the global topology of the worldsheet since it is in fact a boundary term, i.e. the integrand in (6.66) is a total derivative.

The action (6.66) is also invariant under Weyl rescalings for a worldsheet without a boundary. In the presence of a boundary an additional boundary term is needed [5].

The action $S_1 + \Phi\chi$, where Φ is a constant, looks like the Hilbert–Einstein action for the 2-dimensional metric $h_{\mu\nu}$ coupled to D massless scalar field X_μ propagating on the worldsheet. These scalar fields also define the embedding of the worldsheet in a background target spacetime with metric $g_{\mu\nu}(X)$.

6.2.7 Beta function: finiteness and Weyl invariance

We start by considering the action S_1. This is a nonlinear sigma model in the conformal gauge $h_{\alpha\beta} = \eta_{\alpha\beta}$. In this gauge one must also impose the Virasoro conditions $T_{\alpha\beta} = 0$. In the critical dimension the two conditions $T_{++} = T_{--} = 0$ are sufficient to define the Hilbert space of physical states without negative norm states. There remains the condition $T_{+-} = 0$ which actually holds classically due to the invariance under rescalings. The goal now is to check whether or not there is an anomaly in T_{+-}.

The breakdown of scale invariance is due to the fact that we cannot regularize the theory in a way which maintains scale or conformal invariance. Even dimensional regularization violates scale invariance. The breakdown of scale invariance can be described by the beta function of the theory which is related to the UV behavior of Feynman diagrams. The more fundamental question here is whether or not the nonlinear sigma model $S_1 |_{h=\eta}$ is Weyl invariant. Indeed, Weyl invariance implies global scale invariance which in turn implies the vanishing of the beta function which is equivalent to UV finiteness. Also, we note that the beta function is in fact the trace of the energy–momentum tensor which shows explicitly why finiteness is equivalent to Weyl invariance.

Furthermore, we note that the quantum mechanical perturbation theory is an expansion around small α', while the coupling constants of the theory are given by the metric components $g_{\mu\nu}(X)$ with corresponding beta functions $\beta_{\mu\nu}(X)$. The couplings $g_{\mu\nu}$ are actually functions and thus their associated beta functions are in fact functionals. In the explicit calculation, we will use dimensional regularization in $2 + \epsilon$ dimensions.

We expand X^μ around inertial coordinates X_0^μ, which do not depend on the worldsheet coordinates σ and τ, with fluctuations given by Riemann normal coordinates x^μ as

$$X^\mu = X_\mu^0 + x^\mu. \tag{6.71}$$

The metric $g_{\mu\nu}$ then starts as the flat metric $\eta_{\mu\nu}$ with corrections given in terms of the Riemann tensor of the target spacetime by

$$g_{\mu\nu} = \eta_{\mu\nu} - \frac{1}{3} R_{\mu\alpha\nu\beta} x^\alpha x^\beta + \cdots \tag{6.72}$$

The action S_1 organizes as an expansion in powers of x. The first term, which is quadratic, is the classical action. The second term, which is quartic, yields the one-loop correction by contracting two of the x's. The beta function is related to the poles in dimensional regularization which originate from logarithmically divergent integrals. The counter term that must be subtracted from S_1 to cancel the logarithmic divergence of the theory is found to be given in terms of the Ricci tensor of the target spacetime by

$$S = -\frac{1}{12\pi\epsilon}\int d^2\sigma\partial_\alpha X^\mu\partial^\alpha X^\nu R_{\mu\nu}(X). \tag{6.73}$$

The $1/\epsilon$ pole is due to the logarithmic divergence of the propagator in two dimensions which behaves as

$$\langle x^\mu(\sigma)x^\nu(\sigma')\rangle = \frac{1}{2\epsilon}\eta^{\mu\nu}, \quad \sigma' \longrightarrow \sigma. \tag{6.74}$$

The beta function is extracted precisely from the $1/\epsilon$ pole. We get the beta functions

$$\beta_{\mu\nu}(X) = -\frac{1}{4\pi}R_{\mu\nu}(X). \tag{6.75}$$

The condition for the vanishing of the beta functions is then precisely given by the Einstein equations

$$R_{\mu\nu}(X) = 0. \tag{6.76}$$

We go back to the action S_1 and check directly that the condition of Weyl invariance leads to the same result. Again we work in $2 + \epsilon$ dimensions. We substitute the conformal gauge $h_{\alpha\beta} = \exp(\phi)\eta_{\alpha\beta}$ in S_1. We also substitute the expansion (6.72) in S_1. The ϕ dependence does not vanish in the limit $\epsilon \longrightarrow 0$. Indeed, we find (with $y^\mu = (1 + \epsilon\phi/2)x^\mu$) [58]

$$S = -\frac{1}{2\pi}\int d^2\sigma\phi\partial^\alpha y^\mu\partial_\alpha y_\mu - \frac{1}{4\pi}\int d^2\sigma\phi R_{\mu\nu}\partial^\alpha y^\mu\partial_\alpha y^\nu. \tag{6.77}$$

This is ϕ independent, and hence the theory is Weyl invariant, if and only if $R_{\mu\nu} = 0$.

By computing the two-loop correction to the beta function we can deduce the leading string-theoretic correction to general relativity. We find the beta function and the corrected Einstein equations

$$\begin{aligned}\beta_{\mu\nu}(X) = -\frac{1}{4\pi}\left(R_{\mu\nu} + \frac{\alpha'}{2}R_{\mu\lambda\alpha\beta}R_\nu^{\ \lambda\alpha\beta}\right) &\Rightarrow R_{\mu\nu} \\ &+ \frac{\alpha'}{2}R_{\mu\lambda\alpha\beta}R_\nu^{\ \lambda\alpha\beta} = 0.\end{aligned} \tag{6.78}$$

We generalize this result to the nonlinear sigma model $S_1 + S_2 + S_3$. Again we work in $2 + \epsilon$ dimensions in the conformal gauge $h_{\alpha\beta} = \exp(\phi)\eta_{\alpha\beta}$. Now the conditions for Weyl invariance in the limit $\epsilon \longrightarrow 0$ are precisely equivalent to the conditions of vanishing of the one-loop beta functions, associated with the background fields $g_{\mu\nu}$, $B_{\mu\nu}$ and Φ, and they are given explicitly by

$$0 = R_{\mu\nu} + \frac{1}{4} H_{\mu}^{\lambda\rho} H_{\nu\lambda\rho} - 2D_{\mu}D_{\nu}\Phi \tag{6.79}$$

$$0 = D_{\lambda}H^{\lambda}_{\mu\nu} - 2D_{\lambda}\Phi H^{\lambda}_{\mu\nu} \tag{6.80}$$

$$0 = 4(D_{\mu}\Phi)^2 - 4D_{\mu}D^{\mu}\Phi + R + \frac{1}{2} H_{\mu\nu\rho}H^{\mu\nu\rho} \tag{6.81}$$

The D_{μ} is the covariant derivative with respect to the metric $g_{\mu\nu}$ of the target spacetime manifold, and $H_{\mu\nu\rho}$ is the third rank antisymmetric tensor field strength associated with the gauge field $B_{\mu\nu}$ defined by $H_{\mu\nu\rho} = \partial_{\mu}B_{\nu\rho} + \partial_{\rho}B_{\mu\nu} + \partial_{\nu}B_{\rho\mu}$, and therefore it must be invariant under the gauge transformations (6.63). The above conditions are in fact the equations of motion derived from the action

$$S = -\frac{1}{2\kappa^2} \int d^d x \sqrt{-g} \exp^{-2\Phi} \left(R - 4D_{\mu}\Phi D^{\mu}\Phi + \frac{1}{12} H_{\mu\lambda\rho} H^{\mu\lambda\rho} \right). \tag{6.82}$$

6.2.8 String perturbation expansions

The Euler characteristic of a compact Riemann surface M of genus g is given, in terms of the number n_h of handles, the number n_b of boundaries and the number of cross-caps n_c of the surface, by

$$\chi = 2(1 - g), \quad 2g = 2n_h + n_b + n_c. \tag{6.83}$$

If we choose Φ to be a constant, viz $\Phi = \lambda =$ constant, then the Euclidean path integral with action $S_1 + S_3$ on a surface of genus g becomes

$$Z_g = \lambda^{-2(g-1)} \int [dXdh] \exp(-S_1). \tag{6.84}$$

The full partition function is then given by the sum over worldsheet

$$Z = \sum_g Z_g. \tag{6.85}$$

The coefficient λ is the string coupling constant controlled by the Euler action or more precisely by the expectation value of the dilaton field, viz

$$\kappa \equiv g_0^2 \sim \exp(\lambda). \tag{6.86}$$

This is not a free parameter in the theory. Obviously the number of distinct topologies characterized by g is very small compared to the number of Feynman diagrams.

We would like to show this result in a different way for its great importance. We consider a tree level scattering process involving M gravitons. The M external gravitons contribute $M - 2$ interaction vertices while each loop contributes two vertices κ (see figure 3.9 of [3]). The process is described by a Riemann surface M of genus g. The genus g of the surface is precisely the number of loops. This is because

$$(\kappa)^{M-2}(\kappa^2)^g = \kappa^M \kappa^{-2(1-g)}. \tag{6.87}$$

The factor κ^M can be absorbed in the normalization of external vertex operators. Thus we obtain for the diagram or the Riemann surface the factor

$$\kappa^{-\chi}. \tag{6.88}$$

This is exactly the factor appearing in the partition function (6.84).

There are four possible string theory perturbation expansions, corresponding to different sums over worldsheet, depending on whether the fundamental strings are oriented or unoriented and whether or not the theory contains open strings in addition to closed strings. Thus, we cannot have a consistent string theory perturbation expansion with open strings alone. Strings can be unoriented because of the presence of the so-called orientifold plane, whereas strings can be open because of the presence of D-branes. We list the four consistent string theories with their massless spectra and their allowed worldsheet topologies:

- **Closed oriented strings: $g_{\mu\nu}$, $B_{\mu\nu}$, Φ. All oriented surfaces without boundaries.** These are found in type II superstring theories and heterotic string theories. Here we can only have closed and oriented worldsheets which have $n_b = n_c =$ and hence $g = n_h$. The genus is precisely the number of loops. Since $g = n_h$ there exists one single topology (one string Feynman graph) at each order of perturbation theory. This string Feynman graph contains all the field theory Feynman diagrams which are generated, with their enormous number, in the singular limit where handles become too long compared to their circumferences and thus can be approximated by lines. The perturbation expansion of closed oriented strings are UV finite.
- **Closed unoriented strings: $g_{\mu\nu}$, Φ. All surfaces without boundaries.**
- **Closed and open oriented strings: $g_{\mu\nu}$, $B_{\mu\nu}$, Φ, A_μ. All oriented surfaces with any number of boundaries.**
- **Closed and open unoriented strings: $g_{\mu\nu}$, Φ. All surfaces with any number of boundaries.** These are found in type I superstring theories. In this case worldsheets are Riemann surfaces with boundaries and cross-caps as well as handles. These theories are also UV finite but the cancellation between string Feynman diagrams of the same Euler class is more delicate.

6.2.9 Spectrum of type II string theory

For the closed superstring we have left-movers and right-movers. The left-movers can be in the Ramond (R) sector or the Neveu–Schwarz (NS) sector and similarly for the right-movers. In other words, we have four sectors: R–R, NS–NS, R–NS and NS–R. In the NS sector the GSO condition means that we project on states with positive G-parity in order to eliminate the tachyon, whereas in the R sector we project on states with positive or negative G-parity depending on the chirality of the ground state. Thus, we can obtain two different theories depending on whether the left- and right-movers in the R sectors have the same or opposite G-parity.

These two theories are:

1. **Type IIA**: In this case the R sector of the left-movers and the R sector of the right-movers are distinct, i.e. they have opposite chiralities. Again, the ground state of the left-handed-movers in the R sector is a Majorana–Weyl spinor with positive chirality $|0, k, +\rangle_R$ while in the NS sector the ground state is the vector field $b^i_{-1/2}|0, k\rangle_{NS}$. Then, the ground states of the right-handed-movers are given by $|0, k, -\rangle_R$ and $b^i_{-1/2}|0, k\rangle_{NS}$. The ground states of the closed RNS superstring in its four sectors are then given by

$$|0, k, +\rangle_R \otimes |0, k, -\rangle_R, \quad R - R \tag{6.89}$$

$$b^i_{-1/2}|0, k\rangle_{NS} \otimes b^i_{-1/2}|0, k\rangle_{NS}, \quad NS - NS \tag{6.90}$$

$$b^i_{-1/2}|0, k\rangle_{NS} \otimes |0, k, -\rangle_R, \quad NS - R \tag{6.91}$$

$$|0, k, +\rangle_R \otimes b^i_{-1/2}|0, k\rangle_{NS}, \quad R - NS. \tag{6.92}$$

We write this tensor product as

$$\begin{aligned}(8_v \oplus 8_{s_+}) \otimes (8_v \oplus 8_{s_-}) = (1 \oplus 28 \oplus 35_v \oplus 8_v \oplus 56_v)_B \\ \oplus (8_{s+} \oplus 8_{s-} \oplus 56_{s+} \oplus 56_{s-})_F.\end{aligned} \tag{6.93}$$

Explicitly, we have

$$\begin{aligned}NS - NS &= (1 \oplus 28 \oplus 35_v)_B \\ &= \phi(\text{scalar dilaton}) \\ &\quad \oplus A_{\mu\nu}(\text{antisymmetric 2–form gauge field}) \\ &\quad \oplus g_{\mu\nu}(\text{symmetric traceless rank–two tensor graviton})\end{aligned} \tag{6.94}$$

$$\begin{aligned}R - R &= (8_v \oplus 56_v)_B \\ &= A_\mu(\text{gauge field}) \\ &\quad \oplus A_{\mu\nu\lambda}(\text{antisymmetric 3–form gauge field})\end{aligned} \tag{6.95}$$

$$\begin{aligned}NS - R &= (8_{s+} \oplus 56_{s+})_F \\ &= \psi_+(\text{spin half dilatino}) \oplus \chi_+(\text{spin three half gravitino})\end{aligned} \tag{6.96}$$

$$\begin{aligned}R - NS &= (8_{s-} \oplus 56_{s-})_F \\ &= \psi_-(\text{spin half dilatino}) \oplus \chi_-(\text{spin three half gravitino}).\end{aligned} \tag{6.97}$$

This is the particle content of type IIA supergravity in 10 dimensions. And it is the particle content obtained from the dimensional reduction of 11-dimensional supergravity. Since we have two dilatinos and two gravitinos we have $\mathcal{N} = 2$ supersymmetry.

2. **Type IIB**: In this case the R sector of the left-movers and the R sector of the right-movers have the same chirality taken + for concreteness. The ground state of the left-handed-movers in the R sector is a Majorana–Weyl spinor with positive chirality $|0, k, +\rangle_{\mathrm{R}}$ while in the NS sector the ground state is the vector field $b^{i}_{-1/2}|0, k\rangle_{\mathrm{NS}}$. Similarly, the ground states of the right-handed-movers are $|0, k, +\rangle_{\mathrm{R}}$ and $b^{i}_{-1/2}|0, k\rangle_{\mathrm{NS}}$. The ground states of the closed RNS superstring in its four sectors are then given by

$$|0, k, +\rangle_{\mathrm{R}} \otimes |0, k, +\rangle_{\mathrm{R}}, \quad \mathrm{R} - \mathrm{R} \tag{6.98}$$

$$b^{i}_{-1/2}|0, k\rangle_{\mathrm{NS}} \otimes b^{i}_{-1/2}|0, k\rangle_{\mathrm{NS}}, \quad \mathrm{NS} - \mathrm{NS} \tag{6.90}$$

$$b^{i}_{-1/2}|0, k\rangle_{\mathrm{NS}} \otimes |0, k, +\rangle_{\mathrm{R}}, \quad \mathrm{NS} - \mathrm{R} \tag{6.100}$$

$$|0, k, +\rangle_{\mathrm{R}} \otimes b^{i}_{-1/2}|0, k\rangle_{\mathrm{NS}}, \quad \mathrm{R} - \mathrm{NS}. \tag{6.92}$$

We write this tensor product as

$$\begin{aligned}(8_v \oplus 8_{s_+}) \otimes (8_v \oplus 8_{s_+}) = &(1 \oplus 28 \oplus 35_v \oplus 1 \oplus 28 \oplus 35_+)_B \\ &\oplus (8_{s+} \oplus 8_{s+} \oplus 56_{s+} \oplus 56_{s+})_F.\end{aligned} \tag{6.102}$$

Explicitly, we have

$$\begin{aligned}\mathrm{NS} - \mathrm{NS} &= (1 \oplus 28 \oplus 35_v)_B \\ &= \phi(\text{scalar dilaton}) \\ &\quad \oplus A_{\mu\nu}(\text{antisymmetric 2–form gauge field}) \\ &\quad \oplus g_{\mu\nu}(\text{symmetric traceless rank–two tensor graviton})\end{aligned} \tag{6.103}$$

$$\begin{aligned}\mathrm{R} - \mathrm{R} &= (1 \oplus 28 \oplus 35_+)_B \\ &= \phi(\text{scalar field}) \\ &\quad \oplus A_{\mu\nu}(\text{antisymmetric 2–form gauge field}) \\ &\quad \oplus A_{\mu\nu\alpha\lambda}(\text{antisymmetric 4–form gauge field} \\ &\quad\quad \text{with self–dual field strength})\end{aligned} \tag{6.104}$$

$$\begin{aligned}\mathrm{NS} - \mathrm{R} &= (8_{s+} \oplus 56_{s+})_F \\ &= \psi_+(\text{spin half dilatino}) \oplus \chi_+(\text{spin three half gravitino})\end{aligned} \tag{6.105}$$

$$\begin{aligned}\mathrm{R} - \mathrm{NS} &= (8_{s+} \oplus 56_{s+})_F \\ &= \psi_-(\text{spin half dilatino}) \oplus \chi_-(\text{spin three half gravitino}).\end{aligned} \tag{6.106}$$

This is the particle content of type IIB supergravity in 10 dimensions which cannot be obtained by dimensional reduction of 11-dimensional supergravity.

6.3 On Dp-branes and T-duality

6.3.1 Introductory remarks

The p-branes are p-dimensional non-perturbative stable configurations which can carry generalized conserved charges. These charges can obviously act as sources for antisymmetric tensor gauge fields with $p + 1$ indices. For example the 0-brane which is a point particle can have an electric and magnetic charges which generate the electromagnetic field. The magnetic dual of a p-brane is a $(D - p - 4)$-brane. For example in $D = 10$ the magnetic dual of a 0-brane is a 6-brane. The Dirichlet p-branes or Dp-branes are p-branes which are characterized by Dirichlet boundary conditions for open strings terminating on them. M-branes, and NS-branes are also p-branes. The NS5-brane is special since it is the magnetic dual of the fundamental string in the heterotic and type II superstring theories. See the short review [31].

We note that the D1-brane configuration is a string which we also call a D-string. D-strings carry an R–R charge and not an NS–NS charge as opposed to the fundamental type IIB superstring which acts as a source for the usual two-form B-field of the NS–NS sector and not as a source for the two-form C-field of the R–R sector. In general D-branes will also carry an R–R charge, i.e. they act as sources for the corresponding R–R $(p + 1)$-forms [32].

The stable Dp-branes in type II superstring theory, which come with even values of p in type IIA and odd values of p in type IIB, preserve 16 supersymmetries and as such they are called half-BPS Dp-branes. They carry conserved R–R charges and the corresponding open string spectrum is free of tachyons.

Let us now consider type II superstring theory which contains only closed strings. The presence of a Dp-brane modifies the allowed boundary conditions of the strings. Both Neumann and Dirichlet boundary conditions are now allowed. Hence in addition to the closed strings we can have open strings whose end points are fixed on the Dp-brane.

These open strings describe therefore the excitation of the Dp-brane and their quantization is clearly identical to the quantization of the ordinary open superstrings. An elementary exposition of this result can be found in [8].

On the other hand, the massless modes of type I open superstrings consist of a Majorana–Weyl spinor ψ coming from the Ramond sector and a gauge field A_a coming from the Neveu–Schwarz sector. Their dynamics are given at low energy by a supersymmetric $U(1)$ gauge theory in 10 dimensions. The field A_a is a function of only the zero modes of the coordinates $x^0, x^1, x^2, \ldots, x^p$ since, by the Dirichlet boundary conditions $x^{p+1} = \cdots = x^9 = 0$ at $\sigma = 0, \pi$, the zero modes of the other coordinates $x^{p+1}, \ldots, x^9$ must vanish. The reduced massless vector field A_a, $a = 0, 1, \ldots, p$, behaves therefore as a $U(1)$ vector field on the p-brane world volume while $X_{a-p} \equiv A_a$ for $a = p + 1, \ldots, 9$ behave as scalar fields normal to the p-brane. Hence, these scalar fields describe fluctuations of the position of the p-brane.

Therefore, at low energy the theory on the $(p + 1)$-dimensional world volume of the Dp-brane is the reduction to $p + 1$ dimensions of 10-dimensional supersymmetric $U(1)$ gauge theory [33]. Generalization of this fundamental result to the case of N coincident Dp-branes is straightforward. At low energy the theory on the $(p + 1)$-dimensional world volume of N coincident Dp-branes is the reduction to $p + 1$ dimensions of 10-dimensional supersymmetric $U(N)$ gauge theory. When the velocities and/or the string coupling are not small the minimal supersymmetric Yang–Mills is replaced by the supersymmetric Born–Infeld action [34].

6.3.2 Coupling to abelian gauge fields

We start by writing Maxwell's equations in the notation of differential forms. See for example the classic and pedagogical exposition [35]. The gauge field A_μ is a one-form $A_1 = A_\mu dx^\mu$, whereas the field strength is a two-form $F_2 = dA_1 = F_{\mu\nu} dx^\mu \wedge dx^\nu/2$. Maxwell's equations in the presence of electric and magnetic charges given respectively by the electric and magnetic currents $J_\mu^e = (\rho_e, \vec{J_e})$ and $J_\mu^m = (\rho_m, \vec{J_m})$ are given by the equations

$$dF = {*J_m}, \quad d{*F} = {*J_e}. \tag{6.107}$$

The J_e and J_m are the electric and magnetic one-form currents defined by $J = J_\mu dx^\mu$. The $*$ is the duality transformation or Hodge star which converts p-forms into $(D - p)$-forms where D is the dimension of spacetime. In the absence of magnetic charges, the first equation in (6.107) is the homogeneous Bianchi identity which is a geometric equation, whereas the second equation is the Euler inhomogeneous equation which is a dynamical equation.

The $U(1)$ gauge transformations in the notation of forms is given by $\delta A_n = d\Lambda_{n-1}$. In our case $n = 1$ and thus Λ_0 is a zero-form (function) and δ is the adjoint of the exterior derivative d which is given on p-forms by the relation $\delta = (-1)^{pD+d+1} * d*$. We have by construction $d^2 = \delta^2 = 0$. The electric and magnetic charges in four dimensions are defined by the integrals

$$e = \int_{S^2} {*F}, \quad g = \int_{S^2} F. \tag{6.108}$$

The electric and magnetic charges are related by the celebrated Dirac quantization condition $eg/2\pi \in \mathbf{Z}$ (Dirac considered the motion of an electric charge in the field of a magnetic monopole and demanded that the wave function can be consistently defined). The electromagnetic duality is given by $F \leftrightarrow {*F}$ and $e \leftrightarrow g$.

Generalization to p-branes in D dimensions is straightforward (the point particle is a 0-brane). The world volume of a Dp-brane is a $(p + 1)$-dimensional spacetime. The gauge field living on this world volume is an n-form A_n where $n = p + 1$ (the gauge field living on the world line of a particle is a one-form A_1). Thus the electric coupling of the n-form gauge field A_n to the world volume of the Dp-brane is given by

$$S = \mu_p \int A_{p+1}. \tag{6.109}$$

For a point particle this is the usual interaction $S = e \int A_1 = e \int d\tau A_\mu dx^\mu / d\tau$. Thus μ_p is the electric charge of the Dp-brane. This electric charge is given in terms of the field strength $(n + 1)$-form $F_{n+1} = dA_n$ by the obvious relation

$$\mu_p = \int *F_{p+2}. \tag{6.110}$$

The dual $*F_{p+2}$ of the $(p + 2)$-form field strength F is a $(D - p - 2)$-form. Thus the integral in the above formula is over a sphere S^{D-p-2} which is the surface that surrounds a p-brane in D dimensions.

The magnetic dual of the electrically charged p-brane will carry a magnetic charge ν_p computed obviously by the integral

$$\nu_p = \int F_{p+2}. \tag{6.111}$$

The integral is over a sphere S^{p+2}. In the same way that the sphere S^{D-p-2} surrounds a p-brane, the sphere $S^{D-q-2} = S^{p+2}$ must surround a q-brane where $q = D - p - 4$. Hence the magnetic dual of a p-brane in D dimensions is a $D - p - 4$ brane. For example, the magnetic dual of a D0-brane is a $D - 4$ brane (in 10 dimensions this is the D6-brane).

The electric and magnetic charges μ_p and $\nu_p = \mu_{D-p-4}$ must also satisfy the Dirac quantization condition, viz $\mu_p \mu_{D-p-4}/2\pi \in \mathbf{Z}$.

In the remainder of this section we will simply follow the presentations [2, 7].

6.3.3 Symmetry under the exchange of momentum and winding

Let us consider bosonic string theory in a spacetime compactified on a circle of radius R. A closed string wrapped around the circle can be contracted to a point and thus it is a topologically stable configuration characterized by a winding number $w \in Z$. The coordinate X^{25} of the string along the circle must then be periodic such that

$$X^{25}(\sigma + \pi, \tau) = X^{25}(\sigma, \tau) + 2\pi R w. \tag{6.112}$$

For the other coordinates X^μ, $\mu = 0, \ldots, 24$, the boundary condition is the usual periodic boundary condition with $w = 0$. Thus the mode expansion along these directions remains unchanged, whereas along the circular 25th direction it becomes given by ($l_s^2 = 2\alpha'$)

$$X^{25} = x^{25} + 2\alpha' p^{25}\tau + 2Rw\sigma + \frac{i}{2} l_s \sum_{n \neq 0} \frac{1}{n} e^{-2in\tau}\left(\alpha_n^{25} e^{2in\sigma} + \tilde{\alpha}_n^{25} e^{-2in\sigma}\right). \tag{6.113}$$

Since the wave function will contain the factor $\exp(ip^{25}x^{25})$, and since x^{25} is compact and periodic, we conclude that the momentum p^{25} must be quantized as

$$p^{25} = \frac{k}{R}, \quad k \in Z. \tag{6.114}$$

The quantum number k is the so-called Kaluza–Klein excitation number. We split the above solution into left- and right-movers as usual, viz (with $\alpha_0^{25} = \tilde{\alpha}_0^{25} = l_s p^{25}/2$ and $\tilde{x}^{25}$ is an arbitrary constant)

$$X_R^{25}(\tau-\sigma)=\frac{1}{2}(x^{25}-\tilde{x}^{25})+\sqrt{2\alpha'}\,\alpha_0^{25}(\tau-\sigma)$$
$$+\frac{i}{2}l_s\sum_{n\neq 0}\frac{1}{n}\alpha_n^{25}\exp(-2in(\tau-\sigma))\tag{6.115}$$

$$X_L^{25}(\tau+\sigma)=\frac{1}{2}(x^{25}+\tilde{x}^{25})+\sqrt{2\alpha'}\,\tilde{\alpha}_0^{25}(\tau+\sigma)$$
$$+\frac{i}{2}l_s\sum_{n\neq 0}\frac{1}{n}\alpha_n^{25}\exp(-2in(\tau-\sigma)).\tag{6.116}$$

The zero modes are given explicitly by

$$\sqrt{2\alpha'}\,\alpha_0^{25}=\alpha'\frac{k}{R}-wR,\quad \sqrt{2\alpha'}\,\tilde{\alpha}_0^{25}=\alpha'\frac{k}{R}+wR.\tag{6.117}$$

The mass relation in the uncompactified 24 + 1 dimensions (where now the winding number k labels different particle species) is given by $M^2=-p^\mu p_\mu$, $\mu=0,\ldots,24$. Thus

$$M^2=\frac{2}{\alpha'}(\alpha_0^{25})^2-p^\mu p_\mu=\frac{2}{\alpha'}(\alpha_0^{25})^2+\frac{4}{\alpha'}(N-1)$$
$$=\frac{2}{\alpha'}(\tilde{\alpha}_0^{25})^2+\frac{4}{\alpha'}(\tilde{N}-1).\tag{6.118}$$

By taking the sum and the difference we get

$$\alpha' M^2=\frac{\alpha' k^2}{R^2}+\frac{w^2R^2}{\alpha'}+2(N+\tilde{N}-2)\tag{6.119}$$

$$N-\tilde{N}=kw.\tag{6.120}$$

These formulas are invariant under the transformations

$$k\leftrightarrow w,\quad R\leftrightarrow\tilde{R}=\alpha'/R.\tag{6.121}$$

Thus the theory on a circle of radius R with momentum modes k and winding modes w is equivalent to the theory on a circle of radius $\tilde{R}$ with momentum modes w and winding modes k. This profound property is called T-duality and it goes beyond perturbative bosonic string theory to a non-perturbative symmetry of supersymmetric string theory.

We remark that in the limit of large R the momentum modes become lighter while winding modes become heavier. Thus, in the strict limit $R\longrightarrow\infty$ only the states with zero winding $w=0$ and all values of momentum k will survive forming a continuum. This is obviously the uncompactified theory.

Let us now consider the limit of small R where momentum modes become heavier while winding modes become lighter. In this case only the states with zero momentum $k=0$ and all values of the winding number w will survive, in the strict limit $R \longrightarrow 0$, forming also a continuum. In other words, we end up with an effective uncompactified direction in the limit $R \longrightarrow 0$ as well. This is not the expected result of dimensional reduction found in field theory and also found in open string theory which should occur in the limit $R \longrightarrow 0$.

Under the above duality we have $\alpha_0^{25} \longrightarrow -\alpha_0^{25}$ and $\tilde{\alpha}_0^{25} \longrightarrow \tilde{\alpha}_0^{25}$. As it turns out, the theory is symmetric under the full exchange of the right-moving and left-moving parts of the compact direction X^{25}, viz

$$X_L^{25}(\tau + \sigma) \longrightarrow X_L^{25}(\tau + \sigma), \quad X_R^{25}(\tau - \sigma) \longrightarrow - X_R^{25}(\tau - \sigma). \tag{6.122}$$

Thus the 25th coordinate is mapped as

$$\begin{aligned} X^{25} &= X_L^{25} + X_R^{25} = x^{25} + 2\alpha'\frac{k}{R}\tau + 2Rw\sigma + \cdots \longrightarrow \\ \tilde{X}^{25} &= X_L^{25} - X_R^{25} = \tilde{x}^{25} + 2\alpha'\frac{k}{R}\sigma + 2Rw\tau + \cdots \end{aligned} \tag{6.123}$$

The coordinate $\tilde{x}^{25}$ parametrizes the dual circle with period $2\pi\tilde{R}$ in the same way that the coordinate x^{25} parametrizes the original circle with period $2\pi R$.

6.3.4 Symmetry under the exchange of Neumann and Dirichlet

We consider now bosonic open string theory in 26 dimensions. As we know, the requirement of Poincaré invariance implies that the ends of the open strings must obey the Neumann boundary conditions

$$\frac{\partial}{\partial\sigma}X^{\mu}|_{\sigma=0,\,\pi} = 0. \tag{6.124}$$

This holds in all directions, i.e. $\mu = 0, \ldots, 25$. We compactify now the theory as before on a circle of radius R. In this case there are no winding modes attached to the open string. Indeed, the general solution for the 25th coordinate satisfies Neumann boundary conditions and thus it must be given by the usual formula

$$X^{25} = x^{25} + l_s^2 p^{25}\tau + il_s \sum_{n\neq 0} \frac{1}{n}\alpha_n^{25} \exp(-in\tau)\cos n\sigma. \tag{6.125}$$

By splitting this solution into left-moving and right-moving parts X_R^{25} and X_L^{25}, and then applying the T-duality transformation $X_R^{25} \longrightarrow \tilde{X}_R^{25} = -X_R^{25}$ and $X_L^{25} \longrightarrow \tilde{X}_L^{25} = X_L^{25}$, we obtain the solution

$$\tilde{X}^{25} = \tilde{x}^{25} + l_s^2 p^{25}\sigma + l_s \sum_{n\neq 0} \frac{1}{n}\alpha_n^{25} \exp(-in\tau)\sin n\sigma. \tag{6.126}$$

We note the following:

- The T-dual theory has no momentum in the 25th direction since there is no τ dependence.

- The T-dual string satisfies Dirichlet boundary conditions in the 25th direction, viz (where $p^{25} = k/R$ and $\tilde{R} = \alpha'/R$)

$$\tilde{X}^{25}|_{\sigma=0} = \tilde{x}^{25}, \quad \tilde{X}^{25}|_{\sigma=\pi} = \tilde{x}^{25} + 2\pi k\tilde{R}. \tag{6.127}$$

  Thus, the ends of the open strings are fixed on the above wall on the dual circle of radius $\tilde{R}$. This corresponds to a hyperplane in spacetime which is precisely the so-called Dirichlet p-brane or Dp-brane for short. In the above case $p = 24$, whereas in the original case where there was no compactified directions $p = 25$. In the general case where there are n compactified directions we have $p = 25 - n$.
- The T-dual string wraps k times around the dual circle of radius $\tilde{R}$ in the 25th direction. The string remains open though since it lives in more dimensions. Again we observe that the momentum k has become winding under T-duality. More importantly, this winding is topologically stable since the ends of the string are fixed.

In summary, under T-duality a bosonic open string (with momentum and no winding) satisfying Neumann boundary conditions on a circle of radius R is transformed into an open string (without momentum and with winding) satisfying Dirichlet boundary conditions on the dual circle of radius $\tilde{R} = \alpha'/R$. The dual strings have their ends fixed on the D24-brane $\tilde{X}^{25} = \tilde{x}^{25}$ and they wrap around the dual circles an integer number of times.

Note also that ordinary open string theory should be thought of as a theory of open strings ending on a D25-brane (a spacetime filling D-brane). T-duality acting (obviously along a parallel direction) on this D25-brane has produced a D24-brane. T-duality acting along the dual circle (which is the perpendicular direction to the D24-brane) will take us back to the D25-brane.

In general, T-duality acting along a parallel direction to a Dp-brane will produce a D$(p-1)$-brane while acting on a perpendicular direction will produce a D$(p+1)$-brane.

6.3.5 Chan–Paton factors

Dp-branes carry background gauge fields on their world volumes. The end points of open strings terminating on Dp-branes are seen as charged particles by these gauge fields. We are thus led to the case of open strings with additional degrees of freedom at their end points which are called Chan–Paton charges.

The Chan–Paton charges are additional degrees of freedom carried by the open string at its end points which preserve spacetime Poincaré invariance and worldsheet conformal invariance. They have zero Hamiltonian and hence they are background degrees of freedom, i.e. non-dynamical.

For example, if we want to describe oriented strings with N additional degrees of freedom at their end points, we should then consider the gauge group $U(N)$. We can place at the end point $\sigma = 0$ the fundamental representation $\mathbf{N}$ of the group $U(N)$, whereas at the end point $\sigma = \pi$ we place the antifundamental representation $\bar{\mathbf{N}}$.

The open string states will then be labeled by the Fock space states ϕ and the momentum k, as usual, but also by two indices i and j running from 1 to N characterizing the Chan–Paton charges at the two ends $\sigma = 0$ and $\sigma = \pi$ of the strings. We have then

$$|\phi, k\rangle \longrightarrow |\phi, k, ij\rangle. \tag{6.128}$$

By construction this state transforms under $U(1)_i$ as a quark of charge +1, whereas it transforms under $U(1)_j$ as an antiquark of charge −1. An arbitrary string state is then described by a linear combination of these states given by means of N^2 Hermitian matrices λ^a_{ij}, $a = 1, \ldots, N^2$ (Chan–Paton matrices) as

$$|\phi, k, a\rangle = \sum_{i,j=1}^{N} |\phi, k, ij\rangle \lambda^a_{ij}. \tag{6.129}$$

These states are called Chan–Paton factors. Since the Chan–Paton charges are non-dynamical, in any open string scattering process, the right end of string number a associated with the matrix λ^a_{ij} is in the same state as the left end of string number b associated with the matrix λ^b_{kl} and so on, and hence summing over all possible values of the indices $i, j, k, l, \ldots$ will produce a trace of the product of Chan–Paton factors, viz $\mathrm{Tr}\,\lambda^a\lambda^b\ldots$. These traces are clearly invariant under the global $U(N)$ worldsheet symmetry

$$\lambda^a \longrightarrow U\lambda^a U^{-1}. \tag{6.130}$$

This shows explicitly that the index i at the end point $\sigma = 0$ of the open oriented string transforms like a quark under the fundamental representation $\mathbf{N}$ of $U(N)$, whereas the index j at the end point $\sigma = \pi$ transforms like an antiquark under the antifundamental representation $\bar{\mathbf{N}}$. Hence string states become $N \times N$ matrices transforming in the adjoint representation of $U(N)$ which can be seen more clearly by going to vertex operators. We have therefore N^2 tachyons, N^2 massless vector fields and so on labeled by the index a. In particular, each of the massless vector fields transforms under the adjoint representation $\mathbf{N} \otimes \bar{\mathbf{N}}$ of $U(N)$ and hence the global $U(N)$ worldsheet symmetry is promoted to a local $U(N)$ spacetime symmetry.

6.3.6 Electromagnetism on a circle and Wislon lines

In this section we will follow mostly [8, 9].

The Schrödinger equation of a particle with mass m and charge q is invariant under the gauge transformations $U = \exp(iq\chi) \in U(1)$ given explicitly by

$$\psi \longrightarrow \psi' = U\psi, \quad A_\mu \longrightarrow A'_\mu = A_\mu - \frac{i}{q}(\partial_\mu U)U^{-1}. \tag{6.131}$$

We assume now that there is a compact spatial direction x which is assumed to be a circle and that the components of the vector potential $\vec{A}$ are all zero except the component A_x along the circle. The Wilson line or holonomy of the gauge field is defined by

$$W = \exp(iw) = \exp\left(iq \oint dx A_x(x)\right). \tag{6.132}$$

The gauge parameter U must be periodic on the circle while the phase χ is quasi-periodic, viz

$$U(x + 2\pi R) = U(x) \Rightarrow q\chi(x + 2\pi R) = q\chi(x) + 2\pi m, \quad m \in \mathbf{Z}. \tag{6.133}$$

Hence

$$w' = w + 2\pi m. \tag{6.134}$$

Thus w is an angle $\theta \in [0, 2\pi]$. The Wilson line $W = \exp(i\theta)$ is then gauge invariant. The solutions χ of (6.133) are not single valued. Among the infinitely many physically equivalent solutions we can take χ to be linear in the compact direction, i.e.

$$q\chi = (2\pi m)\frac{x}{2\pi R}. \tag{6.135}$$

This solves by construction (6.133). The transformation of the gauge field becomes

$$qRA_x \longrightarrow qRA_x' = qRA_x + m. \tag{6.136}$$

Thus in order to obtain a non-trivial Wilson line we can simply choose constant backgrounds. In particular, we choose qRA_x to be exactly equal to the holonomy angle θ, viz

$$qA_x = \frac{\theta}{2\pi R}. \tag{6.137}$$

This is a trivial background since locally it is a pure gauge $A'_x = 0$ or equivalently $(U = \Lambda^{-1})$

$$qA_x = -i\Lambda^{-1}\partial_\mu\Lambda, \quad \Lambda = \exp(iq\chi), \quad q\chi = \frac{\theta x}{2\pi R}. \tag{6.138}$$

In other words, a constant gauge field on the circle can be gauged away by a suitable gauge transformation. However, this is only true locally. Since globally the constant background still has a non-trivial effect due to the compactness of the circle. This is exhibited by the fact that the gauge transformation Λ is not single valued. Indeed, we have

$$\Lambda(x + 2\pi R) = W \cdot \Lambda(x). \tag{6.139}$$

Thus this constant background gauge field which corresponds to a zero magnetic field everywhere (flat potential $F = 0$) and solves the source free Maxwell equations has concrete physical effects (Aharonov–Bohm effect). This effect lies precisely in the holonomy (the Wilson line W) which cannot be set equal to one by a local gauge transformation Λ. Indeed, as the particle loops around the compact direction it picks up the phase factor W due to the trivializing local gauge transformation Λ.

Next we solve the Schrodinger equation on the circle. By demanding periodicity of the wave function $\psi(x + 2\pi R) = \psi(x)$ we arrive at the solutions $\psi \sim \exp(ikx/R)$ where $k \in \mathbf{Z}$. The momentum along the compact direction is found to be fractional given by

$$p = \frac{k}{R} - \frac{\theta}{2\pi R}. \tag{6.140}$$

This spectrum is invariant under $\theta \longrightarrow \theta + 2\pi$ and $k \longrightarrow k + 1$.

This result can also be seen by considering the coupling of a point particle of mass m and charge q to an electromagnetic field A_x on a circle of radius R given by the action

$$S = \int \left(-m\sqrt{-\dot{X}^\mu \dot{X}_\mu} + q\dot{X}^\mu A_\mu\right) d\tau. \tag{6.141}$$

We note that

$$\exp(iS) \sim \exp(iq \oint dx A_x(x)). \tag{6.142}$$

Thus the wave function $\exp(iS)$ is proportional to the Wilson line and hence when the particle moves along the circle the Wilson line calculates the corresponding phase factor. Since $w = q\oint dx A_x(x)$ is periodic we conclude that the wave function $\exp(iS)$ is gauge invariant as we loop around the compact direction a full circle.

In the gauge $\tau = X^0 = t$ the above action reduces to ($v_i = \dot{x}_i$)

$$S = \int \left(-m\sqrt{1 - \vec{v}^2} + q\left(A_0 + \vec{v}\vec{A}\right)\right) d\tau. \tag{6.143}$$

The canonical momentum P associated with the compact direction x is given by

$$P = \frac{\delta S}{\delta \dot{x}} = \frac{m\dot{x}}{\sqrt{1 - \vec{v}^2}} + qA_x = p + \frac{\theta}{2\pi R}. \tag{6.144}$$

On the other hand, the physical wave function $\exp(iPx)$ is periodic and hence we must have $P = k/R$, i.e.

$$\frac{k}{R} = p + \frac{\theta}{2\pi R}. \tag{6.145}$$

6.3.7 The D-branes on the dual circle

We return now to the case of string theory and we consider a constant $U(N)$ background on the compact direction. More precisely we assume that only the component of the gauge potential along the circle takes a non-zero constant value. As we have seen in the case of the point particle, a flat potential on a compact direction can still have a non-trivial effect analogous to the Aharanov–Bohm effect. Indeed, a constant background on the circle is locally trivial (since it can be removed by a gauge transformation) but globally it is non-trivial (since the trivializing gauge transformation is not single valued). Such a topologically non-trivial background can be characterized by its Wilson line.

Let us then consider the constant $U(N)$ background along the 25th direction (bosonic open oriented stings) given by the pure gauge $A'_{25} = 0$ or equivalently

$$A_{25} = \frac{1}{2\pi R}\text{diag}(\theta_1, \ldots, \theta_N) = -i\Lambda^{-1}\partial_{25}\Lambda. \tag{6.146}$$

The wave function picks up a factor Λ^{-1} under this gauge transformation. The trivializing local $U(N)$ gauge transformation is given by

$$\Lambda = \text{diag}\left(\exp\left(\frac{iX^{25}\theta_1}{2\pi R}\right), \ldots, \exp\left(\frac{iX^{25}\theta_N}{2\pi R}\right)\right). \tag{6.147}$$

This gauge transformation is not single valued since

$$\Lambda(X^{25} + 2\pi R) = W\Lambda(X^{25}). \tag{6.148}$$

The Wilson line is given explicitly by

$$W = \exp\left(i\int_0^{2\pi R} dX^{25}A_{25}\right) = \text{diag}(\exp i\theta_1, \ldots, \exp i\theta_N). \tag{6.149}$$

Hence fields must pick a phase factor given precisely by the Wilson line W as we loop around the compact direction.

The above $U(N)$ background gauge field configuration breaks the Chan–Paton $U(N)$ gauge symmetry at the end points of the open strings down to some abelian subgroup of $U(N)$ such as the maximal subgroup $U(1)^N$. The underlying reason is already what we have said which is the fact that we cannot trivialize the Wilson line by setting it equal to one by a local gauge transformation. The $U(N)$ group is broken to its maximal abelian subgroup $U(1)^N$ when all the holonomy angles θ_a are distinct.

Let us now consider a string in the Chan–Paton state $|\phi, k, ij\rangle$. The end point i (fundamental rep) of the string picks up a factor $\exp(-i\theta_i X^{25}/2\pi R)$ under the effect of the gauge transformation Λ, whereas the end point j (antifundamental rep) picks up a factor $\exp(i\theta_j X^{25}/2\pi R)$. The string wave function then acquires a phase $\exp(-i(\theta_i - \theta_j)X^{25}/2\pi R)$ and hence as we loop around the compact direction $X^{25} \longrightarrow X^{25} + 2\pi R$ it will pick up the Wilson line

$$|\phi, k, ij\rangle \longrightarrow \exp(i(\theta_j - \theta_i))|\phi, k, ij\rangle. \tag{6.150}$$

On the other hand, this wave function contains the plane wave $\exp(iPX^{25})$ and thus under the rotation $X^{25} \longrightarrow X^{25} + 2\pi R$ it will acquire a phase equal to $\exp(iP2\pi R)$. Thus

$$P = \frac{k}{R} - \frac{\theta_i - \theta_j}{2\pi R}, \quad k \in \mathbf{Z}. \tag{6.151}$$

The momentum number is then fractional. When we apply now T-duality these fractional Kaluza–Klein excitation numbers on the circle will be mapped to fractional winding numbers on the dual circle. A fractional winding number means that the open string partially winds around the dual circle since it is connecting two separated D24-branes i and j. The angles θ_i are thus interpreted as the angular positions of N D24-branes on the dual circle. The D24-branes i and j are coincident only when $\theta_i = \theta_j$ in which case we get an integer winding.

The mode expansions of the open string and its dual can now be found to be given by

$$X^{25} = x^{25} + \theta_i \frac{\alpha'}{R} + 2\frac{\alpha'}{R}\left(k - \frac{\theta_i - \theta_j}{2\pi}\right)\tau + \cdots \tag{6.152}$$

$$\tilde{X}_{ij}^{25} = \tilde{x}^{25} + \theta_i \tilde{R} + 2\tilde{R}\left(k - \frac{\theta_i - \theta_j}{2\pi}\right)\sigma + \cdots \tag{6.153}$$

The two end points of the dual string are at $\tilde{x}^{25} + \theta_i \tilde{R}$ and $\tilde{x}^{25} + \theta_j \tilde{R}$. These are precisely the locations of the i and j D24-branes respectively. Indeed, the Dirichlet boundary conditions become

$$\tilde{X}_{ij}^{25}|_{\sigma=\pi} - \tilde{X}_{ij}^{25}|_{\sigma} = 0 = \tilde{R}(2\pi k + \theta_j - \theta_i). \tag{6.154}$$

Thus the end points of the open strings in the gauge state i are located on the hyperplanes (D24-branes) located at the positions

$$\tilde{X}_{ij}^{25} = 2\pi\alpha'(A_{25})_{ii}. \tag{6.155}$$

The mass relation in the uncompactified 24 + 1 dimensions for the open string is given by $M^2 = k^2/R^2 + (N - 1)/\alpha'$. In the presence of Wilson lines the spectrum of the ij open string becomes

$$M_{ij}^2 = \frac{1}{R^2}\left(k - \frac{\theta_i - \theta_j}{2\pi}\right)^2 + \frac{N - 1}{\alpha'}. \tag{6.156}$$

The main observation here is that only diagonal strings (strings starting and ending on the same D24-branes) contain in their spectrum a massless vector field. Hence, if all the angles θ_i are different (no D24-branes coincide), the gauge group is $U(1)^N$, while if all the D24-branes coincide (the angles are all equal), the gauge group is $U(N)$.

6.4 Quantum gravity in two dimensions

The standard references for the matrix models of $D = 0$ and $D = 1$ string theories are the systematic reviews [14, 44]. However, the short review [36] is an extremely useful concise description of the relevant points.

6.4.1 Dynamical triangulation

The full string action includes the Hilbert–Einstein term as well as a cosmological constant for the worldsheet metric. The action then defines a theory of D scalar fields X^μ coupled minimally to the worldsheet metric h_{ab}. Explicitly we have the Euclidean action

$$S = \int d^2\sigma\sqrt{h}\left(\frac{1}{4\pi\alpha'}h^{ab}\partial_a X^\mu \partial_b X_\mu + \frac{\lambda}{4\pi}R + \Lambda\right). \tag{6.157}$$

The Hilbert–Einstein action in two dimensions is a topological term that equals the Euler character, viz

$$\chi = \int d^2\sigma\sqrt{h}\left(\frac{1}{4\pi}R\right) = 2 - 2n, \tag{6.158}$$

where n is the genus of the worldsheet surface $\mathcal{M}_n$ which is a sphere with n handles. The parameter λ is related to the expectation value of the dilaton field and thus it is determined by the string coupling constant, i.e.

$$\exp(\lambda) = g_s = g_o^2. \tag{6.159}$$

The partition function is then given by

$$Z = \sum_n g_s^{2n-2} \int [dX^\mu][dh_{ab}] \exp\left(-\int_{\mathcal{M}_n} d^2\sigma\sqrt{h}\left(\frac{1}{4\pi\alpha'}h^{ab}\partial_a X^\mu \partial_b X_\mu + \Lambda\right)\right). \tag{6.160}$$

A lattice-like regularization of this theory consists in discretizing the worldsheet geometries by dynamical triangulations, i.e. we replace the integration over the worldsheet metrics by a summation over worldsheet triangulations [37–39]. Another approach, which we will not pursue here, is given by Liouville theory [40, 41]. See also [14] and the extensive list of references therein.

Let us for simplicity consider the $D = 0$ theory which is a pure theory of surfaces with no conformal matter. We have

$$\begin{aligned} Z &= \sum_n g_s^{2n-2} \int [dh_{ab}] \exp\left(-\int_{\mathcal{M}_n} d^2\sigma\sqrt{h}\Lambda\right) \\ &= \sum_n \int [dh] \exp(-\Lambda A - \lambda\chi). \end{aligned} \tag{6.161}$$

A random triangulation of a given surface is a discretization of the surface by equilateral triangles. On the plane at each vertex i we find $N_i = 0$ triangles meeting since there is no curvature. On a general curved surface at each vertex i we find N_i triangles. If $N_i = 6$ there is no curvature at the vertex i, if $N_i < 6$ there is a positive curvature, and if $N_i > 6$ there is a negative curvature. Indeed, the Ricci scalar at the vertex i is given by

$$R_i = 2\pi\frac{6 - N_i}{N_i}. \tag{6.162}$$

This can be verified as follows. Each triangulation is characterized by a number of vertices V, a number of edges E and a number of faces F. Thus $V = \sum_i N_i$ by construction. However, topologically since each edge is shared by two vertices we must have $2E = V$, and since each face has three edges and each edge is shared by two faces we must have $3F = 2E$. The area of the triangulation is given by

$$\int d^2\sigma\sqrt{h} = \sum_i \frac{N_i}{3} = \frac{V}{3} = \frac{2E}{3} = F. \tag{6.163}$$

This is the total number of triangles. The Hilbert–Einstein term is given by

$$\begin{aligned}\int d^2\sigma\sqrt{h}\,R &= \sum_i \frac{N_i}{3}2\pi\frac{6-N_i}{N_i} = 4\pi\left(V - \frac{F}{2}\right)\\ &= 4\pi(V - E + F) = 4\pi\chi.\end{aligned} \tag{6.164}$$

This is then the Euler character as it should be. The above partition function becomes in the discrete given by

$$Z_{DT} = \sum_{n=0} g_s^{2n-2} \sum_{T_n} \exp(-\Lambda F), \tag{6.165}$$

where the summation is over the dynamical triangulations T_n of the surfaces $\mathcal{M}_n$ which are explicitly constructed above. The continuum limit is defined by

$$\Lambda \longrightarrow \Lambda_c, \tag{6.166}$$

where Λ_c is independent of the genus n. The total number of graphs with a fixed genus n and a fixed number of triangles F increases with F as $\exp(\Lambda_c F)/F^{b_n}$ [42]. Thus, at $\Lambda = \Lambda_c$ the contribution from entropy (degeneracy increases greatly, i.e. the number of graphs with fixed F becomes too large) dominates over the contribution from energy (the exponential convergent Boltzmann weight) and as a consequence the partition function diverges in a second order phase transition. This behavior can be characterized by the behavior of the string susceptibility given by

$$f = \frac{\partial^2}{\partial\Lambda^2}Z_{DT} = \sum_{n=0} g_s^{2n-2}(\Lambda - \Lambda_c)^{-\gamma_n}, \quad \gamma_n = -b_n + 3. \tag{6.167}$$

6.4.2 Matrix models of $D = 0$ string theory

We consider now the cubic matrix model

$$Z_\alpha = \int [d\Phi]\exp(-N\,\mathrm{Tr}\,V(\Phi)), \quad V = \frac{1}{2}\Phi^2 - \frac{\alpha}{3}\phi^3. \tag{6.168}$$

The propagator is given by

$$\langle\Phi_{ij}\Phi_{kl}\rangle = \frac{1}{Z_0}\frac{(2\pi)^{N^2/2}}{N}\delta_{il}\delta_{jk}. \tag{6.169}$$

Thus, the propagator is represented by a double line. These two lines carry arrows in opposite directions because we are dealing with Hermitian matrices which will correspond to orientable surfaces. Clearly, three such propagators come together in a matrix 3-point vertex. A typical Feynman diagram is then an oriented two-dimensional surface formed by polygons which are bounded by index loops. The so-called dual diagram is constructed by drawing lines through the centers of the polygons. It is seen that the dual diagram can also be obtained by placing the matrix 3-point vertices inside triangles and thus by construction the dual diagram is composed of triangles. In other words, the dual diagram is a dynamical triangulation of some Riemann surface.

Following 't Hooft [43] we can organize the diagrammatic expansion in powers of $1/N$ where each order corresponds to a distinct topology. A given Feynman diagram of the cubic matrix model is characterized by V vertex, E propagators (edges) and F loops (faces). The vertex is associated with a factor of N, the propagator is associated with a factor of $1/N$ and the loop is associated with a factor of N. Thus the Feynman diagram is of order $N^{V-E+F} = N^{\chi}$ where χ is the Euler character, viz $\chi = 2 - 2n$. The free energy admits then the $1/N$ expansion

$$F_\alpha = \log Z_\alpha = \sum_{n=0} N^{2-2n} F_n. \tag{6.170}$$

The free energy F_n is given by the sum of the connected Feynman diagrams which can be drawn on a sphere with n handles (obviously Z_α generates both connected and disconnected diagrams).

Since each matrix 3-point vertex is placed inside a triangle the number of vertices v_G in a connected Feynman diagram G of the cubic matrix model is equal to the number of triangles F of the corresponding dynamical triangulation. Hence the area is $A = F = v_G$ (the area of each triangle is one). This Feynman diagram G is obviously proportional to α^{v_G}. We must also divide by the appropriate symmetry factor, i.e. $\alpha^{v_G} \longrightarrow \alpha^{v_G}/S_G$ where S_G is the order of the discrete symmetry group of the diagram. Indeed, the symmetry group of the Feynman diagram or the dynamical triangulation is exactly the analogue of the isometry group of continuum manifolds [44]. The free energy F_n is then given by

$$F_n = \sum_G \frac{\alpha^{v_G}}{S_G}. \tag{6.171}$$

Thus

$$F_\alpha = \log Z_\alpha = \sum_{n=0} N^{2-2n} \sum_G \frac{\alpha^{v_G}}{S_G}. \tag{6.172}$$

By comparing with (6.165) we get

$$N = \exp(-\lambda) = \frac{1}{g_s}, \quad \exp(-\Lambda) = \alpha. \tag{6.173}$$

In other words,

$$F_\alpha = Z_{DT}. \tag{6.174}$$

The continuum limit is a double scaling limit defined by sending $N \longrightarrow \infty$ and $\alpha \longrightarrow \alpha_c$ keeping fixed the string coupling constant [45–47]

$$\lambda = \frac{1}{N(\alpha_c - \alpha)^{5/4}}. \tag{6.175}$$

In this limit the partition function diverges signaling a second order phase transition. Indeed, the planar partition function $Z_\alpha = N^2 Z_\alpha^{(0)} + \cdots$ behaves in the limit $\alpha \longrightarrow \alpha_c$ as

$$Z_\alpha^{(0)} \sim (\alpha_c - \alpha)^{2-\gamma} \sim -A^{\gamma-2}, \tag{6.176}$$

where A is the expectation value of the area, viz $A = \langle F \rangle = \langle v_G \rangle$ and γ is the string susceptibility exponent. For pure quantum gravity $\gamma = -1/2$ (see below).

Generalization of the above construction is straightforward. Instead of random triangulation by means of a cubic matrix model we can have random polygonulations by means of a general potential of the form

$$V = \sum_{j>1} \alpha_j \Phi^j. \tag{6.177}$$

6.4.3 Matrix models of $D = 1$ string theory

We can generalize the above construction to strings in higher dimensions by considering multi-matrix models. For strings in the $0 < D \leqslant 1$ dimension we consider the q-matrix model

$$Z = \int \prod_{i=1}^{q} [d\Phi_i] \exp\left(-N \sum_i \mathrm{Tr}\, V(\Phi_i) + N \sum_i \mathrm{Tr}\, \Phi_i \Phi_{i+1}\right). \tag{6.178}$$

The diagrammatic expansion of this model generates discretized surfaces with q different states Φ_i existing at the vertices [44]. More precisely, it describes bosonic strings in $0 < D \leqslant 1$ or two-dimensional quantum gravity coupled to conformal matter in D dimensions where D is identified with the central charge of the Virasoro algebra. For example, in the unitary discrete series of conformal field theories which are labeled by an integer $m \geqslant 2$ we have [14]

$$D = c = 1 - \frac{6}{m(m+1)}. \tag{6.179}$$

The case $m = 2$ gives $D = c = 0$ (pure gravity), whereas the case $m = \infty$ gives $D = c = 1$ (one boson). Fractional dimensions start with $m = 3$ which gives $D = 1/2$ (half boson!) corresponding to the Ising model.

As it turns out, the model (6.178) describes also a scalar field on a one-dimensional lattice when $q \longrightarrow \infty$. The coupling term is a nearest neighbor interaction and hence in the limit $q \longrightarrow \infty$ the partition function becomes

$$Z = \int [d\Phi(t)] \exp\left(-N \int dt \left(\frac{1}{2}\dot{\Phi}^2 + \frac{m^2}{2}\Phi^2 + V_{\text{int}}(\Phi)\right)\right). \quad (6.180)$$

This model has been solved in [48]. For a cubic interaction the free energy F_n, given by the sum of the connected Feynman diagrams which can be drawn on a sphere with n handles, is now given by (compare with (6.171))

$$F_n = \sum_G \frac{\alpha^{v_G}}{S_G} F_G. \quad (6.181)$$

The Feynman integral F_G is given explicitly by [44, 36]

$$F_G = \int \prod_i \frac{dX_i}{2m} \exp\left(-m \sum_{\langle ij \rangle} |X_i - X_j|\right). \quad (6.182)$$

The variables X_i are the values of the string coordinate X at the vertex i and the summation is over links $\langle ij \rangle$ between vertices. Thus the one-dimensional inverse propagator yields precisely the kinetic term for the bosonic field X.

In the continuum limit the partition function diverges as before with the leading singular behavior given by $Z(\alpha) \sim (\alpha_c - \alpha)^{2-\gamma}$ where the string susceptibility exponent is given by $\gamma = 0$.

In general, the string susceptibility exponent γ for two-dimensional quantum gravity coupled with conformal matter is given in terms of the central charge $c = D$ by the formula

$$\gamma = \frac{1}{12}\left(D - 1 - \sqrt{(D-1)(D-25)}\right) = -\frac{1}{m}. \quad (6.183)$$

Obviously $D = 1$ is a barrier since the quantity under the square root becomes negative for $1 < D < 25$. The existence of this barrier is also related to the presence of a state with a mass squared proportional to $1 - D$ in the string spectrum which becomes negative for $D > 1$ (the tachyon). As a consequence the string phase is absent for $D > 1$.

6.4.4 Preliminary synthesis

In these notes we will replace the $D = 0$ matrix model of string theory with Type IIB matrix model (IKKT matrix model), whereas the $D = 1$ matrix model of string theory will be replaced with M-(atrix) theory (BFSS matrix quantum mechanics). Generalization to higher dimensions beyond the $D = 1$ barrier exists and it starts with matrix string theory (DVV matrix quantum gauge theory).

Another generalization beyond the $D = 1$ barrier and even beyond two-dimensional quantum gravity is provided by our recent proposal on emergent quantum gravity from multitrace matrix models and noncommutative geometry [49].

It seems also that a very natural generalization of the theory (6.180) is provided by the matrix and noncommutative scalar field theories considered in [50–52].

6.5 Exercise

Exercise: We consider the following Hermitian matrix model

$$S = \frac{N}{g}\,\mathrm{Tr}\left(-\phi^2 + \frac{1}{4}\phi^4\right). \tag{6.184}$$

This model was identified in [53] as a dual description of $\hat{c} = 0$ noncritical string theory with worldsheet supersymmetry (pure supergravity).

Write a Monte Carlo code which simulates the quantum dynamics of this model. Construct the double scaling limit of this model thus providing a complete constructive formulation of string theory.

Modify the above action by adding for example the doubletrace term $(\mathrm{Tr}\,\phi^2)^2$. This model can be considered as a model for string theories with $c > 1$ and additional curvature term such as $\int d^2x\sqrt{g}R^2$ since the two theories share the same phase diagram [56, 57]. Construct the double scaling limit of this model.

Solution: See appendix for the description of various possible algorithms that can be used. However, in this particular case you can diagonalize the matrix ϕ to avoid ergodicity problems and use ordinary Metropolis algorithm. You can compare the results with the numerical investigation of [55] and with the analytical exact solution of [54].

References

[1] Ydri B 2017 Quantum black holes (arXiv:1708.00748 [hep-th]).

[2] Becker K, Becker M and Schwarz J H 2006 *String Theory and M-theory: A Modern Introduction* (Cambridge: Cambridge University Press)

[3] Di Francesco P, Ginsparg P H and Zinn-Justin J 1995 2-D gravity and random matrices *Phys. Rep.* **254** 1

[4] Ginsparg P H and Moore G W 1993 Lectures on 2-D gravity and 2-D string theory, Yale Univ. New Haven - YCTP-P23-92 (92,rec.Apr.93) Los Alamos Nat. Lab. - LA-UR-92-3479 (92,rec.Apr.93) e: LANL hep-th/9304011 [hep-th/9304011].

[5] Green M B, Schwarz J H and Witten E 1987 *Superstring Theory. Vol. 1: Introduction* (Cambridge Monographs On Mathematical Physics) (Cambridge: Cambridge University Press)

[6] Green M B, Schwarz J H and Witten E 1987 *Superstring Theory. Vol. 2: Loop Amplitudes, Anomalies and Phenomenology* (Cambridge Monographs On Mathematical Physics) (Cambridge: Cambridge University Press)

[7] Polchinski J 2005 *String Theory. Vol. 1: An Introduction to the Bosonic String* (Cambridge: Cambridge University Press)

[8] Polchinski J 2005 *String Theory. Vol. 2: Superstring Theory and Beyond* (Cambridge: Cambridge University Press)
[9] Johnson C V 2003 *D-branes* (Cambridge: Cambridge University Press)
[10] Zwiebach B 2009 *A First Course in String Theory* (Cambridge: Cambridge University Press)
[11] Szabo R J 2004 *An Introduction to String Theory and D-Brane Dynamics* (London: Imperial College Press)
[12] Schellekens A N Conformal field theory. 2016 (Downloaded from author website)
[13] Lundholm D 2005 The Virasoro algebra and its representations in physics. *Report for the course "Lie algebras and quantum groups" at KTH.*
[14] Goddard P, Kent A and Olive D I 1986 Unitary representations of the Virasoro and supervirasoro algebras *Commun. Math. Phys.* **103** 105
[15] Friedan D, Shenker S H and Qiu Z A 1986 Details of the nonunitarity proof for highest weight representations of the Virasoro algebra *Commun. Math. Phys.* **107** 535
[16] Ginsparg P H and Moore G W 1993 Lectures on 2-D gravity and 2-D string theory Yale Univ. New Haven - YCTP-P23-92 (92,rec.Apr.93) Los Alamos Nat. Lab. - LA-UR-92-3479 (92,rec.Apr.93) e: LANL hep-th/9304011 [hep-th/9304011].
[17] Maldacena J M 1999 The large N limit of superconformal field theories and supergravity *Int. J. Theor. Phys.* **38** 1113 [*Adv. Theor. Math. Phys.* **2** 231 (1998)]
[18] Natsuume M 2005 *AdS/CFT Duality User Guide* Lecture Notes in Physics vol. 903, p 1
[19] Nastase H 2007 Introduction to AdS-CFT (arXiv:0712.0689 [hep-th]).
[20] Gubser S S, Klebanov I R and Polyakov A M 1998 Gauge theory correlators from noncritical string theory *Phys. Lett.* B **428** 105
[21] Witten E 1998 Anti-de Sitter space and holography *Adv. Theor. Math. Phys.* **2** 253
[22] Banks T, Fischler W, Shenker S H and Susskind L 1997 M theory as a matrix model: a conjecture *Phys. Rev.* D **55** 5112
[23] Polchinski J 2016 The black hole information problem (arXiv:1609.04036 [hep-th]).
[24] Hawking S W 1975 Particle creation by black holes *Commun. Math. Phys.* **43** 199; Erratum: [*Commun. Math. Phys.* **46** 206 (1976)].
[25] Hawking S W 1976 Breakdown of predictability in gravitational collapse *Phys. Rev.* D **14** 2460
[26] Susskind L and Lindesay J 2005 *An Introduction to Black Holes, Information and the String Theory Revolution: The Holographic Universe* (Hackensack, NJ: World Scientific)
[27] Page D N 1993 Black hole information (hep-th/9305040).
[28] Harlow D 2016 Jerusalem lectures on black holes and quantum information *Rev. Mod. Phys.* **88** 015002
[29] Jacobson T 2003 Introduction to quantum fields in curved space-time and the Hawking effect (gr-qc/0308048).
[30] Mukhanov V and Winitzki S 2007 *Introduction to Quantum Effects in Gravity* (Cambridge: Cambridge University Press)
[31] Carroll S M 2004 *Spacetime and Geometry: An Introduction to General Relativity* (San Francisco, CA: Addison-Wesley)
[32] Schwarz J H 2010 Status of superstring and M-theory *Int. J. Mod. Phys.* A **25** 4703 [*Subnucl. Ser.* **46** 335 (2011)]
[33] Polchinski J 1995 Dirichlet Branes and Ramond-Ramond charges *Phys. Rev. Lett.* **75** 4724
[34] Witten E 1996 Bound states of strings and p-branes *Nucl. Phys.* B **460** 335

[35] Leigh R G 1989 Dirac-Born-Infeld action from Dirichlet sigma model *Mod. Phys. Lett.* A **4** 2767

[36] Eguchi T, Gilkey P B and Hanson A J 1980 Gravitation, gauge theories and differential geometry *Phys. Rep.* **66** 213

[37] Zarembo K L and Makeenko Y M 1998 An introduction to matrix superstring models *Phys. Usp.* **41** 1 [*Usp. Fiz. Nauk* **168** 3 (1998)].

[38] Kazakov V A 1985 Bilocal regularization of models of random surfaces *Phys. Lett.* **150B** 282

[39] David F 1985 Planar diagrams, two-dimensional lattice gravity and surface models *Nucl. Phys.* B **257** 45

[40] Ambjorn J, Durhuus B and Frohlich J 1985 Diseases of triangulated random surface models, and possible cures *Nucl. Phys.* B **257** 433

[41] Polyakov A M 1981 Quantum geometry of bosonic strings *Phys. Lett.* **103B** 207

[42] D'Hoker E and Phong D H 1988 The geometry of string perturbation theory *Rev. Mod. Phys.* **60** 917

[43] Koplik J, Neveu A and Nussinov S 1977 Some aspects of the planar perturbation series *Nucl. Phys.* B **123** 109

[44] 't Hooft G 1974 A planar diagram theory for strong interactions *Nucl. Phys.* B **72** 461

[45] Brezin E and Kazakov V A 1990 Exactly solvable field theories of closed strings *Phys. Lett.* B **236** 144

[46] Douglas M R and Shenker S H 1990 Strings in less than one-dimension *Nucl. Phys.* B **335** 635

[47] Gross D J and Migdal A A 1990 Nonperturbative two-dimensional quantum gravity *Phys. Rev. Lett.* **64** 127

[48] Brezin E, Itzykson C, Parisi G and Zuber J B 1978 Planar diagrams *Commun. Math. Phys.* **59** 35

[49] Ydri B, Soudani C and Rouag A 2017 Quantum gravity as a multitrace matrix model (arXiv:1706.07724 [hep-th]).

[50] Ferretti G 1995 On the large N limit of 3-d and 4-d Hermitian matrix models *Nucl. Phys.* B **450** 713

[51] Nishigaki S 1996 Wilsonian approximated renormalization group for matrix and vector models in $2 < d < 4$ *Phys. Lett.* B **376** 73

[52] Ydri B and Ahmim R 2013 Matrix model fixed point of noncommutative ϕ^4 theory *Phys. Rev.* D **88** 106001

[53] Klebanov I R, Maldacena J M and Seiberg N 2004 Unitary and complex matrix models as 1-d type 0 strings *Commun. Math. Phys.* **252** 275

[54] Cicuta G M, Molinari L and Montaldi E 1986 Large *N* phase transitions in low dimensions *Mod. Phys. Lett.* A **1** 125

[55] Kawahara N, Nishimura J and Yamaguchi A 2007 Monte Carlo approach to nonperturbative strings - demonstration in noncritical string theory *J. High Energy Phys.* **0706** 076

[56] Korchemsky G P 1992 Matrix model perturbed by higher order curvature terms *Mod. Phys. Lett.* A **7** 3081

[57] Gubser S S and Klebanov I R 1994 A modified c = 1 matrix model with new critical behavior *Phys. Lett.* B **340** 3

[58] Green M B, Schwarz J H and Witten E 1987 Superstring Theory. Vol. 1: Introduction (Cambridge Monographs on Mathematical Physics) (Cambridge: Cambridge University Press)

IOP Publishing

Badis Ydri

Chapter 7

M-(atrix) theory and matrix string theory

In this chapter we will introduce the matrix quantum mechanics model discovered by Banks, Fischler, Shenker and Susskind [1]. This is obtained by the reduction to one dimension of $\mathcal{N} = 1$ super Yang–Mills theory in $D = 10$ dimensions. This matrix quantum mechanics model describes the low energy dynamics of a system of N type IIA D0-branes [2]. Banks, Fischler, Shenker, and Susskind conjectured that the large N limit of this model describes precisely M-theory in the infinite momentum frame ($P_z \longrightarrow \infty$) in the uncompactified limit ($R_s \longrightarrow \infty$). We remark that the infinite momentum limit is equivalent to the light-cone quantization only in the limit $N \longrightarrow \infty$.

This BFSS model can also be derived from compactification as follows.

We can start with the IKKT or IIB matrix model discovered in 1996 by Ishibashi, Kawai, Kitazawa, and Tsuchiya [3] which is the reduction to 0 dimensions of $\mathcal{N} = 1$ super Yang–Mills theory in D dimensions. Next, it is seen that the Euclidean IKKT model compactified on a circle gives the BFSS model at finite temperature [23].

In this chapter we will also discuss the so-called matrix string theory [5] which can be thought of as a matrix gauge theory in the same way that M-(atrix) theory or the BFSS model is a matrix quantum mechanics. Furthermore, in the same way that the Euclidean IKKT matrix model decompactified on a circle S^1 gives the BFSS model at finite temperature, M-(atrix) theory decompactified on a circle S^1 should give the above matrix string theory [6].

Two main applications are considered in detail: (1) the black-hole/black-string phase transition and its connection with the confinement/deconfinement phase transition [13] and (2) the black hole evaporation process and gauge/gravity duality tests [7, 12].

7.1 The quantized membrane

We start by writing down the action of a p-brane in D-dimensional spacetime with metric $g_{\mu\nu} = (-1, +1, \dots, +1)$. The p-brane is a p-dimensional object moving in spacetime with $p < D$. Thus the local coordinates will be denoted by σ^α, $\alpha = 0, \ldots, p$, $\sigma^0 = \tau$, with a local metric denoted by $h_{\alpha\beta}$. The p-brane will sweep a $(p+1)$-dimensional hyper-volume called the world hyper-volume. The 0-brane is a point, the 1-brane is a string, the 2-brane is a membrane,...which sweep a worldline, a worldsheet, a worldvolume,...respectively. The coordinates of the p-brane will be denoted by

$$X^\mu = X^\mu(\sigma^0, \sigma^1, \ldots, \sigma^p). \tag{7.1}$$

The induced metric is immediately given by

$$G_{\alpha\beta} = g_{\mu\nu}\partial_\alpha X^\mu \partial_\beta X^\nu. \tag{7.2}$$

The Lorentz invariant infinitesimal hyper-volume element is given by

$$d\mu_p = \sqrt{-\det G_{\alpha\beta}}\, d^{p+1}\sigma. \tag{7.3}$$

The action of the p-brane is then given by (with T_p the p-brane tension)

$$S_p = -T_p \int d\mu_p = -T_p \int \sqrt{-\det G_{\alpha\beta}}\, d^{p+1}\sigma. \tag{7.4}$$

The case of the membrane is given by

$$S_2 = -T_2 \int d\mu_2 = -T_2 \int \sqrt{-\det G_{\alpha\beta}}\, d^3\sigma. \tag{7.5}$$

This is the Nambu–Goto action. The Polyakov action is given by (using the same symbol)

$$S_2 = -T_2' \int d^3\sigma \sqrt{-h}\,(h^{\alpha\beta}G_{\alpha\beta} - \Lambda). \tag{7.6}$$

The addition of the cosmological term is due to the fact that the membrane, as opposed to the string, is not scale invariant. The quantized supermembrane exists in 11 dimensions in the same sense that the quantized superstring exists in 10 dimensions [14–16]. The equation of motion with respect to $h_{\alpha\beta}$ is (with $\delta\sqrt{-h} = -\sqrt{-h}\, h_{\alpha\beta}\delta h^{\alpha\beta}/2$)

$$G_{\alpha\beta} = \frac{1}{2} h_{\alpha\beta}(h^{\alpha\beta}G_{\alpha\beta} - \Lambda). \tag{7.7}$$

By tracing we get

$$h^{\alpha\beta}G_{\alpha\beta} = 3\Lambda \Leftrightarrow G_{\alpha\beta} = \Lambda h_{\alpha\beta}. \tag{7.8}$$

Substituting this solution in the Polyakov we get the Nambu–Goto with the identification

$$2T_2' = \sqrt{\Lambda}\, T_2. \tag{7.9}$$

The metric $h_{\alpha\beta}$ contains six independent components and the membrane action is invariant under three diffeomorphisms $\sigma^\alpha \longrightarrow \sigma'^\alpha = f^\alpha(\sigma)$. Thus three components of the metric can be fixed by a suitable gauge choice. If we suppose now that the topology of the membrane worldvolume is $\mathbf{R} \times \Sigma$ where the Riemann surface Σ is of fixed topology, then we can fix the components h_{00} and h_{0i} as [17]

$$h_{0i} = 0, \quad h_{00} = -\frac{4}{\rho^2 \Lambda} \det G_{ij}. \tag{7.10}$$

With this gauge choice the constraints become

$$g_{\mu\nu}\partial_0 X^\mu \partial_i X^\nu = 0, \quad g_{\mu\nu}\partial_0 X^\mu \partial_0 X^\nu = -\frac{4}{\rho^2} \det G_{ij}. \tag{7.11}$$

We get then

$$\sqrt{-\det G_{\alpha\beta}} = -\frac{\rho}{2} g_{\mu\nu}\partial_0 X^\mu \partial_0 X^\nu = \frac{2}{\rho} \det G_{ij}. \tag{7.12}$$

Thus the membrane action becomes

$$S_2 = \frac{T_2 \rho}{4} \int d^2\sigma dt \left(g_{\mu\nu}\partial_0 X^\mu \partial_0 X^\nu - \frac{4}{\rho^2} \det G_{ij} \right). \tag{7.13}$$

We introduce a canonical Poisson bracket on the membrane defined by (with $\epsilon^{12} = 1$)

$$\{f, g\} = \epsilon^{\alpha\beta} \partial_\alpha f \partial_\beta g. \tag{7.14}$$

Then it is not difficult to show that

$$\det G_{ij} = \frac{1}{2}\{X^\mu, X^\nu\}\{X_\mu, X_\nu\}. \tag{7.15}$$

The action becomes

$$S_2 = \frac{T_2 \rho}{4} \int d^2\sigma dt \left(g_{\mu\nu}\partial_0 X^\mu \partial_0 X^\nu - \frac{2}{\rho^2}\{X^\mu, X^\nu\}\{X_\mu, X_\nu\} \right). \tag{7.16}$$

The second constraint becomes

$$g_{\mu\nu}\partial_0 X^\mu \partial_0 X^\nu = -\frac{2}{\rho^2}\{X^\mu, X^\nu\}\{X_\mu, X_\nu\}. \tag{7.17}$$

The first constraint $g_{\mu\nu}\partial_0 X^\mu \partial_i X^\nu = 0$ leads immediately to

$$g_{\mu\nu}\{\partial_0 X^\mu, X^\nu\} = 0. \tag{7.18}$$

The equation of motion deriving from the above action reads

$$\partial_0^2 X^\mu = \frac{4}{\rho^2}\{\{X^\mu, X^\nu\}, X_\nu\}. \tag{7.19}$$

As in the case of the string, there is a residual invariance which allows us to fix the gauge further. We choose the light-cone gauge

$$X^+(\tau, \sigma_1, \sigma_2) = \tau. \tag{7.20}$$

The light-cone coordinates are defined by

$$X^\pm = \frac{X^0 \pm X^{D-1}}{\sqrt{2}}. \tag{7.21}$$

In this gauge the number of degrees of freedom reduce from D to $D - 2$ since X^+ is fixed by the above condition while X^- is obtained by solving the constraints which take the form

$$\partial_0 X^- = \frac{1}{2}(\partial_0 X_a)^2 + \frac{1}{\rho^2}\{X_a, X_b\}^2, \quad \partial_i X^- = \frac{1}{2}\partial_0 X^a \partial_i X^a. \tag{7.22}$$

The indices a and b run from 1 to $D - 2$. The Hamiltonian of the remaining transverse degrees of freedom is computed as follows:

$$\mathcal{L}_2 = \frac{T_2\rho}{4}\left(\partial_0 X^- + \frac{1}{2}(\partial_0 X_a)^2 - \frac{1}{\rho^2}\{X_a, X_b\}^2\right) \tag{7.23}$$

$$X^- \longrightarrow P^+ = \frac{\delta \mathcal{L}_2}{\delta(\partial_0 X^-)} = \frac{T_2\rho}{4}, \quad X^a \longrightarrow P^a = \frac{\delta \mathcal{L}_2}{\delta(\partial_0 X^a)} = \frac{T_2\rho}{4}\partial_0 X_a \tag{7.24}$$

$$\begin{aligned}\mathcal{H}_2 &= P^+\partial_0 X^- + P^a \partial_0 X_a - \mathcal{L}_2\\ &= \frac{T_2\rho}{4}\left(\frac{1}{2}(\partial_0 X_a)^2 + \frac{1}{\rho^2}\{X_a, X_b\}^2\right)\end{aligned} \tag{7.25}$$

$$H = \int d^2\sigma\left(\frac{2}{T_2\rho}P_a^2 + \frac{T_2}{4\rho}\{X_a, X_b\}^2\right). \tag{7.26}$$

The remaining constraint is the statement

$$\{P^a, X^a\} = 0. \tag{7.27}$$

This light-cone theory, as opposed to the case of string theory, is still very difficult to quantize. A solution due to Nicolai and Hoppe was found in the case of a spherical membrane $\xi_1^2 + \xi_2^2 + \xi_3^2 = 1$ in [14] where functions on the sphere are mapped to $N \times N$ matrices, the Poisson brackets are replaced by Dirac commutation rules, the integral is replaced by an appropriately normalized trace, derivations by adjoint commutators, and the coordinates ξ_i are mapped to $SU(2)$ generators L_i in the irreducible representation of spin $s = (N - 1)/2$. The total number of degrees of freedom N^2 is equal to the number of linearly independent polarization tensors T_{lm} with $l \leqslant N - 1$ and they (these tensors) go in the large N limit to the usual spherical harmonics Y_{lm}. Functions on the spherical membrane are expanded in terms of Y_{lm}, whereas functions on the regularized (fuzzy) spherical membrane are expanded in terms of T_{lm}. This is essentially the philosophy of fuzzy spaces and fuzzy physics. In summary, the dictionary for passing to the regularized theory is

$$\xi_i \longrightarrow \frac{2}{N}L_i, \quad \{.,.\} \longrightarrow -\frac{iN}{2}[.,.],$$
$$\mathcal{L}_i = -i\epsilon_{ijk}x_j\partial_k \longrightarrow [L_i,.], \quad \int d^2\sigma \longrightarrow \frac{4\pi}{N}\,\mathrm{Tr}. \tag{7.28}$$

This can be generalized to membranes of arbitrary topology. But by supposing a spherical membrane for concreteness we obtain the Hamiltonian (with $\rho = N$ and $\pi T_2 = 1/2\pi l_p^3$)

$$H = \frac{1}{2\pi l_p^3}\,\mathrm{Tr}\left(\frac{1}{2}(\partial_0 X_a)^2 - \frac{1}{4}[X_a, X_b]^2\right). \tag{7.29}$$

In terms of the momentum $P_a = \pi T_2 \partial_0 X_a$ we get

$$H = 2\pi l_p^3\,\mathrm{Tr}\left(\frac{1}{2}P_a^2\right) - \frac{1}{2\pi l_p^3}\,\mathrm{Tr}\left(\frac{1}{4}[X_a, X_b]^2\right). \tag{7.30}$$

The constraint reads

$$[P^a, X^a] = 0. \tag{7.31}$$

This is the Gauss constraint, i.e. observables must be $U(N)$ invariant. Thus, the remaining invariance of the un-regularized Hamiltonian, which is time-independent area-preserving diffeomorphisms, is replaced in the regularized theory by $U(N)$ invariance.

The quantization of this finite system is straightforward precisely because it is finite although it remains non-trivial in practice.

Three other points are worth mentioning:

- A quantum supermembrane with 16 supercharges exists only in $D = 11$ dimensions. The light-cone Hamiltonian of the regularized supermembrane can be found to be of the form [17, 18]

$$H = \frac{1}{2\pi l_p^3}\,\mathrm{Tr}\left(\frac{1}{2}(\partial_0 X_a)^2 - \frac{1}{4}[X_a, X_b]^2 + \frac{1}{2}\psi^T\gamma^a[X^a, \psi]\right). \tag{7.32}$$

Recall that $a, b = 1, \ldots, D - 2$. The γ^a are 16×16 Euclidean $SO(9)$ gamma matrices and ψ is a 16-component Majorana spinor of $SO(9)$.

- The κ-symmetry of the classical supermembrane action guarantees that the background geometry solves the equations of motion of 11-dimensional supergravity [16, 18].
- The regularized supermembrane suffers from an instability due to flat directions which corresponds to a continuous spectrum [124]. This problem is absent in string theory (where the spectrum is discrete) and also is absent in the bosonic regularized membrane where despite the presence of flat directions the spectrum is discrete. This means that we can interpret the states of the theory as a discrete particle spectrum. This issue and its proposed resolution in terms of viewing the quantum theory as a second quantized theory from the point of view of the target space is nicely discussed in [17].

7.2 The IKKT model or type IIB matrix model

The IKKT model is equivalent to Connes' approach to geometry!

A commutative/noncommutative space in Connes' approach to geometry is given in terms of a spectral triple $(\mathcal{A}, \Delta, \mathcal{H})$ rather than in terms of a set of points [19]. $\mathcal{A}$ is the algebra of functions or bounded operators on the space, Δ is the Laplace operator or, in the case of spinors, the Dirac operator, and $\mathcal{H}$ is the Hilbert space on which the algebra of bounded operators and the differential operator Δ are represented.

In the IKKT model the geometry is in a precise sense emergent. And thus from this point of view it is obviously equivalent to Connes' noncommutative geometry. The algebra $\mathcal{A}$ is given, in the large N limit, by Hermitian matrices with smooth eigenvalue distributions and bounded square traces [20]. The Laplacian/Dirac operator is given in terms of the background solutions while the Hilbert space $\mathcal{H}$ is given by the adjoint representation of the gauge group $U(N)$.

We start immediately by presenting to you the fundamental model:

$$S_{\text{IKKT}} = \frac{1}{g^2}\,\text{Tr}\left(\frac{1}{4}[X_\mu, X_\nu][X^\mu, X^\nu] + \frac{1}{2}\bar{\Psi}^\alpha\Gamma^\mu_{\alpha\beta}[X_\mu, \Psi^\beta]\right). \tag{7.33}$$

This is the IKKT or IIB matrix model discovered in 1996 by Ishibashi, Kawai, Kitazawa, and Tsuchiya [3].

This has $\mathcal{N} = 2$ supersymmetry between the Hermitian $N \times N$ bosonic matrices X^μ, $\mu = 1, \ldots, D$, and the Hermitian $N \times N$ fermionic matrices Ψ_α, $\alpha = 1, \ldots, 2^{[D/2]}$. The first supersymmetry is inherited from the D-dimensional super Yang–Mills theory, while the second supersymmetry is a $U(1)$-shift of $U(1)$-components of fermionic matrices, which originates in the non-independence of the action on $\text{Tr}\,\Psi$ [21]. These two supersymmetries are given explicitly by

$$\delta\Psi^{IJ}_\alpha = \frac{i}{2}[X^\mu, X^\nu]^{IJ}(\Gamma_{\mu\nu}\epsilon)_\alpha + \delta^{IJ}\xi_\alpha, \quad \delta X^{IJ}_\mu = i\bar{\epsilon}\Gamma_\mu\Psi^{IJ}. \tag{7.34}$$

Another term which is invariant under the above supersymmetry, and which is important in the limit $N \longrightarrow \infty$, is given by

$$S_2 = \gamma_{\mu\nu} \operatorname{Tr} [X^\mu, X^\nu]. \tag{7.35}$$

In the above action, we have assumed an implicitly Euclidean signature, viz $\eta^{\mu\nu} = \delta^{\mu\nu}$, and that Γ are the Dirac matrices in D-dimensions in the Weyl representation. Indeed, the fermion Ψ is a complex Weyl spinor which satisfies also the Majorana reality condition. The isometry group is $SO(D)$, whereas the gauge group is obviously $U(N)$. In other words, the field/matrix Ψ provides a Majorana–Weyl representation of spin (D) whereas X^μ provides a vector representation of $SO(D)$.

This model is the reduction to 0-dimensions of $\mathcal{N} = 1$ super Yang–Mills theory in D dimensions. $\mathcal{N} = 1$ super Yang–Mills theories only exists in $D = 10, 6, 4, 3$. However, it can also be obtained from the Green–Schwarz action for IIB closed string theory, after gauge fixing in the Schild gauge, and with matrix regularization. In this latter case clearly D must be equal to 10. Alternatively, the IIB superstring action can be obtained from the IKKT model in the double scaling limit $N \longrightarrow \infty$, $g^2 \longrightarrow 0$, keeping Ng^2 fixed.

The theory is given by the partition function/path integral

$$Z = \int dX d\Psi \, \exp(-S_{\text{IKKT}}). \tag{7.36}$$

The configuration space (X, Ψ) defines a complex supermanifold, the action S_{IKKT} is a holomorphic function on this space, while physical observables are given by integrals with the weight $\exp(-S_{\text{IKKT}})$.

The convergence properties of this path integral in various dimensions are studied in [22]. The integral exists in $D = 10, 6, 4$.

But in Euclidean signature we have an important technical problem.

The fermion Ψ is a complex Weyl spinor which satisfies the Majorana reality condition and thus it contains only $2^{[D/2]-2}$ independent degrees of freedom. However, Majoranan–Weyl fermions do not exist in Euclidean signature. The absence of Majorana–Weyl fermions means that there is no $SO(D, C)$-invariant real cycle or slice (reality condition) in the space of Weyl spinors, which is required to perform the integral over the complex supermanifold. Fortunately, this is irrelevant since the result of the integration does not depend on the choice of cycle for odd variables (spinors). This very technical point is discussed nicely in [23, 24].

Next, it is said that the Euclidean IKKT model compactified on a circle gives the BFSS model at finite temperature [23]. This is the matrix quantum mechanics model discovered by Banks, Fischler, Shenker and Susskind [1].

We define a restriction of the IKKT action functional to the subspace where some gauge equivalence relation holds. This subspace consists of all points (X, Ψ) which remain in the same gauge class after a shift by a real number $2\pi R_0$ in the direction X_0 given explicitly in terms of unitary matrix U by [23, 24]

$$\begin{aligned} UX_0U^{-1} &= X_0 + 2\pi R_0 \mathbf{1} \\ UX_IU^{-1} &= X_I, \quad I \neq 0 \\ U\Psi^\alpha U^{-1} &= \Psi^\alpha. \end{aligned} \tag{7.37}$$

This cannot be satisfied for finite matrices.

We consider then the infinite-dimensional Hilbert space $\mathcal{H} = L_2(S^1) \otimes \mathcal{E}$ where $\mathcal{E}$ is some other Hilbert space which may or may not be finite-dimensional. The Hilbert space $\mathcal{H}$ is infinite-dimensional because of the factor $L_2(S^1)$. Thus, $\mathcal{H}$ is the space of functions f on the circle, i.e. $f = f(s)$, which take values in the Hilbert space $\mathcal{E}$, i.e. they are states in this vector space. Thus $\mathcal{H}$ is a direct sum of an infinite number of copies of $\mathcal{E}$.

We assume now that X and Ψ are operators in this infinite-dimensional Hilbert space $\mathcal{H}$.

A solution of the above gauge equivalence relation (7.37) is

$$\begin{aligned} X_0 &= 2\pi i R_0 \frac{\partial}{\partial\sigma} + A_0(\sigma) \\ X_I &= A_I(\sigma), \quad I \neq 0 \\ \Psi^a &= \chi^a(\sigma) \\ (Uf)(\sigma) &= e^{i\sigma} f(\sigma). \end{aligned} \tag{7.38}$$

Indeed, we have

$$UX_0U^{-1} = 2\pi i R_0 \frac{\partial}{\partial\sigma} + UA_0U^{-1} + 2\pi i R_0 U \frac{\partial}{\partial\sigma} U^{-1}. \tag{7.39}$$

Obviously, $0 \leqslant \sigma \leqslant 2\pi$ is a coordinate on $S^1 = R/2\pi Z$, and $A_0(\sigma)$, $A_i(\sigma)$ are functions on the circle taking values in the space of operators which act on $\mathcal{E}$. It can be shown that all other solutions to the gauge equivalence relation (7.37) are equivalent to the solution (7.38). See [23] for the elegant proof.

We want to return to finite-dimensional matrices. Hence, we assume that $\mathcal{E}$ is of finite dimension N, and also regularize the circle by a lattice of spacing a, and thus $\partial/\partial\sigma$ must be understood as a finite difference operator. However, under these conditions, the solution (7.38) becomes an approximate solution to (7.37).

By substituting the above approximate solution in the IKKT action we get (with the scaling $A \longrightarrow 2\pi R_0 A$, $X_i \longrightarrow 2\pi R_0 A_i$, $\chi \longrightarrow i\sqrt{2}(2\pi R_0)^{3/2}\chi$, and defining $D_0 = \partial/\partial\sigma - iA_0$) the action

$$S_{\mathrm{IKKT}} = \frac{(2\pi R_0)^4}{g^2} \mathrm{Tr}\left(\frac{1}{4}[A_I, A_J]^2 + \frac{1}{2}(D_0 A_I)^2 + \bar{\chi}\Gamma^I[A_I, \chi] - i\chi^\dagger D_0 \chi\right). \tag{7.40}$$

By the Weyl and Majorana conditions we can rewrite the 32-component spinor χ in terms of a 16-component spinor ψ satisfying $\bar{\psi} = \psi^+ = \psi^T C_9$ where C_9 is the charge conjugation in nine dimensions. Also by an appropriate choice of the 10-dimensional Dirac matrices Γ in terms of the 9-dimensional Dirac matrices γ (see next section) we can rewrite the above action as

$$S_{\mathrm{IKKT}} = \frac{(2\pi R_0)^4}{g^2} \mathrm{Tr}\left(\frac{1}{4}[A_I, A_J]^2 + \frac{1}{2}(D_0 A_I)^2 + \bar{\psi}\gamma^I[A_I, \psi] - i\bar{\psi} D_0 \psi\right). \tag{7.41}$$

Of course, the trace over the circle is a sum over the lattice sites in the σ direction. Finally we obtain, by taking the lattice spacing $a \longrightarrow 0$ while keeping the dimension N of the Hilbert space $\mathcal{E}$ fixed, the action

$$S_{\rm BFSS} = \frac{(2\pi R_0)^4}{g^2}\int d\sigma\, {\rm Tr}\left(\frac{1}{4}[A_I, A_J]^2 + \frac{1}{2}(D_0A_I)^2 + \bar{\psi}\gamma^I[A_I, \psi] - i\bar{\psi}D_0\psi\right). \quad (7.42)$$

By Wick rotation we obtain the BFSS quantum mechanics also known as the M-(atrix) theory. Thus the circle is actually a compact time direction (finite temperature).

This exercise is an example of T-duality.

However, I should say that compactification which is an analogue of dimensional reduction does not really describe what has just happened. We know that dimensional reduction can be achieved by taking one dimension to be a circle and then keeping only the zero modes, whereas in compactification we keep all modes in the lower dimensional theory. In a very clear sense the IKKT is the lower dimensional theory of the BFSS and thus the process of compactification should really be taking us from the BFSS to IKKT where the circle becomes a point in the infinite temperature limit. The description given above which escalated the IKKT to the BFSS is strictly speaking a decompactification where one extra dimension or coordinate has emerged from imposing an equivalence relation among gauge configurations along the X^0 direction.

7.3 The BFSS model from dimensional reduction

But what exactly is the BFSS model? The above derivation in terms of compactification on the circle is not mathematically clean since it involves escalation of the problem to infinite-dimensional matrices.

A closely related model to the IKKT matrix model is obtained by the reduction to one dimension of $\mathcal{N} = 1$ super Yang–Mills theory in $D = 10$ dimensions. This is a matrix quantum mechanics model which describes the low energy dynamics of a system of N type IIA D0-branes [2]. After this discovery, Banks, Fischler, Shenker, and Susskind conjectured that the large N limit of this model describes precisely M-theory in the infinite momentum frame ($P_z \longrightarrow \infty$) in the uncompactified limit ($R_s \longrightarrow \infty$). We remark that the infinite momentum limit is equivalent to the light-cone quantization only in the limit $N \longrightarrow \infty$. This is the same model derived above from compactification.

Let us perform now the dimensional reduction explicitly [25].

We start from $D = 10$ with metric $\eta^{MN} = (-1, +1, \ldots, +1)$. The Clifford algebra is 32-dimensional given by $\{G^M, G^N\} = 2\eta^{MN}\mathbf{1}_{32}$. The basic object of $\mathcal{N} = 1$ SUSY in 10 dimensions is a 32-component complex spinor Λ which satisfies the Majorana reality condition and the Weyl condition given by

$$\bar{\Lambda} = \Lambda^{+}G^{0} \equiv \Lambda^{T}C_{10}, \quad C_{10}G^{M}C_{10}^{-1} = (G^{M})^{T}$$
$$G_{11}\Lambda = \Lambda, \quad G_{11} = G^{0}\ldots G^{9}. \tag{7.43}$$

The $\mathcal{N} = 1$ supersymmetric action in $d = 10$ dimensions is given by

$$S = \frac{1}{g^2}\int d^{10}x \ \mathrm{Tr}\left[\left(-\frac{1}{4}F_{MN}F^{MN}\right)|_{d=10} + \frac{i}{2}(\bar{\Lambda}G^{M}D_{M}\Lambda)|_{d=10}\right]. \tag{7.44}$$

By using $D_M = \partial_M - i[A_M,..]$, $A_i = X_i$, $\partial_i = 0$ we can immediately compute the reduction of the fermionic action to $p + 1$ dimensions to be given by

$$-\frac{1}{2}(\bar{\Lambda}G^{M}D_{M}\Lambda)|_{d=10} = -\frac{1}{2}(\bar{\Lambda}G^{\mu}D_{\mu}\Lambda)|_{p+1} + \frac{i}{2}\bar{\Lambda}G^{i}[X_i, \Lambda]. \tag{7.45}$$

The reduction of the bosonic action to $p + 1$ dimensions is given by

$$\left(-\frac{1}{4}F_{MN}F^{MN}\right)|_{d=10} = \left(-\frac{1}{4}F_{\mu\nu}F^{\mu\nu}\right)_{p+1} + \frac{1}{4}[X_i, X_j]^2 - \frac{1}{2}(D_{\mu}X_i)(D^{\mu}X_i)$$
$$F_{\mu\nu} = \partial_{\mu}A_{\nu} - \partial_{\nu}A_{\mu} - i[A_{\mu}, A_{\nu}]. \tag{7.46}$$

The index μ runs over the values 0,1, ..., p, whereas the index i runs over the values $p + 1, p + 2, \ldots, 9$. As an example we write down the action on the 1-dimensional world-volume of N coincident D0-branes, i.e. $p = 0$. This is given by

$$S = \frac{1}{g^2}\int dt \ \mathrm{Tr}\left(\frac{1}{4}[X_I, X_J]^2 - \frac{1}{2}(D_0X_I)(D^0X_I) + \frac{1}{2}\bar{\Lambda}G^{I}[X_I, \Lambda] - \frac{i}{2}\Lambda^{+}D_0\Lambda\right). \tag{7.47}$$

We choose the following representation

$$G^0 = i\sigma_2 \otimes \mathbf{1}_{16}, \quad G^I = \sigma_1 \otimes \Gamma^I, \quad I = 1, \ldots, 9. \tag{7.48}$$

The charge conjugation decomposes as

$$C_{10} = \sigma_1 \otimes C_9. \tag{7.49}$$

The charge conjugation in nine dimensions satisfies $C_9\Gamma^I C_9^{-1} = (\Gamma^I)^T$. By the Weyl and Majorana conditions the 32-component spinor Λ can be rewritten in terms of a 16-component spinor Ψ as follows

$$\Lambda = \sqrt{2}\begin{pmatrix}\Psi\\0\end{pmatrix}, \quad \Psi^{+} = \Psi^{T}C_9. \tag{7.50}$$

The action becomes

$$S = \int dt\ L$$
$$L = \frac{1}{g^2}\,\mathrm{Tr}\left(\frac{1}{4}[X_I, X_J]^2 + \frac{1}{2}(D_0X_I)^2 + \Psi^T C_9 \Gamma^I [X_I, \Psi] - i\Psi^T C_9 D_0 \Psi\right). \tag{7.51}$$

The Clifford algebra in $D = 9$ dimensions can be taken to be given by [26]

$$\Gamma^a = \begin{pmatrix} -\sigma_a \otimes \mathbf{1}_4 & 0 \\ 0 & \sigma_a \otimes \mathbf{1}_4 \end{pmatrix}, \quad \Gamma^i = \begin{pmatrix} 0 & \mathbf{1}_2 \otimes \rho_i \\ \mathbf{1} \otimes (\rho_i)^+ & 0 \end{pmatrix}. \tag{7.52}$$

In the above $a = 1, \ldots, 3$, $i = 4, \ldots, 9$ and

$$\rho_i(\rho_j)^+ + \rho_j(\rho_i)^+ = (\rho_i)^+\rho_j + (\rho_j)^+\rho_i = 2\delta_{ij}\mathbf{1}_4. \tag{7.53}$$

The charge conjugation in $D = 9$ is

$$C_9 = \begin{pmatrix} 0 & -i\sigma_2 \otimes \mathbf{1}_4 \\ i\sigma_2 \otimes \mathbf{1}_4 & 0 \end{pmatrix}. \tag{7.54}$$

We can effectively work with a charge conjugation operator C_9 equal to $\mathbf{1}_8$ [123]. Integration over Ψ leads to the Pfaffian $\mathrm{Pf}(C_9\mathcal{O})$ where $\mathcal{O} = -iD_0 + \Gamma^I[X_I, \ldots]$. Hence the operator C_9 will drop from calculations. Thus we obtain the action

$$S = \int dt\ L$$
$$L = \frac{1}{g^2}\,\mathrm{Tr}\left(\frac{1}{4}[X_I, X_J]^2 + \frac{1}{2}(D_0X_I)^2 + \Psi^T \Gamma^I [X_I, \Psi] - i\Psi^T D_0 \Psi\right). \tag{7.55}$$

Now we must have $\Psi^+ = \Psi^T$. In the gauge $A_0 = 0$ we get the Lagrangian

$$L = \frac{1}{g^2}\,\mathrm{Tr}\left(\frac{1}{4}[X_I, X_J]^2 + \frac{1}{2}(\partial_0X_I)^2 + \Psi^T \Gamma^I [X_I, \Psi] - i\Psi^T \partial_0 \Psi\right). \tag{7.56}$$

The momentum conjugate corresponding to X_I is $P_I = \partial_0 X_I$, whereas the momentum conjugate corresponding to Ψ_α is $\Pi_\alpha = -i\Psi_\alpha$. The Hamiltonian is given by

$$H = \frac{1}{g^2}\,\mathrm{Tr}(P_I\partial_0X_I + \Pi^T\partial_0\Psi - L)$$
$$= \frac{1}{g^2}\,\mathrm{Tr}\left(\frac{1}{2}P_I^2 - \frac{1}{4}[X_I, X_J]^2 - \Psi^T\Gamma^I[X_I, \Psi]\right). \tag{7.57}$$

As opposed to the IKKT model, the BFSS model does not enjoy the full Lorentz group $SO(1,9)$. It is invariant only under $SO(9)$ which rotates the X_I among each other.

In order to obtain Euclidean signature we perform the Wick rotation $x^0 = t \longrightarrow -ix^0 = -it$. Thus $\partial_0 \longrightarrow i\partial_0$, $A_0 \longrightarrow iA_0$ and $D_0 \longrightarrow iD_0$. Also we change $L \longrightarrow -L$. The action becomes

$$S = \int dt \ L$$
$$L = \frac{1}{g^2} \mathrm{Tr}\left(-\frac{1}{4}[X_I, X_J]^2 + \frac{1}{2}(D_0 X_I)^2 - \Psi^T \Gamma^I [X_I, \Psi] - \Psi^T D_0 \Psi\right). \tag{7.58}$$

7.4 Introducing gauge/gravity duality

We will initially follow the excellent presentations of [27–29].

7.4.1 Dimensional reduction to $p + 1$ dimensions

The starting point is the action of $\mathcal{N} = 1$ supersymmetric Yang–Mills in $D = 10$ dimensions given by the functional

$$S = \frac{1}{g^2} \int d^D x \left[-\frac{1}{4} F_{\mu\nu} F^{\mu\nu} + \frac{i}{2} \bar{\Psi} \Gamma^\mu D_\mu \Psi\right]. \tag{7.59}$$

The g^2 is the Yang–Mills coupling. Obviously, Γ^μ is the 10-dimensional Clifford algebra which is 32-dimensional. This theory consists of a gauge field A_μ, $\mu = 0, \ldots, 9$, with field strength $F_{\mu\nu} = \partial_\mu A_\nu - \partial_\nu A_\mu - i[A_\mu, A_\nu]$ coupled in a supersymmetric fashion to a Majorana–Weyl spinor Ψ in the adjoint representation of the gauge group. Since Ψ is a Majorana–Weyl it can be rewritten as[1] (see also previous section)

$$\Psi = \begin{pmatrix} \psi \\ 0 \end{pmatrix}. \tag{7.60}$$

Hence the above action takes the form

$$S = \frac{1}{g^2} \int d^D x \left[-\frac{1}{4} F_{\mu\nu} F^{\mu\nu} + \frac{i}{2} \bar{\psi} \gamma^\mu D_\mu \psi\right]. \tag{7.61}$$

Now, γ^μ are the left-handed sector of the 10-dimensional gamma matrices[2]. The gauge field A and the Majorana–Weyl spinor ψ are $N \times N$ Hermitian matrices which are represented by complex and Grassmann numbers, respectively, in the measure of the path integral. Since Ψ is in the adjoint representation the covariant derivative $D_\mu \psi$ is defined via commutator, i.e. $D_\mu \psi = \partial_\mu \psi - i[A_\mu, \psi]$. The above action enjoys thus $U(N)$ gauge invariance given by

---

[1] In general a spinor in D (even) dimension will have $2^{D/2}$ components. The Majorana condition is a reality condition which reduces the number of components by a factor of two, whereas the Weyl condition is a chirality condition (left-handed or right-handed) which reduces further the number of independent components down by another factor of two. In four dimensions these two conditions are not mutually consistent because there is no a real irreducible representation of the spinor bundle, whereas in 10 dimensions we can impose both conditions at once. Hence in 10 dimensions we have $2^5/4 = 16$ complex components or 16 real components ψ_α.

[2] In particular $\gamma^0 = -1$, γ^a, $a = 1, \ldots, 9$, are the Dirac matrices in $D = 9$, and $\bar{\psi} = \psi^T C_9$. See the previous section.

$$D_\mu \longrightarrow g D_\mu g^\dagger, \quad \psi \longrightarrow g \psi g^\dagger. \tag{7.62}$$

But it does also enjoy $\mathcal{N} = 1$ supersymmetry given by

$$A_\mu \longrightarrow A_\mu + i\bar{\epsilon}\gamma_\mu\psi, \quad \psi \longrightarrow \psi + \frac{1}{2}F_{\mu\nu}\gamma^{\mu\nu}\epsilon. \tag{7.63}$$

Hence, there are 16 real supercharges corresponding to the 16 components of ϵ.

Dimensional reduction to $D = p + 1$ dimensions consists in keeping only the coordinates $x_0, x_1, \ldots, x_p$ and dropping all dependence on the coordinates $x_{p+1}, \ldots x_9$. The gauge fields A_a in the directions $a = p + 1, \ldots, 9$ become scalar fields X_I where $I = a - p = 1, 2, \ldots, 9 - p$. The action becomes given immediately by

$$\begin{aligned} S = \frac{V_{9-p}}{g^2} \int d^{p+1}x \Big[&-\frac{1}{4}F_{\mu\nu}F^{\mu\nu} - \frac{1}{2}(D_\mu X_I)(D^\mu X_I) \\ &+ \frac{1}{4}[X_I, X_J]^2 + \frac{i}{2}\bar{\psi}\gamma^\mu D_\mu\psi + \frac{1}{2}\bar{\psi}\gamma^I[X_I, \psi]\Big]. \end{aligned} \tag{7.64}$$

In the above action the indices μ and ν run from 0 to p, whereas the indices I and J run from 1 to $D - 1 - p$ where $D = 10$. This action is maximally supersymmetric enjoying 16 supercharges. More importantly, it was established more than 20 years ago that this action gives an effective description of Dp-branes in the low energy limit [2, 31, 32].

7.4.2 Dp-branes revisited

But what are Dp-branes?

In the words of Polchinski [31]: '..Dirichlet(p)-branes (are) extended objects defined by mixed Dirichlet–Neumann boundary conditions in string theory (which) break half of the supersymmetries of the type II superstring and carry a complete set of electric and magnetic Ramond–Ramond charges...'.

Thus a Dp-brane is an extended object with p spatial dimensions evolving in time. Let us consider type II superstring theory which contains only closed strings. The presence of a stable supersymmetric Dp-brane with p even for type IIA and with p odd for type IIB modifies the allowed boundary conditions of the strings. Both Neumann and Dirichlet boundary conditions are now allowed. Hence in addition to the closed strings, which propagate in the bulk (1 + 9)-dimensional spacetime, we can have open strings whose endpoints are fixed on the Dp-brane.

Recall that for worldsheets with boundaries there is a surface term in the variation of the Polyakov action which vanishes iff:

- $$\partial^\sigma x^\mu(\tau, 0) = \partial^\sigma x^\mu(\tau, l) = 0. \tag{7.65}$$

  These are Neumann boundary conditions, i.e. the ends of the open string move freely in spacetime, and the component of the momentum normal to the boundary of the worldsheet vanishes.

•
$$x^\mu(\tau, 0) = x^\mu(\tau, l). \tag{7.66}$$

The fields are periodic which corresponds to a closed string.

These are the only boundary conditions which are consistent with D-dimensional Poincaré invariance.

However, in the presence of a Dp-brane defined for example by the condition

$$x^{p+1} = \cdots = x^9 = 0 \tag{7.67}$$

the open strings must necessarily end on this surface and thus they must satisfy Dirichlet boundary conditions in the directions $p + 1$,..,9 while in the directions 0,..., p they must satisfy the usual Neumann boundary conditions. In other words, we must have

$$\begin{aligned} \partial_\sigma x^0 &= \cdots = \partial_\sigma x^p = 0, \quad \text{at } \sigma = 0, \pi \\ x^{p+1} &= \cdots = x^9 = 0, \quad \text{at } \sigma = 0, \pi. \end{aligned} \tag{7.68}$$

Thus, the coordinates x^0, x^1, ..., x^p obey Neumann boundary conditions while the coordinates x^{p+1}, ..., x^9 obey Dirichlet boundary conditions. As it turns out the reduced spacetime defined by the longitudinal coordinates (x^0, x^1, ..., x^p) corresponds to the worldvolume of the Dp-brane, i.e. the $U(N)$ gauge theory (7.64) is living on the Dp-brane. However, this is really a theory of N coincident branes and not a single brane. Indeed, the diagonal elements (X_1^{ii}, X_2^{ii}, ... , X_{9-p}^{ii}) define the position of the ith Dp-brane along the transverse directions (x^{p+1}, x^{p+2}, ..., x^9), whereas the off diagonal elements (X_1^{ij}, X_2^{ij}, ... , X_{9-p}^{ij}) define open string excitations between the ith and jth Dp-branes. See for example [32].

As we have already discussed in the previous chapter, this fundamental result can also be seen as follows.

In addition to the closed strings of the type II superstring theory we can have one class of open strings starting and ending on the Dp-brane. The quantization of these open strings is equivalent to the quantization of the ordinary open superstrings. See for example [33].

On the other hand we know that the massless modes of type I open superstrings consist of a Majorana–Weyl spinor ψ and a gauge field A_a with dynamics given at low energy by a supersymmetric $U(1)$ gauge theory in 10 dimensions. The gauge field A_a depends only on the zero modes of the coordinates x^0, x^1, x^2, ..., x^p since the zero modes of the other coordinates vanish due to the Dirichlet boundary condition $x^{p+1} = \cdots = x^9 = 0$ at $\sigma = 0, \pi$. The reduced gauge field A_a, $a = 0, \ldots, p$, behaves as a $U(1)$ vector on the p-brane worldvolume while the other components A_a with $a = p + 1, \ldots, 9$ behave as scalar fields normal to the p-brane. These scalar fields describe therefore fluctuations of the position of the p-brane.

Therefore, at low energy the theory on the (p+1)-dimensional worldvolume of the Dp-brane is the reduction to $p + 1$ dimensions of 10-dimensional supersymmetric $U(1)$ gauge theory [2]. When the velocities and/or the string coupling are not small the minimal supersymmetric Yang–Mills is replaced by the supersymmetric Born–Infeld action [34].

Let us consider now two parallel Dp-branes located at $X^a = 0$ and $X^a = Y^a$, respectively, where $a = p + 1, \ldots, 9$. In addition to the closed strings of the type II superstring theory we can have four different classes of open strings. The first class consists of open strings which begin and end on the first Dp-brane $X^a = 0$, $a = p + 1, \ldots, 9$. The second class consists of open strings which begin and end on the second Dp-brane $X^a = Y^a$, $a = p + 1, \ldots, 9$. The third and fourth classes consist of open strings which begin on one Dp-brane and end on the other Dp-brane. These are stretched strings. These four classes can be labeled by pairs $[ij]$ where $i, j = 1,2$. These are the Chan–Paton indices. The open strings in the $[ij]$ sector extend from the Dp-brane i to the Dp-brane j. As we have already seen the open string sectors [11] and [22] give rise to $U(1) \times U(1)$ gauge theory where the first $U(1)$ lives on the first brane and the second $U(1)$ lives on the second brane. The massless open string bosonic ground state includes for each factor of $U(1)$ a massless $(p{+}1)$-vector field and $9{-}p$ massless normal scalars propagating on the $(p{+}1)$-dimensional worldvolumes of the Dp-branes.

The stretched strings in the sectors [12] and [21] will have a massive bosonic ground state. The mass being equal to $T|Y|$ where T is the string tension. First, let us assume that $Y^a = 0$. Then the normal coordinates to the brane will correspond to $9{-}p$ normal scalar fields while the tangent coordinates to the brane give rise to $(p{+}1) - 2 = p - 1$ states. However, because $Y^a \neq 0$ we will have instead a massive $(p{+}1)$-vector field with p (and not $p - 1$) independent components and $8{-}p$ massive normal scalars. This is because the normal scalar field parallel to the unique normal direction Y^a which takes us from one brane to the other is in fact a part of the massive gauge boson. Let us also say that these $(p{+}1)$-dimensional fields live on a fixed $(p{+}1)$-dimensional space not necessarily identified with any of the Dp-branes. Alternatively, we can say that these fields live on the two Dp-branes at the same time. These fields have clearly non-local interactions since the two Dp-branes are separated. The spacetime interpretation of these fields seem to require noncommutative geometry [33].

Now we take the limit in which the separation between the two Dp-branes goes to zero. The two Dp-branes remain distinguishable. In this case the open string bosonic ground state which represents strings starting on one brane and ending on the other becomes massless. Therefore, the open string bosonic ground state includes from each of the four sectors $[ij]$ a massless $(p{+}1)$-vector field and $9{-}p$ massless normal scalar fields. In other words, when the two Dp-branes coincide we get four (and not two) massless $(p{+}1)$-vector fields giving a $U(2)$ Yang–Mills gauge theory on the $(p{+}1)$-worldvolume of the coincident Dp-branes. In addition, we will have $9{-}p$ massless normal scalar fields in the adjoint representation of $U(2)$ which will play the role of position coordinates of the coincident Dp-branes. The coordinates became noncommuting matrices. At low energy the theory on the $(p{+}1)$-dimensional worldvolume of the two coincident Dp-branes is therefore the reduction to $p + 1$ dimensions of 10-dimensional supersymmetric $U(2)$ gauge theory [2]. As before when the velocities and/or the string coupling are not small this minimal supersymmetric Yang–Mills is replaced by a non-abelian generalization of the supersymmetric Born–Infeld action which is still not known.

Generalization to N coincident Dp-branes is obvious: at low energy (i.e. small velocities and small string coupling) the theory on the (p+1)-dimensional worldvolume of N coincident Dp-branes is the reduction to $p + 1$ dimensions of 10-dimensional supersymmetric $U(N)$ gauge theory. This will be corrected by higher dimension operators when the velocities and/or the string coupling are not small. In some sense the use of Yang–Mills action instead of the Born–Infeld action is equivalent to a nonrelativistic approximation to the dynamics of the coincident Dp-branes.

7.4.3 The corresponding gravity dual

The gauge/gravity duality in the words of [27] is the statement that 'Hidden within every non-abelian gauge theory, even within the weak and strong nuclear interactions, is a theory of quantum gravity'.

Let us then consider a system of N coincident Dp-branes described by the $U(N)$ gauge theory (7.64). This is a theory characterized by the gauge coupling constant $g_{\mathrm{YM}}^2 \equiv g^2/V_{9-p}$ and the number of colors N. A gauge theory with an infinite number of degrees of freedom, which is the one relevant to supergravity and superstring and which must live in more dimensions by the holographic principle [35, 36] in order to avoid the Weinberg–Witten no-go theorem [37], is given by the 't Hooft planar limit in which N is taken large and g_{YM}^2 taken small keeping fixed the 't Hooft coupling λ given by [38]

$$\lambda = g_{\mathrm{YM}}^2 N. \tag{7.69}$$

There is another parameter in the gauge theory which is the temperature $T = 1/\beta$. This is obtained by considering the finite temperature path integral over periodic Euclidean time paths with period β, i.e. by compactifying the Euclidean time direction on a circle of length equal β. See for example [39].

We will study the Dp-branes in the field theory limit defined by

$$g_{\mathrm{YM}}^2 = (2\pi)^{p-2} g_s (\alpha')^{\frac{p-3}{2}} = \text{fixed}, \quad \alpha' \longrightarrow 0. \tag{7.70}$$

The $\alpha' = l_s^2$ is the inverse of the string tension and g_s is the string coupling constant related to the dilaton field by the expression

$$g_s = \exp(\Phi). \tag{7.71}$$

The system of N coincident Dp-branes is obviously a massive object which will curve the spacetime around it. As such it is called a black p-brane and is a higher-dimensional analogue of the black hole. The singularity in a black p-brane solution is extended in p-spatial directions. The radial (transverse) distance from the center of mass of the black p-brane to a given point in spacetime will be denoted by U. More precisely, $U = r/\alpha'$. From the gauge theory point of view U is the energy scale.

The limit (7.70) will be taken in such a way that we keep U fixed, i.e. we are interested in finite energy configurations in the gauge theory [40].

It is known that type II supergravity solution which describes N coincident extremal ($T = 0$) Dp-branes is given by the metric [40–42]

$$ds^2 = \frac{1}{\sqrt{f_p}}(-dt^2 + dx_1^2 + \cdots + dx_p^2) + \sqrt{f_p}(dx_{p+1}^2 + \cdots + dx_9^2), \tag{7.72}$$

where

$$f_p = 1 + \frac{d_p g_{\mathrm{YM}}^2 N}{\alpha'^2 U^{7-p}}, \quad d_p = 2^{7-2p}\pi^{\frac{9-3p}{2}}\Gamma\left(\frac{7-p}{2}\right). \tag{7.73}$$

The near extremal p-brane solutions in the limit (7.70) keeping the energy density on the brane finite, and which corresponds to the above finite temperature gauge theory, are given by the metric [40–42]

$$\begin{aligned} ds^2 = \alpha' \Bigg[& \frac{U^{\frac{7-p}{2}}}{g_{\mathrm{YM}}\sqrt{d_p N}}\left(-\left(1 - \frac{U_0^{7-p}}{U^{7-p}}\right)dt^2 + dy_{\|}^2\right) \\ & + \frac{g_{\mathrm{YM}}\sqrt{d_p N}}{U^{\frac{7-p}{2}}} \frac{dU^2}{\left(1 - \frac{U_0^{7-p}}{U^{7-p}}\right)} + g_{\mathrm{YM}}\sqrt{d_p N}\, U^{\frac{p-3}{2}} d\Omega_{8-p}^2 \Bigg]. \end{aligned} \tag{7.74}$$

This is the corresponding metric of the curved spacetime around the black p-brane. The $(t, y_{\|})$ are the coordinates along the p-brane worldvolume, Ω_{8-p} is the transverse solid angle associated with the transverse radius U. The U_0 corresponds to the radius of the horizon and it is given in terms of the energy density of the brane E (which is precisely the energy density of the gauge theory) by

$$U_0^{7-p} = a_p g_{\mathrm{YM}}^4 E, \quad a_p = \frac{\Gamma\left(\frac{9-p}{2}\right)2^{11-2p}\pi^{\frac{13-3p}{2}}}{9-p}. \tag{7.75}$$

The string coupling constant in this limit is given by

$$\exp(\phi) = (2\pi)^{2-p} g_{\mathrm{YM}}^2 \left(\frac{d_p g_{\mathrm{YM}}^2 N}{U^{7-p}}\right)^{\frac{3-p}{4}}. \tag{7.76}$$

In order to determine the Hawking temperature we introduce the Euclidean time $t_E = it$ and define the proper distance from the horizon by the relation

$$d\rho^2 = \frac{g_{YM}\sqrt{d_p N}}{U^{\frac{7-p}{2}}} \frac{dU^2}{\left(1 - \frac{U_0^{7-p}}{U^{7-p}}\right)}. \tag{7.77}$$

Near the horizon we write $U = U_0 + \delta$ then

$$\rho^2 = \frac{4g_{YM}\sqrt{d_p N}}{(7-p)U_0^{\frac{5-p}{2}}}\delta = 4r_s\delta. \tag{7.78}$$

The relevant part of the metric becomes

$$ds^2 = \alpha'\left[\frac{\rho^2}{4r_s^2}dt^2 + d\rho^2 + \cdots\right]. \tag{7.79}$$

The first two terms correspond to 2-dimensional flat space, viz

$$X = \rho\cos\frac{t_E}{2r_s}, \quad Y = \rho\sin\frac{t_E}{2r_s}. \tag{7.80}$$

Hence in order for the Euclidean metric to be smooth the Euclidean time must be periodic with period $\beta = 4\pi\, r_s$ otherwise the metric has a conical singularity at $\rho = 0$. We then get the temperature

$$T = \frac{1}{4\pi r_s} = \frac{(7-p)U_0^{\frac{5-p}{2}}}{4\pi g_{YM}\sqrt{d_p N}}. \tag{7.81}$$

The entropy density is given by the Bekenstein–Hawking formula

$$\mathcal{S} = \frac{A}{4}, \tag{7.82}$$

where A is the area density of the horizon. This can be calculated from the first law of thermodynamics as follows

$$\begin{aligned} d\mathcal{S} &= \frac{1}{T}dE \\ &= \frac{4\pi g_{YM}\sqrt{d_p N}}{(7-p)U_0^{\frac{5-p}{2}}}dE \\ &= \frac{4\pi g_{YM}\sqrt{d_p N}}{(7-p)\left(a_p g_{YM}^4\right)^{\frac{5-p}{2(7-p)}}} \frac{dE}{E^{\frac{5-p}{2(7-p)}}} \\ &= \frac{4\pi g_{YM}\sqrt{d_p N}}{(7-p)\left(a_p g_{YM}^4\right)^{\frac{5-p}{2(7-p)}}} \frac{dE^{\frac{9-p}{2(7-p)}}}{\frac{9-p}{2(7-p)}}. \end{aligned} \tag{7.83}$$

We get then the entropy

$$\mathcal{S} = \frac{8\pi\sqrt{d_pN}}{(9-p)a_p^{\frac{5-p}{2(7-p)}}} g_{\rm YM}^{\frac{p-3}{7-p}} E^{\frac{9-p}{2(7-p)}}. \tag{7.84}$$

In summary, we have found that maximally supersymmetric (p+1)-dimensional $U(N)$ gauge theory is equivalent to type II superstring theory around black p-brane background spacetime. This is the original conjecture of Maldacena that weakly coupled super Yang–Mills theory and weakly coupled type II superstring theory both provide a description of N coincident Dp-branes forming a black p-brane [43]. This equivalence should be properly understood as a non-perturbative definition of string theory since the gauge theory is rigorously defined by a lattice à la Wilson [44]. More precisely, we have [27–29]

- The gauge theory in the limit $N \longrightarrow \infty$ (where extra dimensions will emerge) and $\lambda \longrightarrow \infty$ (where strongly quantum gauge fields give rise to effective classical gravitational fields) should be equivalent to classical type II supergravity around the p-brane spacetime.
- The gauge theory with $1/N^2$ corrections should correspond to quantum loop corrections, i.e. corrections in g_s, in the gravity/string side.
- The gauge theory with $1/\lambda$ corrections should correspond to stringy corrections, i.e. corrections in l_s, corresponding to the fact that degrees of freedom in the gravity/string side are really strings and not point particles.

7.5 Black hole unitarity from M-theory

7.5.1 The black 0-brane

The case of $p = 0$ is of particular interest to us here. The gauge theory in this case is a maximally supersymmetric $U(N)$ quantum mechanics given by the Wick rotation to Euclidean signature of the action (7.64) with $p = 0$, viz

$$S = \frac{1}{g_{\rm YM}^2}\int_0^\beta dt\left[\frac{1}{2}(D_tX_I)^2 - \frac{1}{4}[X_I, X_J]^2 \right. \\ \left. - \frac{1}{2}\bar{\psi}\gamma^0 D_0\psi - \frac{1}{2}\bar{\psi}\gamma^I[X_I, \psi]\right]. \tag{7.85}$$

This is the BFSS quantum mechanics. This describes a system of N coincident D0-branes forming a black 0-brane, i.e. a black hole. The corresponding metric is given by the type IIA supergravity solution

$$ds^2 = \alpha' \left[-\frac{U^{\frac{7}{2}}}{g_{\rm YM}\sqrt{d_0 N}} \left(1 - \frac{U_0^7}{U^7}\right) dt^2 + \frac{g_{\rm YM}\sqrt{d_0 N}}{U^{\frac{7}{2}}} \frac{dU^2}{\left(1 - \frac{U_0^7}{U^7}\right)} + g_{\rm YM}\sqrt{d_0 N}\, U^{\frac{-3}{2}} d\Omega_8^2 \right]. \tag{7.86}$$

We write this as

$$ds^2 = \alpha' \left[-\frac{F(U)}{\sqrt{H}} dt^2 + \frac{\sqrt{H}}{F(U)} dU^2 + \sqrt{H}\, U^2 d\Omega_8^2 \right], \tag{7.87}$$

where

$$F(U) = 1 - \frac{U_0^7}{U^7}, \quad H = \frac{\lambda d_0}{U^7}, \quad d_0 = 240\pi^5. \tag{7.88}$$

The temperature, the energy and the entropy are

$$T = \frac{7 U_0^{\frac{5}{2}}}{4\pi\sqrt{d_0 \lambda}} \tag{7.89}$$

$$E = \frac{U_0^7}{a_0 g_{\rm YM}^4} \Rightarrow \frac{E}{N^2} = \frac{1}{a_0}\left(\frac{4\pi\sqrt{d_0}}{7}\right)^{14/5} \lambda^{-3/5} T^{14/5}. \tag{7.90}$$

$$\mathcal{S} = \frac{8\pi\sqrt{d_0 N}}{9 a_0^{\frac{5}{14}}} g_{\rm YM}^{\frac{-3}{7}} E^{\frac{9}{14}} \Rightarrow \frac{\mathcal{S}}{N^2} = \frac{14}{9 a_0}\left(\frac{4\pi\sqrt{d_0}}{7}\right)^{14/5} \lambda^{-3/5} T^{9/5}. \tag{7.91}$$

This black 0-brane solution is one very important example of matrix black holes [45–54]. We will now show that this dual geometry for the BFSS model can be lifted to a solution to 11-dimensional supergravity.

7.5.2 Supergravity in 11 dimensions and M-wave solution

M-theory is not a field theory on an ordinary spacetime except approximately in the low energy perturbative region and it is not a string theory except via circle compactification. In this theory, understood as M-(atrix)-theory or AdS/CFT, there is the remarkable possibility of dynamically growing new dimensions of space, topology change, spontaneous emergence of geometry from 0 dimensions, and more dramatically we can have emergence of gravitational theories from gauge theories.

The five superstring theories as well as 11-dimensional supergravity are believed to be limits of the same underlying 11-dimensional theory which came to be called M-theory. At low energy M-theory is approximated by 11-dimensional supergravity, i.e. M-theory is by construction the UV-completion of 11-dimensional supergravity.

It was shown in [55] that the largest number of dimensions in which supergravity can exist is 11. This is the largest number of dimensions consistent with a single graviton. Beyond 11 dimensions spinors have at least 64 components which lead to massless fields with spins larger than two which have no consistent interactions.

The field content of 11-dimensional supergravity consists of a vielbein e_M^A (or equivalently the metric g_{MN}), a Majorana gravitino ψ_M of spin 3/2, and a 3-form gauge potential $A_{MNP} \equiv A_3$ (with field strength $F_4 = dA_3$). There is a unique classical action [56]. The bosonic part of this action reads

$$S_{11} = \frac{1}{2k_{11}^2} \int d^{11}x\sqrt{-g}\left(R - \frac{1}{2}\,|F_4|^4\right) - \frac{1}{12k_{11}^2} \int A_3 \wedge F_4 \wedge F_4. \tag{7.92}$$

The form action is given explicitly by

$$\begin{aligned} \int d^{11}x\sqrt{-g}\;|F_4|^2 &= \int d^{11}x\sqrt{-g}\,F_4 \wedge^* \\ F_4 &= \int d^{11}x\sqrt{-g}\,\frac{1}{4!}g^{M_1N_1}\,\ldots\,g^{M_4N_4}F_{M_1\ldots M_4}F_{N_1\ldots N_4}. \end{aligned} \tag{7.93}$$

Newton's constant in 11 dimensions is given by $2k_{11}^2/16\pi$, viz

$$G_{11} = \frac{2k_{11}^2}{16\pi}. \tag{7.94}$$

In D dimensions the gravitational force falls with the distance r as $1/r^{D-2}$ and the Newton's and the Planck constants are related by the relation [33]

$$16\pi G_D = 2k_D^2 = \frac{1}{2\pi}(2\pi l_P)^{D-2}. \tag{7.95}$$

Let us count the number of degrees of freedom [56–58]. The metric is a traceless symmetric tensor. It is contained in the symmetric part of the tensor product of two fundamental vector representations of $SO(9)$. This is because each index of the metric is a vector index which can take only $D - 2 = 11 - 2 = 9$ values. Since the metric tensor must also be a traceless tensor it will contain in total $\frac{9^2-9}{2} + 9 - 1 = 44$ degrees of freedom.

The gauge potential A_3 must be contained in the antisymmetric part of the tensor product of three fundamental vector representations of $SO(9)$. The independent number of A_{MNP} is obviously given by $\frac{9\times8\times7}{3!} = 84$. The total number of bosonic variables is thus given by $44 + 84 = 128$.

In 11 dimensions the Dirac matrices are 32×32. The gravitino is a Majorana spin $\frac{3}{2}$ field which carries vector and spinor indices. The Rarita–Schwinger gravitino field must clearly be contained in the tensor product of the fundamental vector representation and the spinor representation of $SO(9)$. There are only $D - 3 = 11 - 3 = 8$ propagating components for a spin $\frac{3}{2}$ field in the same way that a spin 1 field has only $D - 2 = 11 - 2 = 9$ propagating components. The independent number of fermionic degrees of freedom is therefore $\frac{32 \times 8}{2} = 128$ where the division by two is due to the Majorana condition. The numbers of bosonic and fermionic degrees of freedom match.

Supergravity in 11 dimensions contains a 2-brane solution which is an electrically charged configuration with respect to A_{MNP} and a 5-brane solution which is a magnetically charged configuration with respect to A_{MNP}. These supergravity 2-brane and 5-brane solutions are the low-energy limits of the M2- and M5-branes, respectively, of M-theory.

The equations of motion of 11-dimensional supergravity are given by the following Einstein–Maxwell system

$$R_{MN} - \frac{1}{2}g_{MN}R = \frac{1}{2}F_{MN}^2 - \frac{1}{4}g_{MN}\,|F_4|^2 \tag{7.96}$$

$$d{*}F_4 + \frac{1}{2}F_4 \wedge F_4 = 0, \quad dF_4 = 0. \tag{7.97}$$

Type IIA supergravity theory can be constructed by dimensional reduction of supergravity in 11 dimensions on a circle S^1. This, as it turns out, is a consequence of the deeper fact that M-theory compactified on a circle of radius R is equivalent to type IIA superstring in 10 dimensions with coupling constant $g_s = R/\sqrt{\alpha'}$ (more on this below). Kaluza–Klein dimensional reduction is obtained by keeping only the zero modes in the Fourier expansions along the compact direction, whereas compactification is obtained by keeping all Fourier modes in the lower-dimensional theory.

The dimensional reduction of the metric g_{MN} is specified explicitly in terms of a 10-dimensional metric g_{mn}, a $U(1)$ gauge field A_m and a scalar dilaton field Φ, by [59]

$$ds^2 = g_{MN}dx^M dx^N = e^{-\frac{2\Phi}{3}}g_{mn}dx^m dx^n + e^{\frac{4\Phi}{3}}(dx_{10} + A_m dx^m)^2. \tag{7.98}$$

Thus the distance l_{11} between any two points as viewed in the 11-dimensional theory should be viewed as a distance l_{10} in the 10-dimensional theory which are related by $l_{10} = g_s^{-1/3} l_{11}$ where $g_s = \langle \exp(\Phi) \rangle$ is the string coupling constant. Thus the Planck scale l_P in 11 dimensions is related to the string length[3] $l_s = \sqrt{\alpha'}$ in 10 dimensions by the relation

$$l_P = g_s^{1/3} l_s. \tag{7.99}$$

[3] Note that the string length is sometimes defined by $l_s = \sqrt{\alpha'}$ and sometimes by $l_s = \sqrt{2\alpha'}$.

It is also not difficult to convince oneself that the 11-dimensional coordinate transformation $x_{10} \longrightarrow x_{10} + f(x_{10})$ is precisely equivalent to the 10-dimensional $U(1)$ gauge transformation $A \longrightarrow A - df$.

The three-form A_3 in 11 dimensions gives a three-form C_3 and a two-form B_2 in 10 dimensions given by

$$C_3 = A_{mnq}dx^m \wedge dx^n \wedge dx^q, \quad B_2 = A_{10mn}dx^m \wedge dx^n. \tag{7.100}$$

From the action (7.92) we see that the 10-dimensional Newton's constant is given in terms of the 11-dimensional Newton's constant by the relation

$$G_{10} = \frac{G_{11}}{2\pi R} = \frac{2k_{10}^2}{16\pi}. \tag{7.101}$$

The Newton's constant G_{10}, the gravitational coupling constant k_{10}, the string length l_s and the coupling constant g_s are related by the formula (see below)

$$16\pi G_{10} = 2k_{10}^2 = \frac{1}{2\pi}g_s^2(2\pi l_s)^8. \tag{7.102}$$

We obtain immediately

$$\frac{2k_{11}^2}{2\pi R} = 2k_0^2 g_s^2, \quad 2k_0^2 = \frac{1}{2\pi}(2\pi l_s)^8 \tag{7.103}$$

and

$$R = g_s l_s. \tag{7.104}$$

Thus, the original metric g_{MN} corresponds to the 10-dimensional fields g_{mn}, Φ and A_1, whereas the original three-form A_3 corresponds to the three-form C_3 and the two-form B_2. The leading $\alpha' = l_s^2$ low energy effective action of the bosonic $D = 11$ supergravity action, which is obtained by integrating over the compact direction, is given precisely by the action of type IIA supergravity in $D = 10$ dimensions. This action consists of three terms

$$S = S_{\mathrm{NS}} + S_{\mathrm{R}} + S_{\mathrm{CS}}. \tag{7.105}$$

The action S_{NS} describes the fields of the NS–NS sector, viz (g_{mn}, Φ, B_2). The field strength of B_2 is denoted $H_3 = dB_2$. It is given in the string frame explicitly by

$$S_{\mathrm{NS}} = \frac{1}{2k_0^2}\int d^{10}x\sqrt{-g}\exp(-2\Phi)\left(R + 4\partial_\mu\Phi\partial^\mu\Phi - \frac{1}{2}|H_3|^2\right). \tag{7.106}$$

The action S_{R} describes the fields of the R–R sectors, viz (A_1, C_3). The field strengths are given, respectively, by $F_2 = dA_1$ and $\tilde{F}_4 = dC_3 + A_1 \wedge H_3$. The action is given by

$$S_{\mathrm{R}} = -\frac{1}{4k_0^2}\int d^{10}x\sqrt{-g}(|F_2|^2 + |\tilde{F}_4|^2). \tag{7.107}$$

The Chern–Simons term is given obviously by (where $F_4 = dA_3$)

$$S_{\rm CS} = -\frac{1}{4k_0^2}\int B_2 \wedge F_4 \wedge F_4. \tag{7.108}$$

The solution of 11-dimensional gravity which is relevant for the dual geometry to the BFSS quantum mechanics corresponds the non-extremal M-wave solution. This corresponds to $A_3 = 0$, i.e. $C_3 = 0$ and $B_2 = 0$. The 11-dimensional metric corresponds to [7, 8]

$$\begin{aligned} ds_{11}^2 &= g_{MN}dx^M dx^N \\ &= l_s^4\Bigg(- H^{-1}Fdt^2 + F^{-1}dU^2 + U^2 d\Omega_8^2 \\ &\quad + \left(l_s^{-4}H^{1/2}dx_{10} - \left(\frac{U_+}{U_-}\right)^{7/2} H^{-1/2}dt\right)^2\Bigg) \end{aligned} \tag{7.109}$$

$$F = 1 - \frac{U_+^7 - U_-^7}{U^7} = 1 - \frac{U_0^7}{U^7}, \quad U_0^7 = U_+^7 - U_-^7, \quad U = \frac{r}{\alpha'} \tag{7.110}$$

$$H = l_s^4 \frac{U_-^7}{U^7}. \tag{7.111}$$

By using the basic formula (7.98) we get immediately the 10-dimensional metric

$$\begin{aligned} ds_{10}^2 &= g_{mn}dx^m dx^n \\ &= l_s^2\left(- H^{-1/2}Fdt^2 + H^{1/2}F^{-1}dU^2 + H^{1/2}U^2 d\Omega_8^2\right) \end{aligned}. \tag{7.112}$$

This is the same solution as the non-extremal black 0-brane solution (7.140). The dilaton Φ and the one-form field $A = A_m\, {\rm d}x^m$ are given by the identification

$$\exp(\Phi) = l_s^{-3}H^{3/4}, \quad A = -\left(\frac{U_+}{U_-}\right)^{7/2} l_s^4 H^{-1}dt. \tag{7.113}$$

The mass M and the R_R charge Q of this black 0-brane solution are given by (with S^8 being the volume of S^8)

$$M = \frac{V_{S^8}\alpha'^7}{2k_{10}^2}\left(8U_+^7 - U_-^7\right), \quad Q = \frac{7V_{S^8}\alpha'^7}{2k_{10}^2}(U_+U_-)^{7/2}. \tag{7.114}$$

The charge of N coincident D0-branes in type IIA superstring is quantized as

$$Q = \frac{N}{l_s g_s}. \tag{7.115}$$

We write the radii $r_\pm$ as $r_\pm{}^7 = (1 + \delta)^{\pm 1} R$. Hence (with $V_{S^8} = 2(2\pi)^4/(7.15)$)

$$\frac{N}{l_s g_s} = \frac{7V_{S^8}}{2k_{10}^2}(r_+ r_-)^{7/2} \Rightarrow R = \frac{N}{l_s g_s}\frac{2k_{10}^2}{7V_{S^8}} = 15\pi N l_s^7 g_s (2\pi)^2. \tag{7.116}$$

The above M-wave solution is non-extremal. The extremal solution corresponds to the limit $U_- \longrightarrow U_+$ or equivalently $\delta \longrightarrow 0$. The Schwarzschild limit corresponds to $U_- \longrightarrow 0$ and thus the charge Q vanishes in this limit.

The third, much more important, limit is the near horizon limit $r \longrightarrow 0$ given by [40]

$$U = \frac{r}{\alpha'} = \text{fixed} \tag{7.117}$$

$$U_0 = \left(U_+^7 - U_-^7\right)^{1/7} = U_- \alpha = \text{fixed} \tag{7.118}$$

$$\lambda = g_{\text{YM}}^2 N = \frac{g_s}{(2\pi)^2}\frac{1}{l_s^3} N = \text{fixed} \tag{7.119}$$

α is a dimensionless parameter. Note that the extremal limit corresponds to $\alpha \longrightarrow 0$. The near horizon limit is also equivalent to the limit $\alpha \longrightarrow 0$ since $U_0/U = r_- \alpha/r =$ fixed. Thus the near horizon limit is equivalent to the near extremal limit and it is given by $\alpha \longrightarrow 0$ keeping fixed

$$\frac{U_0}{U} = \frac{r_- \alpha}{r} = \text{fixed}. \tag{7.120}$$

$$\frac{1}{U_0} = \frac{\alpha'}{r_- \alpha} = \text{fixed}. \tag{7.121}$$

$$U_0^3 \frac{g_s^2 N^2}{l_s^6} = \frac{g_s^2 N^2}{(r_- \alpha)^3} = \text{fixed}. \tag{7.122}$$

In summary, the M-wave solution which is a purely geometrical solution in 11 dimensions corresponds to a charged black hole in 10 dimensions smeared along the compact x_{10} direction. In the extremal or near horizon limit the functions F and H characterizing this solution are given by

$$H = \frac{H_0}{U^7}, \quad H_0 = l_s^4 U_-^7 = \frac{1}{1+\delta} 240\pi^5 \lambda. \tag{7.123}$$

$$F = 1 - \frac{U_0^7}{U^7}, \quad U_0^7 = \frac{\delta(2+\delta)}{1+\delta}\frac{15\pi\lambda(2\pi)^4}{l_s^4}. \tag{7.124}$$

By comparing with (A.62) we get that δ is proportional to the energy and that the extremal limit is also characterized by $l_s \longrightarrow 0$ keeping δ/l_s^4 fixed given by

$$\frac{\delta}{l_s^4} = \frac{56\pi^2\lambda}{9}\frac{E}{N^2}. \tag{7.125}$$

The mass and the charge are given by

$$M = \left(1 + \frac{9\delta}{7}\right)Q, \quad Q = \frac{N^2}{(2\pi)^2 l_s^4 \lambda}. \tag{7.126}$$

7.5.3 Type IIA string theory at strong couplings is M-Theory

The theory of 11-dimensional supergravity arises also as a low energy limit of type IIA superstring theory [60]. We will discuss this point in some more detail following [57]. We consider the behaviour of Dp-branes at strong coupling in type IIA string theory. The Dp-branes have tensions given by the Polchinski formula [31, 61]

$$T_p^2 = \frac{2\pi}{2k_{10}^2}(4\pi^2\alpha')^{3-p}. \tag{7.127}$$

The fundamental string tension is $1/(2\pi\alpha')$ and Newton's constant in 10 dimensions is $2k^2{}_{10}/16\pi$. Type IIA string theory contains only Dp-branes with p even while type IIB string theory contains Dp-branes with p odd. These two series are related by T-duality. Indeed, it was shown using T-duality that [62]

$$T_{p-1} = 2\pi\alpha'^{\frac{1}{2}} T_p. \tag{7.128}$$

This is consistent with (7.127). It was also shown using $SL(2, Z)$ duality that the D-string (i.e. $p = 1$) in type IIB string theory has a tension given by [63]

$$T_1 = g^{-1}(2\pi\alpha')^{-1}. \tag{7.129}$$

The perturbative effective string coupling is given by $g = \langle e^{\Phi}\rangle + \cdots$ where Φ is the dilaton field. From (7.127) and (7.129) we get that

$$2k_{10}^2 = (2\pi)^3 g^2 (2\pi\alpha')^4. \tag{7.130}$$

Putting this last equation back in (7.127) we obtain

$$T_p = (2\pi)^{\frac{1-p}{2}} g^{-1} (2\pi\alpha')^{-\frac{p+1}{2}}. \tag{7.131}$$

This tension translates into a mass scale [31]

$$T_p^{\frac{1}{p+1}} = O(g^{-\frac{1}{p+1}}\alpha'^{-\frac{1}{2}}). \tag{7.132}$$

Remark that α'^{-1} has units of mass squared and g is dimensionless and thus $T_p^{\frac{1}{p+1}}$ has units of mass. Clearly, at strong string coupling g the smallest p (i.e. $p = 0$) gives the smallest scale, i.e. the lowest modes. These are the D0-branes. Their mass is

$$T_0 = \frac{1}{g\alpha'^{\frac{1}{2}}}. \tag{7.133}$$

At weak coupling this is very heavy. Bound states of any number n of D0-branes do exist. They have mass

$$nT_0 = \frac{n}{g\alpha'^{\frac{1}{2}}}. \tag{7.134}$$

As $g \longrightarrow \infty$ these states become massless and the above spectrum becomes a continuum identical to the continuum of momentum Kaluza–Klein states which correspond to a periodic (i.e. compact) 11 dimensions of radius

$$R_{10} = g\alpha'^{\frac{1}{2}}. \tag{7.135}$$

Thus, the limit $g \longrightarrow \infty$ is the decompactification limit $R_{10} \longrightarrow \infty$ where an extra 11 dimensions appear. This extra dimension is invisible in string perturbation theory because if $g \longrightarrow 0$ it becomes very small (i.e. $R_{10} \longrightarrow 0$).

In the dimensional reduction of 11-dimensional supergravity to 10-dimensional type IIA supergravity only the states with momentum $p_{10} = 0$ survives. However, type IIA string theory (with type IIA supergravity as its low energy limit) contains also states with $p_{10} \neq 0$ which are the D0-branes and their bound states. In this dimensional reduction the Kaluza–Klein gauge field coupling to the 11th dimension is exactly the Ramond–Ramond gauge field coupling to the D0-branes of type IIA supergravity.

From the action (7.92) we can immediately see that the 11-dimensional and 10-dimensional gravitational couplings k_{11} and k_{10} are related by

$$k_{11}^2 = 2\pi R_{10} k_{10}^2 = \frac{1}{2}(2\pi)^8 g^3 \alpha'^{\frac{9}{2}}. \tag{7.136}$$

The 11-dimensional Planck mass is defined by

$$2k_{11}^2 \equiv (2\pi)^8 M_{11}^{-9}. \tag{7.137}$$

The power of -9 is dictated by the fact that we have (by using the last two equations)

$$M_{11} = g^{-\frac{1}{3}}\alpha'^{-\frac{1}{2}}. \tag{7.138}$$

Since $\alpha'^{-\frac{1}{2}}$ has units of mass the M_{11} will have units of mass. From (7.135) and (7.138) we find that the parameters g and α' of type IIA string theory are related to the parameters R_{10} and M_{11} of the 11-dimensional theory by

$$g = (R_{10}M_{11})^{\frac{3}{2}}, \quad \alpha' = R_{10}^{-1}M_{11}^{-3}. \tag{7.139}$$

7.6 M-theory prediction for quantum black holes

Let us start by writing down the metric of the 10-dimensional black 0-brane solution in type II A supergravity theory:

$$ds^2 = \alpha'\left[-\frac{F(U)}{\sqrt{H}}dt^2 + \frac{\sqrt{H}}{F(U)}dU^2 + \sqrt{H}\,U^2 d\Omega_8^2\right], \tag{7.140}$$

where

$$F(U) = 1 - \frac{U_0^7}{U^7}, \quad H = \frac{240\pi^5\lambda}{U^7}. \tag{7.141}$$

The horizon is located at U_0 where

$$U_0^7 = a_0\lambda^2\frac{E}{N^2}. \tag{7.142}$$

This solution is characterized by two parameters U and λ which correspond to a typical energy scale and 't Hooft coupling constant in the dual gauge theory. These two parameters define also the mass and the charge of the black hole.

By the gauge/gravity duality, classical type IIA supergravity around this black 0-brane solution is equivalent to the limit $\lambda \longrightarrow \infty$, $\lambda \longrightarrow \infty$ of the $(0+1)$-dimensional gauge theory given by the BFSS model

$$\begin{aligned} S = \frac{1}{g_{\rm YM}^2}\int_0^\beta dt\bigg[&\frac{1}{2}(D_t X_I)^2 - \frac{1}{4}[X_I, X_J]^2 \\ &- \frac{1}{2}\bar{\psi}\gamma^0 D_0\psi - \frac{1}{2}\bar{\psi}\gamma^I[X_I, \psi]. \end{aligned} \tag{7.143}$$

The quantum gravity corrections to the above black 0-brane solution were compute analytically in [7], whereas the corresponding non-perturbative gauge theory corrections using Monte Carlo simulations of the above gauge theory were computed in [12].This is an extremely important, very concrete check of the gauge/gravity correspondence. In the following we will outline their main results.

7.6.1 Quantum gravity corrections

In this section we follow closely [7].

The quantum gravity corrections are loop corrections in the string coupling g_s. From the dual gauge theory side this corresponds to corrections in $1/N^2$. The calculation of the effective action is done in 11 dimensions by demanding local supersymmetry. This effective action of M-theory is given by the action of 11-dimensional supergravity with the addition of higher derivative terms

The effective action of M-theory becomes, after dimensional reduction, the one-loop effective action of type IIA superstring theory. Also, since we are only interested in the M-wave solution, which is a purely geometrical object in 11 dimensions, we should only concentrate on the graviton field. The structure of the effective action of type IIA superstring theory must be such that it is consistent with the scattering amplitudes of strings. The leading corrections to type IIA supergravity involve four gravitons as asymptotic states and correspond to quartic terms of the Riemann tensor [11]. The corrections $[eR^4]$ when lifted to 11 dimensions become $B_1 = [eR^4]_7$ where the subscript 7 indicates the number of potentially different contractions. The other corrections are determined by demanding local supersymmetry. In this way the combination of the quartic terms of the Riemann tensor in B_1 is determined uniquely up to an overall factor. The effective action of the M-theory relevant to the graviton scattering is given explicitly by [7, 9, 10]

$$\Gamma_{11} = \frac{1}{2k_{11}^2} \int d^{11}x\sqrt{-g}\left(R + \gamma l_s^{12}\left(t_8 t_8 R^4 - \frac{1}{4!}\epsilon_{11}\epsilon_{11}R^4\right)\right). \tag{7.144}$$

The tensor t_8 is the product of four Kronecker's symbols, whereas the tensor ϵ_{11} is an antisymmetric tensor with 11 indices. The parameter γ is given by

$$\gamma = \frac{\pi^6}{2^7 3^2} \frac{\lambda^2}{N^2}. \tag{7.145}$$

By dimensional reduction we get from Γ_{11} the one-loop effective action of type IIA superstring theory.

Next we derive the equations of motion corresponding to the action Γ_{11} and solve them up to the linear order in γ. The near horizon geometry of the M-wave solution with quantum gravity corrections included takes the form [7]

$$ds_{11}^2 = l_s^4\left(- H_1^{-1}F_1 dt^2 + F_1^{-1}dU^2 + U^2 d\Omega_8^2 + \left(l_s^{-4}H_2^{1/2}dx_{10} - H_3^{-1/2}dt\right)^2\right). \tag{7.146}$$

The functions H_i and F_1 are given by

$$H_i = H + \frac{1}{N^2}\frac{5\pi^{16}\lambda^3}{24U_0^{13}}h_i, \quad F_1 = F + \frac{1}{N^2}\frac{\pi^6\lambda^2}{1152U_0^6}f_1. \tag{7.147}$$

The functions h_i and f are functions of $x = U/U_0$ given by equation (44) of [7]. The dimensional reduction of the above solution gives immediately the near horizon geometry of the black 0-brane solution with quantum gravity corrections. This is given explicitly by

$$ds_{10}^2 = l_s^2\left(- H_1^{-1}H_2^{1/2}F_1 dt^2 + H_2^{1/2}F_1^{-1}dU^2 + H_2^{1/2}U^2 d\Omega_8^2\right). \tag{7.148}$$

$$\exp(\Phi) = l_s^{-3}H_2^{3/4}, \quad A = -l_s^4 H_2^{-1/2}H_3^{-1/2}dt. \tag{7.149}$$

The horizon becomes shifted given by

$$F_{1H} = 0 \Rightarrow \frac{U_H^7}{U_0^7} = 1 - \frac{1}{N^2}\frac{\pi^6\lambda^2}{1152U_0^6}f_1. \tag{7.150}$$

The temperature of the black 0-brane is derived by performing Wick rotation and then removing the conical singularity by imposing periodicity in the Euclidean time direction. We define the proper distance from the horizon by

$$d\rho^2 = F_1^{-1}dU^2 \Rightarrow \rho = 2F'^{-\frac{1}{2}}_{1H}\sqrt{U - U_H}. \tag{7.151}$$

Then

$$F_1 = \frac{\rho^2}{4}F'^2_1|_H. \tag{7.152}$$

The relevant part of the metric now takes the form

$$ds_{10}^2 = l_s^2 H_2^{1/2}\left(\frac{\rho^2}{4r_s^2}dt^2 + d\rho^2 + \cdots\right), \quad \frac{1}{r_s} = H_{1H}^{-1/2}F'_{1H}. \tag{7.153}$$

We get immediately the temperature

$$T = \frac{1}{4\pi r_s} = \frac{1}{4\pi}H_{1H}^{-1/2}F'_{1H}. \tag{7.154}$$

In terms of $\tilde{U}_0 = U_0/\lambda^{1/3}$, $\tilde{T} = T/\lambda^{1/3}$ and $\epsilon = \gamma/\lambda^2$ this formula reads

$$\tilde{T} = a_1\tilde{U}_0^{5/2}\left(1 + \epsilon a_2\tilde{U}_0^{-6}\right). \tag{7.155}$$

The coefficient a_1 is precisely the coefficient computed before in equation (7.89). The coefficient a_2 is given by equation (48) of [7].

Next we compute the entropy of the black 0-brane solution by using Wald's formula [64, 65] in order to maintain the first law of thermodynamics. This entropy is exactly equal to the entropy of the M-wave solution. Wald's formula for the M-wave solution is given in terms of the effective action Γ_{11} and an antisymmetric binormal tensor $N_\mu\nu$ by

$$\mathcal{S} = -2\pi\int_H d\Omega_8 dx_{10}\sqrt{h}\frac{\partial\Gamma_{11}}{\partial R_{\mu\nu\alpha\beta}}N_{\mu\nu}N_{\alpha\beta}. \tag{7.156}$$

The binormal tensor $N_\mu\nu$ satisfies $N_{\mu\nu}N^{\mu\nu} = -2$ with the only non-zero component being $N_{tU} = -l_s^4U_0H_1^{-1/2}$ and where $\sqrt{h} = (l_s^2U)^8H_2^{1/2}/l_s^2$ is the volume form at the horizon. The result of the calculation is [7]

$$\frac{\mathcal{S}}{N^2} = a_3\tilde{T}^{9/5}(1 + \epsilon a_4\tilde{T}^{-12/5}). \tag{7.157}$$

Again the coefficient a_3 is computed before in equation (7.91). The coefficient a_4 is given by equation (53) of [7].

The energy $\tilde{E} = E/\lambda^{1/3}$ is given by the first law of thermodynamics, viz $d\tilde{E} = \tilde{T}d\mathcal{S}$. We get immediately

$$\frac{\tilde{E}}{N^2} = \frac{9a_3}{14}\tilde{T}^{14/5} - \frac{3\epsilon a_3 a_4}{2}\tilde{T}^{2/5}. \tag{7.158}$$

The numerical coefficients are given by [7]

$$\frac{9a_3}{14} = 7.41, \quad \frac{3\epsilon a_3 a_4}{2} = \frac{5.77}{N^2}. \tag{7.159}$$

We also get the specific heat

$$\frac{1}{N^2}\frac{d\tilde{E}}{dT} = \frac{9a_3}{5}\tilde{T}^{9/5} - \frac{3\epsilon a_3 a_4}{5}\tilde{T}^{-3/5}. \tag{7.160}$$

This can become negative at low temperature which means that the black 0-brane behaves as an evaporating Schwarzschild black hole. This instability is obviously removed in the limit $N \longrightarrow \infty$.

7.6.2 Non-perturbative tests of the gauge/gravity duality

The basic prediction coming from the dual gravity is the formula for the energy per one degree of freedom given by

$$\mathcal{E}_{\text{gravity}} = \frac{\tilde{E}}{N^2} = 7.41 \cdot \tilde{T}^{14/5} - \frac{5.77}{N^2} \cdot \tilde{T}^{2/5}. \tag{7.161}$$

This formula includes the quantum gravity (second term) corrections which correspond to loop corrections proportional to g_s or equivalently $1/N^2$. We can also include stringy corrections which are proportional to α' or equivalently $1/\lambda$ following [66, 67]. We get then the formula

$$\begin{aligned}\mathcal{E}_{\text{gravity}} = \frac{\tilde{E}}{N^2} &= (7.41 \cdot \tilde{T}^{2.8} + a \cdot \tilde{T}^{4.6} + \cdots) \\ &+ (-5.77\tilde{T}^{0.4} + b\tilde{T}^{2.2} + \cdots)\frac{1}{N^2} + O\left(\frac{1}{N^4}\right).\end{aligned} \tag{7.162}$$

This result is reproduced non-perturbatively by Monte Carlo simulation of the dual gauge theory, here it is the BFSS quantum mechanics, at the level of classical supergravity (for $N = \infty$ with the result $a = -5.58(1)$) in [67] and at the level of quantum gravity in [12]. The gauge/gravity duality is also tested for D0-branes in [72, 73]. Other non-perturbative original tests of the gauge/gravity duality can be found in [70, 71]. Here we will follow closely [12].

The BFSS quantum mechanics

$$S = \frac{1}{g_{\text{YM}}^2}\int_0^\beta dt\left[\frac{1}{2}(D_t X_I)^2 - \frac{1}{4}[X_I, X_J]^2 - \frac{1}{2}\bar{\psi}\gamma^0 D_0\psi - \frac{1}{2}\bar{\psi}\gamma^I[X_I, \psi]\right] \tag{7.163}$$

is computed along the lines of [67, 68]. The goal is to compute the internal energy

$$\mathcal{E}_{\text{gauge}} = \frac{E}{N^2}, \quad E = -\frac{\partial}{\partial \beta} \ln Z, \quad Z = \int \mathcal{D}X\mathcal{D}A\mathcal{D}\psi \exp(-S). \tag{7.164}$$

The theory can be regularized in the time direction either by a momentum cutoff Λ [12, 69], which incidentally preserves gauge invariance in one dimension, or by a conventional lattice a [71] which really remains the preferred method because of the dire need for parallelization in this sort of calculation. As usual the fermionic Pfaffian is complex and thus in practice only the absolute value can be taken into consideration in the Boltzmann probability which is a valid approximation in this context [12, 73].

In order to be able to compare sensibly the quantum gauge theory corrections on the gauge theory side with the corresponding quantum gravity corrections on the gravity side given by the $1/N^2$ term in equation (7.162) the authors of [12] simulated the gauge theory with small values of N. Here the problem of flat directions (commuting matrices) becomes quite acute which is a problem intimately related to the physical effect of Hawking radiation and the instability of the black hole solution encountered in equation (7.160). In particular, it is observed that the eigenvalues of the bosonic matrices X_I, which describe the positions of the D-particles forming the black hole, start to diverge for small N signifying that these particles are being radiated away out of the black hole due to quantum gravity effects. In other words, the black hole for small N is a metastable bound state of the D0-branes and it becomes really stable only in the large N limit when quantum gravity effects can be suppressed and classical gravity becomes an exact description.

Hence, the black hole for small N is only a metastable bound state of the D0-branes and quantum gravity is acting as a destabilizing effect. The energy of this metastable bound state is measured as follows:

- We introduce the extent of space

$$R^2 = \frac{1}{N\beta} \int_0^\beta dt \sum_{I=1}^{9} X_I(t)^2. \tag{7.165}$$

  The histogram of R^2 presents a peak (bound state) and a tail (Hawking instability). See figure 7.1.
- The instability is tamed numerically by the addition of an appropriate potential.
- We fix T, N and Λ. In the measurement of the internal energy we only consider configurations which satisfy $R^2 < x$. The corresponding estimation of the internal energy is denoted by $\mathcal{E}(x) = E(x)/N^2$. This estimation reaches a plateau in the region of the tail of R^2, i.e. for all values of x in the tail of the histogram of R^2 the energy takes on effectively the same value. This value is the measurement $\mathcal{E}(\Lambda) = E(\Lambda)/N^2$ of the internal energy for that particular set of values of T and N and Λ.
- We repeat the same analysis for other sets of values of T, N and Λ. The internal energy in the continuum limit $\mathcal{E}_{\text{gauge}} = E/N^2$ is obtained by fitting the results $\mathcal{E}(\Lambda) = E(\Lambda)/N^2$ for different Λ, but for the same T and N, using

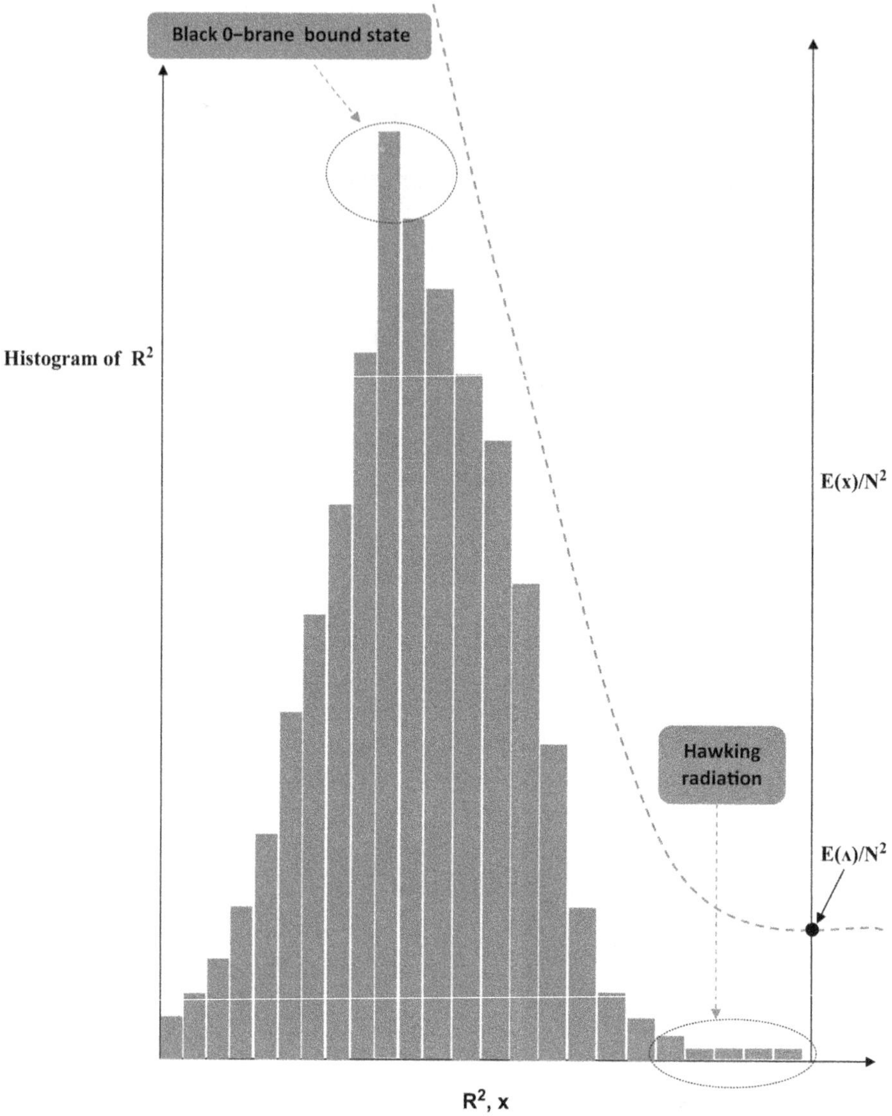

Figure 7.1. Black 0-brane (black hole) and its Hawking radiation.

the ansatz $E(\Lambda) = E + \text{constant}/\Lambda$. We plot the energy $\mathcal{E}_{\text{gauge}}$ as a function of T for some fixed N. The authors of [12] considered $N = 3,4,5$. It is observed that the internal energy increases as T decreases signaling that the specific heat is negative which is consistent with the result (7.160) on the gravity side. See figure 7.2.

- The stringy effects are found to be irrelevant in the temperature range $0.07 \leqslant T \leqslant 0.12$ considered by the author of [12]. The Monte Carlo results

$\mathcal{E}_{\text{gauge}}$ (figure 3 of their paper) should then be compared with the gravity result (7.161). They have provided convincing evidence that

$$\mathcal{E}_{\text{gravity}} - \mathcal{E}_{\text{gauge}} = \frac{\text{constant}}{N^4}. \tag{7.166}$$

See figure 7.3. This implies immediately that

$$\mathcal{E}_{\text{gauge}} = \frac{E}{N^2} = 7.41 \cdot T^{14/5} - \frac{5.77}{N^2} \cdot T^{2/5} + \frac{c_2}{N^4}. \tag{7.167}$$

The T dependence of the coefficient c_2 is also found to be consistent with the prediction from the gravity side given by $c_2 = cT^{-2.6} + \cdots$.

7.7 Matrix string theory

We consider now dimensional reduction to $D = 2$ dimensions, i.e. $p = 1$. Thus, we are dealing with a system of N coincident D1-branes forming a black 1-brane solution with dynamics given by the maximally supersymmetric $U(N)$ gauge action (with $\mu, \nu = 0{,}1$ and $I, J = 1, \ldots, 8$)

$$\begin{aligned} S = \frac{1}{g_{\text{YM}}^2} \int d^2x \Big[&-\frac{1}{4} F_{\mu\nu}F^{\mu\nu} - \frac{1}{2}(D_\mu X_I)(D^\mu X_I) \\ &+ \frac{1}{4}[X_I, X_J]^2 + \frac{i}{2}\bar{\psi}\gamma^\mu D_\mu \psi + \frac{1}{2}\bar{\psi}\gamma^I [X_I, \psi]\Big]. \end{aligned} \tag{7.168}$$

This is equivalent to type IIB superstring theory around the black 1-brane background spacetime. This black 1-brane solution can be mapped via S-duality to a black string solution [40]. This theory is exactly equivalent to the so-called matrix string theory [5] which can be thought of as a matrix gauge theory in the same way that M-(atrix) theory or the BFSS model is a matrix quantum mechanics.

In the same way that the Euclidean IKKT matrix model decompactified on a circle S^1 gives the BFSS model at finite temperature, M-(atrix) theory decompactified on a circle S^1 should give the above matrix string theory. See for example [6]. Indeed, compactification should always take us from a higher-dimensional theory to a lower-dimensional one since it is the analogue of dimensional reduction. Thus, we go from the 2-dimensional matrix gauge theory (the DVV matrix string theory) to the 1-dimensional matrix quantum mechanics (the BFSS M-(atrix) theory) and from the 1-dimensional matrix quantum mechanics to the 0-dimensional type IIB matrix model (the IKKT matrix model) via circle compactifications (assuming also Euclidean signature so there is no difference between timelike and spacelike circles).

As we have discussed previously, the full dynamics of M-theory is captured by the large N limit of the M-(atrix) theory or the BFSS quantum mechanics, whereas compactification of M-theory on a circle gives type IIA string theory. The fundamental degrees of freedom of M-(atrix) theory are D-particles. The D0-branes (D-particles) of type IIA string theory are described by collective coordinates given

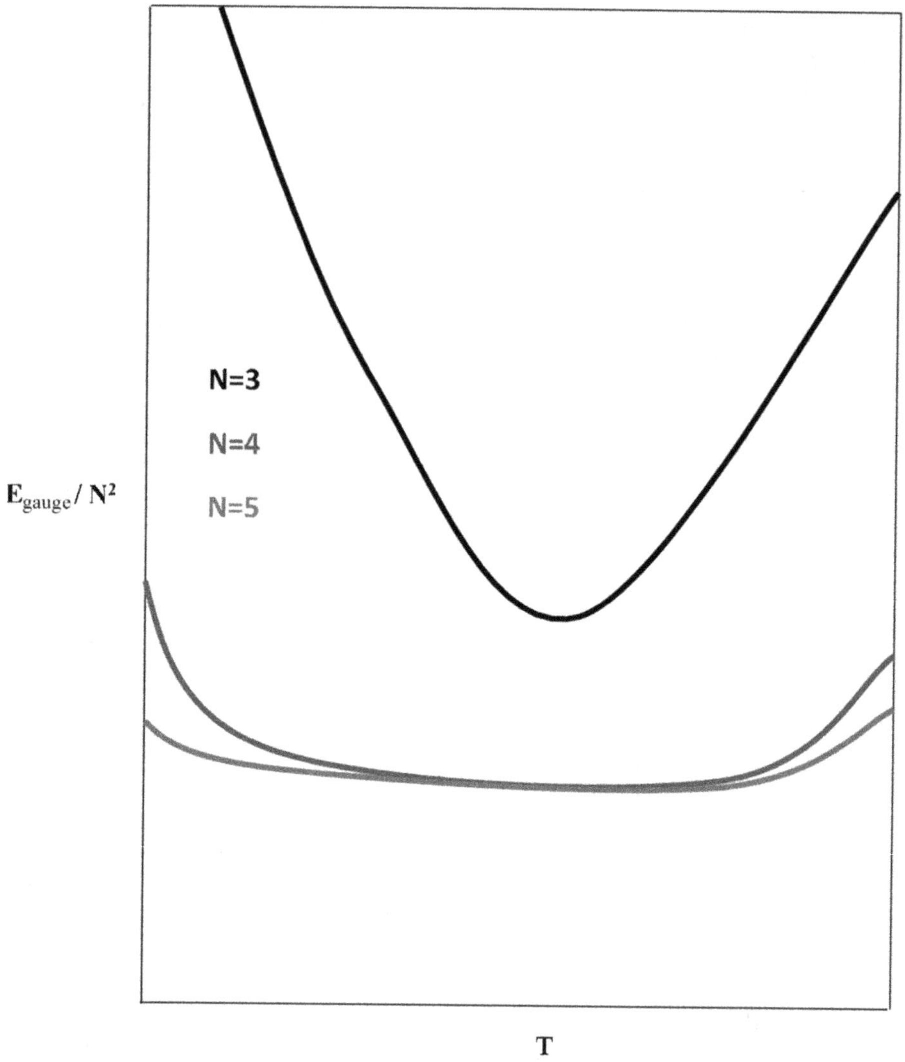

Figure 7.2. Instability of the black hole: the specific heat increases as T decreases.

by $N \times N$ matrices and the limit $N \longrightarrow \infty$ is precisely the compactification limit (recall that $\lambda = g_{YM}^2 N = g_s N/(2\pi)^2 l_s^3$).

In this decompactification, the strong limit of the string coupling constant g_s determines the large radius of the circle by $R = g_s\sqrt{\alpha'}$ while the type IIA D-particles are identified with non-zero Kaluza–Klein modes along the circle.

Similarly, the circle decompactification of the BFSS quantum mechanics is achieved by reinterpreting the large $N \times N$ matrices X_I as covariant derivatives of a $U(N)$ gauge field defined on a circle S^1. Thus, a new compact coordinate along the circle emerges in M-(atrix) theory which then becomes matrix string theory.

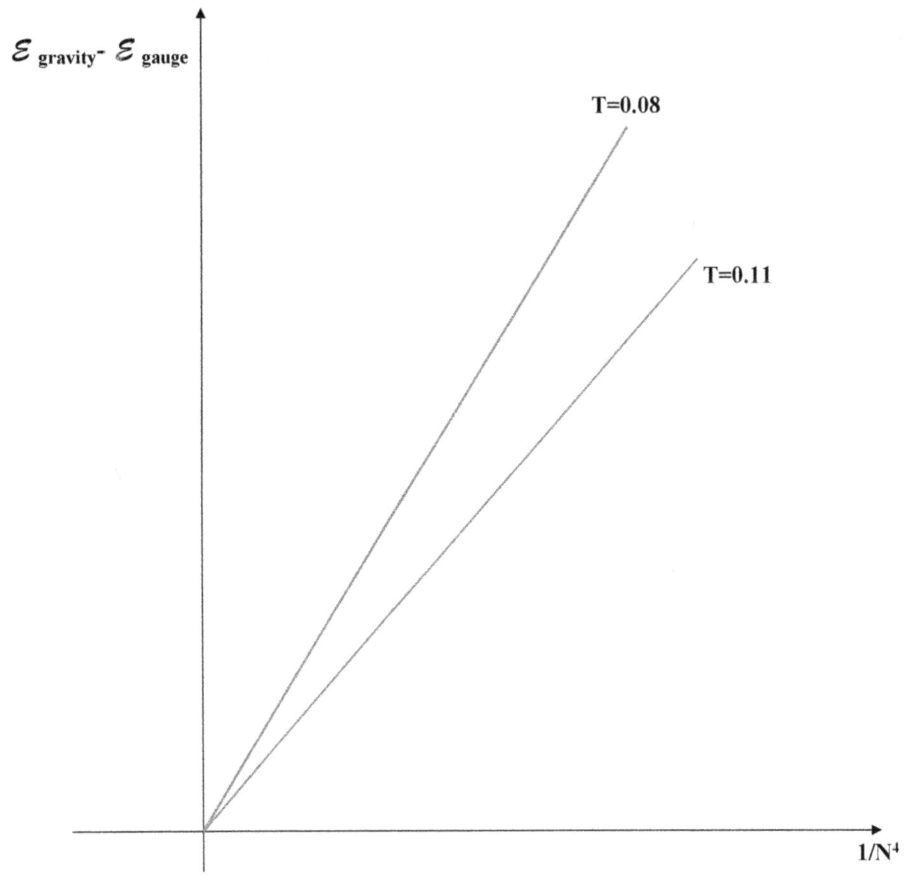

Figure 7.3. Consistency of the gauge and gravity predictions.

In summary,

1. M-theory compactified on a circle gives type IIA string theory.
2. M-theory is given by M-(atrix) theory.
3. Matrix string theory compactified on a circle gives M-(atrix) theory.

Hence, we conclude that dynamics of type IIA string theory must be captured by matrix string theory in the strong coupling region. See [5] and also [74–76].

However, matrix string theory describes D-strings, forming a black string, in type IIB string theory, whereas M-(atrix) theory describes D-particles, forming a black hole, in type IIA string theory. These are related to each other via T-duality which maps type IIA string theory compactified on a circle of radius R to type IIB string theory compactified on a circle of radius α'/R, and also simultaneously exchanging momentum modes with winding numbers.

The mother of all theories M-theory and its relations to other theories is depicted in figure 7.4.

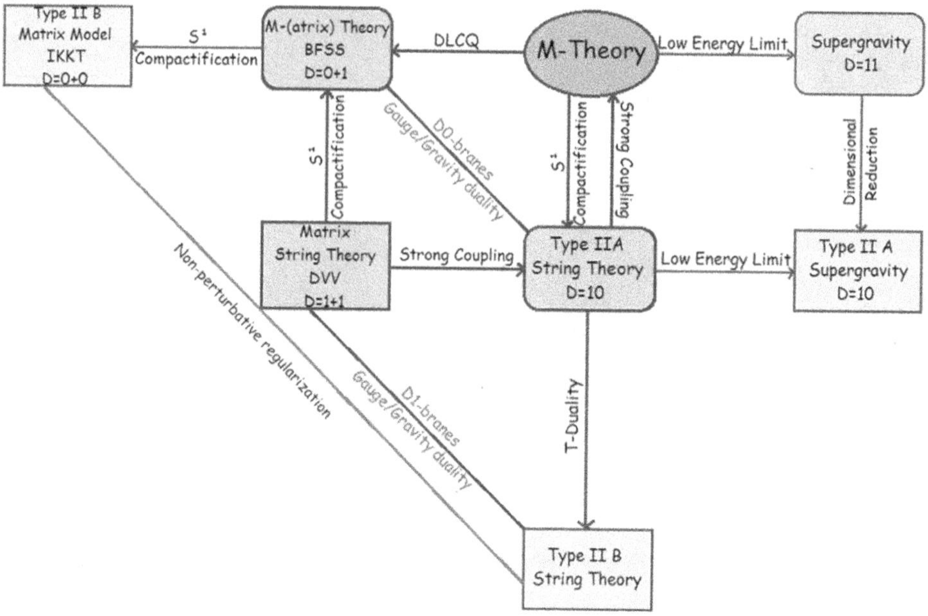

Figure 7.4. M-theory *et al.*

7.8 The black hole and confinement phase transitions

7.8.1 The black-hole/black-string phase transition

In this section we follow the presentations of [13, 29].

We consider the action of 2-dimensional maximally symmetric, i.e. $\mathcal{N} = 8$, $U(N)$ Yang–Mills theory given by the action (7.168). We Wick-rotate to Euclidean signature and denote the time by $\tilde{t}$. The theory is put at finite temperature via compactification of the Euclidean time on a circle of circumference given by the inverse temperature $\tilde{\beta} = 1/\tilde{T}$. The spatial dimension is also compactified on a circle of circumference $\tilde{L}$.

For very large compactification circumference $\tilde{L}$ this theory describes N coincident D1-branes in type IIB string theory which are winding on the circle. When $\tilde{L} \longrightarrow 0$ the appropriate description becomes given by T-duality in terms of N coincident D0-branes in type IIA string theory which are winding on a circle of circumference $\alpha'/\tilde{L}$. The positions of these D-particles on the T-dual circle are given by the eigenvalues of the Wilson loop winding on the circle, i.e. of the holonomy matrix

$$W = \mathcal{P}\exp\left(i \oint dx A_x\right). \tag{7.169}$$

By an appropriate gauge transformation the Wilson line can be diagonalized as

$$W = \text{diag}(\exp(i\theta_1), \dots, \exp(i\theta_N)). \tag{7.170}$$

The phase θ_i is precisely the position of the ith D0-brane on the T-dual circle. If all the angles θ_i accumulate at the same point then we obtain a black hole at that location, whereas if they are distributed uniformly on the circle we obtain a uniform black string. We can also obtain a non-uniform black string phase or a phase with several black holes depending on the distribution of the eigenvalues θ_i.

In the high temperature limit a very nice reduction of this model occurs. In this limit $\tilde{\beta}\longrightarrow 0$ and as a consequence the temporal Kaluza–Klein mode, which have temporal momenta of the form $p = n/\tilde{\beta}$ for $n \in \mathbf{Z}$, become very heavy and thus decouple from the theory. Effectively, the time direction is reduced to a point in the high temperature limit $\tilde{\beta}\longrightarrow 0$ and the theory reduces back to M-(atrix) theory. If we also assume that the fermions obey the anti-periodic boundary condition in the time direction, viz

$$\psi(\tilde{t}+\tilde{\beta}) = -\psi(\tilde{t}). \tag{7.171}$$

Then, in the high temperature limit $\tilde{\beta}\longrightarrow 0$ the fermions decouple and we end up with a bosonic theory, i.e. the bosonic part of the BFSS quantum mechanics given explicitly by

$$S = \frac{1}{g_{\mathrm{YM}}^2}\int_0^{\beta} dt\left[\frac{1}{2}(D_t X_I)^2 - \frac{1}{4}[X_I, X_J]^2\right]. \tag{7.172}$$

However, note that the time direction of this 1-dimensional model is the spatial direction of the 2-dimensional model. More precisely, we have

$$\frac{1}{g_{\mathrm{YM}}^2} = \frac{\tilde{\beta}}{\tilde{g}_{\mathrm{YM}}^2} \Rightarrow \lambda = \tilde{\lambda}\tilde{T}. \tag{7.173}$$

$$\tilde{L} = \beta. \tag{7.174}$$

It is not difficult to see that all physical properties of the 1-dimensional system depend only on the effective coupling constant

$$\lambda_{\mathrm{eff}} = \frac{\lambda}{T^3} = \tilde{\lambda}\tilde{T}\tilde{L}^3. \tag{7.175}$$

The Wilson loop (7.169) winding around the spatial circle in the 2-dimensional theory becomes in the 1-dimensional theory the Polyakov line or holonomy

$$P = \frac{1}{N}\mathrm{Tr}\, U, \quad U = \mathcal{P}\exp\left(i\int_0^{\beta} dt A(t)\right), \tag{7.176}$$

where U is the holonomy matrix.

The high temperature 2-dimensional Yang–Mills theory on a circle was studied in [77, 78] where a phase transition around $\lambda_{\mathrm{eff}} = 1.4$ was observed. This result was made more precise by studying the 1-dimensional matrix quantum mechanics in [13] where two phase transitions were identified of second and third order, respectively, at the values (see next section for detailed discussion)

$$\lambda_{\rm eff} = 1.35(1) \Rightarrow \tilde{T}\tilde{L} = \frac{1.35(1)}{\tilde{\lambda}\tilde{L}^2}. \quad (7.177)$$

$$\lambda_{\rm eff} = 1.487(2) \Rightarrow \tilde{T}\tilde{L} = \frac{1.487(2)}{\tilde{\lambda}\tilde{L}^2}. \quad (7.178)$$

The second order transition separates between the gapped phase and the non-uniform phase, whereas the third order separates between the non-uniform phase and the uniform phase. These phases, in the 2-dimensional phase diagram with axes given by the dimensionless parameters $\tilde{T}\tilde{L}$ and $\tilde{\lambda}\tilde{L}^2$, occur at high temperatures in the region where the 2-dimensional Yang–Mills theory reduces to the bosonic part of the 1-dimensional BFSS quantum mechanics. It is conjectured in [13] that by continuing the above two lines to low temperatures we will reach a triple point where the two lines intersect and as a consequence the non-uniform phase ceases to exist below this tri-critical point.

A phase diagram may look like the one on figure 7.5.

At small $\tilde{T}$ and large $\tilde{\lambda}$ it was shown in [77] that the 2-dimensional Yang–Mills theory exhibits a first order phase transition at the value

$$\tilde{T}\tilde{L} = \frac{2.29}{\sqrt{\tilde{\lambda}\tilde{L}^2}}. \quad (7.179)$$

Note the extra square root. This corresponds in the dual gravity theory side to a transition between the black hole phase (gapped phase) and the black string phase (the uniform phase) [80]. This black hole/black string first order phase transition is intimately related to Gregory–Laflamme instability [79]. Thus, it seems that the first order black hole/black string transition seen at low temperatures splits at the triple point into the second order gapped/non-uniform and the third order non-uniform/uniform transitions seen at high temperatures, i.e. in the bosonic part of the 1-dimensional BFSS quantum mechanics [13]. See the above proposed phase diagram. The black hole/black string transition is a very important example of topology change transitions.

7.8.2 The confinement/deconfinement phase transition

In this section we follow the presentation of [13].

The goal in this section is to describe in some detail the phase structure of the 1-dimensional BFSS quantum mechanics given by the model

$$S = \frac{1}{g_{\rm YM}^2}\int_0^\beta dt\left[\frac{1}{2}(D_tX_I)^2 - \frac{1}{4}[X_I, X_J]^2\right]. \quad (7.172)$$

There is one single effective coupling constant $\lambda_{\rm eff} = \lambda/T^3$ and thus without any loss of generality we can choose the 't Hooft coupling as $\lambda = g_{\rm YM}^2 N = 1$. A Monte Carlo

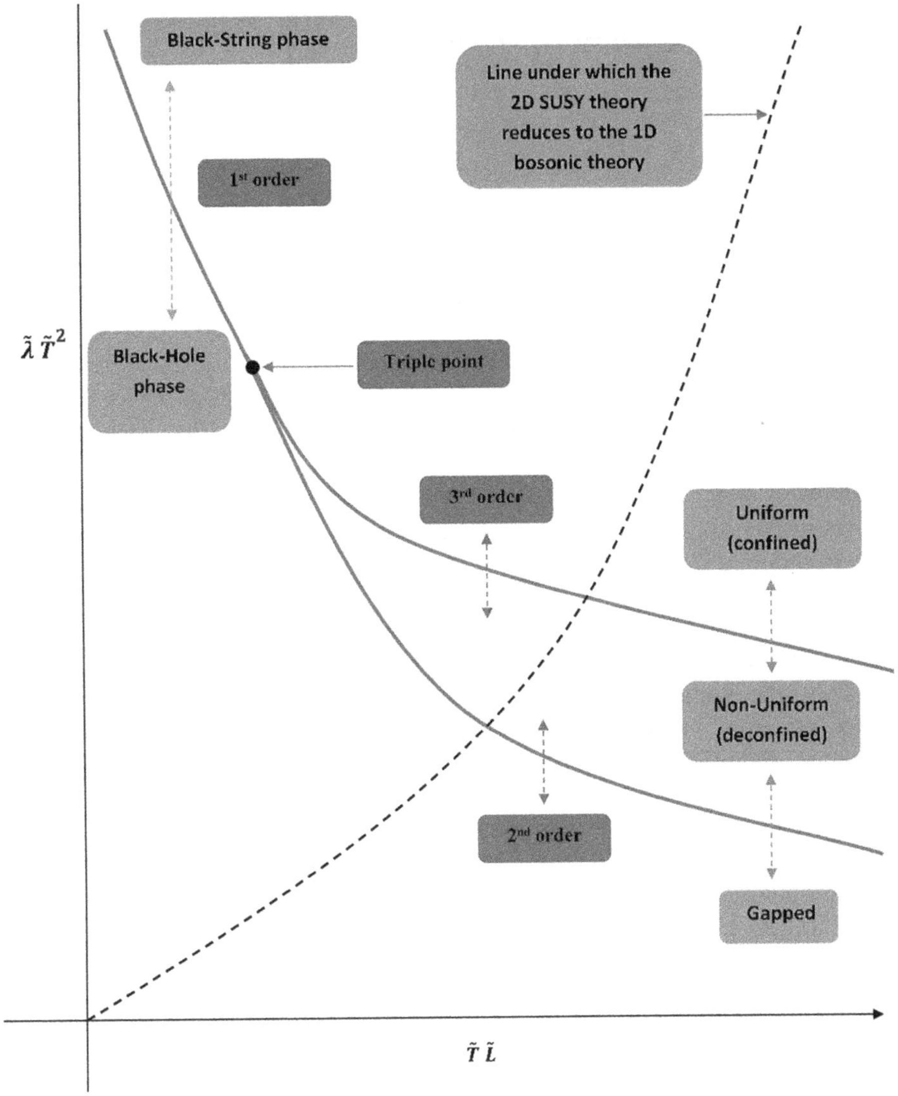

Figure 7.5. The phase diagram of the 2-dimensional Yang–Mills theory on a circle and its relation to 1-dimensional Yang–Mills theory on a circle.

study of the above model using the heat bath algorithm was performed in [13] and using the hybrid Monte Carlo algorithm was performed in [28]. The basic observables which we track in Monte Carlo simulation are the Polyakov line, the energy of the system and the extent of space given, respectively, by

$$\langle |P| \rangle = \left\langle \frac{1}{N} |\text{Tr } U| \right\rangle, \quad U = \mathcal{P}\exp\left(i \int_0 \beta dt A(t) \right). \tag{7.181}$$

$$\frac{E}{N^2} = -\left\langle \frac{3T}{4N} \int_0^{\beta} dt\ \mathrm{Tr}[X_I, X_J]^2 \right\rangle. \tag{7.182}$$

$$R^2 = \left\langle \frac{T}{N} \int_0^{\beta} dt\ \mathrm{Tr}\, X_I^2 \right\rangle. \tag{7.183}$$

The order parameter is the Polyakov line. Some of the main results include:

- The Polyakov line is found to approach one in the deconfined (non-uniform) phase, then it starts changing quite fast at around $T \simeq 0.9$, then it goes to zero in the confined (uniform) phase. The data in the deconfined phase is well reproduced by the high temperature expansion [81] especially for $T \geqslant 2$. In the confined phase the Polyakov line goes to zero as $1/N$ as $T \longrightarrow 0$ which can be reproduced by generating the holonomy matrix U with a probability given by the Haar measure dU.
- Thus $T = 0.9$ marks the transition from the deconfined (non-uniform) to confined (uniform) phase transition. In the confining uniform phase the $U(1)$ symmetry

$$A(t) \longrightarrow A(t) + a \cdot \mathbf{1} \tag{7.184}$$

  is not broken, whereas in the deconfining non-uniform phase it is broken. Thus, the confining/deconfining phase transition at $T = 0.9$ is associated with spontaneous symmetry breaking of the above $U(1)$ symmetry [13, 77, 78, 83–85, 89]. This transition is intimately related to the string theory Hagedorn transition [86–88].
- The energy and the extent of space show a flat behavior in the confined (uniform) phase for $T < 0.9$. This can be interpreted following [13] as due to the Eguchi–Kawai reduction [82] of $U(N)$ gauge theory on a lattice down to a $U(N)$ gauge theory on a point in the 't Hooft limit which is possible because in the planar approximation we find that Wilson loop amplitudes for disconnected diagrams enjoys factorization. Thus in Eguchi–Kawai reduction only global invariance is left and expectation values of single trace operators are independent of the volume in the large N limit if the central $U(1)$ symmetry is not broken[4]. See figure 7.6.
- The Polyakov line, the energy and the extent of space depend continuously on the temperature but their first derivative is discontinuous at the critical temperature

$$T_{c1} = 0.905(2). \tag{7.185}$$

Thus, the transition from confining phase to deconfining phase is second order.

[4] In D dimensions the central symmetry is $U(1)^D$

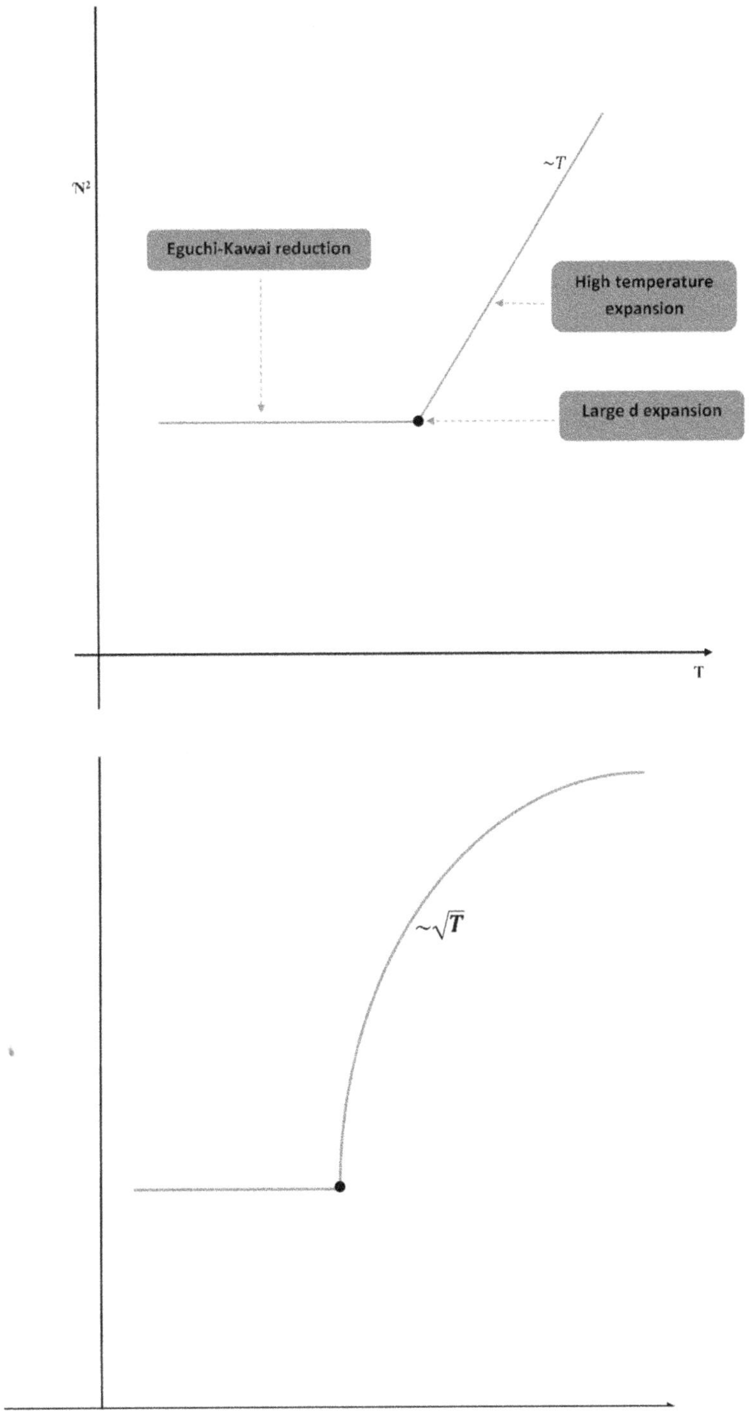

Figure 7.6. The energy and the extent of space.

- A much more powerful order parameter is the eigenvalue distribution $\rho(\theta)$ of the eigenvalues $\exp(i\theta_i)$, $i = 1, \ldots, N$, of the holonomy matrix U. We must have $\theta_i \in [-\pi, \pi]$ and $\sum_i \theta_i = 0$.

  The deconfinement phase where the central $U(1)$ symmetry is spontaneously broken is seen from the behavior of the eigenvalue distribution $\rho(\theta)$ to be divided into two distinct phases: the gapped phase and the non-uniform phase. As it turns out, these phases can be described by the Gross–Witten one-plaquette model [90, 91]

$$Z_{\rm GW} = \int dU \exp\left(\frac{N}{\kappa}({\rm Tr}\, U + {\rm Tr}\, U^{\dagger})\right). \tag{7.186}$$

  The eigenvalue distributions in the large N limit of this model are determined by the solutions

$$\rho_{\rm gapped}(\theta) = \frac{2}{\pi\kappa}\cos\frac{\theta}{2}\sqrt{\frac{\kappa}{2} - \sin^2\frac{\theta}{2}}, \quad |\theta| \leqslant 2\sin^{-1}\sqrt{\kappa/2}, \quad \kappa < 2. \tag{7.187}$$

$$\rho_{\rm non-uniform}(\theta) = \frac{1}{2\pi}\left(1 + \frac{2}{\kappa}\cos\theta\right), \quad |\theta| \leqslant \pi, \quad \kappa \geqslant 2. \tag{7.188}$$

  Thus, there exists in the Gross–Witten one-plaquette model a phase transition between the above two solutions occurring at $\kappa = 2$ which is found to be of third order. This transition in the full model occurs at $T_{1c} = 0.905(2)$ between the gapped and the non-uniform phases and is of second order (not third order) yet the above distributions are still very good fits to the actual Monte Carlo data with some value of κ for each T. The second order phase transition at $T_{1c} = 0.905(2)$ is associated with the emergence of a gap in the spectrum.
- In the confining phase the eigenvalue distribution is uniform. The $U(1)$ symmetry is unbroken and as a consequence the Eguchi–Kawai equivalence holds. Thus, the energy in this phase must be a constant proportional to N^2. The breaking of $U(1)$ symmetry and the resulting Eguchi–Kawai equivalence, as we increase T, is of the order of $1/N^2$. The phase boundary at $T = T_{c2}$ between the confining uniform phase and the deconfining non-uniform phase can thus be determined by means of the Eguchi–Kawai equivalence instead of using the eigenvalue distribution. The behavior of the energy around T_{c2} is found to be of the form [13]

$$\frac{E}{N^2} = \epsilon_0, \quad T \leqslant T_{c2}. \tag{7.189}$$

$$\frac{E}{N^2} = \epsilon_0 + c(T - T_{c2})^p, \quad T > T_{c2}. \tag{7.190}$$

  We find for $N = 32$ [13]

$$\epsilon_0 = 6.695(5), \quad c = 413 \pm 310, \quad T_{c_2} = 0.8758(9), \quad p = 2.1(2). \tag{7.191}$$

Similarly, we get for the extent of space the values $T_{c_2} = 0.8763(4)$, $p = 1.9(2)$. The transition from the confining uniform to the deconfining non-uniform occurs then at the average value

$$T_{c_2} = 0.8761(3). \tag{7.192}$$

The power $p = 2$ suggests that the second derivative of the energy is discontinuous and as a consequence the transition is third order.

- Thus the transition from the confining uniform phase to the deconfining gapped phase goes through the deconfining non-uniform phase. See figure 7.7. There is also the suggestion in [77, 78] that the transition is possibly a direct single first order phase transition which is the behavior observed in the plane-wave BMN matrix model [92–94].

7.8.3 The mass gap and the Gaussian structure

More interesting results concerning the bosonic BFSS quantum mechanics can be found in [28, 95–97].

We only consider the theory at zero temperature $T = 0$ or $\beta = \infty$. On a finite lattice $\beta = a\Lambda$ where a is the lattice spacing and Λ is the number of lattice points. The symmetric static gauge is given by the link variable $V = D^{1/\Lambda}$ where $D = \text{diag}(\exp(i\theta_1), \dots, \exp(i\theta_N))$. Thus, in the zero temperature limit $\Lambda \longrightarrow \infty$ and thus $V \longrightarrow \mathbf{1}$. Hence in this limit the gauge field can be completely gauged away.

The mass gap can be captured at zero temperature by the large time behavior of the correlator

$$\langle \text{Tr}\, X^1(0) X^1(t) \rangle \propto \exp(-mt) + \cdots \tag{7.193}$$

The mass gap is given in terms of the energies of the first excited state E_1 and the ground state E_0 by

$$m = E_1 - E_0. \tag{7.194}$$

For finite temperature the above formula for the correlator is modified as

$$\langle \text{Tr}\, X^1(0) X^1(t) \rangle \propto A(\exp(-mt) + \exp(-m(\beta - t)). \tag{7.195}$$

A measurement of this correlator for $N = 30$, $a = 0.25$ and $\beta = 10$ yields precisely this behavior with the values [95]

$$A = 7.50 \pm 0.2, \quad m = (1.90 \pm 0.01)\lambda^{1/3}. \tag{7.196}$$

The fundamental observation of [95] is that this behavior of the correlator can be obtained from the Gaussian model of d independent scalar fields with effective action

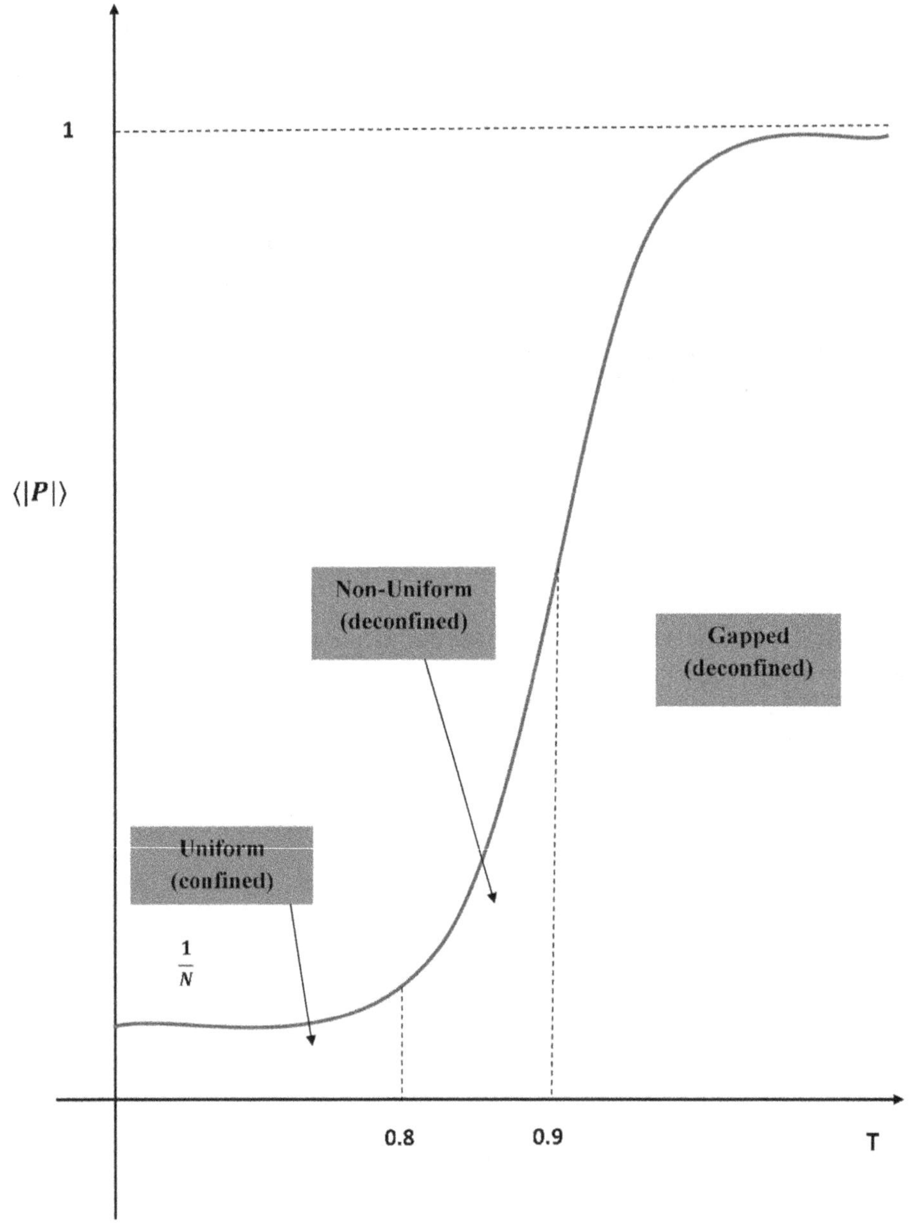

Figure 7.7. The phase structure in terms of the Polyakov line.

$$S = N\int_0^\beta dt\ \mathrm{Tr}\left[\frac{1}{2}(D_tX_I)^2 + \frac{m^2}{2}X_I^2\right]. \tag{7.197}$$

At zero temperature where the gauge field can be set to zero this becomes

$$S = N\int_0^{\infty} dt \,\mathrm{Tr}\left[\frac{1}{2}(\partial_t X_I)^2 + \frac{m^2}{2}X_I^2\right]. \tag{7.198}$$

The eigenvalue distribution of any one of the X_I is given by Wigner semicircle law of radius

$$R_\lambda = \sqrt{\frac{2}{m}}. \tag{7.199}$$

Also, the correlator $\langle \mathrm{Tr}\, X^1(0)X^1(t)\rangle$ in this theory is given exactly by the formula (7.195) with A given by

$$A = \frac{N}{2m(1 - \exp(-\beta m))}. \tag{7.200}$$

Monte Carlo measurement of the eigenvalue distribution of X_1 is indeed found to be given by a Wigner semicircle law with radius given, for the above parameters, by $R_\lambda = 1.01$ [95] which is consistent with the above measured value of m from the correlator.

The results of [95] suggest that for all values of the temperatures the eigenvalue distribution of any one of the X_I is given by a semicircle law with a radius given in terms of the expectation value of the extent of space R^2 by

$$R_\lambda = \frac{4}{d}\langle R^2\rangle. \tag{7.201}$$

Thus the phase transition from the uniform confining phase at $T = T_{c2} = 0.8...$ to the gapped deconfining phase at $T = T_{c1} = 0.9...$, i.e. the emergence of a gap, is associated with a change in the radius of the eigenvalue distribution but not the shape which remains always given by a Wigner semicircle law. The effective action (7.197) works very well at low temperatures which can be motivated by a large d expansion [98] of the bosonic BFSS model in which the quartic commutator term is replaced with a quadratic mass term with a specific value of the mass which depends on the dimension d, i.e. on the number of scalar fields.

7.8.4 The large *d* approximation

To see this in more detail we go back to our original action

$$S = N\int_0^{\beta} dt \,\mathrm{Tr}\left[\frac{1}{2}(D_t X_I)^2 - \frac{1}{4}[X_I, X_J]^2\right]. \tag{7.202}$$

We introduce $SU(N)$ generators t^a satisfying $\mathrm{Tr}\, t^a t^b = \delta^{ab}$ and we expand the matrices X_I as $X_I = t^a X_I^a$. The above action can be rewritten as

$$S = N\int_0^{\beta} dt\frac{1}{2}\,\mathrm{Tr}(D_t X_I)^2 - \frac{N}{4}\lambda^{abcd}\int_0^{\beta} dt X_I^a X_I^b X_J^c I_J^d, \tag{7.203}$$

where λ^{abcd} is some $SU(N)$ tensor (see for example equation (3.28) of [95]). We can add to the action without changing the dynamics any term ΔS depending on X_I and other fields k^{ab} such that $\int \mathcal{D}k^{ab} \exp(-\Delta S) = $ constant. We add

$$\Delta S = \frac{N}{4}\mu_{abcd}\int_0^\beta dt\Big(k^{ab} + \lambda^{abef}X_I^eX_I^f\Big)\Big(k^{cd} + \lambda^{cdgh}X_J^gX_J^h\Big), \tag{7.204}$$

where μ_{abcd} is the inverse kernel of λ^{abcd}. We consider then the action $S' = S + \Delta S$ given by

$$S' = N\int_0^\beta dt\bigg(\frac{1}{2}\mathrm{Tr}(D_tX_I)^2 + \frac{k^{ab}}{2}X_I^aX_I^b\bigg) + \frac{N}{4}\mu_{abcd}\int_0^\beta dt k^{ab}k^{cd}. \tag{7.205}$$

We will assume that the fields k's are time independent, perform a Fourier transformation in the time direction with modes n, integrate out the fields X's, and also choose for k the ansatz

$$k_{ij}P_{ij,\,lm} = P_{ij,\,lm}k_{lm} = t_{ij}^ak^{ab}t_{lm}^b. \tag{7.206}$$

The P is the projector on traceless matrices given by

$$P_{ij,\,lm} = t_{ij}^at_{lm}^a = \delta_{im}\delta_{jl} - \frac{1}{N}\delta_{ij}\delta_{lm}. \tag{7.207}$$

The above ansatz means in particular that $k_{ii} = k_{jj}$ for all i and j and we will also choose k_{ij} to be symmetric. We obtain then the effective action for the fields k given by

$$S_{\mathrm{eff}} = \frac{d}{2}\sum_n\sum_{ij}P_{ij,\,ji}\log W(n)_{ij} + \frac{N\beta}{4}\mu_{abcd}k^{ab}k^{cd}. \tag{7.208}$$

The matrix $W(n)$ is given in terms of the holonomy angles θ_i by

$$W(n)_{ij} = \left(\frac{2\pi n + \theta_i - \theta_j}{\beta}\right)^2 + k_{ij}. \tag{7.209}$$

Next we write down the saddle point equation $\partial S_{\mathrm{eff}}/\partial k_{ij} = 0$ given explicitly by

$$\frac{d}{2}\sum_n\frac{P_{ij,\,ji}}{\left(\frac{2\pi n + \theta_i - \theta_j}{\beta}\right)^2 + k_{ij}} + \frac{\beta N}{2}\mu_{abcd}k^{ab}t_{ij}^ct_{ji}^d. \tag{7.210}$$

It is found that at low temperatures the effect of the holonomy is exponentially suppressed and thus the angles θ_i can be simply set equal to zero. The leading order is given by the ansatz $k_{ij} = m^2$ or equivalently $k^{ab} = m^2\delta^{ab}$ with

$$m = d^{1/3}. \tag{7.211}$$

By substituting this solution in (7.205) we get the Gaussian action (7.197). The radius (7.199) is then given by

$$R_\lambda = \left(\frac{8}{d}\right)^{1/6}. \tag{7.212}$$

Next order corrections in $1/d$ were computed in [99] and they are given by

$$R_\lambda = \left(\frac{8}{d}\right)^{1/6}\left(1 + \frac{1}{d}\left(\frac{7\sqrt{5}}{30} - \frac{9}{32}\right) + \cdots\right). \tag{7.213}$$

The Gaussian model (7.197) enjoys also a phase transition which occurs at the temperature [100]

$$T_c^{\text{gaussian}} = \frac{m}{\ln d} = 0.9467. \tag{7.214}$$

As we can see immediately, this value is in excellent agreement with the critical temperature T_{c1} found in the full model of the phase transition to the gapped phase.

7.8.5 High temperature limit

At high temperatures the bosonic part of the BFSS quantum mechanics reduces to the bosonic part of the IKKT model [81]. The leading behavior of the various observables of interest at high temperatures can be obtained in terms of the corresponding expectation values in the IKKT model by the relations (with $D = d + 1$ and X appropriately scaled)

$$\langle R^2\rangle = \sqrt{T}\left\langle\frac{1}{N}\,\mathrm{Tr}\,X_I^2\right\rangle_{\text{IKKT}} = \sqrt{T}\chi_1. \tag{7.215}$$

$$\langle P\rangle = 1 - \frac{1}{2}T^{-3/2}\left\langle\frac{1}{N}\,\mathrm{Tr}\,X_D^2\right\rangle_{\text{IKKT}} = 1 - \frac{1}{2d}T^{-3/2}\chi_1. \tag{7.216}$$

$$\left\langle\frac{E}{N^2}\right\rangle = -\frac{3}{4}T\left\langle\frac{1}{N}\,\mathrm{Tr}[X_I, X_J]^2\right\rangle_{\text{IKKT}} = \frac{3}{4}T\chi_2,\quad \chi_2 = (d-1)\left(1 - \frac{1}{N^2}\right). \tag{7.217}$$

The coefficient χ_1 for various d and N can be read off from table 1 of [81], whereas the coefficient χ_2 was determined exactly from the Schwinger–Dyson equation. The next-leading order corrections can also be computed from the reduced IKKT model along these lines [81].

7.9 The discrete light-cone quantization (DLCQ) and infinite momentum frame (IMF)

7.9.1 Light-cone quantization and discrete light-cone quantization

In this section we follow mostly [101] but also [1, 102–105]. We consider $D = 11$-dimensional spacetime with coordinates $X^\mu = (t, z, X^1, \ldots, X^{D-2})$ and metric $(1, -1, \ldots, -1)$. The coordinate z is called the longitudinal direction while $X^1, \ldots, X^{D-2}$ are the transverse directions. The line-like coordiante $X^\pm$ are defined by

$$X^\pm = \frac{X^0 \pm X^{D-1}}{\sqrt{2}} = \frac{t \pm z}{\sqrt{2}}. \tag{7.218}$$

We introduce the conjugate momenta

$$P_\pm = \frac{P^0 \mp P^{D-1}}{\sqrt{2}} = \frac{P_t \mp P_z}{\sqrt{2}}. \tag{7.219}$$

The on-shell condition is

$$P_\mu P^\mu = M^2 \Leftrightarrow 2P_+P_- - P_i^2 = M^2. \tag{7.220}$$

In light-cone frame the coordinate X^+ plays the role of time and thus P_+ is the Hamiltonian H. We have

$$H = \frac{P_i^2}{2P_-} + \frac{M^2}{2P_-}. \tag{7.221}$$

This is a non-relativistic expression where P_- plays the role of the mass. This can be made precise as follows.

The Poincaré group is generated by P_i: the transverse translations, $P_+ = H$: the translation in the X^+ direction, $P_- = \mu$: the translation in the X^-direction, $L_{ij} = M_{ij}$: the transverse rotations, $L_{iz} = M_{iz}$: rotations in the (X^i, z) planes, $K_{0z} = M_{0z}$: Lorentz boost along z and $K_{0i} = M_{0i}$: Lorentz boosts along X^i. The Poincaré algebra is

$$\begin{aligned}
[P_\mu, P_\nu] &= 0 \\
[M_{\mu\nu}, P_\rho] &= \eta_{\mu\rho}P_\nu - \eta_{\nu\rho}P_\mu \\
[M_{\mu\nu}, M_{\rho\sigma}] &= \eta_{\mu\rho}M_{\nu\sigma} - \eta_{\mu\sigma}M_{\nu\rho} - \eta_{\nu\rho}M_{\mu\sigma} + \eta_{\nu\sigma}M_{\mu\rho}.
\end{aligned} \tag{7.222}$$

The Galilean group is essentially a subgroup of the Poincaré group defined as follows. We introduce

$$\mu X_i^{C.M} = i\frac{K_{0i} - L_{zi}}{\sqrt{2}}. \tag{7.223}$$

We can immediately compute

$$[P_i, \mu X_j^{C.M}] = i\mu\eta_{ij}. \tag{7.224}$$

The generators $\mu X_i^{C.M}$ are the Galilean boosts. Indeed we can compute

$$e^{-iV(\mu X^i)}P_i e^{iV(\mu X^i)} = P_i + \mu V. \tag{7.225}$$

The generator $P_- = \mu$ commutes with all the generators P_i, P_+, L_{ij} and $\mu X_i^{C.M}$. The mass is therefore a central charge. The Galilean group is isomorphic to the Poincaré subgroup generated by P_i, P_+, L_{ij}, $\mu X_i^{C.M}$ and P_-. Physics in the light-cone frame is Galilean invariant and hence the Hamiltonian must be of the form

$$H = \frac{P_i^2}{2P_-} + H_{\text{internal}}. \tag{7.226}$$

The internal energy H_{internal} is Galilean invariant. The generator K_{0z} is not a part of the Galilean group. We have the commutation relations

$$[K_{0z}, P_\pm] = \pm P_\pm. \tag{7.227}$$

Thus $[K_{0z}, P_+P_-] = 0$, i.e. HP_- and as a consequence $H_{\text{internal}}P_- = M^2/2$ are invaraint under longitudinal Lorentz boosts. We can also show that $[\mu X_i^{C.M}, H_{\text{internal}}P_-] = 0$. Hence M^2 is invariant under Galilean boosts and Lorentz transformations. Note also that H scales as $1/P_-$.

Let us now consider the action

$$\begin{aligned} S &= \int dX^D\left(\frac{1}{2}\partial_\mu\phi\partial^\mu\phi - \frac{1}{2}m^2\phi^2 - \lambda\phi^3\right) \\ &= \int dX^+\mathcal{L}, \quad \mathcal{L} = \int dX^- dX^i\left(\partial_+\phi\partial_-\phi - \frac{1}{2}(\partial_i\phi)^2 - \frac{m^2}{2}\phi^2 - \lambda\phi^3\right). \end{aligned} \tag{7.228}$$

The canonical momentum π to the field ϕ is

$$\pi = \frac{\partial\mathcal{L}}{\partial(\partial_+\phi)} = \partial_-\phi. \tag{7.229}$$

The equal time commutation relations

$$[\phi(X^-, X^i), \partial_-\phi(Y^-, Y^i)] = i\delta(X^- - Y^-)\delta(X^i - Y^i). \tag{7.230}$$

The field can be expanded as follows

$$\phi(X^-, X^i) = \int_0^\infty dk_-\left[\frac{\phi(k_-, X^i)}{\sqrt{2\pi k_-}}e^{-ik_-X^-} + \frac{\phi^*(k_-, X^i)}{\sqrt{2\pi k_-}}e^{ik_-X^-}\right]. \tag{7.231}$$

Thus

$$\partial_{-}\phi(X^{-}, X^{i}) = -i\int_{0}^{\infty} dk_{-}\left[\frac{\phi(k_{-}, X^{i})}{\sqrt{2\pi}}\sqrt{k_{-}}e^{-ik_{-}X^{-}} - \frac{\phi^{*}(k_{-}, X^{i})}{\sqrt{2\pi}}\sqrt{k_{-}}e^{ik_{-}X^{-}}\right]. \quad (7.232)$$

We can then compute the non-relativistic commutation relations

$$\begin{aligned}
&[\phi(k_{-}, X^{i}), \phi(l_{-}, Y^{i})] = [\phi^{*}(k_{-}, X^{i}), \phi^{*}(l_{-}, Y^{i})] = 0 \\
&[\phi(k_{-}, X^{i}), \phi^{*}(l_{-}, Y^{i})] = \delta(k_{-} - l_{-})\delta(X^{i} - Y^{i}).
\end{aligned} \quad (7.233)$$

The Hamiltonian in the light-cone frame takes the non-relativistic form

$$H = H_0 + H_I. \quad (7.234)$$

$$\begin{aligned}
H_0 = &\int_{0}^{\infty} dk_{-} \\
&\int dX^{i}\left[\frac{\partial_i\phi^{*}(k_{-}, X^{i})\partial_i\phi(k_{-}, X^{i}) + m^2\phi^{*}(k_{-}, X^{i})\phi(k_{-}, X^{i})}{2k_{-}} + \text{c. c}\right].
\end{aligned} \quad (7.235)$$

$$\begin{aligned}
H_I = &\frac{3\lambda}{\sqrt{2\pi}}\int_{0}^{\infty} dk_{-} \\
&\int_{0}^{\infty} dl_{-}\int dX^{i}\left[\frac{\phi^{*}(k_{-}, X^{i})}{k_{-}}\frac{\phi^{*}(l_{-}, X^{i})}{l_{-}}\frac{\phi(k_{-} + l_{-}, X^{i})}{k_{-} + l_{-}} + \text{c. c}\right]
\end{aligned} \quad (7.236)$$

We observe that k_{-} can only take positive values and that it is a conserved number at the interaction vertex. This is similar to the positivity and the conservation of the mass in non-relativistic quantum mechanics.

In discrete light-cone quantization we compactify the light-like coordinate X^{-} on a circle of radius R. The spectrum of the corresponding momentum P_{-} becomes discrete given by

$$P_{-} = \frac{N}{R}. \quad (7.237)$$

We have seen that P_{-} plays the role of an invariant mass and that in the field theory it is a conserved quantum number. Hence in the discrete light-cone quantization we get an infinite number of superselection sectors defined by the positive integer N.

The expansion (7.231) becomes (with $\phi(k_{-}, X^{i}) \equiv \sqrt{R}\phi_N(X^{i})$)

$$\phi(X^{i}, X^{i}) = \phi_0(X^{i}) + \sum_{N=1}^{\infty}\frac{\phi_N(X^{i})}{\sqrt{2\pi N}}e^{-i\frac{N}{R}X^{-}} + \text{c. c.} \quad (7.238)$$

The mode $\phi_0(X^{i})$ corresponds to $P_{-} = 0$. The commutation relations (7.233) become

$$[\phi_N(X^i), \phi_M(Y^i)] = [\phi_N^*(X^i), \phi_M^*(Y^i)] = 0$$
$$[\phi_N(X^i), \phi_M^*(Y^i)] = \delta_{NM}\delta(X^i - Y^i). \tag{7.239}$$

The Hamiltonians H_0 and H_I become

$$H_0 = R\sum_{N=1}^{\infty}\int dX^i\left[\frac{\partial_i\phi_N^*(X^i)\partial_i\phi_N(X^i) + m^2\phi_N^*(X^i)\phi_N(X^i)}{2N} + \text{c. c}\right]. \tag{7.240}$$

$$H_I = \frac{3\lambda\sqrt{R}}{\sqrt{2\pi}}R^2\sum_{N=1}^{\infty}\sum_{M=1}^{\infty}\int dX^i\left[\frac{\phi_N^*(X^i)}{N}\frac{\phi_M^*(X^i)}{M}\frac{\phi_{N+M}(X^i)}{N+M} + \text{c. c}\right]. \tag{7.241}$$

At the level of the action the zero mode ϕ_0 can be shown to be non-dynamical because $\partial_-\phi_0 = 0$. Thus it can be integrated out yielding new complicated terms in the Hamiltonian which will (by construction) conserve P_- and the Galilean symmetry.

The limit of physical interest is defined by $N \longrightarrow \infty$ and $R \longrightarrow \infty$ keeping the momentum $P_- = N/R$ fixed.

Note also that H_0 is the Hamiltonian of free particles in non-relativistic quantum mechanics where $\phi_N(X^i)$ is the second quantized Schrödinger field for the Nth type of particle with mass $\mu_N \equiv N/R = P_-$.

7.9.2 Infinite momentum frame and BFSS conjecture

The original BFSS conjecture [1] relates M-theory in the infinite momentum frame (IMF) (and not M-theory in the light-cone frame) to the theory of N D0-branes (7.58) or (7.57). This goes as follows.

(1) In the IMF formulation we boost the system along the longitudinal direction z until longitudinal momenta are much larger than any other scale in the problem. The energy $E = \sqrt{P_z^2 + P^2 + m^2}$ where $\vec{P}$ is the transverse spatial momentum becomes in the limit $P_z \longrightarrow \infty$ given by

$$E = P_z + \frac{\vec{P}^2 + m^2}{2P_z}. \tag{7.242}$$

(2) In the IMF formulation we compactify the spacelike direction z on a circle of radius R_s and hence the momentum P_z becomes quantized as $P_z = N/R_s$ in contrast with the light-cone formulation where a light-like direction X^- is compactified. However, as in the light-cone formulation all objects in the IMF formulation with vanishing and negative P_z decouple. The limit $P_z = NR_s \longrightarrow \infty$ is equivalent to $N \longrightarrow \infty$ and/or $R_s \longrightarrow \infty$.

(3) M-theory with a compactified direction is by definition type IIA string theory in the same way that 11-dimensional supergravity (which is the low energy limit of M-theory) with a compactified direction is by definition type IIA supergravity (which is the low energy limit of type IIA string theory). In the limit $R_s \longrightarrow 0$ the

only objects in M-theory (or equivalently type IIA string theory) which carry P_z are the D0-brane. In a sector with momentum $P_z = N/R_s$ the lowest excitations are N D0-branes. The effective action is given by the supersymmetric Yang–Mills theory reduced to 1 dimension (7.58). The corresponding Hamiltonian is given by the quantum mechanical system (7.57).

(4) Supergravitons carrying Kaluza–Klein momentum $P_z = 1/R_s$ are the elementary D0-branes. They carry the quantum numbers of the basic 11-dimensional supergravity multiplet which contains 44 gravitons G_{MN}, 128 gravitinos ψ_M and 84 independent components of the 3-form field A_3. Supergravitons carrying Kaluza–Klein momentum $P_z = N/R_s$ are bound composites of N D0-branes. Perturbative string states carry $P_z = 0$ while elementary and bound composites of N anti-D0-branes carry negative P_z and as we said they decouple.

Thus the BFSS conjecture relates M-theory in the infinite momentum frame ($P_z \longrightarrow \infty$) in the uncompactified limit ($R_s \longrightarrow \infty$) to the large N limit of the supersymmetric quantum mechanical system (7.57) describing N D0-branes.

The infinite momentum frame and the light-cone quantization are equivalent only in the limit $N \longrightarrow \infty$. For finite N the light-cone quantization is superior since finite N infinite momentum formulation does not have Galilean invariance and also in this formulation negative and vanishing P_z do not decouple for finite N.

A stronger BFSS conjecture [102] relates discrete light-cone quantization (DLCQ) of M-theory to the supersymmetric quantum mechanical system (7.57). The DLCQ of M-theory corresponds to the quantization of M-theory compactified on a light-like circle of radius R in a sector with momentum $P_- = N/R$. This theory is characterized by a finite momentum P_- and it is conjectured to be exactly given by the finite N supersymmetric matrix model (7.58). Since for a light-like circle the value of R cannot be changed via a boost the uncompactified theory is obtained by letting N, $R \longrightarrow \infty$ keeping P_- always fixed.

7.9.3 More on light-like versus spacelike compactifications

In this section we follow [103]. The compactification of the light-like coordinate $X^- = \frac{t-z}{\sqrt{2}}$ on a circle of radius R corresponds to the identification $X^- \sim X^- + l$ where $l = 2\pi R$ or equivalently

$$\begin{aligned} z &\sim z - \frac{l}{\sqrt{2}} \\ t &\sim t + \frac{l}{\sqrt{2}}. \end{aligned} \tag{7.243}$$

Let us consider now the compactification on the spacelike circle

$$\begin{aligned} z &\sim z - \sqrt{\frac{l^2}{2} + l_s^2} \\ t &\sim t + \frac{l}{\sqrt{2}}. \end{aligned} \tag{7.244}$$

For $l_s \ll l$ this spacelike compactification tends to the previous light-like compactification, viz

$$z \sim z - \frac{l}{\sqrt{2}} - \frac{l_s^2}{\sqrt{2}l}$$
$$t \sim t + \frac{l}{\sqrt{2}}. \quad (7.245)$$

Consider now the simpler spacelike compactification

$$z \sim z - l_s$$
$$t \sim t. \quad (7.246)$$

A Lorentz boost is given by

$$z' = \frac{1}{\sqrt{1-\beta^2}}(z - \beta t)$$
$$t' = \frac{1}{\sqrt{1-\beta^2}}(t - \beta z). \quad (7.247)$$

The point $(-l_s, 0)$ is boosted to the point $(-\sqrt{\frac{l^2}{2} + l_s^2}, \frac{l}{\sqrt{2}})$ if we choose the velocity β to be

$$\beta = \frac{l}{\sqrt{l^2 + 2l_s^2}}. \quad (7.248)$$

In other words, the spacelike compactification (7.244) is related by the above boost to the spacelike compactification (7.246). For $l_s \ll l$ the space-like compactification (7.244) becomes the light-like compactification (7.243) and the velocity becomes large given by $\beta = 1 - \frac{l_s^2}{l^2}$. We can conclude that the light-like compactification (7.243) is the $l_s \longrightarrow 0$ limit of the almost light-like compactification (7.244). Equivalently, the light-like compactification (7.243) is the $l_s \longrightarrow 0$ limit of the boosted spacelike compactification (7.246) with a large velocity given by (7.248).

The point $(-\frac{l}{\sqrt{2}}, \frac{l}{\sqrt{2}})$ is boosted under a Lorentz transformation to the point $\sqrt{\frac{1+\beta}{1-\beta}}(-\frac{l}{\sqrt{2}}, \frac{l}{\sqrt{2}})$. In other words, under a longitudinal boost of the light-like compactification (7.243) the radius R of the compactification is rescaled as

$$R' = \sqrt{\frac{1+\beta}{1-\beta}}R. \quad (7.249)$$

The momenta $P_{\pm}$ transform as

$$P'_+ = \sqrt{\frac{1+\beta}{1-\beta}} P_+ = \frac{R'}{R} P_+, \; P'_- = \sqrt{\frac{1-\beta}{1+\beta}} P_- = \frac{R}{R'} P_-. \tag{7.250}$$

It is then obvious that under a longitudinal boost of the light-like compactification (7.243) the light-cone energy P_+ is also rescaled. In fact we see that P_+ is proportional to R. For the spacelike compactification (7.244) the light-cone energy P_+ is also proportional to R in the limit $R_s \longrightarrow 0$ where this spacelike compactification becomes light-like. For $R_s \longrightarrow 0$ the velocity (7.248) becomes large given by $\beta = 1 - R_s^2/R^2$ and hence $\sqrt{\frac{1+\beta}{1-\beta}} = \sqrt{2}\frac{R}{R_s}$. Since P_+ in the almost light-like compactification (7.244) is proportional to R and since this compactification is obtained from the spacelike compactification (7.246) with the above large boost we can immediately conclude that the value of P_+ in the spacelike compactification (7.246) must be proportional to R_s. In other words, the value of P_+ can be made independent of R and of order R_s by an appropriate large boost.

From the above discussion we can see that the light-like compactification of M-theory (i.e. the DLCQ and the stronger BFSS conjecture) is related to the spacelike compactification of M-theory (i.e. the IMF quantization and the original BFSS conjecture).

In the limit $R_s \longrightarrow 0$ the spacelike compactification (7.246) of M-theory yields weakly coupled type IIA string theory where the parameters $g \equiv \tilde{g}_s$ ($\tilde{g}_s$ is the string coupling) and $\alpha' \equiv \tilde{M}_s^{-2}$ ($\tilde{M}_s$ is the string scale) of type IIA string theory are related to the parameters $R_{10} \equiv R_s$ and $M_{11} \equiv \tilde{M}_p$ of the 11-dimensional theory by

$$\tilde{g}_s = (R_s \tilde{M}_p)^{\frac{3}{2}}, \quad \tilde{M}_s^2 = R_s \tilde{M}_p^3. \tag{7.251}$$

Clearly, when $R_s \longrightarrow 0$ we have $\tilde{g}_s \longrightarrow 0$ and $\tilde{M}_s \longrightarrow 0$. This is a complicated theory. Next we apply the boost. As we have said, the energy P_+ which is proportional to R_s becomes proportional to R. Since P_+ has dimension of mass and R_s has dimension of length we conclude that P_+ must be of the order of $R_s \tilde{M}_p^2$ where $\tilde{M}_p^2$ is inserted on dimensional grounds. In order to focus on the modes with these energies we replace the M-theory with parameters $\tilde{M}_p$ and R_s with a new M-theory with parameters M_p and R such that the energy P_+ is kept fixed in the double scaling limit $R_s \longrightarrow 0$, $\tilde{M}_p \longrightarrow \infty$, viz

$$R_s \tilde{M}_p^2 = R M_p^2. \tag{7.252}$$

In this limit the string coupling and the string scale become

$$\begin{aligned} \tilde{g}_s &= (R_s \tilde{M}_p)^{\frac{3}{2}} = R_s^{\frac{3}{4}} (R M_p^2)^{\frac{3}{4}} \longrightarrow 0 \\ \tilde{M}_s^2 &= R_s \tilde{M}_p^3 = R_s^{-\frac{1}{2}} (R M_p^2)^{\frac{3}{2}} \longrightarrow \infty. \end{aligned} \tag{7.253}$$

This is weakly coupled string theory with large string scale which is a very simple theory. Indeed the sector with $P_- = N/R_s$ is the theory of N D0-branes.

In summary, M-theory with Planck scale M_p compactified on a light-like circle of radius R and momentum $P_- = N/R$ can be mapped to the M-theory with Planck scale $\tilde{M}_p \longrightarrow \infty$ compactified on a spacelike circle of radius $R_s = (RM_p^2)/\tilde{M}_p^2 \longrightarrow 0$ which is a theory of N D0-branes.

7.10 M-(atrix) theory in pp-wave spacetimes

7.10.1 The pp-wave spacetimes and Penrose limit

The BFSS model [1] is a matrix model associated with the DLCQ (discrete light-cone quantization) description of the maximally supersymmetric 11-dimensional flat background. The BMN model [106] is a matrix model which describes the DLCQ compactification of M-theory on the maximally supersymmetric pp-wave background of 11-dimensional supergravity.

Supergravity in 11 dimensions admits four types of maximally supersymmetric solutions [107, 108]. These are

1. the 11-dimensional Minkowski space and its toroidal compactifications,
2. the $AdS^7 \times S^4$ (the M5-brane),
3. the $AdS^4 \times S^7$ (the M2-brane) and
4. the Kowalski–Glikman solution [108] which is a pp-wave metric.

Similarly, type IIB supergravity admits three types of solutions

1. 10-dimensional Minkowski space,
2. $AdS^5 \times S^5$ and
3. a pp-wave metric [107].

All maximally supersymmetric pp-wave geometries can arise as Penrose limits of $AdS_p \times S^q$ spaces [109]. The powerful Penrose theorem states that near null geodesics (which are paths of light rays) any spacetime becomes a pp-wave spacetime, i.e. any metric near a null geodesic becomes a pp-wave metric [110].

First we discuss pp-wave geometries a little further. These spaces are solutions of the Einstein equations which correspond to perturbations moving at the speed of light with plane wave fronts. See [39] and references therein. The bosonic content of 11-dimensional supergravity consists of the metric and a 4-form F_4. The pp-wave solutions of 11-dimensional supergravity are given by [39, 107, 111]

$$\begin{aligned} ds^2 &= 2dx^+dx^- + (dx^+)^2 H(x^+, x^i) + \sum_{i=1}^{9}(dx^i)^2 \\ F_4 &= dx^+ \wedge \phi. \end{aligned} \tag{7.254}$$

In the above equation $x^{\pm} = (x \pm t)/\sqrt{2}$, $x \equiv x^{10}$. The ϕ is a 3-form satisfying

$$d\phi = d^*\phi = 0, \quad \partial_i^2 H = \frac{1}{12}|\phi|^2. \tag{7.255}$$

Recall that

$$\phi = \phi_{\mu\nu\rho} dx^{\mu} \wedge dx^{\nu} \wedge dx^{\rho}, \quad |\phi|^2 = \phi_{\mu\nu\rho}\phi^{\mu\nu\rho}, \; (*\phi)_{\mu_1\ldots\mu_8} = \epsilon_{\mu_1\ldots\mu_{11}}\phi^{\mu_9\mu_{10}\mu_{11}}. \tag{7.256}$$

The only non-zero component of the Ricci tensor of the above metric is

$$R_{++} = -\frac{1}{2}\partial_i^2 H(x^+, x^i) = -\frac{1}{24}|\phi|^2. \tag{7.257}$$

An interesting class of solutions is given by

$$H = \sum_{i,j} A_{ij}x^i x^j, \quad A_{ij} = A_{ji}, \quad 2trA = \frac{1}{12}|\phi|^2. \tag{7.258}$$

For generic (A, ϕ) this solution will preserve 1/2 of the supersymmetry. Kowalski–Glikman showed in 1984 that all supersymmetry will be preserved for precisely one non-trivial choice of A_{ij} and ϕ given by

$$H = \sum_{i,j} A_{ij}x^i x^j = -\sum_{i=1}^{3}\frac{\mu^2}{9}x_i^2 - \sum_{a=4}^{9}\frac{\mu^2}{36}x_a^2, \quad \phi = \mu dx^1 \wedge dx^2 \wedge dx^3. \tag{7.259}$$

Similarly, the bosonic content of 10-dimensional type IIB supergravity consists of the metric and a 5-form F_5. The pp-wave solutions of 10-dimensional type IIB supergravity are given by [39]

$$\begin{aligned} ds^2 &= 2dx^+dx^- + (dx^+)^2 H(x^+, x^i) + \sum_{i=1}^{8}(dx^i)^2 \\ F_5 &= dx^+ \wedge (\omega + *\omega). \end{aligned} \tag{7.260}$$

The ω is a 4-form satisfying

$$d\omega = d{*}\omega = 0, \quad \partial_i^2 H = -32|\omega|^2. \tag{7.261}$$

Again the general metric preserves 1/2 of the supersymmetry while all supersymmetry will be preserved for precisely one non-trivial choice of A_{ij} and ω given by

$$H = \sum_{i,j} A_{ij}x^i x^j = \mu^2 \sum_{i=1}^{8} x_i^2, \quad \omega = \frac{\mu}{2}dx^1 \wedge dx^2 \wedge dx^3 \wedge dx^4. \tag{7.262}$$

The above maximally supersymmetric pp-waves are Penrose limits of maximally supersymmetric $AdS_{p+2} \times S^n$ spaces. For 11-dimensional supergravity the pp-wave metric (7.254) arises as limit of $AdS_7 \times S^4$ or $AdS_4 \times S^7$ where both spaces give the same metric. For 10-dimensional type IIB supergravity the pp-wave metric (7.260) arises as limit of $AdS_5 \times S^5$. Let us also say that the near horizon geometry of M2-, M5- and D3-brane solutions is $AdS_{p+2} \times S^{D-p-2}$ [109]. For the M2- and the M5-brane solutions $D = 11$ and $p = 2$ and 5, respectively. For the D3-brane solution $D = 10$ and $p = 3$.

Define $\rho = R_{AdS_{p+2}}/R_{S^{D-p-2}}$ then for the M2-brane $\rho = 1/2$, for the M5-brane $\rho = 2$ while for the D3-brane $\rho = 1$. The metrics for AdS_{p+2}, S^{D-p-2} and $AdS_{p+2} \times S^{D-p-2}$ are given, respectively, by [107]

$$ds^2_{AdS_{p+2}} = R^2_{AdS}\left[-(d\tau)^2 + \sin^2\tau\left(\frac{dr^2}{1+r^2} + r^2 d\Omega_p^2\right)\right]. \tag{7.263}$$

$$ds^2_{S^n} = R^2_S\left[(d\psi)^2 + \sin^2\psi d\Omega^2_{n-1}\right]. \tag{7.264}$$

$$ds^2_{AdS_{p+2}\times S^n} = ds^2_{AdS_{p+2}} + ds^2_{S^n}. \tag{7.265}$$

Above $d\Omega_p^2$ is the p-sphere metric, ψ is the colatitude and $d\Omega^2_{n-1}$ is the metric on the equatorial $(n-1)$-sphere. Introduce the coordinates

$$u = \psi + \rho\tau, \quad v = \psi - \rho\tau. \tag{7.266}$$

The metric becomes

$$\begin{aligned} ds^2_{AdS_{p+2}\times S^n} R^2_S = dudv + \rho^2 \sin^2\frac{u-v}{2\rho}\left(\frac{dr^2}{1+r^2} + r^2 d\Omega_p^2\right) \\ + \sin^2\frac{u+v}{2} d\Omega^2_{n-1}. \end{aligned} \tag{7.267}$$

We consider the Penrose limit along the null geodesic parametrised by u. All coordinates with the exception of u will be scaled to 0. The coordinate v will be scaled to 0 faster than all the other coordinates and hence dependence on the coordinate v will be dropped. We get

$$ds^2_{AdS_{p+2}\times S^n}/R^2_S = dudv + \rho^2 \sin^2\frac{u}{2\rho} ds^2_{E^{p+1}} + \sin^2\frac{u}{2} d\Omega^2_{n-1}. \tag{7.268}$$

Let $y^1, \ldots, y^{p+1}$ be the coordinates of E^{p+1} and $z^{p+2}, \ldots, z^{D-2}$ be the coordinates of S^{n-1}. We introduce

$$\begin{aligned} x^- &= \frac{u}{2} \\ x^+ &= v - \frac{1}{4}\left[\rho\vec{y}^2 \sin\frac{u}{\rho} + \vec{z}^2 \sin u\right] \\ x^a &= \left(\rho \sin\frac{u}{2\rho}\right) y^a, \quad a = 1, \ldots, p+1 \\ x^a &= \left(\sin\frac{u}{2}\right) z^a, \quad a = p+2, \ldots, D-2. \end{aligned} \tag{7.269}$$

We compute

$$\sum_{a=1}^{D-2}(dx^a)^2 = \rho^2 \sin^2\frac{u}{2\rho}(dy^a)^2 + \sin^2\frac{u}{2}(dz^a)^2$$
$$+\frac{1}{4}\left[\vec{y}^2\cos^2\frac{u}{2\rho} + \vec{z}^2\cos^2\frac{u}{2}\right](du)^2 \tag{7.270}$$
$$+\frac{1}{4}\left[\rho\sin\frac{u}{\rho}d(y^a)^2 + \sin u d(z^a)^2\right]du.$$

$$2dx^+dx^- = dudv - \frac{1}{4}\left[\rho\sin\frac{u}{\rho}d(y^a)^2 + \sin u d(z^a)^2\right]$$
$$du - \frac{1}{4}\left[\vec{y}^2\cos\frac{u}{\rho} + \vec{z}^2\cos u\right](du)^2. \tag{7.271}$$

The metric can be rewritten as

$$ds^2_{AdS_{p+2}\times S^n}/R_S^2 = 2dx^+dx^- - \left[\frac{1}{\rho^2}\sum_{a=1}^{p+1}(x^a)^2 + \sum_{a=p+2}^{D-2}(x^a)^2\right]$$
$$(dx^-)^2 + \sum_{a=1}^{D-2}(dx^a)^2. \tag{7.272}$$

This is a pp-wave metric as promised. The two cases $\rho = 2$ and $\rho = 1/2$ are isometric. The corresponding diffeomorphism is

$$x^- \longrightarrow \frac{1}{2}x^-, \; x^+ \longrightarrow 2x^+$$
$$(x^1, \ldots, x^6, x^7, \ldots, x^9) \longrightarrow (x^4, \ldots, x^9, x^1, \ldots, x^3). \tag{7.273}$$

Let us reconsider the case of $AdS_5 \times S^5$ more explicitly. In this case we have $R_{AdS} = R_S = R$ or equivalently $\rho = 1$. The metric is

$$ds^2_{AdS_5\times S^5} = R^2\left[-(d\tau)^2 + \sin^2\tau\left(\frac{dr^2}{1+r^2} + r^2d\Omega_3^2\right) + (d\psi)^2 + \sin^2\psi d\Omega_4^2\right]. \tag{7.274}$$

This can also be put in the form[5]

$$ds^2 = R^2[-dt^2\cosh^2\rho + d\rho^2 + \sinh^2\rho d\Omega_3^2$$
$$+ d\psi^2\cos^2\theta + d\theta^2 + \sin^2\theta d\Omega_3'^2]. \tag{7.275}$$

We consider a particle moving at the speed of light along an equator of S^5 ($\theta = 0$)

[5] **Exercise:** verify this.

while staying in the center of AdS_5 ($\rho = 0$). This is a null geodesic (since it is the path of a light ray) parametrized by ψ. The geometry near this trajectory is given by the metric

$$\begin{aligned} ds^2 = R^2\bigg[& -\left(1 + \frac{\rho^2}{2}\right)dt^2 + d\rho^2 + \rho^2 d\Omega_3^2 \\ & + \left(1 - \frac{\theta^2}{2}\right)d\psi^2 + d\theta^2 + \theta^2 d\Omega_3'^2\bigg]. \end{aligned} \tag{7.276}$$

We define the null coordinates

$$\tilde{x}^{\pm} = \frac{t \pm \psi}{2}. \tag{7.277}$$

Then we take the limit

$$\tilde{x}^+ = x^+,\ \tilde{x}^- = \frac{x^-}{R^2}, \quad \rho = \frac{r}{R}, \quad \theta = \frac{y}{R}. \tag{7.278}$$

The metric becomes

$$\begin{aligned} ds^2 &= -4dx^+dx^- - (r^2 + y^2)(dx^+)^2 + dr^2 + dy^2 + r^2 d\Omega_3^2 + y^2 d\Omega_3'^2 \\ &= -4dx^+dx^- - (\vec{r}^{\,2} + \vec{y}^{\,2})(dx^+)^2 + d\,\vec{r}^{\,2} + d\,\vec{y}^{\,2}. \end{aligned} \tag{7.279}$$

The Penrose limit (7.278) can be understood as follows. We consider the following boost along the equator of S^5 given by

$$t = \cosh\beta t' + \sinh\beta\psi', \qquad \psi = \sinh\beta t' + \cosh\beta\psi'. \tag{7.280}$$

This is equivalent to

$$\tilde{x}^+ = e^{\beta}\tilde{x}^{+'},\ \tilde{x}^- = e^{-\beta}\tilde{x}^{-'}. \tag{7.281}$$

If we make the identification $e^{\beta} = R$ and scale all coordinates t', ψ' and the rest by $1/R$ then we will obtain (7.278).

The D3-brane carries fluxes with respect to a $D - p - 2 = 5$-form field strength. It is obvious that only the components of this 5-form field F_5 with an index + will survive the above Penrose limit, viz

$$F_{+1234} = F_{+5678} = \mu. \tag{7.282}$$

Thus the 5-form field F_5 of the $AdS_5 \times S^5$ solution matches in the Penrose limit the maximally supersymmetric pp-wave solution of IIB supergravity given in (7.260) and (7.262).

The M2- and M5-brane solutions carry fluxes with respect to $D - p - 2 = 7$ and 4-form field strengths, respectively.

7.10.2 The BMN matrix model

As we have seen, the BFFS model is postulated to describe DLCQ quantization of M-theory in flat background spacetime. It is given by

$$S_{\rm BFFS} = \frac{1}{g^2}\,{\rm Tr}\bigg(-\frac{1}{4}[X_I, X_J]^2 + \frac{1}{2}(D_0X_I)^2 - \frac{1}{2}\psi^T C_9 \Gamma^I [X_I, \psi] - \frac{1}{2}\psi^T C_9 D_0 \psi\bigg). \tag{7.283}$$

Similarly, the BMN model is postulated to describe DLCQ quantization of M-theory in pp-wave background spacetime. It is in a precise sense a mass-deformation of the BFSS model given by the action [106]

$$S_{\rm BMN} = S_{\rm BFSS} + \Delta S_{\rm defor}. \tag{7.284}$$

$$\Delta S_{\rm defor} = \frac{\mu^2}{2g^2}\,{\rm Tr}\left(\sum_{i=1}^{3} X_i^2 + \frac{1}{4}\sum_{a=4}^{9} X_a^2\right) - \frac{i\mu}{2g^2}\epsilon_{ijk}\,{\rm Tr}[X_i, X_j]X_k - \frac{3i\mu}{8g^2}\,{\rm Tr}\,\psi^T C_9 \gamma^{123}\psi. \tag{7.285}$$

This model as opposed to the BFSS model has as a solution a fuzzy sphere solution $X_i = \mu J_i$, $X_a = 0$, $A_0 = 0$, $\psi = 0$, where $[J_i, J_j] = i\epsilon_{ijk}J_k$, which preserves full supersymmetry. Non-perturbative studies of the BMN and its relation to the gauge/gravity duality can be found in [67, 68, 71, 95, 112, 113, 120–122]. For a concise summary of the results obtained for the BMN model and future prospects see [29].

7.10.3 Construction of the BMN matrix model

The original derivation of the BMN model consisted in showing that the $N = 1$ mass-deformed BFSS model is given by the action of a superparticle in the above pp-wave background, then they obtained the $N > 1$ mass-deformed BFSS model by extending this result in a way consistent with supersymmetry.

The BMN model can be derived in a more direct way as follows [114]. We start from the pp-wave solutions of 11-dimensional supergravity given by

$$\begin{aligned} ds^2 &= 2dx^+dx^- + H(dx^+)^2 + (dx^I)^2 \\ &= -\left(1 - \frac{H}{2}\right)dt^2 + \left(1 + \frac{H}{2}\right)dx^2 + Hdxdt + (dx^I)^2. \end{aligned} \tag{7.286}$$

$$H = -\sum_{i=1}^{3} \frac{\mu^2}{9}x_i^2 - \sum_{a=4}^{9} \frac{\mu^2}{36}x_a^2. \tag{7.287}$$

$$F_4 = -\frac{\mu}{\sqrt{2}} dx^1 \wedge dx^2 \wedge dx^3 \wedge dx + \frac{\mu}{\sqrt{2}} dt \wedge dx^1 \wedge dx^2 \wedge dx^3. \tag{7.288}$$

If we start with $\tilde{M}$-theory compactified on a spacelike circle of radius R_s then we apply a boost with velocity $\beta = R\sqrt{R^2 + 2R_s^2}$ and take the limit $R_s \longrightarrow 0$ we obtain M-theory compactified on an almost light-like circle of radius R. We are interested in the DLCQ of this M-theory on the above pp-wave background with N units of momentum, viz $P_- = N/R$.

The energy P_+ in the compactification on the almost light-like circle of radius R is proportional to R. The momentum P_-^s and the energy P_+^s in the compactification on the spacelike circle of radius R_s are given through

$$P_- = \sqrt{\frac{1-\beta}{1+\beta}} P_-^s, \quad P_+ = \sqrt{\frac{1+\beta}{1-\beta}} P_+^s, \tag{7.289}$$

with

$$\sqrt{\frac{1-\beta}{1+\beta}} = \sqrt{2}\frac{R}{R_s}, \quad R_s \longrightarrow 0. \tag{7.290}$$

In other words, $P_-^s = \sqrt{2}\frac{N}{R_s}$ and P_+^s is proportional to $\frac{R_s}{\sqrt{2}}$ or equivalently

$$P_t^s = \frac{N}{R_s} + \frac{R_s}{2}, \quad P_x^s = \frac{N}{R_s} - \frac{R_s}{2}. \tag{7.291}$$

In particular, we see that the momentum P_x^s goes to ∞ when $R_s \longrightarrow 0$. Furthermore, the string coupling and the string scale in the $\tilde{M}$-theory are $\tilde{g}_s = (R_s\tilde{M}_p)^{\frac{3}{2}} = R_s^{\frac{3}{4}}(RM_p^2)^{\frac{3}{4}} \longrightarrow 0$ and $\tilde{M}_s^2 = R_s\tilde{M}_p^3 = R_s^{-\frac{1}{2}}(RM_p^2)^{\frac{3}{2}} \longrightarrow \infty$ when $R_s \longrightarrow 0$ and $\tilde{M}_p \longrightarrow \infty$ keeping $R_s\tilde{M}_p^2$ fixed, viz $R_s\tilde{M}_p^2 = RM_p^2$.

The compactification on the spacelike circle of radius $R_s \longrightarrow 0$ corresponds therefore to the quantization of $\tilde{M}$-theory in IMF with N units of longitudinal momentum which we know is weakly coupled type IIA string theory. The DLCQ of M-theory with N units of momentum is exactly mapped to the theory of N D0-branes. The M-theory on the above pp-wave background with mass μ corresponds to $\tilde{M}$-theory on the same pp-wave background with mass μ_s given through

$$\mu = \sqrt{\frac{1-\beta}{1+\beta}}\mu_s = \sqrt{2}\frac{R}{R_s}\mu_s. \tag{7.292}$$

In other words, $\mu_s = \frac{\mu}{\sqrt{2}}\frac{R_s}{R}$. Recall that the energies $E_s = P_t^s - \frac{N}{R_s}$ of the D0-brane states are proportional to $\frac{R_s}{2}$ so that they go to 0 as $R_s \longrightarrow 0$. The light-cone energies P_+ are proportional to R. In order that the energies of the D0-brane states match the light-cone energies we must multiply E_s by $2\frac{R}{R_s}$. In other words, $E_s' = R$. Multiplying μ_s by $2\frac{R}{R_s}$ we get $\mu_s' = \sqrt{2}\mu$ or equivalently $\mu = \frac{\mu_s'}{\sqrt{2}}$.

To take this rescaling into account we make the replacement $\mu \longrightarrow \mu/\sqrt{2}$ so that the 4-form field becomes

$$F_4 = -\frac{\mu}{2}dx^1 \wedge dx^2 \wedge dx^3 \wedge dx + \frac{\mu}{2}dt \wedge dx^1 \wedge dx^2 \wedge dx^3. \tag{7.293}$$

We also take this rescaling into account by replacing $\mu \longrightarrow \mu/\sqrt{2}$ in H. In other words, we replace H by $H/2$ in the metric. The metric becomes (with $H \equiv -F^2$)

$$ds^2 = -\left(1 + \frac{F^2}{4}\right)dt^2 + \left(1 - \frac{F^2}{4}\right)dx^2 - \frac{F^2}{2}dxdt + (dx^i)^2. \tag{7.294}$$

The most general metric which is invariant under translations in the 10-direction $x^{10} \equiv x$ is of the form [57]

$$ds^2 = G^{10}_{\mu\nu}(x^\mu)dx^\mu dx^\nu + e^{2\sigma(x^\mu)}(dx + A_\nu(x^\mu)dx^\nu)^2, \quad \mu, \nu = 0, \ldots, 9. \tag{7.295}$$

A_μ is the RR one-form. We can immediately find that

$$\begin{aligned} e^{2\sigma} &= 1 - \frac{F^2}{4} \\ A_0 &= \frac{-\frac{F^2}{4}}{1 - \frac{F^2}{4}}, \quad A_i = 0 \\ G^{10}_{00} &= -e^{-2\sigma}, \quad G^{10}_{0i} = 0, \quad G^{10}_{ij} = \delta_{ij}. \end{aligned} \tag{7.296}$$

The dilaton field is defined through

$$\sigma = \frac{2\Phi}{3}. \tag{7.297}$$

The corresponding type IIA background is obtained by the redefinition

$$ds^2(\text{new}) = e^\sigma ds^2. \tag{7.298}$$

In other words

$$G^{10}_{00}(\text{new}) = -e^{-\sigma}, \quad G^{10}_{0i}(\text{new}) = 0, \quad G^{10}_{ij}(\text{new}) = e^\sigma \delta_{ij}. \tag{7.299}$$

The 11-dimensional pp-wave metric has zero scalar curvature. The curvature of the 10-dimensional metric $e^{\sigma(x^\mu)}G^{10}_{\mu\nu}(x^\mu)dx^\mu dx^\nu$ is not zero given by

$$\mathcal{R} \propto -\frac{\mu^2}{8}\left(1 - \frac{F^2}{4}\right)^{-\frac{3}{2}}. \tag{7.300}$$

This means that we must always have $F^2 \leqslant 4$.

The NS–NS and R–R three-form fields are given by

$$H_{123} \equiv (F_4)_{12310} = -\frac{\mu}{2}, \quad F_{0123} \equiv (F_4)_{0123} = \frac{\mu}{2}. \tag{7.301}$$

For small F^2 we have

$$\begin{aligned} \Phi &\sim -\frac{3F^2}{16} \\ A_0 &\sim -\frac{F^2}{4} \\ G_{00}^{10}(\text{new}) &= -1 - \frac{F^2}{8} \\ G_{ij}^{10}(\text{new}) &= \delta_{ij} - \frac{F^2}{8}\delta_{ij}. \end{aligned} \tag{7.302}$$

Alternatively we can write the metric as

$$\begin{aligned} ds^2(\text{new}) &= \eta_{\mu\nu}dx^\mu dx^\nu + (dx^{10})^2 \\ &+ h_{\mu\nu}dx^\mu dx^\nu + h_{1010}(dx^{10})^2 + 2h_{010}dx^0 dx^{10}. \end{aligned} \tag{7.303}$$

$$h_{\mu\nu} = \sigma\delta_{\mu\nu} = -\frac{F^2}{8}\delta_{\mu\nu}, \quad h_{1010} = 2\Phi = -\frac{3F^2}{8}, \quad h_{010} = A_0 = -\frac{F^2}{4}. \tag{7.304}$$

The matrix model corresponding to the flat metric $ds^2(\text{new}) = \eta_{\mu\nu}dx^\mu dx^\nu + (dx^{10})^2$ is given by the BFSS model in Minkowski signature given by the equation (7.51), namely

$$\begin{aligned} S_0 &= \int dt \; L_0 \\ L_0 &= \frac{1}{g^2}\,\mathrm{Tr}\bigg(\frac{1}{4}[X_I, X_J]^2 + \frac{1}{2}(D_0X_I)^2 + \psi^T C_9 \Gamma^I [X_I, \psi] \\ &\quad - i\psi^T C_9 D_0 \psi\bigg). \end{aligned} \tag{7.305}$$

above $D_0 = \partial_0 - i[A_0,..]$ and $I, J = 1, \ldots, 9$. The coupling constant g^2 is of dimension L^2, the matrices X_I are of dimension L, the operator D_0 is of dimension L and the spinor ψ is of dimension $L^{\frac{3}{2}}$.

In the rest of this section we will follow the derivation from D0-brane dynamics on compactified reduced pp-waves outlined in [115] but given in detail in [114]. The correction to the matrix model L_0 corresponding to the metric h_{MN} is given by terms of the form [114] (see also [116])

$$\Delta S_0[h] = \int dt \;\; \Delta L_0[h]. \tag{7.306}$$

$$\Delta L_0[h] = \frac{1}{2}\sum_{n=0}^{\infty}\sum_{I_1,\ldots,I_n}\frac{1}{n!}T^{MN(I_1\ldots I_n)}\partial_{I_1}\ldots\partial_{I_n}h_{MN}(0)$$
$$= \frac{1}{4}\sum_{I_1,I_2}T^{MN(I_1I_2)}\partial_{I_1}\partial_{I_2}h_{MN}(0) \tag{7.307}$$
$$= -\frac{1}{8}\left(\frac{1}{4}T^{\mu\mu(I_1I_2)} + T^{010(I_1I_2)} + \frac{3}{4}T^{1010(I_1I_2)}\right)\partial_{I_1}\partial_{I_2}F^2.$$

where $T^{MN(I_1,\ldots,I_n)}$ are the matrix theory forms of the multipole moments of the stress–energy tensor of 11-dimensional supergravity which couple to the derivatives of the background supergravity fields. Making use of equation (17) from [114] we have

$$\begin{aligned}T^{\mu\mu(I_1I_2)} &= T^{00(I_1I_2)} + T^{II(I_1I_2)} = T^{++(I_1I_2)} + T^{+-(I_1I_2)} + T^{II(I_1I_2)}\\ T^{010(I_1I_2)} &= T^{++(I_1I_2)}\\ T^{1010(I_1I_2)} &= T^{++(I_1I_2)} - \frac{1}{3}T^{+-(I_1I_2)} - \frac{1}{3}T^{II(I_1I_2)}.\end{aligned} \tag{7.308}$$

We obtain

$$\Delta L_0[h] = -\frac{1}{4}T^{++(I_1I_2)}\partial_{I_1}\partial_{I_2}F^2. \tag{7.309}$$

The zeroeth moment component T^{++} of the stress–energy tensor is given by

$$T^{++} = \frac{1}{g^2}\,\mathrm{Tr}\,(\mathbf{1}). \tag{7.310}$$

This moment obviously corresponds to the momentum $P_- = N/R$. The higher moments $T^{++(I_1I_2)}$ of the stress–energy tensor are defined by [114]

$$T^{++(I_1I_2)} = \mathrm{Sym}(T^{++};\, X_{I_1},\, X_{I_2}) = \frac{1}{g^2}\,\mathrm{Sym}(\mathrm{Tr}\,\mathbf{1};\, X_{I_1},\, X_{I_2}). \tag{7.311}$$

In general

$$T^{IJ(I_1\ldots I_n)} = \mathrm{Sym}(T^{IJ};\, X_{I_1},\, \ldots,\, X_{I_n}) + T^{IJ(I_1\ldots I_n)}_{\mathrm{fermion}}. \tag{7.312}$$

The contributions $\mathrm{Sym}(\mathrm{STr}(Y);\, X_{I_1},\ldots, X_{I_n})$ (where STr indicates a trace which is symmetrized over all orderings of terms under the trace) are the symmetrized average over all possible orderings when the matrices X_{I_k} are inserted into the trace of the product Y. Thus

$$T^{++(I_1I_2)} = \frac{1}{g^2}\,\mathrm{Tr}\,X_{I_1}X_{I_2}. \tag{7.313}$$

The first correction due to the background metric h (the 10-dimensional metric, the dilaton field Φ and the R–R field A) takes then the form

$$\Delta L_0[h] = -\frac{\mu^2}{18g^2}\,\mathrm{Tr}\sum_{i=1}^{3} X_i^2 - \frac{\mu^2}{72g^2}\,\mathrm{Tr}\sum_{a=4}^{9} X_a^2. \qquad (7.314)$$

The other degrees of freedom are the NS–NS and R–R three-form fields $H_{123} = -\frac{\mu}{2}$ and $F_{0123} = \frac{\mu}{2}$. The corresponding potentials are

$$B_{ij} = \frac{\mu}{6}\epsilon_{ijk}x_k, \quad C_{0ij} = \frac{\mu}{6}\epsilon_{ijk}x_k. \qquad (7.315)$$

In the above we have used the fact that $H = -dB$, $F = dC$ or explicitly $H_{kij} = -\partial_k B_{ij} - \partial_i B_{jk} - \partial_j B_{ki}$ and $F_{0kij} = \partial_k C_{0ij} + \partial_i C_{0jk} + \partial_j C_{0ki}$. The correction to the matrix model L_0 given in equation (7.51) arising from the fields B and C are given by terms of the form [114]

$$\begin{aligned}\Delta L_0[B, C] &= \sum_{n=0}^{\infty}\sum_{i_1,\ldots,i_n} \frac{1}{n!}T_B^{\mu\nu(i_1\ldots i_n)}\partial_{i_1}\ldots\partial_{i_n}B_{\mu\nu}(0)\\ &+ \sum_{n=0}^{\infty}\sum_{i_1,\ldots,i_n} \frac{1}{n!}T_C^{\mu\nu\lambda(i_1\ldots i_n)}\partial_{i_1}\ldots\partial_{i_n}C_{\mu\nu\lambda}(0)\\ &= \sum_{i_1} T_B^{\mu\nu(i_1)}\partial_{i_1}B_{\mu\nu} + \sum_{i_1} T_C^{\mu\nu\lambda(i_1)}\partial_{i_1}C_{\mu\nu\lambda}\\ &= \frac{\mu}{2}\epsilon_{ijk}\left(\frac{1}{3}T_B^{ij(k)} + T_C^{0ij(k)}\right).\end{aligned} \qquad (7.316)$$

The $T_B^{MN(I_1,\ldots,I_n)}$ and $T_B^{MN(I_1,\ldots,I_n)}$ (called I_s and I_2 in [114]) are the matrix theory forms of the multipole moments of the membrane current of 11-dimensional supergravity. By using equation (19) of [114] we have the leading behavior

$$\begin{aligned}T_B^{ij(k)} &= 3J^{+ij(k)}\\ T_C^{0ij(k)} &= J^{+ij(k)}.\end{aligned} \qquad (7.317)$$

$$\Delta L_0[B, C] = \mu\epsilon_{ijk}J^{+ij(k)}. \qquad (7.318)$$

We have (equation (37) of [114])

$$J^{+ij} = \frac{i}{6g^2}\,\mathrm{STr}[X_i, X_j]. \qquad (7.319)$$

Although these zeroeth moments are zero for finite N, the higher moments $J^{+ij(k)}$ are not zero given (by using equation (7.311)) precisely by the Chern–Simons action

$$J_B^{+ij(k)} = \frac{i}{6g^2}\,\mathrm{Tr}[X_i, X_j]X_k. \qquad (7.320)$$

This is the Myers effect [117]. The fermionic contribution to $J^{+ij(k)}$ is given by (see the appendix of [114])

$$J_F^{+ij(k)} = \frac{i}{24g^2}\,\mathrm{Tr}\,\psi^T C_9 \gamma^{[ijk]}\psi. \tag{7.321}$$

The correction $\Delta L_0[B, C]$ is then given by

$$\Delta L_0[B, C] = \frac{i\mu}{6g^2}\epsilon_{ijk}\,\mathrm{Tr}[X_i, X_j]X_k + \frac{i\mu}{4g^2}\,\mathrm{Tr}\,\psi^T C_9 \gamma^{123}\psi. \tag{7.322}$$

Putting (7.305), (7.314) and (7.322) together we get the total BMN model, viz

$$\begin{aligned} L &= \frac{1}{g^2}\,\mathrm{Tr}\left(\frac{1}{4}[X_I, X_J]^2 + \frac{1}{2}(D_0X_I)^2 + \psi^T C_9 \Gamma^I [X_I, \psi] - i\psi^T C_9 D_0 \psi\right) \\ &\quad - \frac{\mu^2}{18g^2}\,\mathrm{Tr}\left(\sum_{i=1}^{3} X_i^2 + \frac{1}{4}\sum_{a=4}^{9} X_a^2\right) \\ &\quad + \frac{i\mu}{6g^2}\epsilon_{ijk}\,\mathrm{Tr}[X_i, X_j]X_k + \frac{i\mu}{4g^2}\,\mathrm{Tr}\,\psi^T C_9 \gamma^{123}\psi. \end{aligned} \tag{7.323}$$

In Euclidean signature we get the Lagrangian

$$\begin{aligned} -L &= \frac{1}{g^2}\,\mathrm{Tr}\left(-\frac{1}{4}[X_I, X_J]^2 + \frac{1}{2}(D_0X_I)^2 - \psi^T C_9 \Gamma^I [X_I, \psi] - \psi^T C_9 D_0 \psi\right) \\ &\quad + \frac{\mu^2}{18g^2}\,\mathrm{Tr}\left(\sum_{i=1}^{3} X_i^2 + \frac{1}{4}\sum_{a=4}^{9} X_a^2\right) \\ &\quad - \frac{i\mu}{6g^2}\epsilon_{ijk}\,\mathrm{Tr}\,[X_i, X_j]X_k - \frac{i\mu}{4g^2}\,\mathrm{Tr}\,\psi^T C_9 \gamma^{123}\psi. \end{aligned} \tag{7.324}$$

We go back to the Minkowski signature and perform the scaling $X_I \longrightarrow RX_I$, $\psi \longrightarrow R^{\frac{3}{2}}\psi$ where $g^2 = R^3$. We obtain

$$\begin{aligned} L &= \mathrm{Tr}\left(\frac{R}{4}[X_I, X_J]^2 + \frac{1}{2R}(D_0X_I)^2 + R\psi^T C_9 \Gamma^I [X_I, \psi] - i\psi^T C_9 D_0 \psi\right) \\ &\quad - \frac{\mu^2}{18R}\,\mathrm{Tr}\left(\sum_{i=1}^{3} X_i^2 + \frac{1}{4}\sum_{a=4}^{9} X_a^2\right) \\ &\quad + \frac{i\mu}{6}\epsilon_{ijk}\,\mathrm{Tr}[X_i, X_j]X_k + \frac{i\mu}{4}\,\mathrm{Tr}\,\psi^T C_9 \gamma^{123}\psi. \end{aligned} \tag{7.325}$$

We set $R = 1$, $\mu = m$, $A_0 = X_0$, $X_a = \Phi_a$, $a = 4, \ldots, 9$, and we define

$$\psi = \begin{pmatrix} \theta_\alpha^A \\ (i\sigma_2)_{\alpha\beta}\theta_\beta^{+A}. \end{pmatrix}, \quad \alpha = 1, 2, \quad A = 1, \ldots, 4. \tag{7.326}$$

We also have

$$\gamma^{123} = \begin{pmatrix} -i\mathbf{1}_2 \otimes \mathbf{1}_4 & 0 \\ 0 & i\mathbf{1}_2 \otimes \mathbf{1}_4 \end{pmatrix}. \tag{7.327}$$

The above Lagrangian becomes therefore (changing also the notation as $a, b = 1, 2, 3$ and $i, j = 4, \ldots, 9$)

$$\begin{aligned} L = \mathrm{Tr}\Bigg[& -\frac{1}{2}[X_0, X_a]^2 - \frac{1}{2}[X_0, \Phi_i]^2 + \frac{1}{4}[X_a, X_b]^2 \\ & + \frac{1}{4}[\Phi_i, \Phi_j]^2 + \frac{1}{2}[X_a, \Phi_i]^2 \\ & + \frac{im}{3}\epsilon_{abc}X_aX_bX_c - \frac{m^2}{18}X_a^2 - \frac{m^2}{72}\Phi_i^2 \Bigg] \\ & + \mathrm{Tr}\Bigg[-2\theta^+[X_0, \theta] - 2\theta^+\left(\sigma_a[X_a, \theta] - \frac{m}{4}\theta\right) \\ & + \theta^+ i\sigma_2\rho_i[\Phi_i, (\theta^+)^T] - \theta^T i\sigma_2\rho_i^+[\Phi_i, \theta] \Bigg] \\ & + \mathrm{Tr}\Bigg[\frac{1}{2}(\partial_0X_a)^2 + \frac{1}{2}(\partial_0\Phi_i)^2 - 2i\theta^+\partial_0\theta \\ & - i\partial_0X_a[X_0, X_a] - i\partial_0\Phi_i[X_0, \Phi_i] \Bigg]. \end{aligned} \tag{7.328}$$

7.10.4 Compactification on R × S$_3$

We give another derivation of the BMN model via dimensional reduction on $\mathbf{R} \times \mathbf{S}^3$ following [26, 123].

We start from $D = 10$ with metric $\eta^{\mu\nu} = (-1, +1, \ldots, +1)$. The Clifford algebra is 32-dimensional given by $\{G^M, G^N\} = 2\eta^{MN}\mathbf{1}_{32}$. The basic object of $\mathcal{N} = 1$ SUSY in 10 dimensions is a 32-component complex spinor Λ which satisfies the Majorana reality condition and the Weyl condition. We use the notation $I = 1, \ldots, 9$, $\mu = 0, \ldots, 3$, $a = 1,.., 3$, $i = 4, .., 9$. The Dirac matrices are given by

$$G^0 = i\sigma_2 \otimes \mathbf{1}_{16}, \quad G^I = \sigma_1 \otimes \Gamma^I. \tag{7.329}$$

Explicitly we have

$$\Gamma^a = \begin{pmatrix} -\sigma^a \otimes \mathbf{1}_4 & 0 \\ 0 & \sigma^a \otimes \mathbf{1}_4 \end{pmatrix}, \quad \Gamma^i = \begin{pmatrix} 0 & \mathbf{1}_2 \times \rho^i \\ \mathbf{1} \otimes (\rho^i)^+ & 0 \end{pmatrix}. \tag{7.330}$$

In the above the matrices ρ_i satisfy

$$\rho_i(\rho_j)^+ + \rho_j(\rho_i)^+ = (\rho_i)^+\rho_j + (\rho_j)^+\rho_i = 2\delta_{ij}\mathbf{1}_4. \tag{7.331}$$

The matrices Γ^I provide the Clifford algebra in $d = 9$ dimensions. The Γ^a provide the $SO(3)$ Clifford algebra, whereas Γ^i provide the $SO(6)$ Clifford algebra. The charge conjugation matrix C_{10} in 10 dimensions is related to the charge conjugation matrix C_9 in nine dimensions via the equation

$$C_{10} = \sigma_1 \otimes C_9, \quad C_9 = \begin{pmatrix} 0 & -i\sigma_2 \otimes \mathbf{1}_4 \\ i\sigma_2 \otimes \mathbf{1}_4 & 0 \end{pmatrix}. \tag{7.332}$$

By the Weyl and Majorana conditions the 32-component spinor Λ can be put in the form

$$\Lambda = \sqrt{2}\begin{pmatrix} \psi \\ 0 \end{pmatrix}, \; \psi^+ = \psi^T C_9. \tag{7.333}$$

By the reality condition $\psi^+ = \psi^T C_9$ the 16-component spinor ψ can be written as

$$\psi = \begin{pmatrix} s_\alpha^A \\ t_\alpha^A \equiv (i\sigma_2)_{\alpha\beta} s_\beta^{+A}. \end{pmatrix}, \quad \alpha = 1, 2, \quad A = 1, \ldots, 4. \tag{7.334}$$

By using $D_M = \nabla_M - i[A_M, ..]$, $A_i = \phi_i$, $\partial_i = 0$ where ∇_M are spacetime covariant derivatives we can immediately compute the fermion action in $D = 10$ to be given by

$$\begin{aligned} -\frac{1}{2}(\bar{\Lambda} G^M D_M \Lambda)|_{d=10} &= \frac{1}{2}\Lambda^+ D_0 \Lambda - \frac{1}{2}\Lambda^+ G^0 G^a D_a \Lambda + \frac{i}{2}\Lambda^+ G^0 G^i [\phi_i, \Lambda] \\ &= 2s^{+A} D_0 s^A + 2s^{+A}\sigma_a D_a s^A + is^{+A} i\sigma_2(\rho_i)_{AB}[\phi_i, (s^{+B})^T] \\ &\quad - i(s^A)^T i\sigma_2(\rho_i^+)_{AB}[\phi_i, s^B]. \end{aligned} \tag{7.335}$$

The Yang–Mills action takes the form

$$\begin{aligned} \left(-\frac{1}{4}F_{MN}F^{MN}\right)\Big|_{d=10} &= -\frac{1}{4}F_{\mu\nu}F^{\mu\nu} + \frac{1}{4}[\phi_i, \phi_j]^2 \\ &\quad - \frac{1}{2}(D_\mu \phi_i)(D^\mu \phi_i), \\ F_{\mu\nu} &= \nabla_\mu A_\nu - \nabla_\nu A_\mu - i[A_\mu, A_\nu]. \end{aligned} \tag{7.336}$$

We assume the $\mathbf{R} \times \mathbf{S}^3$ metric given by

$$\begin{aligned} ds^2 = h_{\mu\nu} dx^\mu dx^\nu &= -dt^2 + R^2(d\theta^2 + \sin^2\theta d\psi^2 + \sin^2\theta \, \sin^2\psi \, d\chi^2) \\ &= +d\tau^2 + R^2(d\theta^2 + \sin^2\theta d\psi^2 + \sin^2\theta \sin^2\psi \, d\chi^2), \quad \tau = it. \end{aligned} \tag{7.337}$$

$\mathbf{R} \times \mathbf{S}^3$ is conformally flat because after the scaling $\tau = R \ln r$ the above metric becomes the flat metric on $\mathbf{R}^4$, viz

$$\frac{e^{\frac{2\tau}{R}}}{R^2} ds^2 = dr^2 + r^2(d\theta^2 + \sin^2\theta \, d\psi^2 + \sin^2\theta \sin^2\psi \, d\chi^2), \quad \tau = it. \tag{7.338}$$

In D dimensions the conformally invariant Laplacian is $\nabla_M \nabla^M - \frac{d-2}{4(d-1)}\mathcal{R}$ where $\mathcal{R}$

is the Ricci scalar curvature. For $\mathbf{R} \times \mathbf{S}^3$ we have $\mathcal{R} = 6/R^2$ and hence we replace the scalar quadratic term in the action as follows

$$-\frac{1}{2}(D_\mu\phi_i)(D^\mu\phi_i) \longrightarrow -\frac{1}{2}(D_\mu\phi_i)(D^\mu\phi_i) - \frac{\mathcal{R}}{12}\phi_i^2. \tag{7.339}$$

The $\mathcal{N} = 1$ SYM action in $D = 10$ is given by

$$S = \frac{1}{g^2}\int d^4x\sqrt{h}\ \mathrm{Tr}\left[\left(-\frac{1}{4}F_{MN}F^{MN}\right)|_{d=10} + \frac{i}{2}(\bar{\Lambda}G^M D_M\Lambda)|_{d=10}\right] \tag{7.340}$$

The $\mathcal{N} = 4$ SYM action in $D = 4$ is given by (with $\sigma_0 = \mathbf{1}_2$)

$$\begin{aligned} S = \frac{1}{g^2}\int d^4x\sqrt{h}\ \mathrm{Tr}\Big[& -\frac{1}{4}F_{\mu\nu}F^{\mu\nu} + \frac{1}{4}[\phi_i, \phi_j]^2 \\ & - \frac{1}{2}(D_\mu\phi_i)(D^\mu\phi_i) - \frac{\mathcal{R}}{12}\phi_i^2 \\ & - 2is^{+A}\sigma_\mu D^\mu s^A + s^{+A}i\sigma_2(\rho_i)_{AB}[\phi_i, (s^{+B})^T] \\ & - (s^A)^T i\sigma_2(\rho_i^+)_{AB}[\phi_i, s^B]\Big]. \end{aligned} \tag{7.341}$$

The supersymmetry transformations are given by

$$\begin{aligned} \delta A_0 &= -2i\left(s_\alpha^{+A}\eta_\alpha^A - \eta_\alpha^{+A}s_\alpha^A\right) \\ \delta A_a &= 2i\left(s_\alpha^{+A}(\sigma_a)_{\alpha\beta}\eta_\beta^A - \eta_\alpha^{+A}(\sigma_a)_{\alpha\beta}s_\beta^A\right) \\ \delta\phi_i &= -2i\left(s_\alpha^{+A}\rho_i^{AB}(i\sigma_2\eta^{+B})_\alpha + s_\alpha^A(\rho_i^+)^{BA}(i\sigma^2\eta^B)_\alpha\right). \end{aligned} \tag{7.342}$$

Also

$$\begin{aligned} \delta s_\alpha^A = & \frac{i}{2}\epsilon_{abc}F^{ab}(\sigma^c\eta^A)_\alpha - D_a\phi_i\rho_i^{AB}(i\sigma^a\sigma^2\eta^{+B})_\alpha - \frac{i}{2}[\phi_i, \phi_j] \\ & (\rho_i\rho_j^+)^{AB}\eta_\alpha^B - \frac{1}{2}\phi_i\rho_i^{AB}(i\sigma^a\sigma^2\nabla_a\eta^{+B})_\alpha \\ & + D_aA_0(\sigma^a\eta^A)_\alpha + [A_0, \phi_i]\rho_i^{AB}(\sigma_2\eta^{+B})_\alpha \\ & - \partial_0A_a(\sigma^a\eta^A)_\alpha + \partial_0\phi_i(i\sigma_2)_{\alpha\beta}\rho_i^{AB}\eta_\beta^{+B} + \frac{1}{2}\phi_i(i\sigma_2)_{\alpha\beta}\rho_i^{AB}\partial_0\eta_\beta^{+B}. \end{aligned} \tag{7.343}$$

The 4th and 9th terms will be absent in the case of flat space. In the above the supersymmetry parameters η_α^A have the following dependence on time

$$\eta_B \equiv \eta_B(t) = e^{-i\alpha t}\eta_B(0). \tag{7.344}$$

The supersymmetry parameters are four Weyl spinors η^A which satisfy one of the two conformal Killing spinor equations

$$\nabla_\mu \eta^\pm = \pm \frac{i}{2R} \sigma_\mu \eta^\pm. \tag{7.345}$$

There are four possible solutions and hence we have $\mathcal{N} = 4$ supersymmetry.

Next we expand the fields A_0, A_a, s_α^A, s_α^{+A} and ϕ_i and the supersymmetry parameters η_α^A, η_α^{+A} in terms of spherical harmonics on $\mathbf{S}^3$ and keep only the zero modes as follows

$$\begin{aligned}
\phi_i &= \Phi_i + \cdots \\
A_0 &= X_0 + .. \\
A_a &= \sum_{\hat{a}=1}^{3} X_{\hat{a}} V_a^{\hat{a}} + \cdots \\
s_\alpha^A &= \sum_{\hat{\alpha}=1}^{2} \theta_{\hat{\alpha}}^A S_\alpha^{\hat{\alpha}+} + \cdots \\
s_\alpha^{+A} &= \sum_{\hat{\alpha}=1}^{2} \theta_{\hat{\alpha}}^{+A} (S_\alpha^{\hat{\alpha}+})^+ + \cdots
\end{aligned} \tag{7.346}$$

The supersymmetry parameters are also expanded as

$$\begin{aligned}
\eta_\alpha^A &= \sum_{\hat{\alpha}=1}^{2} \epsilon_{\hat{\alpha}}^A S_\alpha^{\hat{\alpha}+} + \cdots \\
\eta_\alpha^{+A} &= \sum_{\hat{\alpha}=1}^{2} \epsilon_{\hat{\alpha}}^{+A} (S_\alpha^{\hat{\alpha}+})^+ + \cdots
\end{aligned} \tag{7.347}$$

The fields ϕ_i, X_0 are scalar under the isometry group $SU(2)_L \times SU(2)_R$ of $\mathbf{R} \times \mathbf{S}^3$ so they transform as (1,1) and hence the corresponding zero mode is the constant function. The field s_α^A (for a given A) transforms as (2,1) under $SU(2)_L \times SU(2)_L$ and hence the two zero modes $S_\alpha^{\hat{\alpha}+}$ are the lowest spinor spherical harmonics of $\mathbf{S}^3$. They satisfy the Killing spinor equation

$$\nabla_a S^{\hat{\alpha}+} = \frac{i}{2R} \sigma_a S^{\hat{\alpha}+}. \tag{7.348}$$

Similarly, the field A_a transforms as (3,1) under $SU(2)_L \times SU(2)_R$ and hence the three zero modes $V_a^{\hat{a}+}$ are the lowest vector spherical harmonics of $\mathbf{S}^3$. They are given by

$$(S^{\hat{\alpha}+})^+ \sigma_a S^{\hat{\beta}+} = (\sigma_{\hat{a}})^{\hat{\alpha}\hat{\beta}} V_a^{\hat{a}+}. \tag{7.349}$$

The zero modes X_0, $X_{\hat{a}}$, Φ_i, $\theta_{\hat{\alpha}}^A$ and $\theta_{\hat{\alpha}}^{+A}$ are time-dependent matrices in the Lie algebra of the gauge group.

In terms of the zero modes X_0, $X_{\hat{a}}$, Φ_i, $\theta_{\hat{\alpha}}^A$, $\theta_{\hat{\alpha}}^{+A}$ and the parameters $\epsilon_{\hat{\alpha}}^A$, $\epsilon_{\hat{\alpha}}^{+A}$ the supersymmetry transformations (7.342) take the form (by dropping also the hat on indices whenever possible)

$$\begin{aligned}\delta X_0 &= 7 - 2i\left(\theta_\alpha^{+A}\epsilon_\alpha^A - \epsilon_\alpha^{+A}\theta_\alpha^A\right)\\ \delta X_a &= 2i\left(\theta_\alpha^{+A}(\sigma_a\epsilon^A)_\alpha - \epsilon_\alpha^{+A}(\sigma_a\theta^A)_\alpha\right)\\ \delta\Phi_i &= -2i\left(\theta_\alpha^{+A}\rho_i^{AB}(i\sigma_2\epsilon^{+B})_\alpha + \theta_\alpha^A(\rho_i^+)^{BA}(i\sigma_2\epsilon^B)_\alpha\right).\end{aligned} \tag{7.350}$$

In the bove we have used the identities

$$\begin{aligned}&(S_\alpha^{\hat{\alpha}})^+ S_\alpha^{\hat{\beta}} = \delta^{\hat{\alpha}\hat{\beta}},\ (S_\alpha^{\hat{\alpha}})^+(\sigma_a)_{\alpha\beta}S_\beta^{\hat{\beta}} = (\sigma_{\hat{a}})^{\hat{\alpha}\hat{\beta}}V_a^{\hat{a}}\\ &S_\alpha^{\hat{\alpha}}(i\sigma^2)_{\alpha\beta}S_\beta^{\hat{\beta}} = (i\sigma^2)^{\hat{\alpha}\hat{\beta}},\quad S_\alpha^{\hat{\alpha}}(i\sigma^2\sigma^a)_{\alpha\beta}S_\beta^{\hat{\beta}} = (i\sigma^2\sigma^{\hat{a}})^{\hat{\alpha}\hat{\beta}}V^{\hat{a}a}.\end{aligned} \tag{7.351}$$

By using other identities we can show that

$$\begin{aligned}&F_{ab} = -V_a^{\hat{a}}V_b^{\hat{b}}\tilde{F}_{\hat{a}\hat{b}},\quad \tilde{F}_{\hat{a}\hat{b}} = i[X_{\hat{a}}, X_{\hat{b}}] - \frac{2}{R}\epsilon^{\hat{a}\hat{b}\hat{c}}X_{\hat{c}}\\ &D_a\phi_i = -i[X_{\hat{a}}, \Phi_i]V_a^{\hat{a}}.\end{aligned} \tag{7.352}$$

Similarly,

$$\begin{aligned}\delta\theta_{\hat{\alpha}}^A =& -\frac{i}{2}\epsilon^{\hat{a}\hat{b}\hat{c}}\tilde{F}_{\hat{a}\hat{b}}(\sigma_{\hat{c}}\epsilon^A)_{\hat{\alpha}} + i[X^{\hat{a}}, \Phi_i]\rho_i^{AB}(i\sigma_{\hat{a}}\sigma_2)_{\hat{\alpha}\hat{\beta}}\epsilon_{\hat{\beta}}^{+B}\\ &+\frac{3}{4R}\Phi_i(\rho_i)^{AB}(\sigma_2)_{\hat{\alpha}\hat{\beta}}\epsilon_{\hat{\beta}}^{+B}\\ &-\frac{i}{2}[\Phi_i, \Phi_j](\rho_i\rho_j^+)^{AB}\epsilon_{\hat{\alpha}}^B - i[X^{\hat{a}}, X_0](\sigma_{\hat{a}}\epsilon^A)^{\hat{\alpha}}\\ &+[X_0, \Phi_i](\rho_i)^{AB}(\sigma_2)_{\hat{\alpha}\hat{\beta}}\epsilon_{\hat{\beta}}^{+B}\\ &-\frac{\alpha}{2}\Phi_i(\sigma_2)_{\hat{\alpha}\hat{\beta}}\rho_i^{AB}\epsilon_{\hat{\beta}}^{+B} - \partial_0X^a(\sigma_a\epsilon^A)_\alpha + i\partial_0\Phi_i(\sigma_2)_{\alpha\beta}\rho_i^{AB}\epsilon_\beta^{+B}.\end{aligned} \tag{7.353}$$

Also

$$\begin{aligned}\delta\theta_{\hat{\alpha}}^{+A} =& \frac{i}{2}\epsilon^{\hat{a}\hat{b}\hat{c}}\tilde{F}_{\hat{a}\hat{b}}(\epsilon^{+A}\sigma_{\hat{c}})_{\hat{\alpha}} - i[X^{\hat{a}}, \Phi_i]\epsilon_{\hat{\beta}}^B(i\sigma_2\sigma_{\hat{a}})_{\hat{\beta}\hat{\alpha}}(\rho_i^+)^{BA}\\ &+\frac{3}{4R}\Phi_i\epsilon_{\hat{\beta}}^B(\sigma_2)_{\hat{\beta}\hat{\alpha}}(\rho_i^+)^{BA}\\ &-\frac{i}{2}[\Phi_i, \Phi_j]\epsilon_{\hat{\alpha}}^{+B}(\rho_j\rho_i^+)^{BA} - i[X^{\hat{a}}, X_0](\epsilon^{+A}\sigma_{\hat{a}})^{\hat{\alpha}}\\ &-[X_0, \Phi_i]\epsilon_{\hat{\beta}}^B(\sigma_2)_{\hat{\beta}\hat{\alpha}}(\rho_i^+)^{BA}\\ &-\frac{\alpha}{2}\Phi_i(\sigma_2)_{\hat{\beta}\hat{\alpha}}(\rho_i^+)^{BA}\epsilon_{\hat{\beta}}^B - \partial_0X^a(\epsilon^{+A}\sigma_a)_\alpha\\ &-i\partial_0\Phi_i(\rho_i^+)^{BA}\epsilon_\beta^B(\sigma_2)_{\beta\alpha}.\end{aligned} \tag{7.354}$$

The equations of motion obtained by varying A_μ are

$$D_\mu F^{\mu\nu} + i[\phi_i, D^\nu\phi_i] + 2\left\{s_\alpha^{+A}, (\sigma^\nu s^A)_\alpha\right\} = 0. \tag{7.355}$$

The other equations of motion are

$$\begin{aligned} & D^\mu D_\mu\phi - \frac{1}{R^2}\phi_i + [\phi_j, [\phi_i, \phi_j]] \\ & - \left\{s_\alpha^{+A}, \rho_i^{AB}(i\sigma_2(s^{+B})^T)_\alpha\right\} + \left\{s_\alpha^A, (\rho_i^+)^{AB}(i\sigma_2 s^B)_\alpha\right\} = 0 \\ & i\sigma^\mu D_\mu s^A - i\sigma^2\rho_i^{AB}[\phi_i, (s^{+B})^T] = 0. \end{aligned} \tag{7.356}$$

Next we insert (7.346) in the above equations of motion. The fields ϕ_i, A_0, A_a, s_α^A and s_α^{+A} are found to solve these equations of motion provided that the zero modes Φ_i, X_0, X_a, θ_α^A and θ_α^{+A} satisfy the equations of motion

$$\begin{aligned} & [X_a, iD_0X_a] + [\Phi_i, iD_0\Phi_i] - 2\left\{\theta_\alpha^{+A}, \theta_\alpha^A\right\} = 0 \\ & D_0^2X_a + \frac{4}{R^2}X_a - \frac{6i}{R}\epsilon_{abc}X_bX_c - [X_b, [X_a, X_b]] \\ & - [\Phi_i, [X_a, \Phi_i]] - 2\left\{\theta_\alpha^{+A}, (\sigma_a\theta^A)_\alpha\right\} = 0 \\ & D_0^2\Phi_i + \frac{1}{R^2}\Phi_i - [X_a, [\Phi_i, X_a]] - [\Phi_j, [\Phi_i, \Phi_j]] \\ & + \left\{\theta_\alpha^{+A}, \rho_i^{AB}(i\sigma_2(\theta^{+B})^T)_\alpha\right\} - \left\{\theta_\alpha^A, (\rho_i^+)^{AB}(i\sigma_2\theta^B)_\alpha\right\} = 0 \\ & iD_0\theta^A - \frac{3}{2R}\theta^A + [X_a, \sigma_a\theta^A] - i\sigma_2\rho_i^{AB}[\Phi_i, (\theta^{+B})^T] = 0. \end{aligned} \tag{7.357}$$

As it turns out, these equations of motion can be derived from the following quantum mechanical model

$$S = \int dt L, \quad L = L_B + L_F + L_T. \tag{7.358}$$

$$\begin{aligned} L_B &= \mathrm{Tr}\left[-\frac{1}{2}[X_0, X_a]^2 - \frac{1}{2}[X_0, \Phi_i]^2 + \frac{1}{4}[X_a, X_b]^2 + \frac{1}{4}[\Phi_i, \Phi_j]^2\right. \\ &\left. + \frac{1}{2}[X_a, \Phi_i]^2 + \frac{im}{3}\epsilon_{abc}X_aX_bX_c - \frac{m^2}{18}X_a^2 - \frac{m^2}{72}\Phi_i^2\right] \\ L_F &= \mathrm{Tr}\left[-2\theta^+[X_0, \theta] - 2\theta^+\left(\sigma_a[X_a, \theta] - \frac{m}{4}\theta\right)\right. \\ &\left. + \theta^+i\sigma_2\rho_i[\Phi_i, (\theta^+)^T] - \theta^Ti\sigma_2\rho_i^+[\Phi_i, \theta]\right] \\ L_T &= \mathrm{Tr}\left[\frac{1}{2}(\partial_0X_a)^2 + \frac{1}{2}(\partial_0\Phi_i)^2 - 2i\theta^+\partial_0\theta\right. \\ &\left. - i\partial_0X_a[X_0, X_a] - i\partial_0\Phi_i[X_0, \Phi_i]\right]. \end{aligned} \tag{7.359}$$

This is exactly the BMN model (7.328). In the above

$$m = \frac{6}{R}. \tag{7.360}$$

Before we conclude this section we define the Clebsch–Gordan coefficients ρ_i^{AB} and $(\rho_i^+)^{AB}$. Let Γ_i, $i = 4, 5, 6, 7, 8, 9$ be the Clifford algebra in six dimensions, viz $\{\Gamma_i, \Gamma_j\} = 2\delta_{ij}$. We also denote them by $\hat{\Gamma}_a = \Gamma_{a+3}$, $\hat{\Gamma}_{a+3} = \Gamma_{a+6}$, $a = 1, 2, 3$. We work in the representation

$$\hat{\Gamma}^a = (\hat{\Gamma}^a)^+ = \begin{pmatrix} 0 & \hat{\rho}^a \\ (\hat{\rho}^a)^+ & 0 \end{pmatrix}, \hat{\Gamma}^{a+3} = (\hat{\Gamma}^{a+3})^+ = \begin{pmatrix} 0 & \hat{\rho}^{a+3} \\ (\hat{\rho}^{a+3})^+ & 0 \end{pmatrix}. \tag{7.361}$$

We will also introduce

$$\Gamma^{AB} = (\Gamma^{AB})^+ = -\Gamma^{BA} = \begin{pmatrix} 0 & \gamma^{AB} \\ (\gamma^{AB})^+ & 0 \end{pmatrix}, \quad A, B = 1, 2, 3, 4$$
$$(\Gamma^{AB})_{CD} = \delta_{AC}\delta_{BD} - \delta_{AD}\delta_{BC}. \tag{7.362}$$

We define the gamma matrices $\hat{\Gamma}_a$ and $\hat{\Gamma}_{a+3}$ as follows

$$\hat{\Gamma}_a = \frac{1}{2}\epsilon_{abc4}\Gamma_{bc} + \Gamma_{a4}, \hat{\Gamma}_{a+3} = \frac{i}{2}\epsilon_{abc4}\Gamma_{bc} - i\Gamma_{a4}, \quad a = 1, 2, 3. \tag{7.363}$$

We find immediately

$$\hat{\rho}_a = -\hat{\rho}_a^+ = \frac{1}{2}\epsilon_{abc4}\gamma_{bc} + \gamma_{a4}, \hat{\rho}_{a+3} = \hat{\rho}_{a+3}^+ = \frac{i}{2}\epsilon_{abc4}\gamma_{bc} - i\gamma_{a4}, \quad a = 1, 2, 3. \tag{7.364}$$

Explicitly

$$\begin{aligned} \hat{\rho}_1 &= \gamma_{23} + \gamma_{14} = i\sigma_2 \otimes \sigma_1 \\ \hat{\rho}_2 &= -\gamma_{13} + \gamma_{24} = -i\sigma_2 \otimes \sigma_3 \\ \hat{\rho}_3 &= \gamma_{12} + \gamma_{34} = \mathbf{1} \otimes i\sigma_2. \end{aligned} \tag{7.365}$$

$$\begin{aligned} \hat{\rho}_4 &= i(\gamma_{23} - \gamma_{14}) = \sigma_1 \otimes \sigma_2 \\ \hat{\rho}_5 &= i(-\gamma_{13} - \gamma_{24}) = \sigma_2 \otimes \mathbf{1} \\ \hat{\rho}_6 &= i(\gamma_{12} - \gamma_{34}) = -\sigma_3 \otimes \sigma_2. \end{aligned} \tag{7.366}$$

We verify that these matrices satisfy $\rho_i^+\rho_j + \rho_j^+\rho_i = \rho_i\rho_j^+ + \rho_j\rho_i^+ = 2\delta_{ij}\mathbf{1}_4$. We also compute the identities

$$\begin{aligned} \sum_{a=1}^{6} \hat{\rho}_a^{AB}\hat{\rho}_a^{CD} &= \sum_{a=1}^{6}(\hat{\rho}_a^+)^{AB}(\hat{\rho}_a^+)^{CD} = 2\epsilon_{ABCD} \\ \sum_{a=1}^{6} \hat{\rho}_a^{AB}(\hat{\rho}_a^+)^{CD} &= -\gamma_{EF}^{AB}\gamma_{EF}^{CD} = -2(\delta_{AC}\delta_{BD} - \delta_{AD}\delta_{BC}). \end{aligned} \tag{7.367}$$

7.10.5 Dimensional reduction

Supersymmetry at fixed times

In this section we go even further and turn compactification into dimensional reduction by dropping the time dependence of the matrices X_0, X_a, Φ_i and θ_α^A, θ_α^{+A}.

The action S given by equation (7.359) remains supersymmetric under (7.350), (7.353), (7.354) provided we vary the fermion term $-2i\,\mathrm{Tr}\,\theta^+\partial_0\theta$ first and then take into account that the supersymmetry parameters satisfy $\partial_0\epsilon_\alpha^A = -i\alpha\epsilon_\alpha^A$ and $\partial_0\epsilon_\alpha^{+A} = i\alpha\epsilon_\alpha^{+A}$ before we fix time. We use the notation $\theta = \psi$ and we recall that the curvature is given by (with $\nu = 2$)

$$\tilde{F}_{ab} = i[X_a, X_b] - \frac{\nu}{R}\epsilon_{abc}X_c. \tag{7.368}$$

Right from the start we will drop the time dependence of the matrices X_a and Φ_i. Thus the terms containing time derivatives of X_a and Φ_i in the supersymmetric transformations (7.353), (7.354) will be absent. We compute immediately the following variation of the fermion action

$$\begin{aligned}
\delta\Big(-2\,\mathrm{Tr}\,\psi_\alpha^{+A}(\sigma_a)_{\alpha\beta}[X_a, \psi_\beta^A]\Big) = {} & 2(\sigma_a)_{\alpha\beta}\,\mathrm{Tr}\,\delta X_a\Big\{\psi_\alpha^{+A}, \psi_\beta^A\Big\} \\
& - i\,\mathrm{Tr}\,\delta X_b[X_a, \tilde{F}_{ab}] \\
& + \mathrm{Tr}\,\delta\Phi_i[X_a, [X_a, \Phi_i]] \\
& - \mathrm{Tr}\Big((\epsilon^B)^T\sigma_2\sigma_c\psi^A(\rho_i^+)^{BA} \\
& \quad - \psi^{+A}\sigma_c\sigma_2(\epsilon^{+B})^T\rho_i^{AB}\Big) \\
& [\epsilon_{abc}\tilde{F}_{ab}, \Phi_i] \\
& - \mathrm{Tr}\Big((\epsilon^B)^T\sigma_2\sigma_c\psi^A(\rho_i^+)^{BA} \\
& \quad - \psi^{+A}\sigma_c\sigma_2(\epsilon^{+B})^T\rho_i^{AB}\Big) \\
& \left[\frac{4\nu - 3}{2R}X_c, \Phi_i\right] \\
& - i\,\mathrm{Tr}\,(\epsilon^{+A}\sigma_a\psi^B + \psi^{+A}\sigma_a\epsilon^B) \\
& \quad (\rho_j\rho_i^+)^{AB}[X_a, [\Phi_i, \Phi_j]] \\
& - \mathrm{Tr}\,\delta X_0[X_a, [X_a, X_0]] \\
& + \epsilon_{abc}\,\mathrm{Tr}\,(\epsilon^{+A}\sigma_c\psi^A + \psi^{+A}\sigma_c\epsilon^A) \\
& \quad [X_0, [X_a, X_b]] \\
& - 2\,\mathrm{Tr}\Big((\epsilon^B)^T\sigma_2\sigma_a\psi^A(\rho_i^+)^{BA} \\
& \quad + \psi^{+A}\sigma_a\sigma_2(\epsilon^{+B})^T\rho_i^{AB}\Big) \\
& [X_a, [X_0, \Phi_i]] \\
& - \alpha\,\mathrm{Tr}\Big((\epsilon^B)^T\sigma_2\sigma_c\psi^A(\rho_i^+)^{BA} \\
& \quad - \psi^{+A}\sigma_c\sigma_2(\epsilon^{+B})^T\rho_i^{AB}\Big) \\
& [X_c, \Phi_i].
\end{aligned} \tag{7.369}$$

Let $\delta\psi_\alpha^A = -\frac{i}{2}\epsilon_{abc}\tilde{F}_{ab}(\sigma_c\epsilon^A)_\alpha$ then

$$\begin{aligned}
&-\mathrm{Tr}\ \delta\psi_\alpha^A(i\sigma_2)_{\alpha\beta}\left(\rho_i^+\right)^{AB}[\Phi_i,\ \psi_\beta^B]\\
&-\mathrm{Tr}\ \psi_\alpha^A(i\sigma_2)_{\alpha\beta}\left(\rho_i^+\right)^{AB}[\Phi_i,\ \delta\psi_\beta^B] + \mathrm{C.\ C}\\
&= \frac{1}{2}\mathrm{Tr}\left[(\epsilon^B)^T\sigma_2\sigma_c\psi^A\left(\left(\rho_i^+\right)^{BA} - \left(\rho_i^+\right)^{AB}\right) - \psi^{+A}\sigma_c\sigma_2(\epsilon^{+B})^T\left(\rho_i^{AB} - \rho_i^{BA}\right)\right]\\
&[\epsilon_{abc}\tilde{F}_{ab},\ \Phi_i].
\end{aligned} \tag{7.370}$$

If we assume that

$$\rho_i^{AB} = -\rho_i^{BA}. \tag{7.371}$$

Then this variation will cancel the first term in the second line of (7.369).

Let $\delta\psi_\alpha^A = i[X_a,\ \Phi_i]\rho_i^{AB}(i\sigma_a\sigma_2)_{\alpha\beta}\epsilon_\beta^{+B}$ then

$$\begin{aligned}
&-\mathrm{Tr}\ \delta\psi_\alpha^A(i\sigma_2)_{\alpha\beta}\left(\rho_i^+\right)^{AB}\left[\Phi_i,\ \psi_\beta^B\right] - \mathrm{Tr}\ \psi_\alpha^A(i\sigma_2)_{\alpha\beta}(\rho_i^+)^{AB}\left[\Phi_i,\ \delta\psi_\beta^B\right] + \mathrm{C.\ C} =\\
&i\ \mathrm{Tr}\left[\left(\rho_j^T\rho_i^* - \rho_j^T\rho_i^+\right)^{AB}\epsilon^{+A}\sigma_a\psi^B - \left(\rho_i^T\rho_j^* - \rho_i\rho_j^*\right)^{AB}\psi^{+A}\sigma_a\epsilon^B\right]\\
&[\Phi_i,\ [X_a,\ \Phi_j]] =\\
&i\ \mathrm{Tr}\ (\epsilon^{+A}\sigma_a\psi^B + \psi^{+A}\sigma_a\epsilon^B)\left(\rho_j\rho_i^+\right)^{AB}[X_a,\ [\Phi_i,\ \Phi_j]] - \mathrm{Tr}\ \delta X_a[\Phi_i,\ [X_a,\ \Phi_i]].
\end{aligned} \tag{7.372}$$

The first term in this variation cancels the last line of (7.369). In here we have used $\rho_i^T = -\rho_i$ and $\rho_i^+ = -\rho_i^*$ which follow from (7.371), the Jacobi identity and the defining equation of the six 4×4 matrices ρ_i given by $\rho_i\rho_j^+ + \rho_j\rho_i^+ = \rho_i^+\rho_j + \rho_j^+\rho_i = 2\delta_{ij}\mathbf{1}_4$.

Let $\delta\psi_\alpha^A = \frac{3}{4R}\Phi_i(\rho_i)^{AB}(\sigma_2)_{\alpha\beta}\epsilon_\beta^{+B}$ then

$$\begin{aligned}
&-\mathrm{Tr}\ \delta\psi_\alpha^A(i\sigma_2)_{\alpha\beta}\left(\rho_i^+\right)^{AB}\left[\Phi_i,\ \psi_\beta^B\right] - \mathrm{Tr}\ \psi_\alpha^A(i\sigma_2)_{\alpha\beta}\left(\rho_i^+\right)^{AB}\left[\Phi_i,\ \delta\psi_\beta^B\right] + \mathrm{C.\ C} =\\
&\frac{3i}{4R}\mathrm{Tr}\left[\left(\rho_j^T\rho_i^* - \rho_j^T\rho_i^+\right)^{AB}\epsilon^{+A}\psi^B + \left(\rho_i^T\rho_j^* - \rho_i\rho_j^*\right)^{AB}\psi^{+A}\epsilon^B\right][\Phi_i,\ \Phi_j] =\\
&-\frac{3i}{2R}\left(\rho_i\rho_j^+\right)^{AB}\mathrm{Tr}\ (\epsilon^{+A}\psi^B - \psi^{+A}\epsilon^B)[\Phi_i,\ \Phi_j].
\end{aligned} \tag{7.373}$$

Let $\delta\psi_\alpha^A = -\frac{i}{2}[\Phi_i,\ \Phi_j](\rho_i\rho_j^+)^{AB}\epsilon_\alpha^B$ then

$$\begin{aligned}
&-\mathrm{Tr}\ \delta\psi_\alpha^A(i\sigma_2)_{\alpha\beta}\left(\rho_i^+\right)^{AB}\left[\Phi_i,\ \psi_\beta^B\right] - \mathrm{Tr}\ \psi_\alpha^A(i\sigma_2)_{\alpha\beta}\left(\rho_i^+\right)^{AB}\left[\Phi_i,\ \delta\psi_\beta^B\right] + \mathrm{C.\ C} =\\
&-\frac{i}{2}\mathrm{Tr}\left[\left(\left(\rho_k^* - \rho_k^+\right)\rho_i\rho_j^+\right)^{AB}\psi^A i\sigma_2\epsilon^B + \left(\left(\rho_k - \rho_k^T\right)\rho_i^*\rho_j^T\right)^{AB}\psi^{+A}i\sigma_2\epsilon^{+B}\right]\\
&[\Phi_k,\ [\Phi_i,\ \Phi_j]] = \mathrm{Tr}\ \delta\Phi_j[\Phi_i,\ [\Phi_i,\ \Phi_j]].
\end{aligned} \tag{7.374}$$

The first term of (7.369) can be put in the form

$$2(\sigma_a)_{\alpha\beta}\,\mathrm{Tr}\,\delta X_a\left\{\psi_\alpha^{+A},\,\psi_\beta^A\right\} = 8i\,\mathrm{Tr}\,\left\{\psi_\alpha^A,\,\psi_\beta^{+A}\right\}\left(\psi_\alpha^{+B}\epsilon_\beta^B - \epsilon_\alpha^{+B}\psi_\beta^B\right)$$
$$- 4i\,\mathrm{Tr}\,\left\{\psi_\alpha^A,\,\psi_\alpha^{+A}\right\}\left(\psi_\beta^{+B}\epsilon_\beta^B - \epsilon_\beta^{+B}\psi_\beta^B\right). \qquad (7.375)$$

We also compute

$$-\,\mathrm{Tr}\,\psi_\alpha^A(i\sigma_2)_{\alpha\beta}(\rho_i^+)^{AB}[\delta\Phi_i,\,\psi_\beta^B] + \mathrm{C.\,C} = \mathrm{Tr}\,\left\{\psi_\alpha^A,\,\psi_\beta^B\right\}$$
$$(i\sigma_2)_{\alpha\beta}\left(\rho_i^+\right)^{AB}\delta\Phi_i + \mathrm{C.\,C}$$
$$= 8i\,\mathrm{Tr}\,\left\{\psi_1^A,\,\psi_1^{+A}\right\}\left(\psi_2^{+B}\epsilon_2^B - \epsilon_2^{+B}\psi_2^B\right)$$
$$+ 8i\,\mathrm{Tr}\,\left\{\psi_2^A,\,\psi_2^{+A}\right\} \qquad (7.376)$$
$$\left(\psi_1^{+B}\epsilon_1^B - \epsilon_1^{+B}\psi_1^B\right)$$
$$- 8i\,\mathrm{Tr}\,\left\{\psi_1^A,\,\psi_2^{+A}\right\}\left(\psi_1^{+B}\epsilon_2^B - \epsilon_1^{+B}\psi_2^B\right)$$
$$- 8i\,\mathrm{Tr}\,\left\{\psi_2^A,\,\psi_1^{+A}\right\}\left(\psi_2^{+B}\epsilon_1^B - \epsilon_2^{+B}\psi_1^B\right).$$

In the above we have used the identity

$$\epsilon^{ABCD}(i\sigma_2)_{\alpha\beta}(i\sigma_2)_{\mu\nu}\,\mathrm{Tr}\,\psi_\alpha^A\psi_\beta^B\psi_\mu^C\epsilon_\nu^D = 0. \qquad (7.377)$$

Thus we get

$$2(\sigma_a)_{\alpha\beta}\,\mathrm{Tr}\,\delta X_a\left\{\psi_\alpha^{+A},\,\psi_\beta^A\right\}$$
$$+\left[-\mathrm{Tr}\,\psi_\alpha^A(i\sigma_2)_{\alpha\beta}(\rho_i^+)^{AB}[\delta\Phi_i,\,\psi_\beta^B] + \mathrm{C.\,C}\right] = \qquad (7.378)$$
$$4i\,\mathrm{Tr}\,\{\psi_\alpha^A,\,\psi_\alpha^{+A}\}(\psi_\beta^{+B}\epsilon_\beta^B - \epsilon_\beta^{+B}\psi_\beta^B) = -2\,\mathrm{Tr}\,\{\psi_\alpha^A,\,\psi_\alpha^{+A}\}\delta X_0.$$

Let $\delta\psi_\alpha^A = -i[X_a,\,X_0](\sigma_a\epsilon^A)_\alpha$ then

$$-\,\mathrm{Tr}\,\delta\psi_\alpha^A(i\sigma_2)_{\alpha\beta}\left(\rho_i^+\right)^{AB}\left[\Phi_i,\,\psi_\beta^B\right]$$
$$-\,\mathrm{Tr}\,\psi_\alpha^A(i\sigma_2)_{\alpha\beta}\left(\rho_i^+\right)^{AB}\left[\Phi_i,\,\delta\psi_\beta^B\right] + \mathrm{C.\,C} = \qquad (7.379)$$
$$-\,2\,\mathrm{Tr}\left((\epsilon^B)^T\sigma_2\sigma_a\psi^A(\rho_i^+)^{BA} + \psi^{+A}\sigma_a\sigma_2(\epsilon^{+B})^T\rho_i^{AB}\right)[\Phi_i,\,[X_a,\,X_0]].$$

Let $\delta\psi_\alpha^A = [X_0,\,\Phi_i]\rho_i^{AB}(\sigma_2)_{\alpha\beta}\epsilon_\beta^{+B}$ then

$$-\,\mathrm{Tr}\,\delta\psi_\alpha^A(i\sigma_2)_{\alpha\beta}\left(\rho_i^+\right)^{AB}\left[\Phi_i,\,\psi_\beta^B\right]$$
$$-\,\mathrm{Tr}\,\psi_\alpha^A(i\sigma_2)_{\alpha\beta}\left(\rho_i^+\right)^{AB}\left[\Phi_i,\,\delta\psi_\beta^B\right] + \mathrm{C.\,C} = \qquad (7.380)$$
$$+\,i\left(\rho_j\rho_i^+\right)^{AB}\,\mathrm{Tr}\,(\psi^{+A}\epsilon^B + \epsilon^{+A}\psi^B)[X_0,\,[\Phi_i,\,\Phi_j]]$$
$$+\,\mathrm{Tr}\,\delta X_0[\Phi_i,\,[X_0,\,\Phi_i]].$$

Let $\delta\psi_\alpha^A = \frac{i\alpha}{2}\Phi_i(i\sigma_2)_{\alpha\beta}\rho_i^{AB}\epsilon_\beta^{+B}$ then

$$\begin{aligned}
&- \mathrm{Tr}\ \delta\psi_\alpha^A(i\sigma_2)_{\alpha\beta}\left(\rho_i^+\right)^{AB}[\Phi_i, \psi_\beta^B] \\
&- \mathrm{Tr}\ \psi_\alpha^A(i\sigma_2)_{\alpha\beta}(\rho_i^+)^{AB}[\Phi_i, \delta\psi_\beta^B] + \mathrm{C.\ C} = \\
&\quad i\alpha(\rho_i\rho_j^+)^{AB}\ \mathrm{Tr}\ (\epsilon^{+A}\psi^B - \psi^{+A}\epsilon^B)[\Phi_i, \Phi_j].
\end{aligned} \tag{7.381}$$

Next we compute (with $3/R_1 = m/2$)

$$\begin{aligned}
\frac{3}{R_1}\delta(\mathrm{Tr}\ \psi^+\psi) = &- \frac{3}{4R_1}\epsilon_{abc}\ \mathrm{Tr}\ \tilde{F}_{ab}\delta X_c + \frac{3}{4R_1}\left(\frac{3}{2R} - \alpha\right)\mathrm{Tr}\ \Phi_i\delta\Phi_i \\
&+ \frac{3}{R_1}\ \mathrm{Tr}\left((\epsilon^B)^T\sigma_2\sigma_c\psi^A(\rho_i^+)^{BA} - \psi^{+A}\sigma_c\sigma_2(\epsilon^{+B})^T\rho_i^{AB}\right)[X_c, \Phi_i] \\
&+ \frac{3i}{2R_1}(\rho_i\rho_j^+)^{AB}\ \mathrm{Tr}\ (\epsilon^{+A}\psi^B - \psi^{+A}\epsilon^B)[\Phi_i, \Phi_j] \\
&- \frac{3i}{R_1}\ \mathrm{Tr}\ (\epsilon^{+A}\sigma_a\psi^A + \psi^{+A}\sigma_a\epsilon^A)[X_a, X_0] \\
&+ \frac{3}{R_1}\ \mathrm{Tr}\left(\psi^{+A}\sigma_2(\epsilon^{+B})^T\rho_i^{AB} + (\psi^A)^T\sigma_2\epsilon^B(\rho_i^+)^{AB}\right)[X_0, \Phi_i].
\end{aligned} \tag{7.382}$$

Also

$$\begin{aligned}
\delta\left(-2\ \mathrm{Tr}\ \psi_\alpha^{+A}[X_0, \psi_\alpha^A]\right) = &\ i\epsilon_{abc}\ \mathrm{Tr}\ (\epsilon^{+A}\sigma_a\psi^A + \psi^{+A}\sigma_a\epsilon^A) \\
&[X_0, \tilde{F}_{ab}] + \mathrm{Tr}\ \delta X_a[X_0, [X_a, X_0]] - i(\rho_j\rho_i^+)^{AB} \\
&\mathrm{Tr}\ (\psi^{+A}\epsilon^B + \epsilon^{+A}\psi^B)[X_0, [\Phi_i, \Phi_j]] - \frac{3}{2R} \\
&\mathrm{Tr}\left(\psi^{+A}\sigma_2(\epsilon^{+B})^T\rho_i^{AB} + (\psi^A)^T\sigma_2\epsilon^B(\rho_i^+)^{AB}\right)[X_0, \Phi_i] \\
&- 2\ \mathrm{Tr}\left((\epsilon^B)^T\sigma_2\sigma_a\psi^A(\rho_i^+)^{BA} + \psi^{+A}\sigma_a\sigma_2(\epsilon^{+B})^T\rho_i^{AB}\right) \\
&[X_0, [\Phi_i, X_a]] \\
&- \mathrm{Tr}\ \delta\Phi_i[X_0, [X_0, \Phi_i]] \\
&+ \alpha\ \mathrm{Tr}\left(\psi^{+A}\sigma_2(\epsilon^{+B})^T\rho_i^{AB} + (\psi^A)^T\sigma_2\epsilon^B(\rho_i^+)^{AB}\right) \\
&\quad [X_0, \Phi_i].
\end{aligned} \tag{7.383}$$

The full variation of the fermion action is then given by

$$\begin{aligned}\delta S_F \;=\; & -i\,\mathrm{Tr}\,\delta X_b[X_a,\tilde{F}_{ab}]+\mathrm{Tr}\,\delta\Phi_i[X_a,[X_a,\Phi_i]]\\ & -\mathrm{Tr}\,\delta X_a[\Phi_i,[X_a,\Phi_i]]-\frac{3}{4R_1}\epsilon_{abc}\,\mathrm{Tr}\,F_{ab}\delta X_c\\ & +\frac{3}{4R_1}\left(\frac{3}{2R}-\alpha\right)\mathrm{Tr}\,\Phi_i\delta\Phi_i+\mathrm{Tr}\,\delta\Phi_j[\Phi_i,[\Phi_i,\Phi_j]]\\ & +\mathrm{Tr}\,\delta X_0[\Phi_i,[X_0,\Phi_i]]+\mathrm{Tr}\,\delta X_a[X_0,[X_a,X_0]]\\ & -\mathrm{Tr}\,\delta X_0[X_a,[X_a,X_0]]-\mathrm{Tr}\,\delta\Phi_i[X_0,[X_0,\Phi_i]]\\ & +\left(-\frac{4v-3}{2R}-\alpha+\frac{3}{R_1}\right)\mathrm{Tr}\\ & \left((\epsilon^B)^T\sigma_2\sigma_c\psi^A(\rho_i^+)^{BA}-\psi^{+A}\sigma_c\sigma_2(\epsilon^{+B})^T\rho_i^{AB}\right)[X_c,\Phi_i]\\ & +i\left(-\frac{3}{2R}+\frac{3}{2R_1}+\alpha\right)(\rho_i\rho_j^+)^{AB}\,\mathrm{Tr}\,(\epsilon^{+A}\psi^B-\psi^{+A}\epsilon^B)[\Phi_i,\Phi_j]\\ & +i\left(-\frac{3}{R_1}+\frac{2v}{R}\right)\mathrm{Tr}\,(\epsilon^{+A}\sigma_a\psi^A+\psi^{+A}\sigma_a\epsilon^A)[X_a,X_0]\\ & +\left(\frac{3}{R_1}-\frac{3}{2R}+\alpha\right)\\ & \mathrm{Tr}\left(\psi^{+A}\sigma_2(\epsilon^{+B})^T\rho_i^{AB}+(\psi^A)^T\sigma_2\epsilon^B(\rho_i^+)^{AB}\right)[X_0,\Phi_i].\end{aligned} \tag{7.384}$$

Let us now consider the Lagrangian

$$\begin{aligned}L_T=\mathrm{Tr}\Bigg[&\frac{1}{2}(\partial_0X_a)^2+\frac{1}{2}(\partial_0\Phi_i)^2-2i\psi^+\partial_0\psi-i\partial_0X_a[X_a,X_a]\\ &-i\partial_0\Phi_i[X_0,\Phi_i]\Bigg].\end{aligned} \tag{7.385}$$

Since the time dependence of the matrices X_a and Φ_i is already dropped we have

$$L_T=-2i\,\mathrm{Tr}\,\psi^+\partial_0\psi. \tag{7.386}$$

Varying this action under full supersymmetry transformations then fixing time we obtain the variation

$$-2i\,\mathrm{Tr}\left[\psi_\alpha^{+A}(\partial_0\delta\psi_\alpha^A)|_{t=\text{fixed}}-\text{H. C}\right]. \tag{7.387}$$

Since $\partial_0\epsilon_\alpha^A=-i\alpha\epsilon_\alpha^A$ and $\partial_0\epsilon_\alpha^{+A}=i\alpha\epsilon_\alpha^{+A}$ we have

$$\begin{aligned}(\partial_0\delta\psi_{\hat{a}}^A)|_{t=\text{fixed}}=i\alpha\Bigg[&\frac{i}{2}\epsilon^{\hat{a}\hat{b}\hat{c}}\tilde{F}_{\hat{a}\hat{b}}(\sigma_{\hat{c}}\epsilon^A)_{\hat{\alpha}}+i[X_{\hat{a}},\Phi_i]\rho_i^{AB}(i\sigma^{\hat{a}}\sigma^2)_{\hat{\alpha}\hat{\beta}}\epsilon_{\hat{\beta}}^{+B}\\ &+\frac{3}{4R}\Phi_i(\rho_i)^{AB}(\sigma^2)_{\hat{\alpha}\hat{\beta}}\epsilon_{\hat{\beta}}^{+B}+\frac{i}{2}[\Phi_i,\Phi_j](\rho_i\rho_j^+)^{AB}\epsilon_{\hat{\alpha}}^B\\ &+i[X_{\hat{a}},X_0](\sigma_{\hat{a}}\epsilon^A)^{\hat{\alpha}}\\ &+[X_0,\Phi_i](\rho_i)^{AB}(\sigma_2)_{\hat{\alpha}\hat{\beta}}\epsilon_{\hat{\beta}}^{+B}-\frac{\alpha}{2}\Phi_i(\sigma_2)_{\alpha\beta}\rho_i^{AB}\epsilon_\beta^{+B}\Bigg].\end{aligned} \tag{7.388}$$

We compute

$$
\begin{aligned}
-2i\,\mathrm{Tr}\left[\psi_\alpha^{+A}(\partial_0\delta\psi_\alpha^A)|_{t=\mathrm{fixed}} - \mathrm{H.\,C}\right] = & \frac{\alpha}{2}\epsilon_{abc}\,\mathrm{Tr}\,\tilde{F}_{ab}\delta X_c \\
& +\alpha\left(\frac{3}{4R} - \frac{\alpha}{2}\right)\mathrm{Tr}\,\Phi_i\delta\Phi_i \\
& +2\alpha\,\mathrm{Tr}\Big((\epsilon^B)^T\sigma_2\sigma_c\psi^A(\rho_i^+)^{BA} \\
& -\psi^{+A}\sigma_c\sigma_2(\epsilon^{+B})^T\rho_i^{AB}\Big) \\
& [X_c,\,\Phi_i] - i\alpha(\rho_i\rho_j^+)^{AB} \\
& \mathrm{Tr}\,(\epsilon^{+A}\psi^B - \psi^{+A}\epsilon^B)[\Phi_i,\,\Phi_j] \\
& +2i\alpha\,\mathrm{Tr}\,(\epsilon^{+A}\sigma_a\psi^A + \psi^{+A}\sigma_a\epsilon^A) \\
& [X_a,\,X_0] \\
& +2\alpha\,\mathrm{Tr}\,\Big(\psi^{+A}\sigma_2(\epsilon^{+B})^T\rho_i^{AB} \\
& +(\psi^A)^T\sigma_2\epsilon^B(\rho_i^+)^{AB}\Big)[X_0,\,\Phi_i].
\end{aligned}
\tag{7.389}
$$

The only possible solution is given by

$$
\nu = 2,\qquad R_1 = R,\qquad \alpha = -\frac{1}{2R}. \tag{7.390}
$$

The full variation of the fermion action is finally given by

$$
\begin{aligned}
\delta S_F + \delta(-2i\,\mathrm{Tr}\,\psi^+\partial_0\psi)|_{t=\mathrm{fixed}} = & -i\,\mathrm{Tr}\,\delta X_b[X_a,\,\tilde{F}_{ab}] \\
& +\mathrm{Tr}\,\delta\Phi_i[X_a,\,[X_a,\,\Phi_i]] \\
& -\mathrm{Tr}\,\delta X_a[\Phi_i,\,[X_a,\,\Phi_i]] \\
& +\left[\frac{\alpha}{2} - \frac{3}{4R_1}\right]\epsilon_{abc}\,\mathrm{Tr}\,F_{ab}\delta X_c \\
& +\left[\alpha\left(\frac{3}{4R} - \frac{\alpha}{2}\right) + \frac{3}{4R_1}\left(\frac{3}{2R} - \alpha\right)\right] \\
& \mathrm{Tr}\,\Phi_i\delta\Phi_i \\
& +\mathrm{Tr}\,\delta\Phi_j[\Phi_i,\,[\Phi_i,\,\Phi_j]] \\
& +\mathrm{Tr}\,\delta X_0[\Phi_i,\,[X_0,\,\Phi_i]] \\
& +\mathrm{Tr}\,\delta X_a[X_0,\,[X_a,\,X_0]] \\
& -\mathrm{Tr}\,\delta\Phi_i[X_0,\,[X_0,\,\Phi_i]] \\
& -\mathrm{Tr}\,\delta X_0[X_a,\,[X_a,\,X_0]].
\end{aligned}
\tag{7.391}
$$

The variations of the bosonic terms are given by

$$
\begin{aligned}
\delta(\frac{1}{4}\mathrm{Tr}\,[X_a, X_b]^2) &= \mathrm{Tr}\,\delta X_a[X_b, [X_a, X_b]] \\
&= i\mathrm{Tr}\,\delta X_b[X_a, \widetilde{F}_{ab}] - \frac{v}{R}\epsilon_{abc}\,\mathrm{Tr}\,\delta X_c\widetilde{F}_{ab} - \frac{v^2}{R^2}\mathrm{Tr}\,\delta X_a^2 \\
\delta(\frac{1}{4}\mathrm{Tr}\,[\Phi_i, \Phi_j]^2) &= -\,\mathrm{Tr}\,\delta\Phi_j[\Phi_i, [\Phi_i, \Phi_j]] \\
\delta(\frac{1}{2}\mathrm{Tr}\,[X_a, \Phi_i]^2) &= -\,\mathrm{Tr}\,\delta\Phi_i[X_a, [X_a, \Phi_i]] + \mathrm{Tr}\,\delta X_a[\Phi_i, [X_a, \Phi_i]] \\
\delta(-\frac{1}{2}\mathrm{Tr}\,[X_0, \Phi_i]^2) &= \mathrm{Tr}\,\delta\Phi_i[X_0, [X_0, \Phi_i]] - \mathrm{Tr}\,\delta X_0[\Phi_i, [X_0, \Phi_i]] \\
\delta(-\frac{1}{2}\mathrm{Tr}\,[X_0, X_a]^2) &= \mathrm{Tr}\,\delta X_a[X_0, [X_0, X_a]] - \mathrm{Tr}\,\delta X_0[X_a, [X_0, X_a]] \\
\delta(\frac{im}{3}\epsilon_{abc}\mathrm{Tr}\,X_aX_bX_c) &= im\epsilon_{abc}\,\mathrm{Tr}\,\delta X_aX_bX_c = \frac{m}{2}\epsilon_{abc}\,\mathrm{Tr}\,\delta X_c\widetilde{F}_{ab} + \frac{mv}{2R}\mathrm{Tr}\,\delta X_a^2 \\
\delta(-\frac{1}{2}(\frac{m}{6})\,\Phi_i^2) &= -\,(\frac{m}{6})^2\,\mathrm{Tr}\,\Phi_i\delta\Phi_i \\
\delta(-\frac{1}{2}(\frac{m}{3})^2\mathrm{Tr}\,X_a^2) &= -\,(\frac{m}{3})^2\,\mathrm{Tr}\,X_a\delta X_a.
\end{aligned}
\tag{7.392}
$$

It is not difficult to convince ourselves that in order to have supersymmetric invariance we must have

$$
m = \frac{6}{R}. \tag{7.393}
$$

$\mathcal{N} = 4$ time-independent supersymmetry

The aim is to construct an $\mathcal{N} = 4$ supersymmetric action without time dependence. The total action is

$$
S = S_B + S_F + S_T. \tag{7.394}
$$

The bosonic action is given as before, viz

$$
\begin{aligned}
S_B = \mathrm{Tr}\Bigg[&-\frac{1}{2}[X_0, X_a]^2 - \frac{1}{2}[X_0, \Phi_i]^2 + \frac{1}{4}[X_a, X_b]^2 \\
&+ \frac{1}{4}[\Phi_i, \Phi_j]^2 + \frac{1}{2}[X_a, \Phi_i]^2 \\
&+ \frac{im}{3}\epsilon_{abc}X_aX_bX_c - \frac{m^2}{18}X_a^2 - \frac{m^2}{72}\Phi_i^2\Bigg].
\end{aligned}
\tag{7.395}
$$

The fermionic action is given now by the action

$$S_F = \mathrm{Tr}\left[-2\Psi^+_{\dot{a}}[X_0, \Psi_{\dot{a}}] - 2\Psi^+_{\dot{a}}\left(\sigma_a[X_a, \Psi_{\dot{a}}] - \frac{m}{4}\Psi_{\dot{a}}\right)\right.$$
$$\left. + \Psi^+_{\dot{a}} i\sigma_2\rho_i[\Phi_i, (\Psi^+_{\dot{a}})^T] - \Psi^T_{\dot{a}} i\sigma_2\rho_i^+[\Phi_i, \Psi_{\dot{a}}]\right]. \tag{7.396}$$

The extra index $\dot{a}$ takes two values 1 and 2 and therefore the spinor Ψ is given by

$$\Psi^A_{\dot{a};\alpha} = \begin{pmatrix} \psi^A_\alpha \\ \chi^A_\alpha \end{pmatrix}. \tag{7.397}$$

Similarly, the supersymmetric parameter will be given by the spinor

$$\Omega^A_{\dot{a};\alpha} = \begin{pmatrix} \epsilon^A_\alpha \\ \omega^A_\alpha \end{pmatrix}. \tag{7.398}$$

We have the supersymmetry transformations

$$\begin{aligned} \delta X_0 &= -2i(\Psi^+_{\dot{a}}\Omega_{\dot{a}} - \Omega^+_{\dot{a}}\Psi_{\dot{a}}) \\ \delta X_a &= 2i(\Psi^+_{\dot{a}}\sigma_a\Omega_{\dot{a}} - \Omega^+_{\dot{a}}\sigma_a\Psi_{\dot{a}}) \\ \delta\Phi_i &= 2\left(\Psi^+_{\dot{a}}\rho_i\sigma_2(\Omega^+_{\dot{a}})^T + \Psi^T_{\dot{a}}\rho_i^*\sigma_2\Omega_{\dot{a}}\right). \end{aligned} \tag{7.399}$$

$$\begin{aligned} \delta\Psi^A_{\dot{a};\alpha} = &- \frac{i}{2}\epsilon_{abc}\tilde{F}_{ab}(\sigma_c)_{\alpha\beta}\Omega^A_{\dot{a};\beta} + i[X_a, \Phi_i] \\ &\rho_i^{AB}(i\sigma_a\sigma_2)_{\alpha\beta}\Omega^{+B}_{\dot{a};\beta} + \frac{1}{R}\Phi_i(\rho_i)^{AB}(\sigma_2)_{\alpha\beta}\Omega^{+B}_{\dot{a};\beta} \\ &- \frac{i}{2}[\Phi_i, \Phi_j](\rho_i\rho_j^+)^{AB}\Omega^B_{\dot{a};\alpha} - i[X_a, X_0] \\ &(\sigma_a)_{\alpha\beta}\Omega^A_{\dot{a};\beta} + [X_0, \Phi_i](\rho_i)^{AB}(\sigma_2)_{\alpha\beta}\Omega^{+B}_{\dot{a};\beta}. \end{aligned} \tag{7.400}$$

where $R = \frac{6}{m}$ and $\tilde{F}_{ab} = i[X_a, X_b] - \frac{2}{R}\epsilon_{abc}X_c$.

The analogue of the action $-2i\,\mathrm{Tr}\,\psi^+\partial_0\psi$ is given here by the term S_T defined by the equation

$$S_T = \frac{1}{R}\,\mathrm{Tr}\,\Psi^+_{\dot{b}}(\tau_2)_{\dot{b}\dot{a}}\Psi_{\dot{a}}, \quad \tau_2 = \begin{pmatrix} 0 & -i \\ i & 0 \end{pmatrix}. \tag{7.401}$$

Under supersymmetry this action changes by the amount

$$\delta S_T = \frac{1}{R}\,\mathrm{Tr}\left[\Psi^+_{\dot{b}}(\tau_2)_{\dot{b}\dot{a}}\delta\Psi_{\dot{a}} + \mathrm{H.\,C}\right]. \tag{7.402}$$

The analogue of the two conditions $i\partial_0\epsilon^A_\alpha = \alpha\epsilon^A_\alpha$ and $i\partial_0\epsilon^{+A}_\alpha = -\alpha\epsilon^{+A}_\alpha$ will be given by the conditions

$$(\tau_2)_{b\dot{a}}\Omega^{A}_{\dot{a};\beta} = \Omega^{A}_{\dot{b};\beta}$$
$$(\tau_2)_{b\dot{a}}\Omega^{+B}_{\dot{a};\beta} = -\,\Omega^{+B}_{\dot{b};\beta}. \tag{7.403}$$

This means that the two supersymmetry parameters ϵ and ω are related by the equation

$$\omega = i\epsilon. \tag{7.404}$$

Thus we compute

$$\begin{aligned}(\tau_2)_{b\dot{a}}\delta\Psi^{A}_{\dot{a};\alpha} = &-\frac{i}{2}\epsilon_{abc}\tilde{F}_{ab}(\sigma_c)_{\alpha\beta}\Omega^{A}_{\dot{b};\beta} - i[X_a, \Phi_i]\\ &\rho_i^{AB}(i\sigma_a\sigma_2)_{\alpha\beta}\Omega^{+B}_{\dot{b};\beta} - \frac{1}{R}\Phi_i(\rho_i)^{AB}(\sigma_2)_{\alpha\beta}\Omega^{+B}_{\dot{b};\beta}\\ &-\frac{i}{2}[\Phi_i, \Phi_j](\rho_i\rho_j^{+})^{AB}\Omega^{B}_{\dot{b};\alpha} - i[X_a, X_0]\\ &(\sigma_a)_{\alpha\beta}\Omega^{A}_{\dot{b};\beta} - [X_0, \Phi_i](\rho_i)^{AB}(\sigma_2)_{\alpha\beta}\Omega^{+B}_{\dot{b};\beta}.\end{aligned} \tag{7.405}$$

We can immediately compute

$$\begin{aligned}\delta S_T = &-\frac{1}{4R}\epsilon_{abc}\,\mathrm{Tr}\,\tilde{F}_{ab}\delta X_c - \frac{1}{2R^2}\,\mathrm{Tr}\,\Phi_i\delta\Phi_i\\ &-\frac{1}{R}\,\mathrm{Tr}\left(\Omega^{T}_{\dot{b}}\rho_i^{+}\sigma_2\sigma_a\Psi_{\dot{b}} - \Psi^{+}_{\dot{b}}\rho_i\sigma_a\sigma_2(\Omega^{+}_{\dot{b}})^{T}\right)[X_a, \Phi_i]\\ &+\frac{i}{2R}\,\mathrm{Tr}\left(\Omega^{+}_{\dot{b}}\rho_i\rho_j^{+}\Psi_{\dot{b}} - \Psi^{+}_{\dot{b}}\rho_i\rho_j^{+}\Omega_{\dot{b}}\right)[\Phi_i, \Phi_j]\\ &-\frac{i}{R}\,\mathrm{Tr}\left(\Omega^{+}_{\dot{b}}\sigma_a\Psi_{\dot{b}} + \Psi^{+}_{\dot{b}}\sigma_a\Omega_{\dot{b}}\right)[X_a, X_0]\\ &-\frac{1}{R}\,\mathrm{Tr}\left(\Psi^{+}_{\dot{b}}\rho_i\sigma_2(\Omega^{+}_{\dot{b}})^{T} + \Psi^{T}_{\dot{b}}\rho_i^{+}\sigma_2\Omega_{\dot{b}}\right)[X_0, \Phi_i].\end{aligned} \tag{7.406}$$

This variation is exactly equal to the variation of the term $-2i\,\mathrm{Tr}\,\psi^{+}\partial_0\psi$ in the original time-dependent theory. Hence it must be equal to minus the variation of the action $S_B + S_F$. In other words

$$\delta S_T = -\delta S_B - \delta S_F. \tag{7.407}$$

This establishes the $\mathcal{N} = 4$ supersymmetry invariance.

7.11 Other matrix models

We will just mention two more matrix models which are relevant to the gauge/gravity duality in lower dimensions:

- The BD (Berkooz–Douglas) theory which is relevant to M2- and M5-branes [118]. The M5-branes appear as additional fundamental hypermultiplets added to the BFSS model. A recent non-perturbative study of this model is given by [97].

- The ABJM theory in (2 + 1)-dimension which is relevant to the discussion of D2-branes (IIA superstring) and M2-branes (M-theory) [119]. For a concise summary of the results obtained for the ABJM model and future prospects see [29].

7.12 Exercises

Exercise: We consider the BFSS and BMN matrix models given, respectively, by the following actions

$$\begin{aligned} S_{\text{BFFS}} = \frac{1}{g^2}\,\text{Tr}\Bigg(&-\frac{1}{4}[X_I, X_J]^2 + \frac{1}{2}(D_0X_I)^2 \\ &- \frac{1}{2}\psi^T C_9 \Gamma^I [X_I, \psi] - \frac{1}{2}\psi^T C_9 D_0 \psi\Bigg). \end{aligned} \tag{7.408}$$

$$S_{\text{BMN}} = S_{\text{BFSS}} + \Delta S_{\text{defor}}. \tag{7.409}$$

$$\begin{aligned} \Delta S_{\text{defor}} = &\frac{\mu^2}{2g^2}\,\text{Tr}\left(\sum_{i=1}^{3} X_i^2 + \frac{1}{4}\sum_{a=4}^{9} X_a^2\right) \\ &- \frac{i\mu}{2g^2}\epsilon_{ijk}\,\text{Tr}[X_i, X_j]X_k - \frac{3i\mu}{8g^2}\,\text{Tr}\,\psi^T C_9 \gamma^{123}\psi. \end{aligned} \tag{7.410}$$

Construct perturbation theory of the BMN model around the fuzzy sphere solution $X_i = \mu J_i$, $X_a = 0$, $A_0 = 0$, $\psi = 0$ where $[J_i, J_j] = i\epsilon_{ijk}J_k$ which preserves full supersymmetry.

Write a Monte Carlo code for the BMN model (the BFSS is a special case and therefore it is automatically included).

Verify numerically the results discussed in the main text regarding the black-hole/black-string phase transition and the black hole evaporation process for the BFSS matrix model.

Solution: See appendix for the description of various possible algorithms that can be used.

References

[1] Banks T, Fischler W, Shenker S H and Susskind L 1997 M theory as a matrix model: a conjecture *Phys. Rev.* D **55** 5112

[2] Witten E 1996 Bound states of strings and p-branes *Nucl. Phys.* B **460** 335

[3] Ishibashi N, Kawai H, Kitazawa Y and Tsuchiya A 1997 A large N reduced model as superstring *Nucl. Phys.* B **498** 467

[4] Connes A, Douglas M R and Schwarz A S 1998 Noncommutative geometry and matrix theory: compactification on tori *J. High Energy Phys.* **9802** 003

[5] Dijkgraaf R, Verlinde E P and Verlinde H L 1997 Matrix string theory *Nucl. Phys.* B **500** 43

[6] Kawahara N and Nishimura J 2005 The large N reduction in matrix quantum mechanics: a bridge between BFSS and IKKT *J. High Energy Phys.* **0509** 040

[7] Hyakutake Y 2014 Quantum near-horizon geometry of a black 0-brane *Progr. Theor. Exp. Phys.* **2014** 033B04

[8] Hyakutake Y 2014 Quantum M-wave and Black 0-brane *J. High Energy Phys.* **1409** 075

[9] Hyakutake Y and Ogushi S 2006 Higher derivative corrections to eleven dimensional supergravity via local supersymmetry *J. High Energy Phys.* **0602** 068

[10] Hyakutake Y 2007 Toward the Determination of R**3 F**2 Terms in M-theory *Prog. Theor. Phys.* **118** 109

[11] Gross D J and Witten E 1986 Superstring modifications of Einstein's equations *Nucl. Phys.* B **277** 1

[12] Hanada M, Hyakutake Y, Ishiki G and Nishimura J 2014 Holographic description of quantum black hole on a computer *Science* **344** 882

[13] Kawahara N, Nishimura J and Takeuchi S 2007 Phase structure of matrix quantum mechanics at finite temperature *J. High Energy Phys.* **0710** 097

[14] Hoppe J 1982 *Quantum theory of a massless relativistic surface and a two-dimensional bound state problem* PhD Thesis *Massachusetts Institute of Technology (MIT)* (Cambridge, MA, USA) http://dspace.mit.edu/handle/1721.1/15717

[15] de Wit B, Hoppe J and Nicolai H 1988 On the quantum mechanics of supermembranes *Nucl. Phys.* B **305** 545

[16] Bergshoeff E, Sezgin E and Townsend P K 1987 Supermembranes and eleven-dimensional supergravity *Phys. Lett.* B **189** 75

[17] Taylor W 2000 The M-(atrix) model of M theory *NATO Sci. Ser.* C **556** 91

[18] Duff M J 1996 Supermembranes [hep-th/9611203]

[19] Connes A 1996 Gravity coupled with matter and foundation of noncommutative geometry *Commun. Math. Phys.* **182** 155

[20] Sochichiu C 2000 M[any] vacua of IIB *J. High Energy Phys.* **0005** 026

[21] Zarembo K L and Makeenko Y M 1998 An introduction to matrix superstring models *Phys. Usp.* **41** 1 [Usp. Fiz. Nauk **168** 3 (1998)]

[22] Krauth W, Nicolai H and Staudacher M 1998 Monte Carlo approach to M theory *Phys. Lett.* B **431** 31

[23] Connes A, Douglas M R and Schwarz A S 1998 Noncommutative geometry and matrix theory: compactification on tori *J. High Energy Phys.* **9802** 003

[24] Konechny A and Schwarz A S 2002 Introduction to M-(atrix) theory and noncommutative geometry *Phys. Rept.* **360** 353

[25] Brink L, Schwarz J H and Scherk J 1977 Supersymmetric Yang–Mills theories *Nucl. Phys.* B **121** 77

[26] Kim N, Klose T and Plefka J 2003 Plane wave matrix theory from N=4 superYang–Mills on R x S**3 *Nucl. Phys.* B **671** 359

[27] Horowitz G T and Polchinski J 2006 Gauge/gravity duality *Approaches to Quantum Gravity* ed D Oriti (Cambridge: Cambridge University Press) 169–86

[28] O'Connor D and Filev V G 2016 Membrane matrix models and non-perturbative checks of gauge/gravity duality *PoS CORFU* **2015** 111

[29] Hanada M 2016 What lattice theorists can do for superstring/M-theory *Int. J. Mod. Phys.* A **31** 1643006

[30] Dai J, Leigh R G and Polchinski J 1989 New connections between string theories *Mod. Phys. Lett.* A **4** 2073

[31] Polchinski J 1995 Dirichlet branes and Ramond-Ramond charges *Phys. Rev. Lett.* **75** 4724

[32] Azeyanagi T, Hanada M, Hirata T and Shimada H 2009 On the shape of a D-brane bound state and its topology change *J. High Energy Phys.* **0903** 121

[33] Zwiebach B 2009 *A First Course in String Theory* (Cambridge: Cambridge University Press)

[34] Leigh R G 1989 Dirac-Born-Infeld action from Dirichlet sigma model *Mod. Phys. Lett.* A **4** 2767

[35] 't Hooft G 1993 Dimensional reduction in quantum gravity arXiv:gr-qc/9310026.

[36] Susskind L 1995 The World as a hologram *J. Math. Phys.* **36** 6377

[37] Weinberg S and Witten E 1980 Limits on massless particles *Phys. Lett.* **96B** 59

[38] 't Hooft G 1974 A planar diagram theory for strong interactions *Nucl. Phys.* B **72** 461

[39] Nastase H 2007 Introduction to AdS-CFT arXiv:0712.0689

[40] Itzhaki N, Maldacena J M, Sonnenschein J and Yankielowicz S 1998 Supergravity and the large N limit of theories with sixteen supercharges *Phys. Rev.* D **58** 046004

[41] Gibbons G W and Maeda K I 1988 Black holes and membranes in higher dimensional theories with dilaton fields *Nucl. Phys.* B **298** 741

[42] Horowitz G T and Strominger A 1991 Black strings and P-branes *Nucl. Phys.* B **360** 197

[43] Maldacena J M 1999 The large N limit of superconformal field theories and supergravity *Int J. Theor. Phys.* **38** 1113 [*Adv. Theor. Math. Phys.* **2** 231 (1998)]

[44] Wilson K G 1974 Confinement of quarks *Phys. Rev.* D **10** 2445

[45] Horowitz G T and Martinec E J 1998 Comments on black holes in matrix theory *Phys. Rev.* D **57** 4935

[46] Li M and Martinec E J 1998 Probing matrix black holes arXiv:hep-th/9801070.

[47] Englert F and Rabinovici E 1998 Statistical entropy of Schwarzschild black holes *Phys. Lett.* B **426** 269

[48] Banks T, Fischler W and Klebanov I R 1998 Evaporation of Schwarzschild black holes in matrix theory *Phys. Lett.* B **423** 54

[49] Liu H and Tseytlin A A 1998 Statistical mechanics of D0-branes and black hole thermodynamics *J. High Energy Phys.* **9801** 010

[50] Das S R, Mathur S D, Kalyana Rama S and Ramadevi P 1998 Boosts, Schwarzschild black holes and absorption cross-sections in M theory *Nucl. Phys.* B **527** 187

[51] Li M 1998 Matrix Schwarzschild black holes in large N limit *J. High Energy Phys.* **9801** 009

[52] Banks T, Fischler W, Klebanov I R and Susskind L 1998 Schwarzschild black holes from matrix theory *Phys. Rev. Lett.* **80** 226

[53] Banks T, Fischler W, Klebanov I R and Susskind L 1998 Schwarzschild black holes in matrix theory. 2 *J. High Energy Phys.* **9801** 008

[54] Klebanov I R and Susskind L 1998 Schwarzschild black holes in various dimensions from matrix theory *Phys. Lett.* B **416** 62

[55] Nahm W 1978 Supersymmetries and their representations *Nucl. Phys.* B **135** 149

[56] Cremmer E, Julia B and Scherk J 1978 Supergravity theory in eleven-dimensions *Phys. Lett.* **76B** 409

[57] Polchinski J 2005 *String Theory. Vol. 2: Superstring Theory and Beyond* (Cambridge: Cambridge University Press)

[58] Miemiec A and Schnakenburg I 2006 Basics of M-theory *Fortsch. Phys.* **54** 5

[59] Becker K, Becker M and Schwarz J H 2006 *String Theory and M-theory: A Modern Introduction* (Cambridge: Cambridge University Press)
[60] Witten E 1995 String theory dynamics in various dimensions *Nucl. Phys.* B **443** 85
[61] de Alwis S P 1996 A note on brane tension and M theory *Phys. Lett.* B **388** 291
[62] Green M B, Hull C M and Townsend P K 1996 D-brane Wess-Zumino actions, t duality and the cosmological constant *Phys. Lett.* B **382** 65
[63] Schwarz J H 1995 An SL(2,Z) multiplet of type IIB superstrings *Phys. Lett.* B **360** 13 Erratum: [*Phys. Lett. B* **364**, 252 (1995)]
[64] Wald R M 1993 Black hole entropy is the Noether charge *Phys. Rev.* D **48** R3427
[65] Iyer V and Wald R M 1994 Some properties of Noether charge and a proposal for dynamical black hole entropy *Phys. Rev.* D **50** 846
[66] Green M B, Russo J G and Vanhove P 2007 Non-renormalisation conditions in type II string theory and maximal supergravity *J. High Energy Phys.* **0702** 099
[67] Hanada M, Hyakutake Y, Nishimura J and Takeuchi S 2009 Higher derivative corrections to black hole thermodynamics from supersymmetric matrix quantum mechanics *Phys. Rev. Lett.* **102** 191602
[68] Anagnostopoulos K N, Hanada M, Nishimura J and Takeuchi S 2008 Monte Carlo studies of supersymmetric matrix quantum mechanics with sixteen supercharges at finite temperature *Phys. Rev. Lett.* **100** 021601
[69] Hanada M, Nishimura J and Takeuchi S 2007 Non-lattice simulation for supersymmetric gauge theories in one dimension *Phys. Rev. Lett.* **99** 161602
[70] Kabat D N, Lifschytz G and Lowe D A 2001 Black hole thermodynamics from calculations in strongly coupled gauge theory *Int. J. Mod. Phys.* A **16** 856 [*Phys. Rev. Lett.* **86** 1426 (2001)]
[71] Catterall S and Wiseman T 2008 Black hole thermodynamics from simulations of lattice Yang–Mills theory *Phys. Rev.* D **78** 041502
[72] Hanada M, Hyakutake Y, Ishiki G and Nishimura J 2016 Numerical tests of the gauge/gravity duality conjecture for D0-branes at finite temperature and finite N *Phys. Rev.* D **94** 086010
[73] Hanada M, Nishimura J, Sekino Y and Yoneya T 2011 dimensional Direct test of the gauge-gravity correspondence for matrix theory correlation functions *J. High Energy Phys.* **1112** 020
[74] Banks T and Seiberg N 1997 Strings from matrices *Nucl. Phys.* B **497** 41
[75] Motl L 1997 Proposals on nonperturbative superstring interactions arXiv:hep-th/9701025
[76] Sethi S and Susskind L 1997 Rotational invariance in the M-(atrix) formulation of type IIB theory *Phys. Lett.* B **400** 265
[77] Aharony O, Marsano J, Minwalla S and Wiseman T 2004 Black hole-black string phase transitions in thermal 1+1 dimensional supersymmetric Yang–Mills theory on a circle *Class. Quant. Grav* **21** 5169
[78] Harmark T and Obers N A 2004 New phases of near-extremal branes on a circle *J. High Energy Phys.* **0409** 022
[79] Gregory R and Laflamme R 1993 Black strings and p-branes are unstable *Phys. Rev. Lett.* **70** 2837
[80] Susskind L 1997 Matrix theory black holes and the Gross-Witten transition arXiv:hep-th/9805115

[81] Kawahara N, Nishimura J and Takeuchi S 2007 High temperature expansion in supersymmetric matrix quantum mechanics *J. High Energy Phys.* **0712** 103
[82] Eguchi T and Kawai H 1982 Reduction of dynamical degrees of freedom in the large N gauge theory *Phys. Rev. Lett.* **48** 1063
[83] Janik R A and Wosiek J 2001 Towards the matrix model of M theory on a lattice *Acta Phys. Polon.* B **32** 2143
[84] Bialas P and Wosiek J 2002 Towards the lattice study of M theory. 2 *Nucl. Phys. Proc* Suppl.**106** 968
[85] Kawahara N, Nishimura J and Takeuchi S 2007 Exact fuzzy sphere thermodynamics in matrix quantum mechanics *J. High Energy Phys.* **0705** 091
[86] Aharony O, Marsano J, Minwalla S, Papadodimas K and Van Raamsdonk M 2004 The Hagedorn - deconfinement phase transition in weakly coupled large N gauge theories *Adv. Theor. Math. Phys.* **8** 603
[87] Aharony O, Marsano J, Minwalla S, Papadodimas K and Van Raamsdonk M 2005 A first order deconfinement transition in large N Yang–Mills theory on a small S**3 *Phys. Rev.* D **71** 125018
[88] Atick J J and Witten E 1988 The Hagedorn transition and the number of degrees of freedom of string theory *Nucl. Phys.* B **310** 291
[89] Azuma T, Morita T and Takeuchi S 2014 Hagedorn instability in dimensionally reduced large-N gauge theories as Gregory-Laflamme and Rayleigh-plateau instabilities *Phys. Rev. Lett.* **113** 091603
[90] Gross D J and Witten E 1980 Possible third order phase transition in the large N lattice gauge theory *Phys. Rev.* D **21** 446
[91] Wadia S R 1980 $N =$ infinity phase transition in a class of exactly soluble model lattice gauge theories *Phys. Lett.* **93B** 403
[92] Furuuchi K, Schreiber E and Semenoff G W 2003 Five-brane thermodynamics from the matrix model arXiv:hep-th/0310286
[93] Semenoff G W 2004 Matrix model thermodynamics arXiv:hep-th/0405107
[94] Kawahara N, Nishimura J and Yoshida K 2006 Dynamical aspects of the plane-wave matrix model at finite temperature *J. High Energy Phys.* **0606** 052
[95] Filev V G and O'Connor D 2016 The BFSS model on the lattice *J. High Energy Phys.* **1605** 167
[96] Filev V G and O'Connor D 2016 A computer test of holographic flavour dynamics *J. High Energy Phys.* **1605** 122
[97] Asano Y, Filev V G, Kovik S and O'Connor D 2017 The flavoured BFSS model at high temperature *J. High Energy Phys.* **1701** 113
[98] Hotta T, Nishimura J and Tsuchiya A 1999 Dynamical aspects of large N reduced models *Nucl. Phys.* B **545** 543
[99] Mandal G and Morita T 2011 Phases of a two dimensional large N gauge theory on a torus *Phys. Rev.* D **84** 085007
[100] Mandal G, Mahato M and Morita T 2010 Phases of one dimensional large N gauge theory in a 1/D expansion *J. High Energy Phys.* **1002** 034
[101] Bigatti D and Susskind L Review of matrix theory *Strings, Branes and Dualities* (Berlin: Springer) pp 277–318
[102] Susskind L 1997 Another conjecture about M(atrix) theory arXiv:hep-th/9704080
[103] Seiberg N 1997 Why is the matrix model correct? *Phys. Rev. Lett.* **79** 3577

[104] Taylor W 2001 M(atrix) theory: matrix quantum mechanics as a fundamental theory *Rev. Mod. Phys.* **73** 419

[105] Sen A 1998 D0-branes on T**n and matrix theory *Adv. Theor. Math. Phys.* **2** 51

[106] Berenstein D E, Maldacena J M and Nastase H S 2002 Strings in flat space and pp waves from N=4 superYang–Mills *J. High Energy Phys.* **0204** 013

[107] Blau M, Figueroa-O'Farrill J M, Hull C and Papadopoulos G 2002 A new maximally supersymmetric background of IIB superstring theory *J. High Energy Phys.* **0201** 047

[108] Kowalski-Glikman J 1984 Vacuum states in supersymmetric Kaluza–Klein theory *Phys. Lett.* **134B** 194

[109] Blau M, Figueroa-O'Farrill J M, Hull C and Papadopoulos G 2002 Penrose limits and maximal supersymmetry *Class. Quant. Grav* **19** L87

[110] Penrose R 1976 Any spacetime has a plane wave as a limit *Differential Geometry and Relativity* (Dordrecht: Reidel) pp 271–5

[111] Hull C M 1984 Exact *pp* wave solutions of eleven-dimensional supergravity *Phys. Lett.* **139B** 39

[112] Kadoh D and Kamata S 2015 Gauge/gravity duality and lattice simulations of one dimensional SYM with sixteen supercharges arXiv:1503.08499

[113] Berkowitz E, Rinaldi E, Hanada M, Ishiki G, Shimasaki S and Vranas P 2016 Precision lattice test of the gauge/gravity duality at large-N *Phys. Rev.* D **94** 094501

[114] Taylor W and Van Raamsdonk M 1999 Multiple D0-branes in weakly curved backgrounds *Nucl. Phys.* B **558** 63

[115] Dasgupta K, Sheikh-Jabbari M M and Van Raamsdonk M 2002 Matrix perturbation theory for M theory on a PP wave *J. High Energy Phys.* **0205** 056

[116] Taylor W and Van Raamsdonk M 1999 Supergravity currents and linearized interactions for matrix theory configurations with fermionic backgrounds *J. High Energy Phys.* **9904** 013

[117] Myers R C 1999 Dielectric branes *J. High Energy Phys.* **9912** 022

[118] Berkooz M and Douglas M R 1997 Five-branes in M(atrix) theory *Phys. Lett.* B **395** 196

[119] Aharony O, Bergman O, Jafferis D L and Maldacena J 2008 $N = 6$ superconformal Chern-Simons-matter theories, M2-branes and their gravity duals *J. High Energy Phys.* **0810** 091

[120] Catterall S and Wiseman T 2010 Extracting black hole physics from the lattice *J. High Energy Phys.* **1004** 077

[121] Catterall S, Joseph A and Wiseman T 2010 Thermal phases of D1-branes on a circle from lattice super Yang–Mills *J. High Energy Phys.* **1012** 022

[122] Hanada M, Miwa A, Nishimura J and Takeuchi S 2009 Schwarzschild radius from Monte Carlo calculation of the Wilson loop in supersymmetric matrix quantum mechanics *Phys. Rev. Lett.* **102** 181602

[123] Kim N and Plefka J 2002 On the spectrum of PP wave matrix theory *Nucl. Phys.* B **643** 31

[124] de Wit B, Luscher M and Nicolai H 1989 *Nucl. Phys.* B **320** 135

IOP Publishing

Matrix Models of String Theory

Badis Ydri

Chapter 8

Type IIB matrix model

In this chapter we will discuss in some detail two very interesting physical applications based on the type IIB matrix model which is also known as the IKKT matrix model [2]. In the first application, the (Lorentzian) type IIB matrix model allows us to study the real time dynamics of the emergence of (1 + 3)-dimensional Minkowski spacetime, the spontaneous symmetry breaking of $SO(9)$ down to $SO(3)$, as well as providing a mechanism for avoiding the Big Bang singularity, and allowing us to obtain the expansion (exponential at early times and power law, i.e. $\sqrt{t}$, at late times) of the Universe [19, 21–27].

In the second application, the idea of emergent gravity from a noncommutative $U(1)$ gauge theory which was put forward first by Rivelles [48] is discussed. See also Yang [52–54]. A semi-classical derivation of Einstein equations from the IKKT model around fuzzy $\mathbf{S}^4$ background due to Steinacker is the main result discussed here [49, 51, 127].

8.1 The IKKT model in the Gaussian expansion method

We start by considering the IKKT matrix model in the Gaussian expansion method following mainly [1]. As we have already discussed, the IKKT model also called the type IIB matrix model is the zero-volume limit, i.e. dimensional reduction to a point, of supersymmetric $U(N)$ Yang–Mills gauge theory in 10 dimensions where the components of the 10-dimensional gauge field reduce to 10 bosonic Hermitian matrices A_μ. This is a Euclidean $SO(10)$ model given by the action

$$S = -\frac{1}{4g^2}\,\mathrm{Tr}[A_\mu, A_\nu]^2 - \frac{1}{2g^2}\,\mathrm{Tr}\,\psi_\alpha(\mathcal{C}\gamma^\mu)_{\alpha\beta}[A_\mu, \psi_\beta]. \tag{8.1}$$

The A_μ (vector) and ψ_α (Majorana–Weyl spinor) are $N \times N$ traceless Hermitian matrices with complex and Grassmannian entries, respectively. The γ are the 16×16 Dirac matrices in 10 dimensions after Weyl projection and the charge conjugation operator is defined by $\gamma_\mu^T = \mathcal{C}\gamma_\mu\mathcal{C}^\dagger$ and $\mathcal{C}^T = \mathcal{C}$. We can without any loss of generality

doi:10.1088/978-0-7503-1726-9ch8

choose the coupling constant (scale parameter) g such that $g^2N = 1$. This model is supposed to give a non-perturbative regularization of type IIB superstring theory in the Schild gauge [2]. The size of the matrices N plays in this regularization the role of the cutoff. It is then supposed that in the limit $N \longrightarrow \infty$ (continuum limit) the 10 bosonic Hermitian matrices A_μ reproduce the 10-dimensional target space of the string [3].

The above model corresponds to a Yang–Mills matrix model in $D = 10$ dimensions. The dimensional reduction to a point of supersymmetric $U(N)$ Yang–Mills gauge theory in D dimensions gives a Yang–Mills matrix model in D dimensions. We can also have supersymmetric models in $D = 6$ and $D = 4$. These three models $D = 10, 6, 4$ enjoy a convergent partition function [4–6]. The $D = 3$ supersymmetric partition function is divergent while the bosonic one is convergent. The determinant in $D = 4$ is real positive while the determinants in $D = 6$ and $D = 10$ are complex, which means in particular that in $D = 10$ and $D = 6$ we can have spontaneous symmetry breaking of rotational invariance [7–10] while in $D = 4$ there is no spontaneous symmetry breaking [11–13].

The main question we want to ask in the Gaussian expansion method is whether or not the rotational $SO(10)$ is spontaneously broken in the continuum large N limit. Towards answering this question the authors of [1] studied the $SO(d)$ symmetric vacua for all values of d in the range between two and seven. They also impose on the shrunken $(10 - d)$ directions an extra Σ_d symmetry and thus the full symmetry imposed is actually $SO(d) \times \Sigma_d \in SO(10)$ which is much more stronger than simply $SO(d)$ which allowed them to reduce the number of free parameters in the Gaussian expansion method considerably (they exhausted all possible extra symmetries which leave not more than five free parameters). This spontaneous symmetry breaking corresponds therefore to a dynamical compactification to d dimensions.

The order parameter of the spontaneous symmetry breaking of $SO(10)$ is given by the set of nine eigenvalues λ_i of the moment of inertia tensor

$$T_{\mu\nu} = \frac{1}{N} \mathrm{Tr}\, A_\mu A_\nu. \tag{8.2}$$

The expectation values $\langle \lambda_i \rangle$ in the large N limit are all equal if $SO(10)$ is not spontaneously broken. In the current Euclidean model it is found that the expectation values $\langle \lambda_1 \rangle$, $\langle \lambda_2 \rangle$ and $\langle \lambda_3 \rangle$ become much larger than the other ones in the large N limit and hence $SO(10)$ is spontaneously broken down to $SO(3)$ due precisely to the phase of the Pfaffian [14, 15]. In the Lorentzian model as we will see, the spontaneous symmetry breaking is due to the noncommutativity of the space coordinates [17].

The Gaussian expansion method is a nonperturbative scheme in which mostly perturbative calculations are performed [16]. We start by introducing a Gaussian action S_0 and split the action S as

$$S = (S - S_0) + S_0. \tag{8.3}$$

The Gaussian action S_0 is arbitrary since it contains many free parameters. It is expected that at any finite order the result of the Gaussian expansion method will depend on the parameters of the Gaussian action S_0. However, it is also known that there exist regions in the parameter space (plateau regions) for which the result at finite order will not depend on the choice of S_0 [1].

By requiring $SO(10)$ rotational invariance the most general $U(N)$-invariant Gaussian action S_0 takes the form

$$S_0 = S_{0b} + S_{0f}, \tag{8.4}$$

$$S_{0b} = \frac{N}{2}\sum_{\mu=1}^{10} M_\mu \,\mathrm{Tr}\, A_\mu^2, \qquad S_{0f} = \frac{N}{2}\sum_{\alpha,\beta=1}^{16} \mathcal{A}_{\alpha\beta}\,\mathrm{Tr}\,\psi_\alpha\psi_\beta. \tag{8.5}$$

There are 10 parameters M_μ and 120 parameters $\mathcal{A}_{\alpha\beta}$. The 16×16 complex matrix $\mathcal{A}$ can be given in terms of gamma matrices. The goal is to compute the free energy

$$F = -\log Z, \quad Z = Z_0\langle \exp(-(S - S_0))\rangle_0, \quad Z_0 = \int dA d\psi \exp(-S_0). \tag{8.6}$$

This will be done by truncating the perturbative series at some finite order and as a consequence the free energy will depend on the parameters M_μ and $\mathcal{A}_{\alpha\beta}$. We look for the regions in the parameter space for which the free energy F is stationary in the sense that it solves the self-consistency conditions

$$\frac{\partial}{\partial M_\mu} F = 0, \qquad \frac{\partial}{\partial \mathcal{A}_{\alpha\beta}} F = 0. \tag{8.7}$$

The number of self-consistency conditions is equal to the number of parameters. The values of the free energy are obtained at the solutions of these conditions. Clearly, the number of solutions increases with the order. The sought after plateau in the parameter space is determined by the existence of multiple solutions with almost the same value of the free energy.

The above perturbative series corresponds to an ordinary loop expansion where the insertion of the 2-point vertex $-S_0$ acts as if it is coming from a one-loop counter term. Furthermore, due to the large N limit, only planar Feynman diagrams are of relevance.

It is expected that $SO(10)$ symmetry will be spontaneously broken dynamically down to some $SO(d)$. Thus only $SO(d)$-symmetric vacua will be left and the Gaussian action in this case must only be required to be $SO(d)$-symmetric. As it turns out, the number of free parameters will be reduced considerably by imposing $SO(d)$ symmetry on the Gaussian action. Indeed, we get the total number of parameters to be 5, 9, 16, 27, 44, 73 for $d = 7, 6, 5, 4, 3, 2$ respectively, by imposing $SO(d)$ symmetry on the Gaussian action [1]. The Gaussian expansion method ceases to work properly for $d \geqslant 8$ but in these cases there exist also spontaneous symmetry breaking and $SO(d)$-symmetric vacua.

However, the self-consistency conditions can be solved only for not more than five parameters. Therefore, an extra symmetry Σ_d is imposed on the shrunken dimensions $x_{d+1}, \ldots, x_{10}$ in order to reduce the number of parameters to five or less.

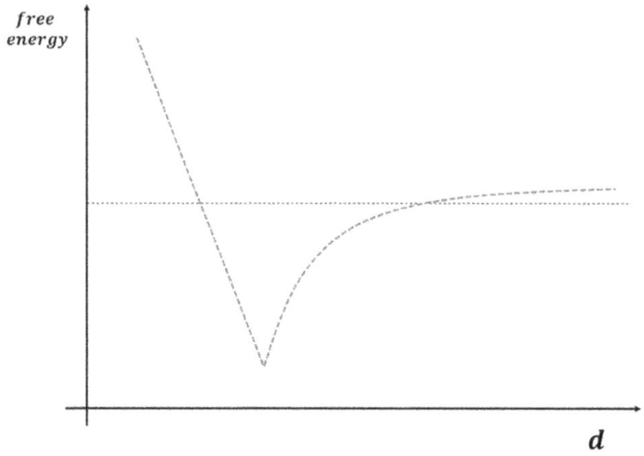

Figure 8.1. The free energy of the IKKT model in the Gaussian expansion method. The minimum is at $d = 3$.

The extra symmetry Σ_d is a subgroup of $SO(10)$ formed out of cyclic permutations and reflections [1]. For each $SO(d)$ there are several choices Σ_d.

We fix then the ansatz $SO(d) \times \Sigma_d$ and we compute the free energy up to the third order as a function of the parameters of the Gaussian action. We differentiate the energy with respect to these parameters to obtain the self-consistency conditions and then solve these equations numerically. By substituting back in the free energy with the solution(s) we get the value(s) of the free energy at the solution(s). A given ansatz may correspond to several solutions. We locate the plateau region by the solutions for the various ansatz which have the same value of the free energy. By averaging over these physical solutions we get the value of the free energy for that particular d.

The main result of [1] is the following: the free energy F (averaged over the plateau region of physical solutions) takes its minimum value at $d = 3$. See figure 8.1. This shows more or less explicitly that in the Euclidean model (and due to the phase of the Pfaffian) the stringy rotational symmetry $SO(10)$ must be spontaneously broken down to the physical rotational symmetry $SO(3)$.

The extent of space in the extended directions $R^2 = \langle \lambda_1 \rangle = \cdots = \langle \lambda_d \rangle$ and the extent of space in the shrunken directions $r^2 = \langle \lambda_{d+1} \rangle \cdots = \langle \lambda_{10} \rangle$ can also be computed in the Gaussian expansion method with very illuminating results [1]. They found that r remains almost constant for all d, and thus it is indeed a universal compactification scale, while R becomes larger for smaller d. See figure 8.2. This behavior is consistent with the so-called constant volume property given by [18]

$$R^d r^{10-d} = l^{10}. \tag{8.8}$$

8.2 Yang–Mills matrix cosmology

8.2.1 Lorentzian type IIB matrix model

The Lorentzian type IIB matrix model allows us to study the real time dynamics of the emergence of (1 + 3)-dimensional Minkowski spacetime, the spontaneous symmetry breaking of $SO(9)$ down to $SO(3)$, as well as providing a mechanism

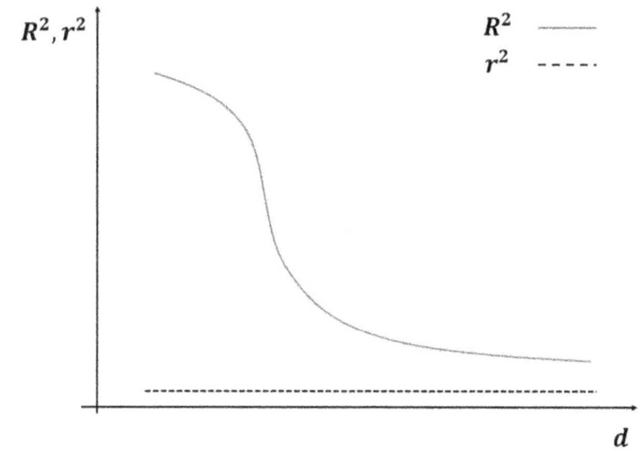

Figure 8.2. The extent of space R and the compactification scale r for the IKKT model.

for avoiding the Big Bang singularity, and allowing us to obtain the expansion (exponential at early times and power law, i.e. $\sqrt{t}$, at late times) of the Universe [17, 19–25]. The model is given by (with $F_{\mu\nu} = i[A_\mu, A_\nu]$ and $\bar{\psi} = \psi^T C$, $\mu, \dots \nu = 0, \dots, 9$, $\alpha, \dots, \beta = 1, \dots, 16$ and the 16×16 matrices γ^μ are the 9-dimensional Dirac matrices)

$$S = -\frac{1}{4g^2} \operatorname{Tr} F_{\mu\nu} F^{\mu\nu} - \frac{1}{2g^2} \operatorname{Tr} \bar{\psi}_\alpha (\gamma^\mu)_{\alpha\beta} [A_\mu, \psi_\beta]. \tag{8.9}$$

The A_μ and ψ_α are $N \times N$ traceless Hermitian matrices with complex and Grassmannian entries, respectively. Because of the signature $\eta = (-1, +1, +1, \dots)$ the Yang–Mills term is not positive definite, i.e. the action is not bounded from below. As opposed to the Euclidean case, the Pfaffian in the Lorentzian case is real (in fact it is positive definite at large N). The path integral is given by

$$Z = \int dA d\psi \exp(iS) = \int dA \ \operatorname{Pf}\mathcal{M}(A) \ \exp(iS_B). \tag{8.10}$$

The Dirac operator is given by $\mathcal{M}(A) = (C\gamma^\mu)_{\alpha\beta}[A_\mu, \dots]$.

This path integral is not finite as it stands (when A_0 diverges the action S_B goes to $-\infty$). We introduce the $SO(9,1)$ symmetric IR cutoff in the temporal direction as [17]

$$\frac{1}{N} \operatorname{Tr} A_0^2 \leqslant \kappa \frac{1}{N} \operatorname{Tr} A_a^2. \tag{8.11}$$

This condition is reminiscent to what happens in causal dynamical triangulation.

The oscillating phase factor in the path integral is also regularized in the usual way by adding a damping factor $\exp(-\epsilon|S_B|)$ to the action where ϵ is some small positive number. We get then

$$Z = \int dA \ \operatorname{Pf}\mathcal{M}(A) \ \exp(iS_B - \epsilon|S_B|). \tag{8.12}$$

The IR cutoff (8.11) is explicitly implemented by inserting in the path integral the expression

$$\int_0^\infty dr\ \delta\left(\frac{1}{N}\mathrm{Tr}\,A_a^2 - r\right)\theta\left(\kappa r - \frac{1}{N}\mathrm{Tr}\,A_0^2\right). \tag{8.13}$$

The variable r is effectively the scale factor. We scale the field as $A_\mu \longrightarrow \sqrt{r}A_\mu$ and perform the integral over r (with $D = 10$ and $d_F = 8$). We find

$$\int_0^\infty dr\ r^{\frac{D(N^2-1)+d_F(N^2-1)}{2}-1}\exp(-r^2(\epsilon|x| - ix)) \sim 1/|x|^{(D+d_F)(N^2-1)/4}. \tag{8.14}$$

This is a divergent integral which can be regularized by introducing a second IR cutoff

$$\frac{1}{N}\mathrm{Tr}\,A_a^2 \leqslant L^2. \tag{8.15}$$

We insert now in the path integral the condition

$$\int_0^{L^2} dr\ \delta\left(\frac{1}{N}\mathrm{Tr}\,A_a^2 - r\right)\theta\left(\kappa r - \frac{1}{N}\mathrm{Tr}\,A_0^2\right). \tag{8.16}$$

We get

$$\begin{aligned} Z = &\int dA \int_0^{L^2} dr\ (r^{1/2})^{D(N^2-1)+d_F(N^2-1)}\frac{1}{r}\delta\left(\frac{1}{N}\mathrm{Tr}\,A_a^2 - 1\right) \\ &\times \theta\left(\kappa - \frac{1}{N}\mathrm{Tr}\,A_0^2\right)\mathrm{Pf}\mathcal{M}(A) \\ &\times \exp(-r^2(\epsilon|S_B| - iS_B)). \end{aligned} \tag{8.17}$$

We use the result [23]

$$\int_0^{L^2} dr\ r^{\frac{D(N^2-1)+d_F(N^2-1)}{2}-1}\exp(-r^2(\epsilon|x| - ix)) = \delta(x), \quad N, L\longrightarrow\infty. \tag{8.18}$$

We get then the path integral (by reinserting the scale parameter L for later convenience)

$$\begin{aligned} Z = &\int dA\ \delta\left(\frac{1}{N}\mathrm{Tr}\,F_{\mu\nu}F^{\mu\nu}\right)\delta\left(\frac{1}{N}\mathrm{Tr}\,A_a^2 - L^2\right) \\ &\theta \times \left(\kappa L^2 - \frac{1}{N}\mathrm{Tr}\,A_0^2\right)\mathrm{Pf}\mathcal{M}(A). \end{aligned} \tag{8.19}$$

In this formulation N plays the role of inverse lattice spacing and $\sqrt{\kappa}L^2$ plays the role of the volume (L is the spacelike length and $\sqrt{\kappa}L$ is the time-like length). These two constraints can be removed in the continuum limit $N\longrightarrow\infty$ and the infinite volume limit $L\longrightarrow\infty$ and only one scale parameter g remains (string coupling constant). See figure 8.3.

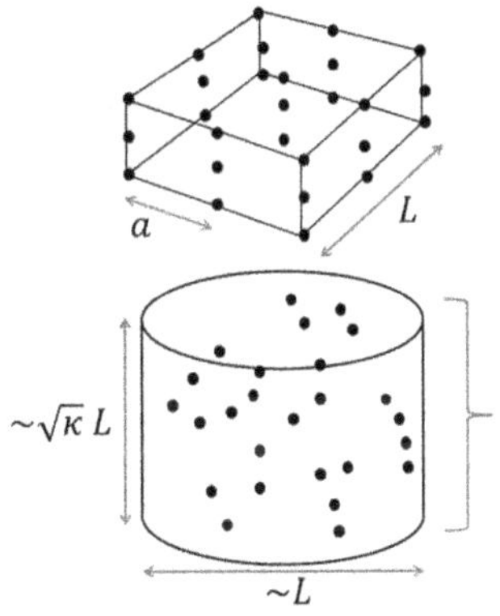

Figure 8.3. The IR cutoffs L and κ.

The problem is then converted into a potential problem of the form

$$Z = \int dA \, \exp(-V_{\text{pot}}) \, \text{Pf}\mathcal{M}(A). \tag{8.20}$$

$$\begin{aligned} V_{\text{pot}} = &\frac{1}{2}\gamma_C\left(\frac{1}{N}\,\text{Tr}\,F_{\mu\nu}F^{\mu\nu}\right)^2 + \frac{1}{2}\gamma_L\left(\frac{1}{N}\,\text{Tr}\,A_a^2 - L^2\right)^2 \\ &+ \frac{1}{2}\gamma_\kappa\left(\kappa L^2 - \frac{1}{N}\,\text{Tr}\,A_0^2\right)^2 \theta\left(\frac{1}{N}\,\text{Tr}\,A_0^2 - \kappa L^2\right). \end{aligned} \tag{8.21}$$

This theory enjoys $SO(1,9)$ Lorentz symmetry, $SO(9)$ rotational symmetry, $U(N)$ gauge symmetry, $\mathcal{N} = 1$ supersymmetry, translation symmetry given by the shift symmetry $A_\mu \longrightarrow A_\mu + \alpha_\mu \mathbf{1}$. But it also enjoys an extended $\mathcal{N} = 2$ supersymmetry and hence it includes implicitly gravity since $\mathcal{N} = 1$ is the maximal supersymmetry without gravity.

The rotational symmetry $SO(9)$ will be spontaneously broken which is the main goal in this model. Also, the shift symmetry $A_\mu \longrightarrow A_\mu + \alpha_\mu \mathbf{1}$ will be spontaneously broken in the dynamics, which causes problems in determining the origin of the time coordinate. This issue can be avoiding by adding a potential of the form

$$V_{\text{sym}} = \frac{1}{2}\gamma_{\text{sym}}\left(\frac{1}{N}\,\text{Tr}\,A^2|_{\text{left}} - \frac{1}{N}\,\text{Tr}\,A^2|_{\text{right}}\right)^2. \tag{8.22}$$

$$\text{Tr}\,A^2|_{\text{left}} = \sum_{i=1}^{d}\sum_{a+b<N+1}|(A_i)_{ab}|^2, \quad \text{Tr}\,A^2|_{\text{right}} = \sum_{i=1}^{d}\sum_{a+b>N+1}|(A_i)_{ab}|^2. \tag{8.23}$$

The parameters $\gamma_{C,\,L,\,\kappa,\,\text{sym}}$ are chosen as [23]

$$\gamma_C = N^2, \quad \gamma_L = \gamma_\kappa = 100N^2, \quad \gamma_{\text{sym}} = 100. \tag{8.24}$$

8.2.2 Spontaneous symmetry breaking and continuum and infinite volume limits

We will now employ $SU(N)$ invariance to diagonalize the time-like matrix A_0, producing also a Vandermonde determinant, as

$$A_0 = \mathrm{diag}(t_1, \ldots, t_N), \quad t_1 < \cdots < t_N. \tag{8.25}$$

Thus the measure becomes

$$\int dA = \int dA_a \int \prod_i dt_i \Delta(t)^2, \quad \Delta(t) = \prod_{i>j} (t_i - t_j). \tag{8.26}$$

The effect of the Vandermonde determinant $\Delta(t)$ cancels exactly at one-loop order due to supersymmetry (more on this below). Indeed, at one-loop the repulsive effective action of the eigenvalues t_i is given by

$$S_{\mathrm{eff}} = (D - 2 - d_F) \sum_{i \neq j} \ln(t_i - t_j)^2 = 0. \tag{8.27}$$

Thus the spectrum of A_0, i.e. time, extends to infinity even for finite N. Locality in time is also guaranteed as follows. Instants of time will be defined by

$$t = \frac{1}{n} \sum_{i=1}^{n} t_{\nu+i}. \tag{8.28}$$

In the eigenbasis of A_0 the spacelike matrices A_a have a band diagonal structure. This highly non-trivial property is determined dynamically. This band diagonal structure means in particular that the off diagonal elements $(A_a)_{ab}$ for $|a - b| \geqslant n$ are very small for some integer n. But n should also be sufficiently large that it includes non-negligible off diagonal elements. Thus we can consider $n \times n$ block matrices

$$(\bar{A}_a)_{IJ}(t) = (A_a)_{I+\nu,\, J+\nu}. \tag{8.29}$$

The indices I, J run from 1 to n and thus the index ν runs from 0 to $N - n$. The time t appearing in this equation is the one defined in equation (8.28). The matrices $\bar{A}_a$ represent the state of the Universe at time t. The progression of t is encoded in the index ν. See figure 8.4.

The block size n is determined as follows [19]. We take N even and consider the $N \times N$ matrix $(Q)_{IJ} = (A_1^2)_{IJ}$. We plot for a fixed $L = (I + J)/2$ the quantity $\sqrt{|Q_{IJ}/Q_{N/2N/2}|}$ as a function of $I - J = 2(I - L)$. It is sufficient to consider only the values $L = 2, 4, 6, \ldots, N/2$. It is found that the quantity $\sqrt{|Q_{IJ}/Q_{N/2N/2}|}$ decreases exponentially with $|I - J|$ and that the half width is maximum for $L = N/2$. The block size n is given precisely by the largest half width.

At time t the square of the extent of space should be defined in terms of $\bar{A}_a$ as follows (where tr is the trace taken over the $n \times n$ block)

$$R^2(t) = \left\langle \frac{1}{n} \mathrm{tr}\, \bar{A}_a^2(t) \right\rangle. \tag{8.30}$$

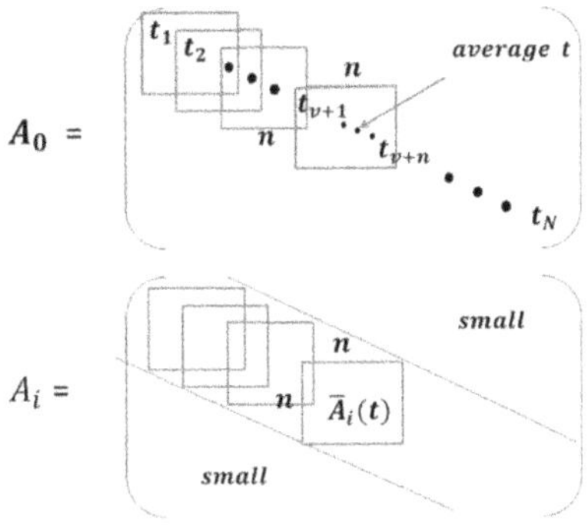

Figure 8.4. The band structure.

The order parameter for the spontaneous symmetry breaking (SSB) of rotational symmetry $SO(9)$ is given by the moment of inertia tensor (symmetric under $t \longrightarrow -t$ and thus we may only take $t < 0$)

$$T_{ab}(t) = \frac{1}{n}\text{tr } \bar{A}_a(t)\bar{A}_b(t). \tag{8.31}$$

This is a real symmetric 9×9 matrix with eigenvalues denoted by $\lambda_i(t)$ where $\lambda_1(t) > \cdots > \lambda_9(t)$. We have then the behavior at early times in the large N and n limits

$$\langle\lambda_1(t)\rangle = \cdots = \langle\lambda_9(t)\rangle, \quad \text{Exact } SO(9). \tag{8.32}$$

For late times we have instead

$$\langle\lambda_1(t)\rangle = \langle\lambda_2(t)\rangle = \langle\lambda_3(t)\rangle \gg \langle\lambda_i(t)\rangle, \quad i \neq 1, 2, 3, \quad \text{SSB of } SO(9). \tag{8.33}$$

The spontaneous symmetry breaking of $SO(9)$ down to $SO(3)$ occurs at some critical time t_c. See figure 8.5. The mechanism behind this breaking is noncommutative geometry and not the complex Pfaffian as is the case in Euclidean signature.

The scaling limit is achieved as follows. The square of the extent of space $R^2(t)$ is given by the sum of the eigenvalues of the moment of inertia tensor $\langle T_{ab}(t)\rangle$. The extent of space $R(t)$ is found to scale in the large N limit by sending the infrared cutoffs κ and L to infinity in a prescribed way [17]. This is done in two steps:

- Continuum limit: We send $\kappa \longrightarrow \infty$ as $\kappa = \beta N^p$, $p = 1/4$. The extent of space $R(t)$ for different values of N is then seen to collapse to a single curve depending on β.
- Infinite volume limit: We fix N then we send $\beta \longrightarrow \infty$ (equivalently κ) with $L \longrightarrow \infty$. The extent of space $R(t)$ for various values of β is then seen to collapse to a single curve.

Thus the two IR cutoffs κ and L are then removed in the large N limit and the theory depends only on one single parameter κ (which should be thought of as the string coupling constant). The only scale parameter is the size of the Universe $R(t_c)$ at the critical time.

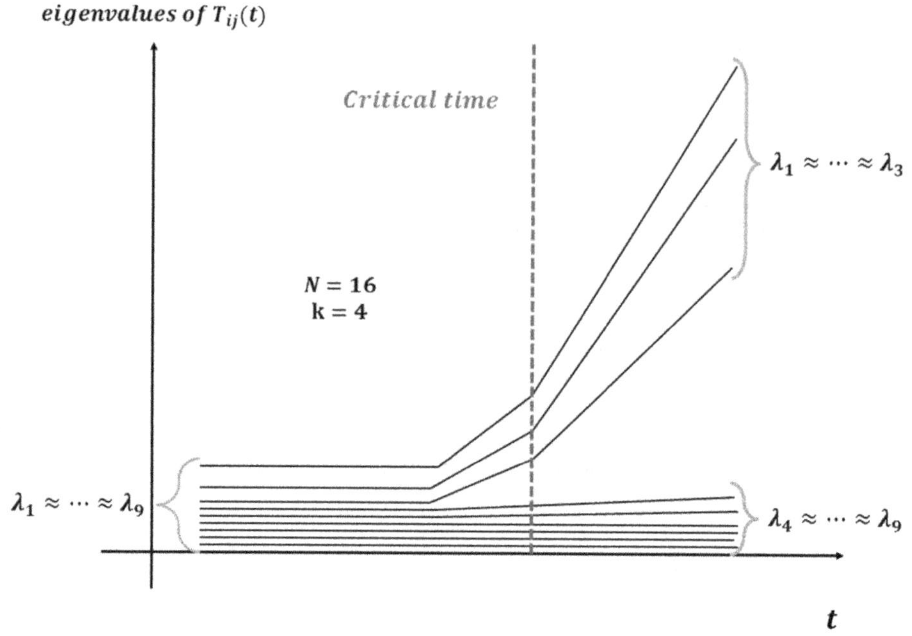

Figure 8.5. SSB of $SO(9)$.

We would like to discuss the infinite volume limit further following [17]. We fix N and κ and calculate the extent of space $R(t)$ for $L = 1$. They observe that the constraint (8.15) is saturated for the dominant configurations. Thus they only need to scale $A_\mu \longrightarrow LA_\mu$ to reinstate the IR cutoff L. Then they choose L such that the extent of space at the critical time t_c is one, viz $R(t_c) = 1$. They repeat for different values of β and each time they determine implicitly the corresponding value of L in this way. Thus increasing values of β is equivalent to increasing values of L and the extent of space $R(t)$ is seen to scale with β.

To summarize, it is seen that the extent of space $R(t)$ for different values of N and κ converge to a single scaled curve. This achieves the non-trivial continuum limit and infinite volume limit of the theory.

8.2.3 Expansion

The first fundamental observation is that the birth of the Universe at the critical time t_c emerges without any singularity, i.e. the problem of the initial singularity is completely avoided, and the underlying mechanism behind it can be determined to be the noncommutativity of the space.

After the birth of the Universe three coordinates start to expand and the other six shrink. It can be verified from the measurement of the extent of space $R(t)$ that for the very early times after t_c the expansion of the three spatial coordinates is indeed exponential (inflation). See figure 8.6. However, at later times the expansion is expected to become a power-law $\sqrt{t}$ behavior (radiation-dominated FLRW Universe) which is a fact that has been explicitly checked in the bosonic model in [25].

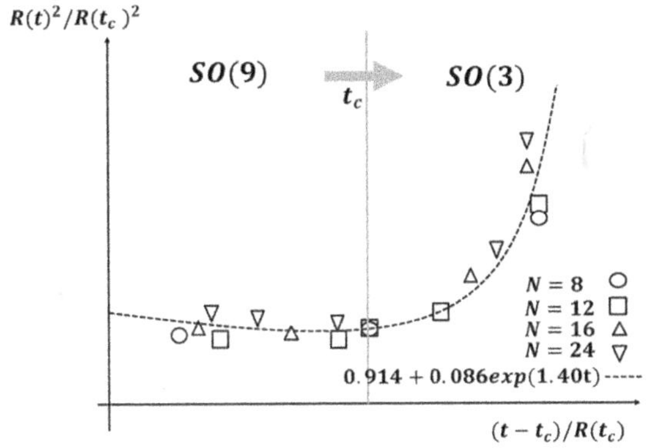

Figure 8.6. Exponential expansion in the type IIB model.

The early and late times can be approximated by the Vandermonde and bosonic models, respectively, as follows. We write the fermion action as

$$S_F = \mathrm{Tr}\ \bar{\Psi}\Gamma^{\mu}[A_{\mu}, \Psi] = \mathrm{Tr}\ \bar{\Psi}\Gamma^{0}[A_0, \Psi] + \mathrm{Tr}\ \bar{\Psi}\Gamma^{a}[A_a, \Psi]. \tag{8.34}$$

Thus for early times $A_0 \gg A_a$ and the first term dominates while at late time $A_0 \ll A_i$ and it is the second time that dominates. The bosonic and fermionic actions can be expanded as follows

$$S_B = -\frac{1}{4g^2}(t_i - t_j)^2|(A_a)_{ij}|^2 + \cdots \tag{8.35}$$

$$S_F = -\frac{1}{2g^2}(t_i - t_j)(\Psi_a)_{ji}(C\Gamma^0)_{\alpha\beta}(\Psi_\beta)_{ij} + \cdots \tag{8.36}$$

The subleading terms are small for large $|t_i - t_j|$. The one-loop integration over bosonic degrees of freedom gives $\Delta(t)^{-2(D-1)}$, whereas the integration over fermionic degrees of freedom gives $\Delta(t)^{2d_F}$. Thus the effective potential at one-loop vanishes since $D - 2 - d_F = 0$ and the spectrum of A_0 extends due to supersymmetry to infinity even for finite N in the limit $\kappa \longrightarrow \infty$. In the bosonic model the eigenvalues t_i are attracted to each other and the spectrum has a finite extent without any cutoff.

At early times we quench the model by including the repulsive force between the eigenvalues t_i given by the fermion determinant $\Delta(t)^{2d_F}$. The Pfaffian is then approximated by

$$\mathrm{Pf}\mathcal{M}(A) = \Delta(t)^{2d_F} = \prod_{i>j}(t_i - t_j)^{2d_F}. \tag{8.37}$$

The corresponding model is called the Vandermonde (VDM) model in [23]. It shares with the original supersymmetric model some crucial features such as spontaneous symmetry breaking and exponential expansion at very early times. It can also be accessed via Monte Carlo simulation as easily as the bosonic model which is valid at late times when it is possible to fully quench the fermions and the Pfaffian is approximated by

$$\mathrm{Pf}\mathcal{M}(A) = 1. \tag{8.38}$$

In this bosonic model the IR cutoff (8.11) is not required. We can observe the emergence of an exponentially expanding Universe only after some critical value of N given by $N_c = 110$ [25]. Also, by studying this bosonic model we find the extent of space behavior

$$R^2(t) \sim t. \tag{8.39}$$

In the VDM model with $D = 6$ the extent of space $R(t)$ was found for very early times to be given by the exponential fit [23]

$$\frac{R^2(t)}{R^2(t_c)} = C + (1 - C)\exp(-bx), \quad x = \frac{t - t_c}{R(t_c)}. \tag{8.40}$$

This is the inflationary behavior seen also in the Lorentzian type IIB matrix model [26].

The calculation time caused by the Pfaffian in the supersymmetric model is of order N^5, whereas in the quenched model it is of order N^3.

The late time behavior in the VDM model can still be studied (in fact very carefully) by using the renormalization group method. First, we note that the late time behavior is described by the inner part of the matrices A_μ. By integrating out the outer part of the matrices A_μ corresponding to early times we get thus a renormalized theory with a smaller number of degrees of freedom which can be studied more efficiently by means of Monte Carlo. This ingenious idea with very interesting results for the late time behavior of the Vandermonde model is reported in [23].

It is also found in the Lorentzian type IIB matrix model that the space–time noncommutativity (given by the double commutator $[A_0, A_a]^2$) is of order $O(1)$ only at $t = 0$ (end of expansion) then it decreases at $|t|^{-1.7}$ at large t. The space–space noncommutativity plays a crucial role in the SSB of $SO(9)$ at early times and dynamically disappears at later times (possibly marking the end of inflation) [19].

The extent of time is defined by

$$\Delta = \frac{t_p - t_c}{R(t_c)}, \tag{8.41}$$

where t_p is the instant at which the extent of space becomes maximum which is by the symmetry $t \longrightarrow -t$ ($A_0 \longrightarrow -A_0$) must be zero. It is a dynamical question to show whether or not $\Delta \longrightarrow \infty$ (no big crunch) in the limit $N \longrightarrow \infty$. If Δ does not diverge in the continuum limit then the extent of space has a genuine maximum and as a consequence there will be a recollpase of the Universe.

8.2.4 Role of noncommutativity

The fundamental role played by noncommutativity in the spontaneous symmetry breaking of rotational symmetry, the emergence of an expanding Universe, and the end of inflation can be found discussed in great detail in [17, 19, 21]. This is done by writing down explicit Lie algebra solutions of the classical equations of motion and studying their properties. Here we will mainly follow this discussion.

We start by the simpler situation of the large κ limit. Due to the IR cutoff $\mathrm{Tr}\, A_0^2/N \geqslant \kappa L^2$, the eigenvalues of the configuration A_0 in the limit $\kappa \longrightarrow \infty$ tend to become larger and thus the first term in $\mathrm{Tr}\, F_{\mu\nu}F^{\mu\nu} = -2\,\mathrm{Tr}\, F_{0i}^2 + \mathrm{Tr}\, F_{ij}^2$ becomes a very large negative quantity. As a consequence, the first term $\mathrm{Tr}\, F_{ij}^2$ must become a large positive quantity in order to maintain the condition $\mathrm{Tr}\, F_{\mu\nu}F^{\mu\nu} = 0$[1]. But we also have to remember the other IR cutoff $\mathrm{Tr}\, A_i^2/N = L^2$. Thus we should maximize $\mathrm{Tr}\, F_{ij}^2$ with the constraint $\mathrm{Tr}\, A_i^2/N = L^2$. This should be done more efficiently at $t = 0$ where the first term in $\mathrm{Tr}\, F_{\mu\nu}F^{\mu\nu}$ has its least value (this is why the peak of $R(t)$ at $t = 0$ grows with κ).

The problem now is to minimize the Lagrangian (we set $L = 1$ and λ is a Lagrange multiplier)

$$L = -\frac{1}{4N}\,\mathrm{Tr}\, F_{ij}^2 + \frac{\lambda}{2}\left(\frac{1}{N}\,\mathrm{Tr}\, A_i^2 - 1\right). \tag{8.42}$$

The equations of motion with respect to A_i are

$$[A_j, [A_j, A_i]] = \lambda A_i. \tag{8.43}$$

We consider Lie algebra solutions given by the ansatz

$$A_i = \chi L_i, \quad i \leqslant d, \quad A_j = 0, \quad j \geqslant d + 1. \tag{8.44}$$

The L_i are the generators of a compact semi-simple d-dimensional Lie algebra in a unitary representation. The Jacobi identity guarantees the solution of the equations of motion (it determines the value of the Lagrange multiplier λ). The coefficient χ is determined as $\chi = \sqrt{N/\,\mathrm{Tr}\, L_i^2}$ and as a consequence

$$\mathrm{Tr}\, F_{ij}^2 = \frac{N^2}{(\mathrm{Tr}\, L_i^2)^2}\,\mathrm{Tr}(f_{ijk}\, L_k)^2. \tag{8.45}$$

The maximum of this quantity is achieved for $SU(2)$ Lie algebra (L_i are angular momentum operators in the representation given by the direct sum of spin 1/2 representations and $N - 2$ copies of the trivial representation and $f_{ijk} = \epsilon_{ijk}$) [17]

$$\frac{1}{N}\,\mathrm{Tr}\, F_{ij}^2 = \frac{N}{(\mathrm{Tr}\, L_i^2)^2}\,\mathrm{Tr}(f_{ijk}\, L_k)^2 \leqslant \frac{N}{(\mathrm{Tr}\, L_i^2)^2}\,\mathrm{Tr}(\epsilon_{ijk} L_k)^2 = \frac{4N}{3}. \tag{8.46}$$

However, it can be checked that the spectrum of the $n \times n$ matrix $Q(t) = \sum_i \bar{A}_i^2(t)$ is continuous and therefore the space is actually not a sphere. This classical picture is confirmed in the quantum theory where κ goes to ∞ as $N^{1/4}$.

[1] If we compare here between Lorentzian and Euclidean we find that $\mathrm{Tr}\, F_{\mu\nu}F^{\mu\nu} = 0 \Rightarrow 2\,\mathrm{Tr}\, F_{0i}^2 = \mathrm{Tr}\, F_{ij}^2 \neq 0$ (Lorentzian, noncommutative) and $\mathrm{Tr}\, F_{\mu\nu}F^{\mu\nu} = 0 \Rightarrow 2\,\mathrm{Tr}\, F_{0i}^2 = \mathrm{Tr}\, F_{ij}^2 = 0$ (Euclidean, commutative).

In general we should consider both A_i and A_0 and extremize the action S_B with the constraints Tr A_i^2/N = fixed and Tr A_0^2/N = fixed. We consider then two Lagrange multipliers λ and $\tilde{\lambda}$ and the Lagrangian

$$L = -\frac{1}{4N}\,\mathrm{Tr}[A_\mu, A_\nu][A^\mu, A^\nu] - \frac{\lambda}{2}\left(\frac{1}{N}\,\mathrm{Tr}\,A_i^2 - L^2\right) + \frac{\tilde{\lambda}}{2}\left(\frac{1}{N}\,\mathrm{Tr}\,A_0^2 - \kappa L^2\right). \quad (8.47)$$

The equations of motion are now given by

$$-[A_0, [A_0, A_i]] + [A_j, [A_j, A_i]] - \lambda A_i = 0, \quad [A_j, [A_j, A_0]] - \tilde{\lambda} A_0 = 0. \quad (8.48)$$

Again we look for a solution given by the generators of a compact semi-simple Lie algebra in unitary representations. We look for a d-dimensional solution which is of the form

$$A_i \neq 0, \quad i \leqslant d, \quad A_i = 0, \quad i \geqslant d + 1. \quad (8.49)$$

We assume spatially commutative solutions, viz

$$[A_i, A_j] = 0. \quad (8.50)$$

We will denote the time–space commutator by

$$[A_0, A_i] = iE_i. \quad (8.51)$$

We need to compute $[A_0, E_i] = iF_i$ and $[A_i, E_j] = iG_{ij}$. By substituting in the first equation of motion we get immediately $F_i = \lambda A_i$. By substituting in the second equation of motion we get $G_{jj} = \tilde{\lambda} A_0$. Also, by using the Jacobi identity between A_i, A_j and A_0 we can show that G_{ij} is a symmetric tensor. We split it then into a symmetric traceless tensor M_{ij} and a diagonal part $\delta_{ij}H/d$ with $H = \tilde{\lambda} A_0$. Furthermore, we will assume that M_{ij} is diagonal, viz $M_{ij} = M_i\delta_{ij}$ where $\sum_i M_i = 0$. We have then the commutation relations

$$[A_0, E_i] = i\lambda A_i, \quad [A_i, E_j] = i\delta_{ij}\left(M_i + \frac{\tilde{\lambda}}{d}A_0\right). \quad (8.52)$$

By computing the commutator $[A_0, G_{ij}]$ we get $\delta_{ij}[A_0, M_i] = [E_i, E_j]$. Thus we must have

$$[A_0, M_i] = 0, \quad [E_i, E_j] = 0. \quad (8.53)$$

Next from the two identities $[A_i, G_{jk}] = [A_j, G_{ki}]$ and $[E_i, G_{jk}] = [E_k, G_{ji}]$ and by employing Jacobi identities we derive for $k = j \neq i$ the two commutators

$$[A_i, M_j] = i\frac{\tilde{\lambda}}{d}E_i, \quad [E_i, M_j] = i\frac{\lambda\tilde{\lambda}}{d}A_i. \quad (8.54)$$

These commutators for any i and j are given by

$$[A_i, M_j] = i\frac{\tilde{\lambda}}{d}(1 - d\delta_{ij})E_i, \quad [E_i, M_j] = i\frac{\lambda\tilde{\lambda}}{d}(1 - d\delta_{ij})A_i. \quad (8.55)$$

We can then check that for $i \neq j$ and by using once again the Jacobi identity we have the commutator $[A_k, [E_i, M_j]] = -i\delta_{ik}[M_j, M_k] = 0$. Thus we have the commutator

$$[M_i, M_j] = 0. \tag{8.56}$$

Our spatially commutative solution is then given by the generators A_i, A_0, E_i and M_i satisfying the Lie algebra (8.50)–(8.53), (8.55) and (8.56).

The solution for $d = 2$ corresponds either to $SO(2,2)$ or $SO(4)$ [21]. There is a single solution for $d = 3$ corresponding to the unique 4-dimensional real Lie algebra with $SO(3)$ symmetry [27]. In this solution $\tilde{\lambda} = M_i = 0$ and $E_i = \pm\sqrt{\lambda}A_i$. However, $\lambda \neq 0$ is crucial for this solution to describe the expanding behavior [19]. The solutions with $\lambda = \tilde{\lambda} = 0$ corresponding to a Minkowski spacetime noncommuting with extra dimensions given by fuzzy spheres are given in [28]. The solutions of the corresponding Euclidean equations of motion are given in [29].

Here we consider as a very illustrative explicit solution of the above Lie algebra the simple case of $d = 1$. In this case $M_1 = 0$ and we get the Lie algebra

$$[A_0, A_1] = iE, \quad [A_0, E] = i\lambda A_1, \quad [A_1, E] = i\tilde{\lambda}A_0. \tag{8.57}$$

This is either an $SU(1,1)$ Lie algebra or an $SU(2)$ Lie algebra depending on the signs of λ and $\tilde{\lambda}$. This very simple $d = 1$ solution can also be used to construct higher dimensional solutions. We rotate by an $SO(9)$ transformation the above solution into a solution with only the ith spatial matrix non-zero given by $r_i A_1$. Obviously we must have $r_i^2 = 1$. Let us then consider the general solution [21]

$$A_0' = A_0 \otimes \mathbf{1}_K, \quad A_i' = A_1 \otimes \mathrm{diag}(r_i^{(1)}, \ldots, r_i^{(K)}), \quad \vec{r}^{(j)} \cdot \vec{r}^{(j)} = 1. \tag{8.58}$$

We can make this solution $SO(D)$ symmetric by requiring the vectors $\vec{r}(j)$ to lie on a sphere S^{D-1}, $d = 1, \ldots, 9$. In other words, $A_i' \neq 0$ only for $i \leqslant 4$. The geometry of spacetime is then $R \times S^{D-1}$. We are interested obviously in $SO(4)$ symmetric solutions.

In the $SU(1, 1)$ ($SL(2, R)$) Lie algebra with $\lambda < 0$ and $\tilde{\lambda} < 0$ ($d = 1$ solution (b) of [21]) we have $A_0 = aT_0$, $A_1 = bT_1$, $E = cT_2$ with $a^2 = -\lambda$, $b^2 = -\tilde{\lambda}$ and $ab = c$. Thus

$$[T_0, T_1] = iT_2, \quad [T_2, T_0] = iT_1, \quad [T_1, T_2] = -iT_0. \tag{8.59}$$

We consider the primary unitary series representation (PUSR) of this algebra (with $\epsilon = 0$ and ρ positive) given by [30]

$$\begin{aligned}
(T_0)_{mn} &= n\delta_{mn} \\
(T_1)_{mn} &= -\frac{i}{2}\left(n - i\rho + \frac{1}{2}\right)\delta_{m,\,n+1} - \frac{i}{2}\left(n + i\rho - \frac{1}{2}\right)\delta_{m,\,n-1} \\
(T_2)_{mn} &= -\frac{1}{2}\left(n - i\rho + \frac{1}{2}\right)\delta_{m,\,n+1} - \frac{1}{2}\left(n + i\rho - \frac{1}{2}\right)\delta_{m,\,n-1}
\end{aligned} \tag{8.60}$$

These are infinite dimensional unitary representations (since $SL(2, R)$ is noncompact).

Since $A_0 = aT_0$ is diagonal and $A_1 = bT_1$ is tri-diagonal we can extract the time evolution of the space by considering the $3K \times 3K$ diagonal blocks defined by the 3×3 submatrices $\bar{A}_0(n)$ and $\bar{A}_1(n)$ of A_0 and A_1 given explicitly by

$$\bar{A}_0'(n) = \bar{A}_0(n) \otimes \mathbf{1}_K = a\begin{pmatrix} n-1 & 0 & 0 \\ 0 & n & 0 \\ 0 & 0 & n+1 \end{pmatrix} \otimes \mathbf{1}_K. \tag{8.61}$$

$$\begin{aligned} \bar{A}_i'(n) &= \bar{A}_i(n) \otimes \mathrm{diag}\left(r_i^{(1)}, \dots, r_i^{(K)}\right) \\ &= \frac{ib}{2}\begin{pmatrix} 0 & n + i\rho - \frac{1}{2} & 0 \\ -n + i\rho + \frac{1}{2} & 0 & n + i\rho + \frac{1}{2} \\ 0 & -n + i\rho - \frac{1}{2} & 0 \end{pmatrix} \\ &\otimes \mathrm{diag}\left(r_i^{(1)}, \dots, r_i^{(K)}\right). \end{aligned} \tag{8.62}$$

This is an $SO(4)$ symmetric solution. We think of n as a discrete time and thus the matrices $\bar{A}_i'(n)$ provide the state of the Universe at time n. The space–time noncommutativity disappears in the continuum limit by construction. The extent of space is then defined by

$$R^2(n) = \frac{1}{3K}\,\mathrm{Tr}\,\bar{A}_i'^{\,2}(n). \tag{8.63}$$

We compute immediately

$$R(n) = \sqrt{\frac{b^2}{3}\left(n^2 + \rho^2 + \frac{1}{4}\right)}. \tag{8.64}$$

The continuum limit is defined by $a \longrightarrow 0$ and $\rho \longrightarrow \infty$ such that $t_0 = a\rho$ (present time) and $b/a = \alpha$ are kept fixed. The time t is then defined by $t = na$. We get

$$R(t) = \sqrt{\frac{\alpha^2}{3}(t^2 + t_0^2)}. \tag{8.65}$$

In the continuum limit the above Lie algebra solution becomes commutative $R \times S^3$. The cutoffs L and κ are determined to be $L \sim N/a$ and $\kappa = 1/\alpha$. Thus $L \longrightarrow \infty$ if we send $N \longrightarrow \infty$ faster than $1/a$, whereas κ remains finite in the continuum limit. We identify $R(t)$ with scale factor and compute the Hubble constant

$$\begin{aligned} H(t) &= \frac{\dot{R}(t)}{R(t)} = \frac{t}{t^2 + t_0^2} \\ &= \frac{\alpha}{\sqrt{3}R^2}\sqrt{R^2 - \frac{\alpha^2}{3}t_0^2}. \end{aligned} \tag{8.66}$$

We can also compute the parameter w as

$$\begin{aligned} w &= -\frac{2R}{3}\frac{d\ln H}{dR} - 1 \\ &= -\frac{2t_0^2}{3t^2} - \frac{1}{3}. \end{aligned} \tag{8.67}$$

In particular we compute that w at the present time is given by $w(t_0) = -1$. This value corresponds to a vacuum density, i.e. a cosmological constant, and hence it explains the current observed acceleration of the expansion of the Universe. Since $H = 1/2t_0$ at $t = t_0$ the cosmological constant is given today by $\Lambda \sim H^2 \sim 1/t_0^2$ which explains its smallness. In the future $t \longrightarrow \infty$ we have $w \longrightarrow -1/3$ and $H \longrightarrow 0$. Hence the cosmological constant will vanish in the future according to this model.

8.2.5 Other related work

The main idea behind the Lorentzian matrix model is to use matrix regularization to avoid the Big Bang singularity and to have the Universe with three expanding directions emerge in a phase transition associated with spontaneous symmetry breaking of rotational invariance.

There are so many good ideas out there which try also to use matrices to reproduce cosmology with or without singularity. It is practically impossible to review all these excellent ideas here. But we can mention for example the cosmology from matrix string theory in [36, 37], the cosmology from the BFSS and BFSS-type models in [35, 38–42], the cosmology from IKKT-type models of noncommutative gauge theories and emergent gravity in [43–45], and the cosmology from the IKKT-type models of fuzzy spaces and emergent cosmology in [31–34]. All these intimately related approaches can be called Yang–Mills matrix cosmology or perhaps emergent cosmology.

8.3 Emergent gravity: introductory remarks

8.3.1 Noncommutative electromagnetism is a gravity theory

The idea of emergent gravity from noncommutative gauge theory was first put forward by Rivelles [46]. It was then pursued vigorously by Yang [50, 51] and by Steinacker [47–49]. See also for example [52, 53].

Emergent gravity is one of the most important developments in recent years, if not the most important one, in noncommutative field theory and noncommutative geometry and their underlying matrix models.

In the rest of this chapter we will give a deconstruction, then hopefully a reconstruction, of the emergent gravity approach of Steinacker in which 4-dimensional Einstein (and other) gravity(ies) emerge from the 'mother of all noncommutative geometry': The IKKT Yang–Mills matrix model. Emergent gravity, which was initiated as we just mentioned by Rivelles and Yang, is an approach to quantum gravity, very similar to the AdS/CFT and to the BFSS quantum mechanics, in which we maintain that gravity is equivalent to a gauge theory. More precisely, it states that noncommutative $U(1)$ gauge theory behaves in many respect as a gravitational theory. This is a profound idea which merits systematic pursuing. The work of Steinacker on obtaining the Einstein equation from the IKKT model, which is deconstructed in the remainder of this chapter, is very promising although the calculation is still very semi-classical and not as rigorous as we would like it to be.

In a nutshell, noncommutative Abelian gauge theory, i.e. a noncommutative gauge theory based on a $U(1)$ group, can be reinterpreted as a gravitational theory. In other words, gravity is equivalent to a gauge theory albeit noncommutative. This is exactly in the same spirit as that of the AdS/CFT correspondence which states that gravity is equivalent to a gauge theory albeit conformal and in one lower dimension. In contrast, the equivalence between the emergent gravity and the noncommutative gauge theory occurs in the same number of dimensions (well almost)!

On the other hand, recall that on commutative spaces the cherished Maxwell theory of electromagnetism is a $U(1)$ gauge theory. Thus, in physical terms, the essence of emergent gravity is the statement that gravity is equivalent to noncommutative electromagnetism, and as such it has a dual meaning of being a theory for the (Riemannian, exclusively external) geometry of spacetime as usual, or of being a theory of the (symplectic, internal as well as external) geometry of noncommutative spacetime.

Emergent gravity is not necessarily general relativity but it is a theory of spin 2 field with many interesting effects. And the equivalence between noncommutative electromganetism and emergent gravity is only expected to hold in the semiclassical limit defined here by the limit of noncommutativity of spacetime going to zero. Interestingly enough, if we stick to this limit the emergent metric will also satisfy Einstein equations in vacuum.

There are always issues regarding Lorentz invariance and renormalizability, which can be avoided if we go to two dimensions which might not be very interesting to many people, but the overall picture presented in this scenario is still very compelling.

Thus, electromagnetism on a noncommutative space is actually gravity. We would like to elaborate on this point a little further using the celebrated Seiberg–Witten map. However, before getting into that we will need some background on noncommutative field theory.

First, let us recall that a noncommutative space is a space which implements the Heisenberg uncertainty principle, and thus its coordinates x^μ satisfy the Dirac canonical quantization relations, viz

$$[\hat{x}^\mu, \hat{x}^\nu] = i\theta^{\mu\nu}, \tag{8.68}$$

where $\theta^\mu\nu$ is the noncommutativity parameter. This Heisenberg algebra defines a Hilbert space $\mathcal{H}$ in the usual way.

The plane waves $\exp(ikx)$, the preferred basis in QFT on flat backgrounds, where k is the momentum, becomes plane wave opeartors $\exp(ik\hat{x})$ which satisfy by the Baker–Cambell–Hausdorff formula the relation

$$\exp(ik\hat{x})\exp(ip\hat{x}) = \exp\left(i(k+p)\hat{x} - \frac{i}{2}k\theta p\right). \tag{8.69}$$

This is the torus algebra. This algebra can be mapped back to the commutative plane waves $\exp(ikx)$ by means of the so-called Weyl map, which maps the coordinates operators $\hat{x}$ back to the commutative coordinates x^μ, and by utilizing the so-called Moyal–Weyl star product $*$, in place of the pointwise multiplication of operators, defined for any two functions f and g, by the formula [54]

$$f * g(x) = \exp\left(\frac{i}{2}\theta^{\mu\nu}\partial_\mu^y\partial_\nu^z\right)f(y)g(z)|_{y=z=x}. \tag{8.70}$$

Indeed, we can easily check that the commutative plane waves $\exp(ikx)$ satisfy with the Moyal–Weyl star product the same torus algebra satisfied by the plane wave operators, viz

$$\exp(ikx)* \exp(ipx) = \exp\left(i(k+p)x - \frac{i}{2}k\theta p\right). \tag{8.71}$$

The Weyl map $\hat{x}^\mu \longrightarrow x^\mu$ extends to all operators on the Hilbert space $\hat{O}$ which are mapped to functions O by the relation

$$O(x) = \langle x|\hat{O}|x\rangle, \tag{8.72}$$

where $|x\rangle$ is an appropriate coherent state which provides a basis for the Hilbert space $\mathcal{H}$.

Furthermore, the derivative and the integral in the noncommutative setting are also defined by the almost obvious relations

$$\partial_\mu \longrightarrow \hat{\partial}_\mu = -i(\theta^{-1})_{\mu\nu}\hat{x}^\nu. \tag{8.73}$$

$$\int d^D x \longrightarrow \sqrt{\det(2\pi\theta)}\,\mathrm{Tr}_{\mathcal{H}}. \tag{8.74}$$

We need now to write down dynamics, i.e. action functionals which describe completely the classical behavior of the physical system and also they are crucial ingredients in the description of the corresponding quantum behavior through path integrals.

Towards this end, we can invoke a kind of 'minimal coupling principle', which can be verified explicitly using the above formalism, allowing us to generate correctly how gauge and matter fields couple to the spacetime noncommutativity by the simple rule:

- To obtain noncommutative action functionals and their noncommutative symmetries, we simply replace everywhere in the commutative action functionals and their commutative symmetries, the pointwise multiplication of functions by the Moyal–Weyl star product.

Let us give the example of massless scalar electrodynamics. In other words, electromagnetism coupled to a charged spin 0 particle. The commutative action functional is given by

$$S = -\frac{1}{4g^2}\int d^Dx F_{\mu\nu}F^{\mu\nu} + \frac{1}{2}\int d^Dx (D_\mu\phi)^\dagger(D^\mu\phi). \tag{8.75}$$

The first term is precisely Maxwell action and ϕ is the complex scalar field describing the charged spin 0 particle. The field strength F and the covariant derivative D are given in terms of the electromagnetic (photon) field A^μ by the equations

$$D_\mu = \partial_\mu - iA_\mu, \quad F_{\mu\nu} = i[D_\mu, D_\nu] = \partial_\mu A_\nu - \partial_\nu A_\mu. \tag{8.76}$$

Thus

$$D_\mu\phi = \partial_\mu\phi - iA_\mu\phi. \tag{8.77}$$

This is a gauge theory meaning it is invariant under the following gauge transformations (labeled by the gauge parameter λ)

$$\begin{aligned} A_\mu \longrightarrow A'_\mu &= A_\mu + [D_\mu, \lambda] = A_\mu + \partial_\mu\lambda \\ \phi \longrightarrow \phi' &= \phi + i\lambda\phi \\ \phi^\dagger \longrightarrow \phi'^\dagger &= \phi^\dagger - i\phi^\dagger\lambda. \end{aligned} \tag{8.78}$$

The electromagnetic photon (gauge) field A^μ transforms in the adjoint representation of the gauge group $U(1)$ (similarly for the gauge parameter λ), the complex scalar field ϕ transforms in the fundamental representation of the gauge group $U(1)$ while $\phi^\dagger$ transforms in the anti-fundamental representation.

By invoking the above minimal coupling principle we obtain immediately the noncommutative action functional as (we may also place a hat over the various functions to distinguish them from their commutative counterparts)

$$S = -\frac{1}{4g^2}\int d^Dx \hat{F}_{\mu\nu} * \hat{F}^{\mu\nu} + \frac{1}{2}\int d^Dx (\hat{D}_\mu\hat{\phi})^\dagger * (\hat{D}^\mu\hat{\phi}), \tag{8.79}$$

where

$$\hat{D}_\mu = \partial_\mu - i\hat{A}_\mu * \Rightarrow \hat{D}_\mu\hat{\phi} = \partial_\mu\hat{\phi} - i\hat{A}_\mu * \hat{\phi}. \tag{8.80}$$

$$\hat{F}_{\mu\nu} = i[\hat{D}_\mu, \hat{D}_\nu]_* = \partial_\mu\hat{A}_\nu - \partial_\nu\hat{A}_\mu - i[\hat{A}_\mu, \hat{A}_\nu]_*. \tag{8.81}$$

The last term is very similar to the commutator term in the Yang–Mills gauge theory. As it turns out noncommutative $U(1)$ gauge theory is actually a large $U(N)$ gauge theory in a precise sense [55–57].

The noncommutative gauge transformations are also obtained in the same way, i.e. by replacing everywhere pointwise multiplication of functions by the Moyal–Weyl star product. We get

$$\begin{aligned} \hat{A}_\mu \longrightarrow \hat{A}'_\mu &= \hat{A}_\mu + [\hat{D}_\mu, \hat{\lambda}]_* = \hat{A}_\mu + \partial_\mu\hat{\lambda} - i[\hat{A}_\mu, \hat{\lambda}]_* \\ \hat{\phi} \longrightarrow \hat{\phi}' &= \hat{\phi} + i\hat{\lambda} * \hat{\phi} \\ \hat{\phi}^\dagger \longrightarrow \hat{\phi}'^\dagger &= \hat{\phi}^\dagger - i\hat{\phi}^\dagger * \hat{\lambda}. \end{aligned} \tag{8.82}$$

This minimal coupling principle is the first indication that noncommutative geometry behaves somehow similarly to gravity in the sense that all fields, regardless of their charges, will couple to the noncommutativity parameter in this way, i.e. via the same prescription of replacing pointwise multiplication of functions by the Moyal–Weyl star product. Consider for example a real scalar field Φ in the adjoint representation of the gauge group $U(1)$. In the commutative setting, this field because it is neutral it cannot couple to the gauge field A^{μ}. In the noncommutative setting there is a coupling given by the action

$$S = \frac{1}{2}\int d^D x(\hat{D}_{\mu}\hat{\Phi}) * (\hat{D}^{\mu}\hat{\Phi}). \tag{8.83}$$

Since $\hat{\Phi}$ is in the adjoint representation its covariant derivative and its gauge transformation are given respectively by

$$\hat{D}_{\mu}\hat{\Phi} = \partial_{\mu}\hat{\Phi} - i[\hat{A}_{\mu}, \hat{\Phi}]_{*}. \tag{8.84}$$

$$\hat{\Phi} \longrightarrow \hat{\Phi}' = \hat{\Phi} + i[\hat{\lambda}, \hat{\Phi}]_{*}. \tag{8.85}$$

The star product reflecting spacetime noncommutativity allows therefore all fields to couple to the spacetime symplectic geometry in the same way. This is very reminiscent of the equivalence principle of general relativity.

8.3.2 Seiberg–Witten map

In order to exhibit the hidden gravity in noncommutative $U(1)$ gauge theory we apply the Seiberg–Witten map [58, 59]. We start with noncommutative Moyal–Weyl $U(1)$ gauge theory coupled to a complex scalar field in the fundamental representation given by the action

$$S = -\frac{1}{4g^2}\int d^D x\hat{F}_{\mu\nu} * \hat{F}^{\mu\nu} + \frac{1}{2}\int d^D x(\hat{D}_{\mu}\hat{\phi})^{\dagger} * (\hat{D}^{\mu}\hat{\phi}). \tag{8.86}$$

The star $U(1)$ gauge transformations are given explicitly by

$$\hat{A}_{\mu} \longrightarrow \hat{A}_{\mu}^{'} = \hat{A}_{\mu} + \partial_{\mu}\hat{\lambda} - i[\hat{A}_{\mu}, \hat{\lambda}]_{*}, \quad \hat{\phi} \longrightarrow \hat{\phi}^{'} = \hat{\phi} + i\hat{\lambda} * \hat{\phi}. \tag{8.87}$$

Following Seiberg and Witten we will now construct an explicit map between the noncommutative vector potential $\hat{A}_{\mu}$ and a commutative vector potential A_{μ} which will implement explicitly the perturbative equivalence of the above noncommutative gauge theory to a conventional gauge theory. Clearly, this map must depend both on the gauge parameter as well as on the vector potential in order to be able to achieve equivalence between the physical orbits in the two theories. Also, since the gauge field is coupled to a scalar field, the noncommutative scalar field will also be mapped to a conventional scalar field. We write then

$$\begin{aligned} \hat{A}_{\mu}(A + \delta_{\lambda}A) &= \hat{A}_{\mu}(A) + \delta_{\hat{\lambda}}\hat{A}_{\mu}(A) \\ \hat{\phi}(A + \delta_{\lambda}A, \phi + \delta_{\lambda}\phi) &= \hat{\phi}(A, \phi) + \delta_{\hat{\lambda}}\hat{\phi}(A, \phi). \end{aligned} \tag{8.88}$$

$$\delta_{\hat{\lambda}}\hat{A}_\mu = \partial_\mu\hat{\lambda} - i[\hat{A}_\mu, \hat{\lambda}]_*, \quad \delta_{\hat{\lambda}}\hat{\phi} = i\hat{\lambda} * \hat{\phi}. \tag{8.89}$$

$$\delta_\lambda A_\mu = \partial_\mu\lambda, \quad \delta_\lambda\phi = i\lambda\phi. \tag{8.90}$$

To solve the above equation we write

$$\hat{A}^\mu = A^\mu + A^{\mu'}(A), \quad \hat{\lambda} = \lambda + \lambda'(\lambda, A). \tag{8.91}$$

It reduces then to

$$\begin{aligned} A'_\mu(A + \delta A) - A'_\mu(A) - \partial_\mu\lambda' &= -i[A_\mu, \lambda]_* \\ &= \theta^{\alpha\beta}\partial_\alpha A_\mu \partial_\beta\lambda. \end{aligned} \tag{8.92}$$

We have

$$A^{\mu'}(A + \delta A) = A^{\mu'}(A) + \frac{\delta}{\delta A_\beta}A^{\mu'} \cdot \partial_\beta\lambda + \frac{\delta}{\delta(\partial_\nu A_\beta)}A^{\mu'} \cdot \partial_\nu\partial_\beta\lambda. \tag{8.93}$$

$A\mu'$ and λ' must be both of order θ. Furthermore, by thinking along the lines of a derivative expansion, we know that $A^{\mu\prime}$ is quadratic in A^μ of the form $A\partial A$, whereas λ' is linear in A^μ. The last equation above also suggests that λ' is proportional to $\partial_\alpha\lambda$. For constant A the above condition reduces then to

$$\begin{aligned} A_\mu(A + \delta A) - A_\mu(A) - \partial_\mu\lambda' = 0 \Rightarrow &\frac{\delta}{\delta(\partial_\nu A_\beta)}A'_\mu \cdot \partial_\nu\partial_\beta\lambda \\ &+ \frac{1}{2}\theta^{\alpha\beta}\partial_\mu\partial_\beta\lambda \cdot A_\alpha = 0, \end{aligned} \tag{8.94}$$

where we had set

$$\lambda' = \frac{1}{2}\theta^{\alpha\beta}\partial_\alpha\lambda A_\beta. \tag{8.95}$$

The coefficient 1/2 is fixed by the requirement that the second derivative in λ cancels. The condition becomes

$$\frac{1}{2}\frac{\delta}{\delta(\partial_\nu A_\beta)}A'_\mu + \frac{1}{2}\frac{\delta}{\delta(\partial_\beta A_\nu)}A'_\mu + \frac{1}{4}\theta^{\alpha\beta}\eta^\nu_\mu A_\alpha + \frac{1}{4}\theta^{\alpha\nu}\eta^\beta_\mu A_\alpha = 0. \tag{8.96}$$

A solution is given by

$$A'_\mu = -\frac{1}{2}\theta^{\alpha\beta}A_\alpha(2\partial_\beta A_\mu - \partial_\mu A_\beta). \tag{8.97}$$

It is now very easy to verify that this solves the Seiberg–Witten condition also for non-constant A. We do now the same for the scalar field. We write

$$\hat{\phi} = \phi + \phi'(A, \phi). \tag{8.98}$$

We get immediately the condition

$$\begin{aligned}\phi'(A+\delta A,\phi+\delta\phi)-\phi'(A,\phi)&=i\hat{\lambda}*\hat{\phi}-\delta_\lambda\phi\\&=i\lambda\phi'+\frac{i}{2}\theta^{\mu\nu}\partial_\mu\lambda A_\nu\phi-\frac{1}{2}\theta^{\mu\nu}\partial_\mu\lambda\partial_\nu\phi.\end{aligned}\tag{8.99}$$

Again, we note that ϕ' is of order θ and it must be proportional to $A\partial\phi$. It is not difficult to convince ourselves that the solution is given by

$$\phi'=-\frac{1}{2}\theta^{\mu\nu}A_\mu\partial_\nu\phi.\tag{8.100}$$

We compute the expression of the action in the new variables. We compute first

$$\begin{aligned}\hat{F}_{\mu\nu}&=\partial_\mu\hat{A}_\nu-\partial_\nu\hat{A}_\mu-i[\hat{A}_\mu,\hat{A}_\nu]_*\\&=F_{\mu\nu}+\partial_\mu A'_\nu-\partial_\nu A'_\mu+\theta^{\alpha\beta}\partial_\alpha A_\mu\partial_\beta A_\nu\\&=F_{\mu\nu}+\theta^{\alpha\beta}F_{\mu\alpha}F_{\nu\beta}-\theta^{\alpha\beta}A_\alpha\partial_\beta F_{\mu\nu}.\end{aligned}\tag{8.101}$$

We get immediately the action

$$\begin{aligned}S_F&=-\frac{1}{4g^2}\int d^Dx\hat{F}_{\mu\nu}\hat{F}^{\mu\nu}\\&=-\frac{1}{4g^2}\int d^Dx[F_{\mu\nu}F^{\mu\nu}+2\theta^{\alpha\beta}F_\beta{}^\nu(F_\alpha{}^\sigma F_{\sigma\nu}+\frac{1}{4}\eta_{\nu\alpha}F^{\rho\sigma}F_{\rho\sigma})].\end{aligned}\tag{8.102}$$

Also we compute the covariant derivative

$$\begin{aligned}\hat{D}_\mu\hat{\phi}&=\partial_\mu\hat{\phi}-i\hat{A}_\mu*\hat{\phi}\\&=D_\mu\phi-\frac{1}{2}\theta^{\alpha\beta}A_\alpha\partial_\beta D_\mu\phi-\frac{\theta^{\alpha\beta}}{2}D_\alpha\phi F_{\beta\mu}.\end{aligned}\tag{8.103}$$

$$\begin{aligned}(\hat{D}_\mu\hat{\phi})^\dagger&=\partial_\mu\hat{\phi}^\dagger+i\hat{\phi}^\dagger*\hat{A}_\mu\\&=(D_\mu\phi)^\dagger-\frac{1}{2}\theta^{\alpha\beta}A_\alpha\partial_\beta(D_\mu\phi)^\dagger-\frac{\theta^{\alpha\beta}}{2}(D_\alpha\phi)^\dagger F_{\beta\mu}.\end{aligned}\tag{8.104}$$

The charged scalar action becomes

$$\begin{aligned}S_\phi&=\frac{1}{2}\int d^Dx(\hat{D}_\mu\hat{\phi})^\dagger*(\hat{D}^\mu\hat{\phi})\\&=\frac{1}{2}\int d^Dx\Bigg[(D_\mu\phi)^\dagger(D^\mu\phi)\\&\quad-\frac{1}{2}\left(\theta^{\mu\alpha}F_\alpha{}^\nu+\theta^{\nu\alpha}F_\alpha{}^\mu+\frac{1}{2}\eta^{\mu\nu}\theta^{\alpha\beta}F_{\alpha\beta}\right)(D_\mu\phi)^\dagger(D_\nu\phi)\Bigg].\end{aligned}\tag{8.105}$$

The main observation of Rivelles [46] is that we can rewrite the above θ-expanded actions of the noncommutative $U(1)$ gauge field $\hat{A}_\mu$ and the noncommutative charged scalar field $\hat{\phi}$ as a coupling of a commutative $U(1)$ gauge field A_μ and a commutative charge scalar field ϕ to a metric $g_{\mu\nu} = \eta_{\mu\nu} + h_{\mu\nu} + \eta_{\mu\nu}h$ with $h_{\mu\nu}\eta^{\mu\nu} = 0$. This metric itself is determined by the commutative $U(1)$ gauge field A_μ and the noncommutativity structure $\theta_\mu\nu$. Indeed, the dynamics of a commutative Maxwell field A_μ and the charged scalar field ϕ in a linearized gravitational field is given by the actions

$$\begin{aligned} S_F &= -\frac{1}{4g^2}\int d^Dx\sqrt{-g}F_{\mu\nu}F^{\mu\nu} \\ &= -\frac{1}{4g^2}\int d^Dx[F_{\mu\nu}F^{\mu\nu} + 2h^{\mu\nu}F_\mu{}^\rho F_{\rho\nu}]. \end{aligned} \tag{8.106}$$

$$\begin{aligned} S_\phi &= \frac{1}{2}\int \sqrt{-g}\,d^Dx(D_\mu\phi)^\dagger * (D^\mu\phi) \\ &= \frac{1}{2}\int d^Dx\left[(D_\mu\phi)^\dagger(D^\mu\phi) - h^{\mu\nu}(D_\mu\phi)^\dagger(D_\nu\phi) + 2h(D_\mu\phi)^\dagger(D^\mu\phi)\right]. \end{aligned} \tag{8.107}$$

We get immediately the traceless metric

$$h^{\mu\nu} = \frac{1}{2}(\theta^{\mu\beta}F_\beta{}^\nu + \theta^{\nu\beta}F_\beta{}^\mu) + \frac{1}{4}\theta^{\alpha\beta}F_{\alpha\beta}\eta^{\mu\nu}. \tag{8.108}$$

The gauge field has in this setting a dual role. It couples minimally to the charged scalar field as usual but also it sources the gravitational field. In other words, the gravitational field is not just a background field since it is determined by the dynamical gauge field. We write the interval

$$ds^2 = (1 + \frac{1}{4}\theta^{\alpha\beta}F_{\alpha\beta})dx_\mu dx^\mu + \theta^{\nu\beta}F_{\beta\mu}dx_\mu dx^\nu. \tag{8.109}$$

We can compute the Riemann tensor, the Ricci tensor and the Ricci scalar and find them of order one, two and two in θ, respectively. Thus, to the linear order in θ, the Riemann tensor is non-zero while the Ricci tensor and the Ricci scalar are zero. In other words, the above metric cannot describe a flat spacetime. In fact it describes a gravitational plane wave since at zero order in θ we must have ordinary electromagnetism and thus $F_\mu\nu$ (and hence the metric $h_\mu\nu$) depends on the plane wave $\exp(ikx)$ with $k^2 = 0$. At first order a plane wave solution $A_\mu = F_\mu \exp(ikx)$ with $k^2 = 0$ and $k^\mu F_\mu = 0$ can also be constructed explicitly [46].

On the other hand, we can see that this metric describes a spacetime with a covariantly constant symplectic form θ which is also null, since to the linear order in θ we must have the equations $D_\mu\theta^{\alpha\beta} = 0$ and $\theta_{\mu\nu}\theta^{\mu\nu} = 0$, and as a consequence this spacetime is indeed a pp-wave spacetime [60], which is precisely a gravitational plane wave as we have checked explicitly above to the first order in θ.

8.4 Fuzzy spheres and fuzzy CP^n

The topics of this section are slightly off the main line of development of this chapter and may be skipped.

8.4.1 Co-adjoint orbits

Fuzzy spaces and their field theories and fuzzy physics are discussed for example in [61–68]. Fuzzy spaces are finite dimensional approximations to the algebra of functions on continuous manifolds which preserve the isometries and (super) symmetries of the underlying manifolds. Thus by construction the corresponding field theories contain a finite number of degrees of freedom. The basic and original motivation behind fuzzy spaces is non-perturbative regularization of quantum field theory similar to the familiar technique of lattice regularization. See for example [69, 70].

Another very important motivation lies in the fact that string theory suggests that spacetime may be fuzzy and noncommutative at its fundamental level [71–73]. For older and other more recent motivations see [75–77].

It is well established that the specification of fuzzy spaces requires the language of Connes' noncommutative geometry. In particular, following Connes [79] and Fröhlich and Gawędzki [78], the geometry of a Riemannian manifold $\mathcal{M}$ can be reconstructed from the so-called spectral triple $(\mathcal{A}, \mathcal{H}, \mathcal{D})$ where $\mathcal{A} = C^{\infty}(\infty)$ is the algebra of smooth bounded functions on the manifold, $\mathcal{H}$ is the Hilbert space of square-integrable spinor functions on $\mathcal{M}$, and $\mathcal{D}$ is the Dirac operator on $\mathcal{M}$ which encodes all the information about the metric aspects of the manifold $\mathcal{M}$. In the case of the absence of spinors the Dirac operator can be replaced by the Laplace–Beltrami operator Δ on $\mathcal{M}$. Similarly, a fuzzy space will be given by a sequence of triples

$$(\mathrm{Mat}_{d_L}, \mathcal{H}_L, \Delta_L), \tag{8.110}$$

where Mat_{d_L} is the algebra of $d_L \times d_L$ Hermitian matrices with inner product $(A, B) = \mathrm{Tr}\, A^{+}B/d_L$, $\mathcal{H}_L = \mathbf{C}^{d_L^2}$ is the Hilbert space of the d_L-dimensional matrix algebra Mat_{d_L}, and Δ_L is an appropriate Laplacian acting on these matrices which encodes in a precise sense the metric aspects of the fuzzy space. For example, the dimension of the space is given by the growth of the number of eigenvalues of the Laplacian. In the limit $L \longrightarrow \infty$ we obtain the spectral triple $(\mathcal{A}, \mathcal{H}, \Delta)$ associated with the corresponding commutative manifold. The fuzzy Laplacian Δ_L is typically a truncated version of the commutative Laplacian Δ with the same isometries and symmetries. The fuzzy geometry can be mapped to an algebra of functions with an appropriate star product by constructing the corresponding Weyl map.

Obviously, the corresponding fuzzy field theories are non-perturbatively regularized since they contain a finite number of degrees of freedom given precisely by ${d_L}^2$ with the correct limiting commutative behavior. The fuzzy approach is thus the most natural one in Monte Carlo simulations of noncommutative field theories or in Monte Carlo simulations of commutative field theories where (super)symmetry plays a crucial role.

Construction of fuzzy spaces by quantizing compact symplectic manifolds, via geometric quantization [80–83], is equivalent to the construction of a quantum Hilbert space from a classical phase space. More explicitly, quantization is the construction of a correspondence between the algebra of Poisson brackets, represented by real functions generating canonical transformations on the phase space or symplectic manifold, and the algebra of commutators, represented by Hermitian operators generating irreducible unitary transformations on the Hilbert space. The irreducibility of the representation is equivalent, in the context of geometric quantization, to the holomorphic or polarization condition. For example, in the usual one-dimensional quantum mechanics given by $[\hat{x}, \hat{p}] = i\hbar$ the wave functions depend only on x and not x and p. In general, within geometric quantization where (1) we consider a prequantum line bundle on the phase space with curvature given by the symplectic 2-form and (2) impose a complex structure on the phase space in which the symplectic 2-form is identified as a Kähler form, the resulting Hilbert space is given by sections of a polarized line bundle satisfying a holomorphic condition. This is intimately related to the Borel–Weil–Bott theorem [80–83] which states that all unitary irreducible representations of a compact Lie group G are realized by holomorphic sections of a complex line bundle on a coset space G/T, where T is the maximal torus of G, G/T is a Kähler manifold, and where the group G acts on holomorphic sections by right translations.

Examples of compact symplectic manifolds are the so-called co-adjoint or adjoint orbits of compact semi-simple Lie groups G which can be geometrically quantized by quantizing their underlying symplectic 2-forms when they satisfy the Dirac quantization condition. Co-adjoint orbits are coset spaces $G/H_t = \{gtg^{-1}: g \in G, t \in \underline{G}\}$ where $\underline{G}$ is the Lie algebra of G and $H_t \subset G$ is the stabilizer of t. The fuzzy co-adjoint orbits corresponding to G/H_t are such that their Hilbert spaces consist of holomorphic sections of a complex line bundle over G/H_t associated with unitary irreducible representations of G. The fuzzy coset spaces constructed so far satisfying (8.110) are fuzzy complex projective spaces, which are mostly degenerate co-adjoint orbits with dimensions $\dim G/H_t = \dim G - \dim H_t$, given by

$$\mathbf{CP}^k = G/H_t = SU(k+1)/U(k). \tag{8.111}$$

Another class is given by flag manifolds in which H coincides with the maximal torus H of G and the dimension of the co-adjoint orbits becomes maximal given by $\dim G/H_t = \dim G - \text{rank } G$. See for example [84].

In the remainder of this section we will discuss further fuzzy $\mathbf{CP}^k$ and write down their Yang–Mills matrix models. The four-dimensional case of fuzzy $\mathbf{S}^2 \times \mathbf{S}^2$ is discussed in [85–90] while the case of fuzzy $\mathbf{CP}^2$ is also discussed in [91–97]. The physics on a single fuzzy sphere is studied for example in [103–110, 113–115]. For related topics see also [98–102, 111, 112]. Fuzzy projective spaces are examples of co-adjoint orbits and homogeneous spaces which, among many other properties, admit an underlying symplectic structure. The existence of fuzzy co-adjoint orbits and similar fuzzy spaces satisfying (8.110) relies on the single fact that the corresponding symplectic structure is quantizable. For recent work on fuzzy physics in two and four dimensions see [116–119, 148]. As a concert example here we will consider in the following the case of fuzzy $\mathbf{CP}^2$ (and fuzzy $\mathbf{CP}^k$) and more importantly the case of fuzzy $\mathbf{S}^4$ which is relevant to the emergence of Einstein gravity from the IKKT matrix model.

8.4.2 Fuzzy projective space $\mathbf{CP}^2$

Let T_a, $a = 1, \ldots, 8$, be the generators of $SU(3)$ in the symmetric irreducible representation $(n, 0)$ of dimension $N = \frac{1}{2}(n+1)(n+2)$. They satisfy

$$[T_a, T_b] = if_{abc}T_c \tag{8.112}$$

and

$$T_a^2 = \frac{1}{3}n(n+3) \equiv |n|^2, \quad d_{abc}T_aT_b = \frac{2n+3}{6}T_c. \tag{8.113}$$

Let $t_a = \lambda_a/2$, where λ_a are the usual Gell–Mann matrices, be the generators of $SU(3)$ in the fundamental representation (1,0) of dimension $N = 3$. They also satisfy

$$\begin{aligned} 2t_at_b &= \frac{1}{3}\delta_{ab} + (d_{abc} + if_{abc})t_c \\ tr_3 t_at_b &= \frac{1}{2}\delta_{ab}, \quad tr_3 t_at_bt_c = \frac{1}{4}(d_{abc} + if_{abc}). \end{aligned} \tag{8.114}$$

The N-dimensional generator T_a can be obtained by taking the symmetric product of n copies of the fundamental 3-dimensional generator t_a, viz

$$T_a = (t_a \otimes \mathbf{1} \otimes \cdots \otimes \mathbf{1} + \mathbf{1} \otimes t_a \otimes \cdots \otimes \mathbf{1} + \cdots + \mathbf{1} \otimes \mathbf{1} \otimes \cdots \otimes t_a)_{\text{symmetric}}. \tag{8.115}$$

The commutative $\mathbf{CP}^2$ is the space of all unit vectors $|\psi\rangle$ in $\mathbf{C}^3$ modulo the phase. Thus $e^{i\theta}|\psi\rangle$, for all $\theta \in [0, 2\pi[$, define the same point on $\mathbf{CP}^2$. It is obvious that all these vectors $e^{i\theta}|\psi\rangle$ correspond to the same projector $P = |\psi\rangle\langle\psi|$. Hence $\mathbf{CP}^2$ is the space of all projection operators of rank one on $\mathbf{C}^3$. Let $\mathbf{H}_N$ and $\mathbf{H}_3$ be the Hilbert spaces of the $SU(3)$ representations $(n, 0)$ and $(1, 0)$, respectively. We will define fuzzy $\mathbf{CP}^2$ through the canonical $SU(3)$ coherent states as follows. Let $\vec{n}$ be a vector in $\mathbf{R}^8$ and we define the projector

$$P_3 = \frac{1}{3}\mathbf{1} + n_at_a. \tag{8.116}$$

The requirement $P_3^2 = P_3$ leads to the condition that $\vec{n}$ is a point on $\mathbf{CP}^2$ satisfying the equations

$$[n_a, n_b] = 0, \quad n_a^2 = \frac{4}{3}, \quad d_{abc}n_an_b = \frac{2}{3}n_c. \tag{8.117}$$

We can write

$$P_3 = |\vec{n}, 3\rangle\langle 3, \vec{n}|. \tag{8.118}$$

We think of $|\vec{n}, 3\rangle$ as the coherent state in $\mathbf{H}_3$, level 3×3 matrices, which is localized at the point $\vec{n}$ of $\mathbf{CP}^2$. Therefore, the coherent state $|\vec{n}, N\rangle$ in $\mathbf{H}_N$, level $N \times N$ matrices, which is localized around the point $\vec{n}$ of $\mathbf{CP}^2$ is defined by the projector

$$P_N = |\vec{n}, N\rangle\langle N, \vec{n}| = (P_3 \otimes P_3 \otimes \cdots \otimes P_3)_{\text{symmetric}}. \tag{8.119}$$

We compute that

$$tr_3 t_a P_3 = \langle \vec{n}, 3|t_a|\vec{n}, 3\rangle = \frac{1}{2}n_a, \quad tr_N T_a P_N = \langle \vec{n}, N|T_a|\vec{n}, N\rangle = \frac{n}{2}n_a. \tag{8.120}$$

Hence it is natural to identify fuzzy $\mathbf{CP}^2$ at level $N = \frac{1}{2}(n+1)(n+2)$ by the coordinate operators

$$x_a = \frac{2}{n}T_a. \tag{8.121}$$

They satisfy

$$[x_a, x_b] = \frac{2i}{n}f_{abc}x_c, \quad x_a^2 = \frac{4}{3}\left(1 + \frac{3}{n}\right), \quad d_{abc}x_a x_b = \frac{2}{3}\left(1 + \frac{3}{2n}\right)x_c. \tag{8.122}$$

Therefore, in the large N limit we can see that the algebra of x_a reduces to the continuum algebra of n_a. Hence $x_a \longrightarrow n_a$ in the commutative limit $N \longrightarrow \infty$.

The algebra of functions on fuzzy $\mathbf{CP}^2$ is identified with the algebra of $N \times N$ matrices Mat_N generated by all polynomials in the coordinate operators x_a. Recall that $N = \frac{1}{2}(n+1)(n+2)$. The left action of $SU(3)$ on this algebra is generated by $(n, 0)$, whereas the right action is generated by $(0, n)$. Thus the algebra Mat_N decomposes under the action of $SU(3)$ as

$$(n, 0) \otimes (0, n) = \otimes_{p=0}^{n}(p, p). \tag{8.123}$$

A general function on fuzzy $\mathbf{CP}^2$ is therefore written as

$$F = \sum_{p=0}^{n} F^{(p)}_{I^2, I_3, Y} T^{(p, p)}_{I^2, I_3, Y} \tag{8.124}$$

$T^{(p, p)}_{I^2, I_3, Y}$ are $SU(3)$ polarization tensors in the irreducible representation (p, p). I^2, I_3 and Y are the square of the isospin, the third component of the isospin and the hypercharge quantum numbers which characterize $SU(3)$ representations.

The derivations on fuzzy $\mathbf{CP}^2$ are defined by the commutators $[T_a, ..]$. The Laplacian is then obviously given by $\Delta_N = [T_a, [T_a, \ldots]]$. Fuzzy $\mathbf{CP}^2$ is completely determined by the sequence of spectral triples $(\mathrm{Mat}_N, \Delta_N, \mathbf{H}_N)$.

8.4.3 Tangent projective module on fuzzy $\mathbf{CP}^2$

We will introduce fuzzy gauge fields A_a, $a = 1, \ldots, 8$, through the covariant derivatives D_a, $a = 1, \ldots, 8$, as follows

$$D_a = T_a + A_a. \tag{8.125}$$

The D_a are $N \times N$ Hermitian matrices which transform covariantly under the action of $U(N)$. In order for the field $\vec{A}$ to be a $U(1)$ gauge field on fuzzy $\mathbf{CP}^2$ it must satisfy some additional constraints so that only four of its components are non-zero. These are the tangent components to $\mathbf{CP}^2_n$. The other four components of $\vec{A}$ are normal to $\mathbf{CP}^2_n$ and in general they will be projected out from the model.

Let us go back to the commutative $\mathbf{CP}^2$ and let us consider a gauge field A_a, $a = 1, \ldots, 8$, which is strictly tangent to $\mathbf{CP}^2$. By construction this gauge field must satisfy

$$A_a = P^T_{ab}A_b, \quad P^T = (n_a Adt_a)^2. \tag{8.126}$$

The P^T is the projector which defines the tangent bundle over $\mathbf{CP}^2$. The normal bundle over $\mathbf{CP}^2$ will be defined by the projector $P^N = 1 - P^T$. Explicitly these are given by

$$P^T_{ab} = n_c n_d (Adt_c)_{ae}(Adt_d)_{eb} = n_c n_d f_{cae} f_{dbe}, \quad P^N_{ab} = \delta_{ab} - n_c n_d f_{cae} f_{dbe}. \tag{8.127}$$

In the above we have used the fact that the generators in the adjoint representation (1,1) satisfy $(Adt_a)_{bc} = -if_{abc}$. Note that we have the identities $n_a P^T_{ab} = n_b P^T_{ab} = 0$. Hence the condition (8.126) takes the natural form

$$n_a A_a = 0. \tag{8.128}$$

This is one condition which allows us to reduce the number of independent components of A_a by one. We know that there must be three more independent constraints which the tangent field A_a must satisfy since it has only four independent components. To find them we start from the identity

$$d_{abk}d_{cdk} = \frac{1}{3}\left[\delta_{ac}\delta_{bd} + \delta_{bc}\delta_{ad} - \delta_{ab}\delta_{cd} + f_{cak}f_{dbk} + f_{dak}f_{cbk}\right]. \tag{8.129}$$

Thus

$$n_c n_d d_{abk} d_{cdk} = \frac{2}{3}[n_a n_b - \frac{2}{3}\delta_{ab} + n_c n_d f_{cak} f_{dbk}]. \tag{8.130}$$

By using the fact that $d_{cdk} n_c n_d = \frac{2}{3} n_k$ we obtain

$$d_{abk} n_k = n_a n_b - \frac{2}{3}\delta_{ab} + n_c n_d f_{cak} f_{dbk}. \tag{8.131}$$

Hence it is a straightforward calculation to find that the gauge field A_a must also satisfy the conditions

$$d_{abk} n_k A_b = \frac{1}{3} A_a. \tag{8.132}$$

In the case of $\mathbf{S}^2$ the projector P^T takes the simpler form $P^T_{ab} = \delta_{ab} - n_a n_b$ and hence $P^N_{ab} = n_a n_b$. From equation (8.131) we have on $\mathbf{CP}^2$

$$P^T_{ab} = d_{abc} n_c - n_a n_b + \frac{2}{3}\delta_{ab}, \quad P^N_{ab} = -d_{abc} n_c + n_a n_b + \frac{1}{3}\delta_{ab}. \tag{8.133}$$

As it turns out, the constraint (8.132) already contains (8.128). In other words, it contains exactly the correct number of equations needed to project out the gauge field A_a onto the tangent bundle of $\mathbf{CP}^2$. Let us also say that given any commutative gauge field A_a which does not satisfy the constraints (8.128) and (8.132) we can always make it tangent by applying the projector P^T. Thus we will have the tangent gauge field

$$A_a^T = P_{ab}^T A_b = d_{abc} n_c A_b - n_a(n_b A_b) + \frac{2}{3} A_a. \tag{8.134}$$

Similarly, the fuzzy gauge field must satisfy some conditions which should reduce to (8.128) and (8.132). As it turns out, constructing a tangent fuzzy gauge field using an expression like (8.126) is a highly non-trivial task due to (1) gauge covariance problems and (2) operator ordering problems. However, implementing (8.128) and (8.132) in the fuzzy setting is quite easy since we will only need to return to the covariant derivatives D_a and require them to satisfy the $SU(3)$ identities (8.113), viz

$$\begin{aligned} D_a^2 &= \frac{1}{3} n(n+3) \\ d_{abc} D_a D_b &= \frac{2n+3}{6} D_c. \end{aligned} \tag{8.135}$$

So D_a are almost the $SU(3)$ generators except that they fail to satisfy the fundamental commutation relations of $SU(3)$ given by equation (8.112). This failure is precisely measured by the curvature of the gauge field A_a, namely

$$\begin{aligned} F_{ab} &= i[D_a, D_b] + f_{abc} D_c \\ &= i[T_a, A_b] - i[T_b, A_a] + f_{abc} A_c + i[A_a, A_b]. \end{aligned} \tag{8.136}$$

This has the correct commutative limit which is clearly given by the usual curvature on $\mathbf{CP}^2$, viz by $F_{ab} = i\mathcal{L}_a A_b - i\mathcal{L}_b A_a + f_{abc} A_c + i[A_a, A_b]$.

8.4.4 Yang–Mills matrix models for fuzzy $\mathbf{CP}^k$

Next, we need to write down actions on fuzzy $\mathbf{CP}^2$. The first piece is the usual Yang–Mills action

$$S_{\rm YM} = \frac{1}{4g^2 N} \mathrm{Tr}\, F_{ab}^2. \tag{8.137}$$

By construction it has the correct commutative limit.

The second piece in the action is a potential term which has to implement the constraints (8.135) in some limit. Indeed, we will not impose these constraints rigidly on the path integral but we will include their effects by adding to the action a very special potential term. In other words, we will not assume that D_a satisfy (8.135). To the end of writing this potential term we will introduce the four normal scalar fields on fuzzy $\mathbf{CP}_n^2$ by the equations (see equations (8.135))

$$\Phi = \frac{1}{n}\left(D_a^2 - \frac{1}{3}n(n+3)\right) = \frac{1}{2} x_a A_a + \frac{1}{2} A_a x_a + \frac{1}{n} A_a^2 \longrightarrow n_a A_a, \tag{8.138}$$

and

$$\begin{aligned}\Phi_c &= \frac{1}{n}\left(d_{abc}D_aD_b - \frac{2n+3}{6}D_c\right) \\ &= \frac{1}{2}d_{abc}x_aA_b + \frac{1}{2}d_{abc}A_ax_b - \frac{2n+3}{6n}A_c + \frac{1}{n}d_{abc}A_aA_b \\ &\longrightarrow d_{abc}n_aA_b - \frac{1}{3}A_c.\end{aligned} \tag{8.139}$$

We add to the Yang–Mills action the potential term

$$V = \frac{M_0^2}{N}\,\mathrm{Tr}\,\Phi^2 + \frac{M^2}{N}\,\mathrm{Tr}\,_N\Phi_a^2. \tag{8.140}$$

In the limit where the parameters M_0^2 and M^2 are taken to be very large positive numbers we can see that only configurations A_a, or equivalently D_a, such that $\Phi = 0$ and $\Phi_c = 0$ dominate the path integral which is precisely what we want.

The total action is then given by

$$\begin{aligned}S &= \frac{1}{2g^2}Tr_N F_{ab}^2 + \beta\,\mathrm{Tr}\,_N\Phi^2 + M^2\,\mathrm{Tr}\,_N\Phi_a^2 \\ &= \frac{1}{g^2N}\,\mathrm{Tr}\left[-\frac{1}{4}[D_a, D_b]^2 + if_{abc}D_aD_bD_c\right] \\ &+ \frac{3n}{4g^2N}\,\mathrm{Tr}\,\Phi + \frac{M_0^2}{N}Tr_N\Phi^2 + \frac{M^2}{N}Tr_N\Phi_a^2.\end{aligned} \tag{8.141}$$

This is the desired Yang–Mills matrix model in which fuzzy $\mathbf{CP}^2$ is described as a noncommutative brane solution of the equations of motion. To obtain the corresponding Yang–Mills matrix models for fuzzy $\mathbf{CP}^k$ we simply replace the $SU(3)$ constants f and d by their $SU(k+1)$ values and extend the indices $a, b, c, \ldots$ from 1 to $k(k+2)$. The case of the sphere is much simpler since $d = 0$ for $SU(2)$. We obtain in this case (with $f = \epsilon$)

$$\begin{aligned}S &= \frac{1}{g^2N}\,\mathrm{Tr}\left[-\frac{1}{4}[D_a, D_b]^2 + i\epsilon_{abc}D_aD_bD_c\right] \\ &+ \frac{3n}{4g^2N}\,\mathrm{Tr}\,\Phi + \frac{M_0^2}{N}\,\mathrm{Tr}\,\Phi^2 + \frac{M^2}{N}\frac{(2n+3)^2}{36n}\,\mathrm{Tr}\,\Phi.\end{aligned} \tag{8.142}$$

This action will be studied in great detail in subsequent sections. Extension of this action to fuzzy $\mathbf{S}^2 \times \mathbf{S}^2$ and higher cartesian products of the fuzzy sphere is straightforward.

8.4.5 Coherent states

We will closely follow [99]. Here Latin indices refer to $SU(k+1)$, viz $a, b = 1, \ldots, k(k+2)$. This section can be skipped by experts.

Classical $\mathbf{CP}^k$ can be given by the projector

$$P = \frac{1}{k+1}\mathbf{1} + \alpha_k n^a t_a. \tag{8.143}$$

The requirement that $P^2 = P$ will lead to the three equations

$$\begin{aligned} \vec{n}^2 &= 1 \\ d_{abc}n^a n^b &= \frac{2}{\alpha_k}\frac{k-1}{k+1}n^c, \\ \alpha_k &= \pm\sqrt{\frac{2k}{k+1}}. \end{aligned} \tag{8.144}$$

This defines $\mathbf{CP}^k$ as embedded in $\mathbf{R}^{k+2}$. First, let us specialize the projector (8.143) to the 'north' pole of $\mathbf{CP}^k$:

$$\vec{n}_0 = (0, 0, \dots, 1). \tag{8.145}$$

We have then the projector

$$P_0 = \frac{1}{k+1}\mathbf{1} + \alpha_k t_{k(k+2)}. \tag{8.146}$$

Now, by using the result

$$t_{k(k+2)} = \frac{1}{\sqrt{2k(k+1)}}\mathrm{diag}(1, 1, \dots, 1, -k), \tag{8.147}$$

we get

$$P_0 = \mathrm{diag}(0, 0, \dots, 1), \tag{8.148}$$

if we choose the minus sign for α_N, namely

$$\alpha_k = -\sqrt{\frac{2k}{k+1}}. \tag{8.149}$$

So at the 'north' pole, our projector projects down to the state

$$|\psi_0\rangle = (0, 0, \dots, 1) \tag{8.150}$$

of the Hilbert space $\mathbf{C}^{k+1}$ on which the defining representation of $SU(k+1)$ is acting.

A general point $\vec{n} \in \mathbf{CP}^k$ can be obtained from $\vec{n}_0$ by the action of a certain element $g \in SU(k+1)$

$$\vec{n} = g\vec{n}_0. \tag{8.151}$$

P will then project down to the state

$$|\psi\rangle = g|\psi_0\rangle \tag{8.152}$$

of $\mathbf{C}^{k+1}$. One can show that

$$P = |\psi\rangle\langle\psi| = g|\psi_0\rangle\langle\psi_0|g^+ = gP_0g^+, \tag{8.153}$$

provided

$$gt_{k(k+2)}g^+ = n^a t_a. \tag{8.154}$$

This last equation is the usual definition of $\mathbf{CP}^k$. Under $g \longrightarrow gh$ where $h \in U(k)$ we have $ht_{k(k+2)}h^+ = t_{k(k+2)}$, i.e. $U(k)$ is the stability group of $t_{k(k+2)}$ and hence

$$\mathbf{CP}^k = SU(k+1)/U(k). \tag{8.155}$$

Points $\vec{n}$ of $\mathbf{CP}^k$ are then equivalent classes $[g] = [gh]$, $h \in U(k)$.

In the case of $SU(2)$, fuzzy $\mathbf{S}^2_N$ is the algebra of operators generated by the orbital angular momenta L_i, $i = 1,2,3$, where $[L_i, L_j] = i\epsilon_{ijk}L_k$, and $\sum_{i=1}^3 L_i^2 = l(l+1)$. Since these operators define the IRR l of $SU(2)$, fuzzy $\mathbf{S}^2$ will act on the Hilbert space $H_l^{(2)}$, which is the $d_l^{(2)} = (2l+1)$-dimensional irreducible representation of $SU(2)$, i.e. $N = 2l + 1$. This representation can be obtained from the symmetric product of $2l$ fundamental representations $\mathbf{2}$ of $SU(2)$. Given an element $g \in SU(2)$, its l-representation matrix $U^{(l)}(g)$ is given as follows

$$U^{(l)}(g) = U^{(\mathbf{2})}(g) \otimes_s \cdots \otimes_s U^{(\mathbf{2})}(g), \; 2l\text{–times}. \tag{8.156}$$

$U^{(\mathbf{2})}(g)$ is the spin $\frac{1}{2}$ representation of $g \in SU(2)$.

Similarly, fuzzy $\mathbf{CP}^k$ is the algebra of all operators which act on the Hilbert space $H_l^{(k+1)}$, where $H_l^{(k+1)}$ is the $d_l^{(k)}$-dimensional irreducible representation of $SU(k+1)$ obtained from the symmetric product of 21 fundamental representations $\mathbf{N}$ of $SU(N)$, where

$$d_l^{(k)} = \frac{(k+2l)!}{k!(2l)!}. \tag{8.157}$$

Remark that for $l = \frac{1}{2}$ we have $d_{\frac{1}{2}}^{(k)} = k + 1$ and therefore $H_{\frac{1}{2}}^{(k+1)} = \mathbf{C}^{k+1}$ is the fundamental representation of $SU(k+1)$.

Clearly, the states $|\psi_0\rangle$ and $|\psi\rangle$ of $H_{\frac{1}{2}}^{(k+1)}$, given by equations (8.150) and (8.152), will correspond in $H_l^{(k+1)}$ to the states $|\vec{n}_0, l\rangle$ and $|\vec{n}, l\rangle$ respectively, so that $|\psi_0\rangle = |\vec{n}_0, \frac{1}{2}\rangle$ and $|\psi\rangle = |\vec{n}, \frac{1}{2}\rangle$. Equation (8.152) becomes

$$|\vec{n}, l\rangle = U^{(l)}(g)|\vec{n}_0, l\rangle. \tag{8.158}$$

$U^{(l)}(g)$, where $g \in SU(k+1)$, is the representation given by

$$U^{(l)}(g) = U^{(\mathbf{k+1})}(g) \otimes_s \cdots \otimes_s U^{(\mathbf{k+1})}(g), \; 2l - \text{times}. \tag{8.159}$$

To any operator $\hat{F}$ on $H_l^{(k+1)}$, which can be thought of as a fuzzy function on fuzzy $\mathbf{CP}^k$, we associate a 'classical' function $F_l(\vec{n})$ on a classical $\mathbf{CP}^k$ by

$$F_l(\vec{n}) = \langle\vec{n}, l|\hat{F}|\vec{n}, l\rangle, \tag{8.160}$$

such that the product of two such operators $\hat{F}$ and $\hat{G}$ is mapped to the star product of the corresponding two functions by the relation

$$F_l * G_l(\vec{n}) = \langle \vec{n}, l | \hat{F}\hat{G} | \vec{n}, l \rangle. \tag{8.161}$$

8.4.6 Star product

This is a very long calculation which I would like to do once and for all [99]. First we use the result that any operator $\hat{F}$ on the Hilbert space $H_l^{(N)}$ admits the expansion

$$\hat{F} = \int_{SU(k+1)} d\mu(h) \tilde{F}(h) U^{(l)}(h), \tag{8.162}$$

where $U^{(l)}(h)$ are taken to satisfy the normalization

$$\mathrm{Tr}\, U^{(l)}(h) U^{(l)}(h') = d_l^{(N)} \delta(h^{-1} - h'). \tag{8.163}$$

Using the above two equations, one can derive the value of the coefficient $\tilde{F}(h)$ to be

$$\tilde{F}(h) = \frac{1}{d_l^{(k+1)}} \mathrm{Tr}\, \hat{F} U^{(l)}(h^{-1}). \tag{8.164}$$

Using the expansion (8.162) in (8.160) we get

$$F_l(\vec{n}) = \int_{SU(k+1)} d\mu(h) \tilde{F}(h) \omega^{(l)}(\vec{n}, h), \quad \omega^{(l)}(\vec{n}, h) = \langle \vec{n}, l | U^{(l)}(h) | \vec{n}, l \rangle. \tag{8.165}$$

On the other hand, using the expansion (8.162) in (8.161) will give

$$F_l * G_l(\vec{n}) = \int \int_{SU(k+1)} d\mu(h) d\mu(h') \tilde{F}(h) \tilde{G}(h') \omega^{(l)}(\vec{n}, hh'). \tag{8.166}$$

The computation of this star product boils down to the computation of $\omega^{(l)}(\vec{n}, hh')$. We have

$$\begin{aligned} \omega^{(l)}(\vec{n}, h) &= \langle \vec{n}, l | U^{(l)}(h) | \vec{n}, l \rangle \\ &= \left[\left\langle \vec{n}, \frac{1}{2} \right| \otimes_s \cdots \otimes_s \left\langle \vec{n}, \frac{1}{2} \right| \right] [U^{(k+1)}(h) \otimes_s \cdots \otimes_s U^{(k+1)}(h)] \\ &\quad \left[\left| \vec{n}, \frac{1}{2} \right\rangle \otimes_s \cdots \otimes_s \left| \vec{n}, \frac{1}{2} \right\rangle \right] \\ &= [\omega^{(\frac{1}{2})}(\vec{n}, h)]^{2l}, \end{aligned} \tag{8.167}$$

where

$$\omega^{(\frac{1}{2})}(\vec{n}, h) = \left\langle \vec{n}, \frac{1}{2} \right| U^{(k+1)}(h) \left| \vec{n}, \frac{1}{2} \right\rangle = \langle \psi | U^{(k+1)}(h) | \psi \rangle. \tag{8.168}$$

In the fundamental representation $\mathbf{k+1}$ of $SU(k+1)$ we have $U^{(\mathbf{k+1})}(h)= \exp(im^a t_a) = c(m)\mathbf{1} + is^a(m)t_a$ and therefore

$$\omega^{(\frac{1}{2})}(\vec{n}, h) = \langle\psi|c(m)\mathbf{1} + is^a(m)t_a|\psi\rangle = c(m) + is^a(m)\langle\psi|t_a|\psi\rangle, \tag{8.169}$$

where

$$\begin{aligned}\omega^{(\frac{1}{2})}(\vec{n}, hh') &= \langle\psi|U^{(\mathbf{N})}(hh')|\psi\rangle \\ &= \langle\psi|(c(m)\mathbf{1} + is^a(m)t_a)(c(m')\mathbf{1} + is^a(m')t_a)|\psi\rangle \\ &= c(m)c(m') + i[c(m)s^a(m') + c(m')s^a(m)]\langle\psi|t_a|\psi\rangle \\ &\quad - s^a(m)s^b(m')\langle\psi|t_a t_b|\psi\rangle.\end{aligned} \tag{8.170}$$

Now, it is not difficult to check that

$$\begin{aligned}\langle\psi|t_a|\psi\rangle &= \mathrm{Tr}\ t_a P = \frac{\alpha_k}{2}n^a \\ \langle\psi|t_a t_b|\psi\rangle &= \mathrm{Tr}\ t_a t_b P = \frac{1}{2(k+1)}\delta_{ab} + \frac{\alpha_k}{4}(d_{abc} + if_{abc})n^c.\end{aligned} \tag{8.171}$$

Hence, we obtain

$$\begin{aligned}\omega^{(\frac{1}{2})}(\vec{n}, h) &= c(m) + i\frac{\alpha_k}{2}\vec{s}(m)\cdot\vec{n} \\ \omega^{(\frac{1}{2})}(\vec{n}, hh') &= c(m)c(m') - \frac{1}{2(k+1)}\vec{s}(m)\cdot\vec{s}(m') + i\frac{\alpha_k}{2}[c(m)s^a(m') \\ &\quad + c(m')s^a(m)]n^a \\ &\quad - \frac{\alpha_k}{4}(d_{abc} + if_{abc})n^c s^a(m)s^b(m').\end{aligned} \tag{8.172}$$

These two last equations can be combined to get the pre-final result

$$\begin{aligned}\omega^{(\frac{1}{2})}(\vec{n}, hh') - \omega^{(\frac{1}{2})}(\vec{n}, h)\omega^{(\frac{1}{2})}(\vec{n}, h') = &- \frac{1}{2(k+1)}\vec{s}(m).\ \vec{s}(m') \\ &- \frac{\alpha_k}{4}(d_{abc} + if_{abc})n^c s^a(m)s^b(m') \\ &+ \frac{\alpha_k^2}{4}n^a n^b s_a(m)s_b(m').\end{aligned} \tag{8.173}$$

We can remark that in this last equation, we have got rid of all reference to c's. We would like also to get rid of all reference to s's. This can be achieved by using the formula

$$s_a(m) = \frac{2}{i\alpha_N}\frac{\partial}{\partial n^a}\omega^{(\frac{1}{2})}(\vec{n}, h). \tag{8.174}$$

By using this formula, we get the final result

$$\omega^{(\frac{1}{2})}(\vec{n}, hh') - \omega^{(\frac{1}{2})}(\vec{n}, h)\omega^{(\frac{1}{2})}(\vec{n}, h') = K_{ab}\frac{\partial}{\partial n^a}\omega^{(\frac{1}{2})}(\vec{n}, h)\frac{\partial}{\partial n^b}\omega^{(\frac{1}{2})}(\vec{n}, h'), \quad (8.175)$$

where

$$K_{ab} = \frac{2}{(k+1)\alpha_k^2}\delta_{ab} - n_a n_b + \frac{1}{\alpha_k}(d_{abc} + if_{abc})n^c. \quad (8.176)$$

Therefore

$$\begin{aligned} F_l * G_l(\vec{n}) &= \int\int_{SU(k+1)} d\mu(h)d\mu(h')\tilde{F}(h)\tilde{G}(h')\omega^{(l)}(\vec{n}, hh') \\ &= \int\int_{SU(k+1)} d\mu(h)d\mu(h')\tilde{F}(h)\tilde{G}(h')[\omega^{(\frac{1}{2})}(\vec{n}, hh')]^{2l}. \end{aligned} \quad (8.177)$$

More explicitly

$$\begin{aligned} F_l * G_l(\vec{n}) &= \int_{SU(k+1)} d\mu(h)d\mu(h')\tilde{F}(h)\tilde{G}(h') \\ &\times \sum_{q=0}^{2l} \frac{(2l)!}{q!(2l-q)!} K_{a_1b_1}\ldots K_{a_qb_q}[\omega^{(\frac{1}{2})}(\vec{n}, h)]^{2l-q} \\ &\times [\omega^{(\frac{1}{2})}(\vec{n}, h')]^{2l-q}\frac{\partial}{\partial n^{a_1}}\omega^{(\frac{1}{2})}(\vec{n}, h)\ldots \\ &\times \frac{\partial}{\partial n^{a_q}}\omega^{(\frac{1}{2})}(\vec{n}, h)\frac{\partial}{\partial n^{b_1}}\omega^{(\frac{1}{2})}(\vec{n}, h')\ldots\frac{\partial}{\partial n^{b_q}}\omega^{(\frac{1}{2})}(\vec{n}, h') \\ &= \sum_{q=0}^{2l} \frac{(2l)!}{q!(2l-q)!} K_{a_1b_1}\ldots K_{a_qb_q}\int_{SU(k+1)} d\mu(h)\tilde{F}(h) \\ &\times \left[\omega^{(\frac{1}{2})}(\vec{n}, h)\right]^{2l-q}\frac{\partial}{\partial n^{a_1}}\omega^{(\frac{1}{2})}(\vec{n}, h)\ldots\frac{\partial}{\partial n^{a_q}}\omega^{(\frac{1}{2})}(\vec{n}, h) \\ &\times \int_{SU(k+1)} d\mu(h')\tilde{G}(h')[\omega^{(\frac{1}{2})}(\vec{n}, h')]^{2l-q} \\ &\times \frac{\partial}{\partial n^{b_1}}\omega^{(\frac{1}{2})}(\vec{n}, h')\ldots\frac{\partial}{\partial n^{b_q}}\omega^{(\frac{1}{2})}(\vec{n}, h'). \end{aligned} \quad (8.178)$$

Next, we use the formula

$$\begin{aligned} \frac{(2l-q)!}{(2l)!}\frac{\partial}{\partial n^{a_1}}\ldots\frac{\partial}{\partial n^{a_q}}F_l(\vec{n}) &= \int_{SU(k+1)} d\mu(h)\tilde{F}(h)[\omega^{(\frac{1}{2})}(\vec{n}, h)]^{2l-q} \\ &\times \frac{\partial}{\partial n^{a_1}}\omega^{(\frac{1}{2})}(\vec{n}, h)\ldots\frac{\partial}{\partial n^{a_q}}\omega^{(\frac{1}{2})}(\vec{n}, h) \end{aligned} \quad (8.179)$$

to get the final result

$$F_l * G_l(\vec{n}) = \sum_{q=0}^{2l} \frac{(2l-q)!}{q!(2l)!} K_{a_1b_1} \ldots K_{a_qb_q} \frac{\partial}{\partial n^{a_1}} \ldots \times \frac{\partial}{\partial n^{a_q}} F_j(\vec{n}) \frac{\partial}{\partial n^{b_1}} \ldots \frac{\partial}{\partial n^{b_q}} G_j(\vec{n}). \tag{8.180}$$

8.4.7 Fuzzy derivatives

Derivations on $\mathbf{CP}^k$ are generated by the vector fields

$$\mathcal{L}_a = -if_{abc}n_b\frac{\partial}{\partial n_c}, \quad [\mathcal{L}_a, \mathcal{L}_b] = if_{abc}\mathcal{L}_c. \tag{8.181}$$

The corresponding adjoint action on the Hilbert space $H_l^{(N)}$ is generated by L_a, $[L_a, L_b] = if_{abc}L_c$, and is given by

$$\langle \vec{n}, l|U^{(l)}(h^{-1})\hat{F}U^{(l)}(h)|\vec{n}, l\rangle = \langle \vec{n}_0, l|U^{(l)}(g^{-1}h^{-1})\hat{F}U^{(l)}(hg)|\vec{n}_0, l\rangle, \tag{8.182}$$

where we have used equation (8.158), and such that $U^{(l)}(h)$ is given by $U^{(l)}(h) = \exp(i\eta_a L_a)$. Now if we take η to be small, then one computes

$$\langle \vec{n}, l|U^{(l)}(h)|\vec{n}, l\rangle = 1 + i\eta_a\langle \vec{n}, l|L_a|\vec{n}, l\rangle. \tag{8.183}$$

On the other hand, we know that the representation $U^{(l)}(h)$ is obtained by taking the symmetric product of $2l$ fundamental representations $\mathbf{k+1}$ of $SU(k+1)$, and hence

$$\begin{aligned}\langle \vec{n}, l|U^{(l)}(h)|\vec{n}, l\rangle &= \left(\left\langle \vec{n}, \frac{1}{2}\right| 1 + i\eta_a t_a \left| \vec{n}, \frac{1}{2}\right\rangle\right)^{2l} \\ &= 1 + i(2l)\eta_a\left\langle \vec{n}, \frac{1}{2} \mid t_a \mid \vec{n}, \frac{1}{2}\right\rangle \\ &= 1 + i(2l)\eta_a\frac{\alpha_k}{2}n_a,\end{aligned} \tag{8.184}$$

where we have used the facts, $L_a = t_a \otimes_s \ldots \otimes_s t_a, |\vec{n}, l\rangle = |\vec{n}, \frac{1}{2}\rangle \otimes_s \ldots \otimes_s |\vec{n}, \frac{1}{2}\rangle$, and the first equation of (8.171). Hence we get the important result

$$\langle \vec{n}, l|L_a|\vec{n}, l\rangle = l\alpha_k n_a. \tag{8.185}$$

We should define the fuzzy derivative $[L_a, \hat{F}]$ by

$$(\mathcal{L}_a F)_l(\vec{n}) = \langle \vec{n}, l|[L_a, \hat{F}]|\vec{n}, l\rangle. \tag{8.186}$$

Indeed, we have

$$\begin{aligned}\langle \vec{n}, l|[L_a, \hat{F}]|\vec{n}, l\rangle &= \langle \vec{n}, l|L_a\hat{F}|\vec{n}, l\rangle - \langle \vec{n}, l|\hat{F}L_a|\vec{n}, l\rangle \\ &= l\alpha_k(n_a * F_l(\vec{n}) - F_l * n_a(\vec{n})).\end{aligned} \tag{8.187}$$

But we can compute

$$n_a * F_l(\vec{n}) = n_a F_l(\vec{n}) + \frac{1}{2l} K_{ab} \frac{\partial}{\partial n^b} F_l(\vec{n}), \quad F_l * n_a(\vec{n})$$
$$= n_a F_l(\vec{n}) + \frac{1}{2l} K_{ba} \frac{\partial}{\partial n^b} F_l(\vec{n}). \tag{8.188}$$

In other words,

$$n_a * F_l(\vec{n}) - F_l * n_a(\vec{n}) = \frac{1}{2l}(K_{ab} - K_{ba}) \frac{\partial}{\partial n^b} F_l(\vec{n})$$
$$= \frac{1}{2l}\left(\frac{2i}{\alpha_k} f_{abc} n^c\right) \frac{\partial}{\partial n^b} F_l(\vec{n}). \tag{8.189}$$

Therefore

$$\langle \vec{n}, l | [L_a, \hat{F}] | \vec{n}, l \rangle = i f_{abc} n^c \frac{\partial}{\partial n^b} F_l(\vec{n}) = (\mathcal{L}_a F_l)(\vec{n}). \tag{8.190}$$

8.5 Fuzzy $\mathbf{S}_N^4$: symplectic and Poisson structures

8.5.1 The spectral triple and fuzzy $\mathbf{CP}_N^k$: another look

In order to avoid the string theory landscape it has been argued in [47, 49–52] that gravity should emerge in the IKKT or IIB matrix model from the noncommutative physics of 4-dimensional brane solutions and not from the 10-dimensional physics of the bulk. On the other hand, although noncommutative gauge theories behave similarly to gravity theories they are generically different from Einstein theory [46, 120, 121].

We consider covariant noncommutative spaces such as fuzzy $\mathbf{S}_N^4$ which is a compact Euclidean version of Snyder space [75, 76]. Indeed, there are extra generators here denoted by $\theta^{\mu\nu}$ and $\mathcal{P}^\mu$, and the non-commutativity $\theta^{\mu\nu}$ is not central here as opposed to the DFR theory [77] which is the source of gravity [48]. However, as in the case of DFR quantum spacetime the non-commutativity θ is averaged over the extra dimensions, which is here a fuzzy $\mathbf{S}_N^2$, in order to recover $SO(5)$ invariance.

We will follow [48, 122–124, 127]. Very closely related constructions are found in [70, 99, 128–133]. See also [134–137].

The fuzzy four-sphere $\mathbf{S}_N^4$, similarly to all fuzzy spaces, is specified by a sequence of triples [78, 79]

$$\mathcal{M}_n = (\mathcal{H}, \mathcal{A}, \Delta). \tag{8.191}$$

All fuzzy spaces, the fuzzy four-sphere $\mathbf{S}_N^4$ included, are given in terms of matrix algebras. Thus, the algebra $\mathcal{A}$ is the algebra of $d_n \times d_n$ matrices with the obvious inner product

$$\langle M, N \rangle = \frac{1}{d_n} \operatorname{Tr} M^+ N. \tag{8.192}$$

In other words,

$$\mathcal{A} = \mathrm{Mat}_{d_n}. \tag{8.193}$$

And

$$\mathcal{H} = H_n = \mathbf{C}^{d_n} \tag{8.194}$$

is the Hilbert space on which the algebra of matrices acts in a natural way, whereas $\Delta = \Delta_n$ is an appropriate Laplacian acting the matrices.

The data contained in the above triple, which defines the fuzzy space, can be specified completely by giving a scalar action on the fuzzy space.

Fuzzy $\mathbf{S}^4_N$ is really fuzzy $\mathbf{CP}^3_N$. More precisely, $\mathbf{CP}^3$ is an $\mathbf{S}^2$ bundle over $\mathbf{S}^4$. Since $\mathbf{CP}^3$ is a coadjoint orbit given by $\mathbf{CP}^3 = SU(4)/U(3)$ it can be subjected to fuzzification by quantization in the usual way to obtain a matrix approximation which is fuzzy $\mathbf{CP}^3_N$.

The space $\mathbf{CP}^k = SU(k+1)/U(k)$ can be thought of as a brane surface embedded in $\mathbf{R}^{k(k+2)}$. The fundamental representation of $SU(k+1)$ is $(k+1)$-dimensional and is denoted $\mathbf{k+1}$. Let Λ_μ, $\mu = 1, \ldots, k(k+2)$, be the generators of $SU(k+1)$ in the fundamental representation. These can be given by the Gell–Mann matrices $t_\mu = \lambda_\mu/2$ satisfying

$$\begin{aligned} [t_\mu, t_\nu] = if_{\mu\nu\lambda}t_\lambda, \quad 2t_\mu t_\nu &= \frac{1}{k+1}\delta_{\mu\nu} \\ + (d_{\mu\nu\lambda} + if_{\mu\nu\lambda})t_\lambda, \quad \mathrm{Tr}\, t_\mu t_\nu &= \frac{1}{2}\delta_{\mu\nu}. \end{aligned} \tag{8.195}$$

We take the n-fold symmetric tensor product of the fundamental representation of $SU(k+1)$ to obtain the d_n^k-dimensional irreducible representation of $SU(k+1)$, viz

$$\mathbf{d}_n^k = (\mathbf{k+1} \otimes \cdots \otimes \mathbf{k+1})_{\mathrm{sym}}, \quad n\text{–times}. \tag{8.196}$$

It is not difficult to show that

$$d_n^k = \frac{(k+n)!}{k!n!}. \tag{8.197}$$

The dimension of the space is given by

$$\text{dimension} = 2 \cdot \lim_{n\longrightarrow\infty}\frac{\ln d_n^k}{\ln n} = 2k. \tag{8.198}$$

It is immediately seen that for fuzzy $\mathbf{CP}^1_N = \mathbf{S}^2_N$ we obtain $d_n^1 = n+1$ and thus $n/2$ is the spin quantum number characterizing the irreducible representations of $SU(2)$. Let T_μ be the generators of $SU(k+1)$ in the d_n^k-dimensional irreducible representation and $-T_\mu^R$ be the generators in the complex conjugate representation, viz

$$T_\mu = (t_\mu \otimes \mathbf{1} \otimes \cdots \otimes \mathbf{1} + \mathbf{1} \otimes t_\mu \otimes \cdots \otimes \mathbf{1} + \mathbf{1} \otimes \mathbf{1} \otimes \cdots \otimes t_\mu)_{\mathrm{sym}}, \quad n\text{–times}. \tag{8.199}$$

We know that functions on fuzzy $\mathbf{CP}^k_N$ are matrices in $\mathrm{Mat}_{d_n^k}$ which transforms under the action of $SU(k+1)$ as the tensor product $\mathbf{d}_n^k \otimes \bar{\mathbf{d}}_n^k$ and thus can be expanded in terms of $SU(k+1)$ polarization tensors. These representations can be found as follows. First, we note that

$$\mathbf{d}_n^k = (n, 0, \ldots, 0), \quad \bar{\mathbf{d}}_n^k = (0, 0, \ldots, n). \tag{8.200}$$

Then, we use the result for $SU(k+1)$, from [138], that

$$\begin{aligned} &(n_1, n_2, \ldots, n_k) \otimes (n, 0, \ldots, 0) = \oplus(b_1, b_2, \ldots, b_k) \\ &b_i = n_i + c_i - c_{i+1}, c_{i+1} \leqslant n_i, \quad 1 \leqslant i \leqslant k, \end{aligned} \tag{8.201}$$

where c_i's are non-negative integers satisfying $c_1 + c_2 + \cdots + c_{k+1} = n$. We compute immediately that

$$(0, 0, \ldots, n) \otimes (n, 0, \ldots, 0) = \oplus(c_1, 0, \ldots, 0, c_1). \tag{8.202}$$

Thus for $\mathbf{CP}^3$ or $SU(4)$ we have

$$(0, 0, n) \otimes (n, 0, 0) = \oplus(c_1, 0, c_1). \tag{8.203}$$

These representations exist only for $c_1 = 0, \ldots, n$ with corresponding dimensions [138]

$$\dim(c_1, 0, c_1) = \frac{1}{12}(c_1+1)^2(c_1+2)^2(2c_1+3). \tag{8.204}$$

Hence, functions on fuzzy $\mathbf{CP}^k_N$ are given by $N \times N$ matrices M in Mat_N where

$$N \equiv d_n^k = \frac{(k+n)!}{k!n!}, \tag{8.205}$$

with permitted $SU(k+1)$ representations $(l, 0, \ldots, 0, l)$, with k factors, such that $l \leqslant n$. The commutative limit is given by $n \longrightarrow \infty$ where

The Laplacian on fuzzy $\mathbf{CP}^k_N$ is given by

$$\Delta_{n,k} = (\mathrm{Ad}\, T_\mu)^2 = [T_\mu, [T_\mu, \ldots]]. \tag{8.206}$$

We need also to find the coordinate operators on fuzzy $\mathbf{CP}^k_N$. We consider the tensor product

$$(n, 0, \ldots, 0) \otimes (1, 0, \ldots, 0) = (n+1, 0, \ldots, 0) \oplus (n-1, 1, 0, \ldots, 0). \tag{8.207}$$

Then we consider the intertwiner on the above vector space given by the operator [133]

$$2X = 2T_\mu t_\mu = (T_\mu + t_\mu)^2 - T_\mu^2 - t_\mu^2. \tag{8.208}$$

For simplicity we consider $SU(4)$ first. The Casimir operators and the dimensions of $SU(4)$ irreducible representations with highest weight $\Lambda = (\lambda_1, \lambda_2, \lambda_3)$ are given by

$$C_2(\Lambda) = \frac{1}{8}\lambda_1(3\lambda_1 + 2\lambda_2 + \lambda_3 + 12) + \frac{1}{4}\lambda_2(\lambda_1 + 2\lambda_2 + \lambda_3 + 8) \\ + \frac{1}{8}\lambda_3(\lambda_1 + 2\lambda_2 + 3\lambda_3 + 12). \tag{8.209}$$

$$\dim(\Lambda) = \frac{1}{12}(\lambda_1 + 1)(\lambda_2 + 1)(\lambda_3 + 1)(\lambda_1 + \lambda_2 + 2) \\ \times (\lambda_2 + \lambda_3 + 2)(\lambda_1 + \lambda_2 + \lambda_3 + 3). \tag{8.210}$$

The relevant Casimirs are

$$T_\mu^2 = C_2(n, 0, 0) = \frac{3n(n + 4)}{8}. \tag{8.211}$$

$$t_\mu^2 = C_2(1, 0, 0) = \frac{15}{8}. \tag{8.212}$$

$$(T_\mu + t_\mu)^2 = C_2(n + 1, 0, 0) \\ = \frac{3(n + 1)(n + 5)}{8}, (T_\mu + t_\mu)^2 \\ = C_2(n - 1, 1, 0) = \frac{1}{8}(3n^2 + 10n + 7). \tag{8.213}$$

The eigenvalues of X are therefore given by

$$2X|_{(n+1, 0, 0)} = \frac{3n}{4}, \quad 2X|_{(n-1, 1, 0)} = -\frac{n}{4} - 1. \tag{8.214}$$

The characteristic equation for X is given by

$$\left(2X - \frac{3n}{4}\right)\left(2X + \frac{n}{4} + 1\right) = 0. \tag{8.215}$$

The generalization to $SU(k + 1)$ is immediately given by

$$\left(2X - \frac{kn}{k + 1}\right)\left(2X + \frac{n}{k + 1} + 1\right) = 0 \Rightarrow X^2 = \frac{1}{4}\frac{kn}{k + 1}\left(\frac{n}{k + 1} + 1\right) \\ + \frac{1}{2}(\frac{kn}{k + 1} - \frac{n}{k + 1} - 1)T_\lambda t_\lambda. \tag{8.216}$$

On the other hand, we compute

$$X^2 = \frac{1}{2(k + 1)}T_\mu^2 + \frac{1}{2}(d_{\mu\nu\lambda} + if_{\mu\nu\lambda})T_\mu T_\nu t_\lambda. \tag{8.217}$$

By identification, we get

$$T_\mu^2 = \frac{kn}{2(k+1)}(n+k+1). \tag{8.218}$$

$$d_{\mu\nu\lambda}T_\mu T_\nu + if_{\mu\nu\lambda}T_\mu T_\nu = \frac{1}{k+1}((k-1)n-k-1)T_\lambda. \tag{8.219}$$

The adjoint representation of $SU(4)$ is (1,0,1) and has dimension and Casimir given by 15 and $C_2^{\text{ad}} = 4$, respectively. For $SU(N)$ the adjoint representation is $(1,0,\ldots,0,1)$ with dimension and Casimir equal to $k(k+2)$ and $C_2^{\text{ad}} = k+1$. The generators in the adjoint representation are given by $(\text{ad}\ t_\mu)_{\alpha\beta} = -if_{\mu\alpha\beta}$. Thus we have

$$\begin{aligned} if_{\mu\nu\alpha}T_\mu T_\nu &= -\frac{1}{2}f_{\mu\nu\alpha}f_{\mu\nu\beta}\,T_\beta \\ &= -\frac{1}{2}(\text{ad}\ t_\mu\ \text{ad}\ t_\mu)_{\alpha\beta}T_\beta \\ &= -\frac{1}{2}C_2^{\text{ad}}T_\alpha \Rightarrow if_{\mu\nu\alpha}T_\mu T_\nu = -\frac{1}{2}(k+1)T_\alpha. \end{aligned} \tag{8.220}$$

In other words, the other defining equations of fuzzy $\mathbf{CP}_N^k$ are given by

$$d_{\mu\nu\lambda}T_\mu T_\nu = (k-1)\left(\frac{n}{k+1}+\frac{1}{2}\right)T_\lambda. \tag{8.221}$$

8.5.2 Fuzzy $\mathbf{S}_N^4$

Now we construct fuzzy $\mathbf{S}_N^4$. Although $\mathbf{S}^4$ does not admit a symplectic structure.

Let Γ_i, $i = 1, \ldots, 4$, be the Dirac matrices in four dimensions and let $\Gamma_5 = \Gamma_1\Gamma_2\Gamma_3\Gamma_4$. Collectively we write them Γ_a, $a = 1, \ldots, 5$, and they satisfy $\{\Gamma_a, \Gamma_a\} = 2\delta_{ab}$.These are the gamma matrices associated with $SO(5)$. We define

$$X_a = \frac{R}{\sqrt{5}}\Gamma_a. \tag{8.222}$$

These satisfy

$$X_a^2 = R^2. \tag{8.223}$$

This is a fuzzy $\mathbf{S}_N^4$ with $N = 4$. This corresponds to the 4-dimensional representation (1/2, 1/2) of $SO(5)$ or spin(5) (or equivalently $Sp(2)$). Spin(5) is a 2-to-1 cover of SO(5) and therefore they have the same Lie algebra and the same representation theory. The generators of spin(5) in the fundamental representation are not Γ_a but they are given by

$$\frac{\sigma_{ab}}{2} = \frac{1}{4i}[\Gamma_a, \Gamma_b]. \tag{8.224}$$

Any 4×4 matrix, i.e. any function on $\mathbf{S}_4^4$, can be expanded in terms of the 16 matrices Γ_a and σ_{ab} and the identity as follows

$$M = M_0\mathbf{1} + M_a\Gamma_a + M_{ab}\sigma_{ab}. \tag{8.225}$$

Fuzzy $\mathbf{S}_4^4$ corresponds only to the first two terms in the above expansion.

Higher approximations of the four-sphere $\mathbf{S}^4$ are obtained as follows. We consider the irreducible representation obtained by taking the n-fold symmetric tensor product of the fundamental representation (1/2,1/2) of spin(5). Thus

$$J_a = \frac{1}{2}(\Gamma_a \otimes \mathbf{1} \otimes \cdots \otimes \mathbf{1} + \mathbf{1} \otimes \Gamma_a \otimes \cdots \otimes \mathbf{1} + \mathbf{1} \otimes \mathbf{1} \otimes \cdots \otimes \Gamma_a)_{\text{sym}}, \tag{8.226}$$

n−times.

This corresponds to the spin $(n/2, n/2)$ irreducible representation of spin(5) or $SO(5)$. The corresponding Dynkin labels are $\lambda_1 = n_1 - n_2 = 0$ and $\lambda_2 = 2n_2 = n$. The gamma matrices are taken in the representation

$$\Gamma_i = \begin{pmatrix} 0 & \sigma_i \\ \sigma_i & 0 \end{pmatrix}, \quad \Gamma_4 = \begin{pmatrix} 0 & i \\ -i & 0 \end{pmatrix}, \quad \Gamma_5 = \begin{pmatrix} 1 & 0 \\ 0 & -1 \end{pmatrix}. \tag{8.227}$$

The irreducible representations of $SO(5)$ are characterized by the highest weight vectors $\Lambda = (n_1, n_2)$ with $n_1 \geqslant n_2 \geqslant 0$ with dimensions and Casimirs

$$\dim(\Lambda) = \frac{1}{6}(2n_1 + 3)(2n_2 + 1)(n_1 - n_2 + 1)(n_1 + n_2 + 2). \tag{8.228}$$

$$C_2(\Lambda) = \frac{1}{2}n_1(n_1 + 3) + \frac{1}{2}n_2(n_2 + 1). \tag{8.229}$$

Thus,

$$\dim\left(\frac{n}{2}, \frac{n}{2}\right) = \frac{1}{6}(n + 1)(n + 2)(n + 3), \quad C_2\left(\frac{n}{2}, \frac{n}{2}\right) = \frac{1}{4}n(n + 4). \tag{8.230}$$

This can be given an explicit construction in terms of creation and annihilation operators [129]. We write

$$J_a = \frac{1}{2}a_\alpha^+(\Gamma_a)_{\alpha\beta}a_\beta, \quad [a_\alpha, a_\beta^+] = \delta_{\alpha\beta}. \tag{8.231}$$

We compute

$$a_\alpha^+\left(\frac{\sigma_{ab}}{2}\right)_{\alpha\beta} a_\beta = \frac{1}{i}[J_a, J_b]. \tag{8.232}$$

This is the analogue of (8.224). Thus the generators of spin(5) in the irreducible representation $(n/2, n/2)$ are given by

$$\mathcal{M}_{ab} = a_\alpha^+\left(\frac{\sigma_{ab}}{2}\right)_{\alpha\beta} a_\beta. \tag{8.233}$$

Also we compute

$$[\mathcal{M}_{ab}, J_c] = \frac{1}{2}a_\alpha^+\left[\frac{\sigma_{ab}}{2}, \Gamma_c\right]_{\alpha\beta} a_\beta$$
$$= \frac{i}{2}a_\alpha^+(\delta_{ac}\Gamma_b - \delta_{bc}\Gamma_a)_{\alpha\beta}a_\beta \tag{8.234}$$
$$= i(\delta_{ac}J_b - \delta_{bc}J_a).$$

And

$$[\mathcal{M}_{ab}, \mathcal{M}_{cd}] = a_\alpha^+\left[\frac{\sigma_{ab}}{2}, \frac{\sigma_{cd}}{2}\right]_{\alpha\beta} a_\beta$$
$$= ia_\alpha^+\left(\delta_{ac}\frac{\sigma_{bd}}{2} - \delta_{ad}\frac{\sigma_{bc}}{2} - \delta_{bc}\frac{\sigma_{ad}}{2} + \delta_{bd}\frac{\sigma_{ac}}{2}\right)_{\alpha\beta} a_\beta \tag{8.235}$$
$$= i(\delta_{ac}\mathcal{M}_{bd} - \delta_{ad}\mathcal{M}_{bc} - \delta_{bc}\mathcal{M}_{ad} + \delta_{bd}\mathcal{M}_{ac}).$$

J_a transforms as a vector in the irreducible representation (1,0) under spin(5). Thus $\sum_a J_a^2$ is invariant under $SO(5)$, and since $(n/2, n/2)$ is an irreducible representation, the quantity $\sum_a J_a^2$ must be proportional to the identity [128]. We show this explicitly as follows.

We go on now to the groups spin(6), $SO(5)$ and $SU(4)$. Spin(6) is a 2-to-1 cover of $SO(6)$ and it is locally isomorphic to $SU(4)$. They have the same Lie algebra. Irreducible representations of $SO(6)$ are characterized by the highest weight vectors $\Lambda = (n_1, n_2, n_3)$ with $n_1 \geqslant n_2 \geqslant |n_3| \geqslant 0$ with dimensions and Casimirs

$$\dim(\Lambda) = \frac{1}{12}((n_1+2)^2 - n_3^2)((n_1+2)^2 - (n_2+1)^2)((n_2+1)^2 - n_3^2). \tag{8.236}$$

$$C_2(\Lambda) = \frac{1}{2}n_1(n_1+4) + \frac{1}{2}n_2(n_2+2) + \frac{1}{2}n_3^2. \tag{8.237}$$

The fundamental is (1/2, 1/2, 1/2), whereas the anti-fundamental is $\overline{(1/2, 1/2, 1/2)} = (1/2, 1/2, -1/2)$. We have the identification with $SU(4)$ representations

$$(n, 0, 0) \leftrightarrow \left(\frac{n}{2}, \frac{n}{2}, \frac{n}{2}\right), \quad (0, 0, n) \leftrightarrow \overline{\left(\frac{n}{2}, \frac{n}{2}, \frac{n}{2}\right)} = \left(\frac{n}{2}, \frac{n}{2}, -\frac{n}{2}\right). \tag{8.238}$$

The generators of spin(6) in the fundamental representation (1/2,1/2,1/2) are given by $\sigma_{ab}/2$ and $\sigma_{a6}/2 = -\sigma_{6a}/2 = \Gamma_a/2$. They are written collectively as $\sigma_{AB}/2$, A, $B = 1, \ldots, 6$, and they satisfy

$$\left[\frac{\sigma_{AB}}{2}, \frac{\sigma_{CD}}{2}\right] = i\left(\delta_{AC}\frac{\sigma_{BD}}{2} - \delta_{AD}\frac{\sigma_{BC}}{2} - \delta_{BC}\frac{\sigma_{AD}}{2} + \delta_{BD}\frac{\sigma_{AC}}{2}\right). \tag{8.239}$$

The irreducible representations $(n/2, n/2, n/2)$ of spin(6) are obtained by taking the n-fold symmetric tensor product of the fundamental representation (1/2,1/2,1/2). The corresponding dimension and Casimir are

$$\dim\left(\frac{n}{2},\frac{n}{2},\frac{n}{2}\right)=\frac{1}{6}(n+1)(n+2)(n+3),\quad C_2\left(\frac{n}{2},\frac{n}{2},\frac{n}{2}\right)=\frac{3}{8}n(n+4). \quad (8.240)$$

The corresponding Dynkin labels are $\lambda_1 = n_1 - n_2 = 0$, $\lambda_2 = n_2 - n_3 = 0$, $\lambda_3 = n_2 + n_3 = n$. The generators are exactly given by $J_a = \mathcal{M}_{a6} = -\mathcal{M}_{6a}$ and $\mathcal{M}_{ab}$ which are denoted collectively $\mathcal{M}_{AB}$, $A, B = 1, \ldots, 6$, and they satisfy

$$[\mathcal{M}_{AB}, \mathcal{M}_{CD}] = i(\delta_{AC}\mathcal{M}_{BD} - \delta_{AD}\mathcal{M}_{BC} - \delta_{BC}\mathcal{M}_{AD} + \delta_{BD}\mathcal{M}_{AC}). \quad (8.241)$$

We know now the Casimirs

$$\frac{1}{4}\mathcal{M}_{ab}\mathcal{M}_{ab} = \frac{1}{4}n(n+4). \quad (8.242)$$

$$\frac{1}{4}\mathcal{M}_{AB}\mathcal{M}_{AB} = \frac{1}{4}\mathcal{M}_{ab}\mathcal{M}_{ab} + \frac{1}{2}J_a^2 = \frac{3}{8}n(n+4). \quad (8.243)$$

Thus we get

$$J_a^2 = \frac{1}{4}n(n+4). \quad (8.244)$$

In summary, the defining equations of fuzzy $\mathbf{S}^4_N$ are

$$X_a^2 = R^2. \quad (8.223)$$

$$X_a = rJ_a,\quad r^2 = \frac{4R^2}{n(n+4)}. \quad (8.246)$$

The coordinate operators do not commute, viz

$$[X_a, X_b] = i\Theta_{ab},\quad \Theta_{ab} = r^2\mathcal{M}_{ab}. \quad (8.247)$$

The coordinate operators X_a are covariant under $SO(5)$, viz

$$[\mathcal{M}_{ab}, X_c] = i(\delta_{ac}X_b - \delta_{bc}X_a). \quad (8.248)$$

$$[\mathcal{M}_{ab}, \mathcal{M}_{cd}] = i(\delta_{ac}\mathcal{M}_{bd} - \delta_{ad}\mathcal{M}_{bc} - \delta_{bc}\mathcal{M}_{ad} + \delta_{bd}\mathcal{M}_{ac}). \quad (8.249)$$

The fuzzy $\mathbf{S}^4_N$ is therefore a covariant quantum space very similar to Snyder quantum spacetime [75, 76]. It is also similar to the DFR spacetime [77] but the Θ are not central.

A general function on fuzzy $\mathbf{S}^4_N$ is an $N \times N$ matrix where

$$N = d_N^3 = \frac{1}{6}(n+1)(n+2)(n+3). \quad (8.250)$$

Thus we are dealing with a sequence of matrix algebras Mat_N. The basis is given by the polarization tensors corresponding to the tensor products

$$\left(\frac{n}{2}, \frac{n}{2}\right) \otimes \left(\frac{n}{2}, \frac{n}{2}\right) = \sum_{l=0}^{n} \sum_{k=0}^{l} (l, k), \quad \text{spin}(5). \tag{8.251}$$

$$\left(\frac{n}{2}, \frac{n}{2}, \frac{n}{2}\right) \otimes \left(\frac{n}{2}, \frac{n}{2}, -\frac{n}{2}\right) = \sum_{l=0}^{n} (l, l, 0), \quad \text{spin}(6). \tag{8.252}$$

$$(0, 0, n) \otimes (n, 0, 0) = \sum_{l=0}^{n} (l, 0, l), \quad SU(4). \tag{8.253}$$

Thus functions on fuzzy $\mathbf{S}^4_N$ obviously involve X_a and $\mathcal{M}_{ab}$, and because of the constraints, the dimension of this space is actually six and not four (we are really dealing with fuzzy $\mathbf{CP}^3_N$). We can see from (8.251), (8.252) and (8.253) that the $(l, l, 0)$ representation of $SO(6)$ and the $(l, 0, l)$ representation of $SU(4)$ decomposes in terms of $SO(5)$ representations as

$$(l, l, 0)_{SO(6)} = \sum_{k=0}^{l} (l, k)_{SO(5)}. \tag{8.254}$$

$$(l, 0, l)_{SU(4)} = \sum_{k=0}^{l} (l, k)_{SO(5)}. \tag{8.255}$$

In terms of representation theory functions on fuzzy $\mathbf{S}^4_N$ correspond only to the representation $(l,0)$ in (8.251). This can be seen as follows. A general $N \times N$ matrix M can be expanded in terms of $SO(5)$ polarization tensors $T^{(l,k)}_{a_1, \ldots, a_{l+k}}$ as follows

$$M = \sum_{l=0}^{n} \sum_{k=0}^{l} M^{(l,k)}_{a_1, \ldots, a_{l+k}} T^{(l,k)}_{a_1, \ldots, a_{l+k}}. \tag{8.256}$$

The polarization tensors $T^{(l,k)}_{a_1, \ldots, a_{l+k}}$ are symmetrized nth order polynomials of X_a and $\mathcal{M}_{ab}$ where l is the order of X_a and k is the order of $\mathcal{M}_{ab}$. Thus fuzzy $\mathbf{S}^4_N$ corresponds to $k = 0$, i.e. to the $(l, 0)$ representations.

These symmetric representations $(l,0)$ which occur in the expansion of functions on $\mathbf{S}^4$ are also due to the fact that $\mathbf{S}^4$ is the coadjoint orbit $\mathbf{S}^4 = SO(5)/SO(4)$. Thus the harmonic expansion of functions on $\mathbf{S}^4$ requires irreducible representations of $SO(5)$ which contain singlets of $SO(4)$ under the decomposition $SO(5) \longrightarrow SO(4)$ [122]. This is analogous to the statement that since $\mathbf{CP}^3$ is the coadjoint orbit $\mathbf{CP}^3_N = SU(4)/U(3)$ harmonic expansion of functions on $\mathbf{CP}^3$ requires irreducible representations of $SU(4)$ which contain singlets of $U(3)$ under the decomposition $SU(4) \longrightarrow SU(3) \times U(1)$. But $\mathbf{CP}^3$ is also the coadjoint orbit $\mathbf{CP}^3 = SO(5)/(SU(2) \times U(1))$ and thus harmonic expansion of functions on $\mathbf{CP}^3$ requires irreducible representations of $SO(5)$ which contain singlets of $SU(2) \times U(1)$ under the decomposition $SO(5) \longrightarrow SU(2) \times U(1)$. In the first case the Laplacian on $\mathbf{CP}^3$ is $SO(6)$-invariant, whereas in the second case the Laplacian is $SO(5)$-invariant although not unique. This is in fact why we can extract fuzzy $\mathbf{S}_N^4$ from fuzzy $\mathbf{CP}_N^3$ with a fiber given by fuzzy $\mathbf{S}_N^2$. Indeed, schematically we have

$$\begin{aligned} SO(5)/(SU(2) \times U(1)) &= (SO(5)/SO(4)) \times (SO(4)/(SU(2) \times U(1))) \\ &= (SO(5)/SO(4)) \times (SU(2)/U(1)). \end{aligned} \tag{8.257}$$

This will be given a precise meaning below.

Thus, fuzzy $\mathbf{S}_N^4$ have in some way an internal structure given by fuzzy $\mathbf{S}_N^2$. If we simply project out the unwanted degrees of freedom we obtain a non-associative algebra [130]. Another more elegant approach is due to O'Connor *et al* and goes as follows. The eigenvalues of the spin(5) and spin(6) Casimirs on the polarization tensors $T^{(l,\,k)}$ are

$$C_2^{SO(5)}T^{(l,\,k)} = \frac{1}{2}(l(l+3) + k(k+1))T^{(l,\,k)}. \tag{8.258}$$

$$C_2^{SO(6)}T^{(l,\,k)} = l(l+3)T^{(l,\,k)}. \tag{8.259}$$

In the last line we have used the fact that all $T^{(l,\,k)}$, $k = 0, \ldots, l$, correspond to the $SO(6)$ representation $(l, l, 0)$. We note then immediately that the operator

$$C_I = 2C_2^{SO(5)} - C_2^{SO(6)}, \tag{8.260}$$

has eigenvalues

$$C_I T^{(l,\,k)} = k(k+1)T^{(l,\,k)}. \tag{8.261}$$

Since it only depends on k it can be used to penalize the representations with $k \neq 0$ in order these unwanted zero modes on fuzzy $\mathbf{S}_N^4$. The Laplacian on the fuzzy $\mathbf{S}_N^4$ is then given by

$$\Delta_n = \frac{1}{R^2}(C_2^{SO(6)} + hC_I). \tag{8.262}$$

The parameter h will be taken to ∞ but the theory is stable for all $h \in [-1, \infty]$. We also write explicitly these Laplacians as

$$C_2^{SO(5)} = \frac{1}{4}[\mathcal{M}_{ab}, [\mathcal{M}_{ab}, \ldots]]. \tag{8.263}$$

$$C_2^{SO(6)} = \frac{1}{4}[\mathcal{M}_{AB}, [\mathcal{M}_{AB}, \ldots]] = \frac{1}{4}[\mathcal{M}_{ab}, [\mathcal{M}_{ab}, \ldots]] + \frac{1}{2}[J_a, [J_a, \ldots]]. \tag{8.264}$$

A non-commutative scalar field theory on fuzzy $\mathbf{S}_N^4$ is given by

$$S = \frac{R^4}{N}\,\mathrm{Tr}(\Phi\Delta_n\Phi + V(\Phi)). \tag{8.265}$$

8.5.3 Hopf map

In this section and the rest of this chapter we will mostly follow [48, 127]. There is no symplectic 2-form on $\mathbf{S}^4$ since $\mathbf{H}^2(\mathbf{S}^4) = 0$. Thus, we insist that fuzzy $\mathbf{S}^4_N$ should be viewed as a squashed fuzzy $\mathbf{CP}^3_N$ with degenerate fiber fuzzy $\mathbf{S}^2_N \equiv \mathbf{S}^2_{n+1}$ (see below). We explain this in some detail.

We consider the fundamental representation (1,0,0) of $SU(4)$. This is a 4-dimensional representation which we will view as $\mathbf{C}^4$. Let $z_0 = (1,0,0,0)$ be some reference point in $\mathbf{C}^4$. Obviously, $SU(4)$ will act on this point giving $z = Uz_0$, where $U \in SU(4)$, in such a way that $z^+ z = 1$, i.e. $z \in \mathbf{S}^7 \in \mathbf{R}^8 = \mathbf{C}^4$. Further, starting from the 4×4 gamma matrices Γ_a of $SO(5)$, we define

$$x_a = z_\alpha^*(\Gamma_a)_{\alpha\beta} z_\beta = \langle z|\Gamma_a|z\rangle. \tag{8.266}$$

By using the result [128, 139]

$$\sum_a (\Gamma_a \otimes \Gamma_a)_{\mathrm{sym}} = (\mathbf{1} \otimes \mathbf{1})_{\mathrm{sym}}, \tag{8.267}$$

we can show that $x_a^2 = 1$, i.e. $x_a \in \mathbf{S}^4$. We can then define the Hopf map

$$\begin{aligned} \mathbf{S}^7 &\longrightarrow \mathbf{S}^4 \\ z_a &\longrightarrow x_a. \end{aligned} \tag{8.268}$$

But since the phase of z drops in a trivial way, this Hopf map is actually a map from $\mathbf{CP}^3 = \mathbf{S}^7/U(1)$ into $\mathbf{S}^4$, viz

$$\begin{aligned} x_a \; : \; \mathbf{CP}^3 &\longrightarrow \mathbf{S}^4 \\ |z\rangle\langle z| &\longrightarrow \langle z|\Gamma_a|z\rangle = x_a. \end{aligned} \tag{8.269}$$

Here, we have identified $\mathbf{CP}^3$ with the space of rank one projectors $|z\rangle\langle z|$.

We consider the gamma matrices in the Weyl basis where Γ_5 is diagonal with eigenvalues $+1$ and -1 and degeneracy equal to 2 for each. The reference point in $\mathbf{C}^4$ is $z_0 = (\hat{z}_0, \tilde{z}_0)$ where $\hat{z}_0 = (1, 0)$ and $\tilde{z}_0 = 0$. The coordinates of the point $p_0 \in \mathbf{S}^4 \in \mathbf{R}^5$ at the reference point are

$$x_i = 0, \quad x_5 = 1 = \hat{z}_0^+ \hat{z}_0 = z_0^+ \Gamma_5 z_0. \tag{8.270}$$

This is essentially the north pole. But recall that $\mathbf{CP}^3 = SO(5)/(SU(2) \times U(1))$. Therefore, the stabilizer at p_0 is obviously given by $SO(4) \in SO(5)$, viz

$$H = \{h \in SO(5); [h, \Gamma_5] = 0\} = SO(4) = SU(2)_L \times SU(2)_R. \tag{8.271}$$

This can also be seen from the fact that $\mathbf{S}^4$ is the $SO(4)$ orbit of $SO(5)$ through Γ_5 given by

$$x_a \Gamma_a = g\Gamma_5 g^{-1}, \quad g \in SO(5). \tag{8.272}$$

The $SU(2)_L$ acts on the eigenspace of Γ_5 with eigenvalue $+1$. The fiber over $p_0 \in \mathbf{S}^4$ is clearly given by the condition (with $z = (\hat{z}, 0)$)

$$z^+\Gamma_5 z = 1 \Rightarrow |z_1|^2 + |z_2|^2 = 1. \tag{8.273}$$

This is $\mathbf{S}^3$ and because the phase of $\hat{z}$ drops we get $\mathbf{S}^3/U(1) = \mathbf{S}^2$, i.e. the fiber is $\mathbf{S}^2$. In other words, $\mathbf{CP}^3$ in an $\mathbf{S}^2$-bundle over $\mathbf{S}^4$. On the other hand, the action of $SU(2)_R$ in $\mathbf{CP}^3$ is trivial.

What is the matrix analogue of the above Hopf map?

The quantization of the classical Hopf map x_a: $\mathbf{CP}^3 \longrightarrow \mathbf{S}^4$ is clearly given by $X_a = rJ_a$ where $r^2 = 4R^2/n(n+4)$ and $J_a = a_\alpha^+(\Gamma_a)_{\alpha\beta}a_\beta/2$. The generators of $SO(5)$ are $\mathcal{M}_{ab} = a_\alpha^+(\sigma_{ab})_{\alpha\beta}a_\beta/2$, whereas the generators of $SO(6) = SU(4)$ are $\mathcal{M}_{AB} = \{\mathcal{M}_{ab}, \mathcal{M}_{a6} = J_a\}$ and they act on the Hilbert space $(n, 0, 0)_{SU(4)} = (n/2, n/2, n/2)_{SO(6)} = (n/2, n/2)_{SO(5)}$. The generators of $SO(6) = SU(4)$ can also be given by

$$J_\mu = \frac{1}{2}a_\alpha^+(\Gamma_\mu)_{\alpha\beta}a_\beta, \quad \mu = 1, \dots, 15. \tag{8.274}$$

Essentially J_μ are the T_μ in (8.199) and the correspondence between Γ_μ and $\{\Gamma_a, \sigma_{ab}\}$ is obvious. Strictly speaking, $X_\mu = rJ_\mu$ is the quantization of the classical Hopf map x_a: $\mathbf{CP}^3 \longrightarrow \mathbf{S}^4$ since fuzzy $\mathbf{S}_N^4$ is a squashed $\mathbf{S}^2$-bundle over $\mathbf{S}^4$ given by $\mathbf{CP}^3$ where the fiber is degenerate.

8.5.4 Poisson structure

This then should be viewed as the quantization of the Kirillov–Kostant symplectic form corresponding to the Poisson structure

$$\{J_\mu, J_\nu\} = f_{\mu\nu\lambda} J_\lambda \tag{8.275}$$

giving the commutators

$$[J_\mu, J_\nu] = if_{\mu\nu\lambda}J_\lambda. \tag{8.276}$$

Thus $X_a = rJ_a \sim x_a$, $a = 1, \dots, 5$, is a subset of the quantized embedding functions $X_\mu = rJ_\mu \sim x_\mu$: $\mathbf{CP}^3 \hookrightarrow su(4) = \mathbf{R}^{15}$.

The noncommutativity is given by $\mathcal{M}_{ab}$ since

$$[X_a, X_b] = i\Theta_{ab} = ir^2\mathcal{M}_{ab}, \tag{8.277}$$

and it arises from the Poisson structure on $\mathbf{CP}^3$. In other words, Θ^{ab} is the quantization of the embedding function defined on $\mathbf{CP}^3$ by θ_{ab}: $\mathbf{CP}^3 \hookrightarrow so(5)$ where the antisymmetric tensor θ_{ab} is given by the Poisson bracket $\theta_{ab} = \{x_a, x_b\}$. This embedding is by construction not constant along the fiber, and thus it does not define a Poisson bracket on $\mathbf{S}^4$, and each point on the fiber $\mathbf{S}^2$ corresponds to a different choice of the noncommutativity θ_{ab} on $\mathbf{S}^4$. This kind of averaging over $\mathbf{S}^2$ of the noncommutativity is also what guarantees $SO(5)$ invariance on this noncommutative space. The embedding function θ_{ab} therefore resolves completely the $\mathbf{S}^2$ fiber over $\mathbf{S}^4$.

Thus, the local noncommutativity is $\mathcal{M}_{ij}$ which also generates the local $SO(4)$ rotations. Indeed, by going to the north pole, we find that the stabilizer group is

$SO(4) = SU(2)_L \times SU(2)_R$. We can decompose there the $SO(5)$ generators $\mathcal{M}_{ab}$ into the $SO(4)$ generators $\mathcal{M}_{ij}$ and the translations

$$P_i = \frac{1}{R}\mathcal{M}_{i5}. \tag{8.278}$$

They satisfy

$$[P_i, X_j] = \frac{i}{R}\delta_{ij}X_5, \quad [P_i, P_j] = \frac{i}{R^2}\mathcal{M}_{ij}, \quad [P_i, X_5] = -\frac{i}{R}X_i. \tag{8.279}$$

X_i and P_i are vectors, whereas X_5 is a scalar under the rotations $\mathcal{M}_{ij}$. The P_i reduce to ordinary translations if we set $X_5 = R$, since we are at the north pole, and take the limit $R \longrightarrow \infty$. This is an Inonu–Wigner contraction of $SO(5)$ yielding the full Poincaré group in four dimensions. Thus the generators P_i will allow us to move around $\mathbf{S}^4$. The local noncommutativity is given by

$$[X_i, X_j] = ir^2\mathcal{M}_{ij}, \quad [X_i, X_5] = ir^2RP_i. \tag{8.280}$$

Obviously, $r^2\mathcal{M}_{ij}$ is the noncommutativity parameter Θ_{ij} and $r^2 \longrightarrow 0$ in the limit since $\mathcal{M} \sim n$. In other words, the noncommutativity scale is given by

$$L_{\rm NC}^2 = r^2 n = \frac{4R^2}{n}. \tag{8.281}$$

8.5.5 Coherent state

This can also be seen as follows. The coherent state on $\mathbf{CP}^3 = SO(5)/(SU(2) \times U(1))$ is given by the orbit $P = gP_0g^+ = |x, \xi\rangle\langle x, \xi|$, where $g \in SO(5)$, $P_0 = |\Lambda\rangle\langle\Lambda|$ where $|\Lambda\rangle$ is the highest weight state of the irreducible representation $(n/2, n/2)$ of $SO(5)$, and $|x, \xi\rangle$ is the coherent state with $x \in \mathbf{S}^4$ and $\xi \in \mathbf{S}^2$. The north pole corresponds to the highest weight state. We have then

$$\mu x^a = \langle x, \xi|X^a|x, \xi\rangle = \langle X^a\rangle. \tag{8.282}$$

This is consistent with

$$\frac{x^a}{R} = \langle x, \xi|\Gamma^a|x, \xi\rangle \tag{8.283}$$

with $\mu = 1/\sqrt{1 + 4/n}$. The spread is given by

$$\Delta_x^2 = \langle (X^a - \langle X^a\rangle)^2\rangle = \langle (X^a)^2\rangle - \langle X^a\rangle^2 = R^2 - \mu^2R^2 = L_{\rm NC}^2. \tag{8.284}$$

The coherent state $|x, \xi\rangle$ is therefore optimally localized since it minimizes the uncertainty relation. Functions on $\mathbf{CP}^3$ are associated with operators on $(n/2, n/2)$ by means of the coherent state by the usual formula

$$\phi(x, \xi) = \langle x, \xi|\hat{\Phi}|x, \xi\rangle. \tag{8.285}$$

The coherent states are in one-to-one correspondence with point on $\mathbf{CP}^3$ up to a $U(1)$ factor. Thus, the coherent state is a $U(1)$ bundle over $\mathbf{CP}^3$. The curvature of the corresponding connection is the symplectic form ω on $\mathbf{CP}^3$ associated with the Poisson structure

$$\theta_{ab}(x, \xi) = \langle x, \xi|[X_a, X_b]|x, \xi\rangle = i\langle x, \xi|\Theta_{ab}|x, \xi\rangle. \tag{8.286}$$

8.5.6 Local flatness

The matrices X_a describe a matrix or quantized membrane 4-sphere which appears locally to be an L5-brane in matrix theory with the correct charge [128]. The rotational invariance under $SO(5)$ is given by the condition

$$R_{ab}X_b = UX_aU^{-1}. \tag{8.287}$$

Thus, the noncommutativity tensor will transform correspondingly as

$$R_{aa'}R_{bb'}\Theta_{ab} = U\Theta_{a'b'}U^{-1}. \tag{8.288}$$

Rotations are then implemented by gauge transformations, i.e. as local transformations, in the spirit of gravity.

The matrices J_a satisfy, among other things, the so-called local flatness condition given by

$$\epsilon_{abcde}J_aJ_bJ_cJ_d = \alpha J_e. \tag{8.289}$$

By rotational invariance this needs only to be checked for $e = 5$. We compute

$$\epsilon_{abcd5}J_aJ_bJ_cJ_d = \{[J_1, J_2], [J_3, J_4]\} - \{[J_1, J_3], [J_2, J_4]\} + \{[J_1, J_4], [J_2, J_3]\}. \tag{8.290}$$

For the smallest possible representation $J_a = \Gamma_a/2$ we get $3J_5$ and thus $\alpha = 3$. For any J_a in the irreducible representation $(n,0,0)$ of $SU(4)$ we obtain instead [139]

$$\epsilon_{abcde}J_aJ_bJ_cJ_d = (n + 2)J_e. \tag{8.291}$$

By using $J_a^2 = n(n + 4)/4$ this can be rewritten as

$$\epsilon_{abcde}J_aJ_bJ_cJ_dJ_e = \frac{1}{4}n(n + 2)(n + 4). \tag{8.292}$$

We can also rewrite this as (remember that $J_e = \mathcal{M}_{e6}$)

$$-\frac{1}{4}\epsilon_{abcde6}\mathcal{M}_{ab}\mathcal{M}_{cd}\mathcal{M}_{e6} = \frac{1}{4}n(n + 2)(n + 4). \tag{8.293}$$

Or

$$-\frac{1}{24}\epsilon_{ABCDEF}\mathcal{M}_{AB}\mathcal{M}_{CD}\mathcal{M}_{EF} = \frac{1}{4}n(n + 2)(n + 4). \tag{8.294}$$

We recognize this to be the cubic Casimir of $SU(4)$, viz

$$\begin{aligned} d_{\mu\nu\lambda}T_\mu T_\nu T_\lambda &= -\frac{1}{32}\epsilon_{ABCDEF}\mathcal{M}_{AB}\mathcal{M}_{CD}\mathcal{M}_{EF} \\ &= \frac{3}{16}n(n+2)(n+4). \end{aligned} \tag{8.295}$$

This with the quadratic Casimir

$$T_\mu^2 = \frac{1}{4}\mathcal{M}_{AB}\mathcal{M}_{AB} = \frac{3}{8}n(n+4) \tag{8.296}$$

provide the defining equations of $\mathbf{CP}_N^3$ as we know.

8.5.7 Noncommutativity scale

The noncommutativity $\Theta_{ij} = r^2\mathcal{M}_{ij}$, in the semi-classical limit, is a self-dual antisymmetric tensor $\theta_{ij}(x, \xi)$, transforming as (1,0) under the local group $SO(4) = SU(2)_L \times SU(2)_R$. It corresponds to a bundle of self-dual frames over $\mathbf{S}^4$ which averages out over $\mathbf{S}^2$ [48]. The generators of $SU(2)_L$ denoted by $J_{\hat{i}}^L$ (the index $\hat{i}$ runs from 1 to 3) generate the local fiber given by the fuzzy sphere $\mathbf{S}_N^2 \equiv \mathbf{S}_{n+1}^2$. In other words, the self-dual antisymmetric tensor $\mathcal{M}_{ij}$ is the flux of a $U(n+1)$-valued noncommutative gauge field given by the non-trivial instanton configuration

$$\mathcal{M}_{ij} = \epsilon_{ij}^{\hat{k}} J_{\hat{k}}^L. \tag{8.297}$$

Thus, locally (north pole) the four-sphere $\mathbf{S}^4$ is characterized by the four coordinates X_i, whereas the fiber $\mathbf{S}^2$ is characterized by two of the three components of the self-dual tensor $\mathcal{M}_{ij}$.

This can also be interpreted as arising from a twisted stack of $n+1$ noncommutative spherical branes carrying $U(n+1)$-valued gauge field [140, 141]. We identify these branes by choosing $n+1$ coherent states $|i\rangle$ on $\mathbf{S}_{n+1}^2$. Thus every point on $\mathbf{S}^4$ is covered by $n+1$ sheets and these sheets are connected by the modes $|i\rangle\langle j| \in U(n+1)$. Also, we can assign a Poisson structure to these sheets or leaves by the usual formula $\langle i|\Theta_{ij}|i\rangle$. We will apply the semi-classical formula

$$tr \sim \frac{1}{\pi^2}\int \frac{1}{2}\omega \wedge \omega \tag{8.298}$$

for symplectic 4-dimensional spaces to the $n+1$ leaves. We consider the local flatness condition (8.292) at the north pole which takes in the semi-classical limit the form

$$\begin{aligned} \frac{1}{4}n(n+2)(n+4) &\sim \frac{1}{4r^5}\epsilon_{abcde}\theta_{ab}\theta_{cd}x_e \\ &\sim \frac{R}{2r^5}\text{Pf}\,\theta. \end{aligned} \tag{8.299}$$

The Pfaffian is defined by

$$\mathrm{Pf}\,\theta = \frac{1}{2}\epsilon_{ijkl}\theta_{ij}\theta_{kl}. \tag{8.300}$$

Thus,

$$\epsilon_{abcde}trJ_aJ_bJ_cJ_dJ_e \sim \frac{1}{24}n^6. \tag{8.301}$$

On the other hand, by applying the trace to the $n + 1$ leaves we obtain

$$\begin{aligned}\epsilon_{abcde}trJ_aJ_bJ_cJ_dJ_e &\sim \frac{n+1}{\pi^2}\int\frac{1}{2}\omega^2(\frac{R}{2r^5}\mathrm{Pf}\,\theta)\\ &\sim \frac{n+1}{\pi^2}\frac{R}{2r^5}V_{\mathbf{S}^4}\\ &\sim \frac{n+1}{\pi^2}\frac{R}{2r^5}\frac{8\pi^2}{3}R^4.\end{aligned} \tag{8.302}$$

By comparing we get $r = 2R/n$ which is correct. Then, fuzzy $\mathbf{S}^4_N$ consists of $n^3/6$ cells of volume L^4_{NC}. The volume of these cells is equal to $n + 1$ (due to the fiber structure given by fuzzy $\mathbf{S}^2_{n+1}$) times the volume of the $\mathbf{S}^4$, viz

$$\frac{n^3}{6}L^4_{\mathrm{NC}} = (n+1)\cdot\frac{8\pi^2}{3}R^4 \Rightarrow L^2_{\mathrm{NC}} = \frac{4\pi R^2}{n}. \tag{8.303}$$

This is consistent with our previous estimate.

8.5.8 Matrix model

We have already discussed the harmonic expansion of functions on fuzzy $\mathbf{CP}^3_N$ given by the $SO(5)$ tensor product

$$\left(\frac{n}{2},\frac{n}{2}\right)\otimes\left(\frac{n}{2},\frac{n}{2}\right) = \sum_{l=0}^{n}\sum_{k=0}^{l}(l,k). \tag{8.304}$$

In terms of the Dynkin labels the representations (l, k) read $(\lambda_1, \lambda_2) = (l - k, 2k)$. Fuzzy $\mathbf{S}^4_N$ is given by the symmetric representations $(l,0)$ which correspond to totally symmetric polynomials of degree l in J_a. The space of functions on fuzzy $\mathbf{S}^4_N$ is written as

$$C_N(\mathbf{S}^4) = \sum_{l=0}^{n}(l,0). \tag{8.305}$$

The representations $(l$–k,2k) with $k \neq 0$ correspond to bosonic higher spin modes of $SO(5)$ on S^4_N which are spanned by the polynomials

$$(l-k, 2k) = \mathrm{span}\{P_{l-k}(J_a)P_k(\mathcal{M}_{ab})\}. \tag{8.306}$$

Indeed, the bosonic modes $(l-2k, 2k)$ are functions on fuzzy $\mathbf{CP}^3_N$, which have a non-trivial dependence along the fiber $\mathbf{S}^2$, i.e. they transform non-trivially under the local group $SU(2)_L \times SU(2)_R$, and thus they are higher spin modes on $\mathbf{S}^4_N$ and not Kaluza–Klein modes. The low spin modes correspond to small k and the spin 2 sector is precisely the graviton [48]. It is also known that projecting out these modes leads to a non-associative algebra [130]. Furthermore, the fermionic higher spin modes will involve representations (l, k) with k odd [127].

The $SO(5)$ generators $\mathcal{M}_{ab}$ are tangential to the 4-sphere since

$$i\mathcal{M}_{ab}J_b + iJ_b\mathcal{M}_{ab} = [J_a, J_b^2] = 0. \tag{8.307}$$

Furthermore, we have

$$\Box J_b = [J_a, [J_a, J_b]] = i[J_a, \mathcal{M}_{ab}] = 4J_b. \tag{8.308}$$

And

$$X_aX_dX_a = R^2X_d - \frac{1}{2}[X_a, [X_a, X_d]] = (R^2 - 2r^2)X_d. \tag{8.309}$$

A crucial property is given by

$$-g_{ac}\{\Theta^{ab}, \Theta^{cd}\} = r^2(\{X^b, X^d\} - 2R^2g^{bd}). \tag{8.310}$$

The proof goes as follows. The ambient metric is given by

$$g_{ab} = \delta_{ab}. \tag{8.311}$$

By using (8.309) we get immediately

$$\begin{aligned} -r^4g_{ac}\{\mathcal{M}^{ab}, \mathcal{M}^{cd}\} &= (R^2 - 4r^2)\{X_b, X_d\} - X_a\{X_b, X_d\}X_a \\ &= -4r^2\{X_b, X_d\} - [X_a, \{X_b, X_d\}]X_a. \end{aligned} \tag{8.312}$$

Then (using Jacobi identity and (8.248) and (8.308))

$$\begin{aligned} [X_a, \{X_b, X_d\}]X_a &= [[X_a, X_d], X_bX_a] + [[X_a, X_b], X_dX_a] \\ &= [X_d, [X_b, X_a]]X_a + [X_b, [X_d, X_a]]X_a \\ &\quad - X_b[X_a, [X_a, X_d]] - X_d[X_a, [X_a, X_b]] \\ &= r^2(2\delta_{bd}R^2 - \{X_b, X_d\}) - 4r^2\{X_b, X_d\}. \end{aligned} \tag{8.313}$$

This leads to the desired result (8.310). The semi-classical limit of this equation (8.310) is given immediately by

$$g_{ac}\theta^{ab}\theta^{cd} = -r^2R^2(P^T)^{bd} \sim -\frac{1}{4}\Delta_x^4(P^T)^{bd}, \tag{8.314}$$

where $\Delta_x = L_{\mathrm{NC}}$ is the noncommutativity scale and $(P^T)^{ab}$ is the tangent projector

$$(P^T)^{ab} = g^{ab} - \frac{x^ax^b}{R^2}. \tag{8.315}$$

We remark that (8.310), as opposed to the noncommutativity itself, is a tensor living on fuzzy $\mathbf{S}_N^4$. As it turns out, the effective background metric around any point $p \in \mathbf{S}^4$ (north pole) is given exactly by [48]

$$\gamma^{jl} = g_{ik}\theta^{ij}\theta^{kl} = -\frac{1}{4}\Delta_x^4 g^{jl}. \tag{8.316}$$

This is a very important result and can also be seen alternatively as follows. (The minus sign is an error due to our correspondence $\theta_{ij} \longrightarrow i\Theta_{ij}$ which should be corrected.)

The result (8.308) allows us to immediately write down a five matrix model with a ground state given by fuzzy $\mathbf{S}_N^4$. This is given explicitly by [127]

$$S = \frac{1}{g^2}\,\mathrm{Tr}(-[D_a, D_b][D^a, D^b] + \mu^2 D_a D^a). \tag{8.317}$$

The equations of motion read

$$[D_b, [D^b, D^a]] + \frac{\mu^2}{2} D^a = 0. \tag{8.318}$$

Clearly, $D_a = J_a$ is a solution if $\mu^2 = -8$.

The above matrix model involves the Laplacian

$$\bar{\square} = C_2^{SO(6)} - C_2^{SO(5)} = \frac{1}{2}[J_a, [J^a, \ldots]] \tag{8.319}$$

with eigenvalues given by

$$\frac{1}{2}(l(l+3) - k(k+1)), \quad k \leqslant l. \tag{8.320}$$

We recall the $SO(5)$ Casimir and its eigenvalues (with $\mathcal{M}_{i5} = RP_i$)

$$\begin{aligned} C_2^{SO(5)} = \frac{1}{4}\mathcal{M}_{ab}^2 &= \frac{R^2}{2}\left(P_i^2 + \frac{1}{2R^2}\mathcal{M}_{ij}^2\right) \\ &\longrightarrow \frac{1}{2}(l(l+3) + k(k+1)). \end{aligned} \tag{8.321}$$

At the north pole $p \in \mathbf{S}^4$ we can neglect the angular momentum contribution compared the translational contribution to get

$$C_2^{SO(5)} \simeq \frac{R^2}{2} P_i^2 = \frac{1}{8r^2}\Delta_x^4 P_i^2. \tag{8.322}$$

In the local frame at p we can also replace $P_i = i\partial_i$. Thus

$$C_2^{SO(5)} \simeq -\frac{1}{2r^2}\gamma^{ij}\partial_i\partial_j. \tag{8.323}$$

Also we note that for the low spin modes $m = 0, 1, 2$ we can make the approximation

$$\bar{\Box} \simeq C_2^{SO(5)} \longrightarrow \frac{1}{2} l(l+3). \tag{8.324}$$

We have then

$$\Box = 2r^2 \bar{\Box} = [X_a, [X_a, \ldots]] \simeq 2r^2 C_2^{SO(5)} = -\gamma^{ij}\partial_i\partial_j. \tag{8.325}$$

Thus, γ^{ij} is indeed the effective background metric.

There are other important results that can be derived from (8.316). First, we introduce a frame $\theta^a_{\mu\nu}$ on the bundle of self-dual tensors $\theta_{\mu\nu}$ over $\mathbf{S}^4$ normalized such that

$$\theta^{\mu\nu}_a \theta^b_{\mu\nu} = 4\delta^b_a. \tag{8.326}$$

This must be a self-dual frame and therefore

$$\theta^a_{\alpha\beta} = \frac{1}{2}\epsilon_{\mu\nu\alpha\beta}\theta^{\mu\nu a}. \tag{8.327}$$

This can be shown by using $\epsilon_{\mu\nu\alpha\beta}\epsilon^{\mu\nu\rho\sigma} = 2(\delta^\rho_\alpha\delta^\sigma_\beta - \delta^\rho_\beta\delta^\sigma_\alpha)$. We write $\theta_{\mu\nu}$ in terms of the frame $\theta^a_{\mu\nu}$ and in terms of the generators J^a of the internal fuzzy sphere $\mathbf{S}^2_{n+1}$ as

$$\theta^{\mu\nu} = r^2\theta^{\mu\nu}_a J^a. \tag{8.328}$$

The normalization is fixed by (8.316) or equivalently $\theta_{\mu\nu}\theta^{\mu\nu} = \Delta^4_x \sim n^2 r^4$. This background flux θ is a function on $\mathbf{CP}^3$, viz $\theta = \theta(x, \xi)$. We want to compute the average over the sphere, i.e. over $\xi \in \mathbf{S}^2$, of various objects constructed from θ. For example, the average of the product $\theta^{\mu\nu}\theta^{\rho\sigma}$ over $\mathbf{S}^2$ obviously can only depend on the two constant tensors on $\mathbf{S}^4$: the Levi–Civita tensor $\epsilon^{\mu\nu\rho\sigma}$ and the generalized Kronecker delta $\delta^{\mu\rho}\delta^{\nu\sigma} - \delta^{\nu\rho}\delta^{\mu\rho}$. Thus

$$[\theta^{\mu\nu}\theta^{\rho\sigma}]_{\mathbf{S}^2} = a(\delta^{\mu\rho}\delta^{\nu\sigma} - \delta^{\nu\rho}\delta^{\mu\rho}) + b\epsilon^{\mu\nu\rho\sigma}. \tag{8.329}$$

By contracting with $\epsilon_{\mu\nu\rho\sigma}$ and using $\epsilon_{\mu\nu\rho\sigma}\epsilon^{\mu\nu\rho\sigma} = 4!$ we find $b = \Delta^4_x/12$, while by contracting with $g_{\mu\rho}$ we obtain $a = \Delta^4_x/12$. Thus we get

$$[\theta^{\mu\nu}\theta^{\rho\sigma}]_{\mathbf{S}^2} = \frac{\Delta^4_x}{12}(\delta^{\mu\rho}\delta^{\nu\sigma} - \delta^{\nu\rho}\delta^{\mu\sigma} + \epsilon^{\mu\nu\rho\sigma}). \tag{8.330}$$

From the fundamental result (8.310) we obtain by substituting $b = d = 5$ and using the semi-classical result $X^\mu \longrightarrow \mu x^\mu$ where $\mu^2 = 1/(1 + 4/n)$ the result

$$P^\mu P_\mu = \frac{4}{\Delta^2_x}. \tag{8.331}$$

This gives us the normalization of the P^a, which should be thought of as function on $\mathbf{CP}^3$, although they vanish in the semi-classical limit at the north pole p. We get then immediately the average on the fiber $\mathbf{S}^2$ given by

$$[P^\mu P^\nu]_{S^2} = \frac{1}{\Delta_x^2} g^{\mu\nu}. \tag{8.332}$$

Since the semi-classical limit of $g_{ac}\{\Theta^{a\mu}, \Theta^{c5}\}$ is zero we get also the average

$$[P^\mu \mathcal{M}^{\alpha\beta}]_{S^2} = 0. \tag{8.333}$$

8.6 Emergent matrix gravity

In this section we will follow mainly [48]. We will also use [47, 49, 127].

8.6.1 Fluctuations on fuzzy S^4_N

We return to the matrix model

$$S[D] = \frac{1}{g^2}\,\mathrm{Tr}(-[D_a, D_b][D^a, D^b] + \mu^2 D_a D^a). \tag{8.334}$$

We will allow now $a = 1, \ldots, 10$. The equations of motion read

$$[D_b, [D^b, D^a]] + \frac{\mu^2}{2} D^a = 0. \tag{8.318}$$

We will expand around the background $D_a = J_a$, $a = 1, \ldots, 5$ and $D_a = J_a = 0$, $i = 6, \ldots, 10$, with fluctuations $\mathcal{A}_a$ as

$$D_a = J_a + \mathcal{A}_a. \tag{8.336}$$

The corresponding fluctuations of the flux are given by

$$\begin{aligned} r^2[D^a, D^b] &= i\Theta^{ab}_{(D)} = iF^{ab} + i\Theta^{ab}, \quad iF^{ab} \\ &= r[J^a, \mathcal{A}^b] - r[J^b, \mathcal{A}^a] + r^2[\mathcal{A}^a, \mathcal{A}^b]. \end{aligned} \tag{8.337}$$

The definition of $\Theta^{ab}_{(J)} = \Theta^{ab}$ is obvious. Obviously, this will lead to a noncommutative gauge theory on the noncommutative brane $\mathcal{M}$ defined by the J_a or equivalently to a geometric deformation of x_a: $\mathcal{M} \hookrightarrow \mathbf{R}^{10}$.

We expand the fluctuation $\mathcal{A}_a$, keeping only tensors of rank up to 3, into tangential and radial components as

$$\mathcal{A}^a = \hat{\kappa}\frac{X^a}{R} + \hat{\xi}^a + i\Theta^{ab}\hat{A}_b. \tag{8.338}$$

The fields $\hat{\kappa}$ (radial fluctuation) and $\hat{\xi}^a$, $\hat{A}_b$ (tangential deformations) are given by

$$\hat{\kappa} = \kappa + \kappa_{bc}\mathcal{M}^{bc} + \cdots. \tag{8.339}$$

$$\hat{\xi}^a = \xi^a + \xi^{abc}\mathcal{M}_{bc} + \cdots. \tag{8.340}$$

$$\hat{A}_b = A_b + A_{bcd}\mathcal{M}^{cd} + \cdots. \tag{8.341}$$

The tensor fields κ, κ_{bc}, ξ^a, ξ^{abc}, A_b and A_{bcd} are functions on fuzzy $\mathbf{S}^4_N$, i.e. $\in C_N(\mathbf{S}^4)$. The modes ξ^a are redundant with the trace sector of the A_{bcd} modes. The tangential deformation $\hat{A}_a$ corresponds to noncommutative gauge theory. Indeed, $\hat{A}_a$ is a $u(1) \times so(5)$-valued gauge field corresponding to symplectomorphisms on the bundle $\mathbf{CP}^3$. More generally, the full expansion into higher spin modes is captured by allowing $\hat{A}_a$, $\hat{\xi}^a$ and $\hat{\kappa}$ to take value in the universal enveloping algebra $U(so(5))$.

Let us consider these fields near the north pole $p \in \mathbf{S}^4$. We will change notation here so that Greek indices refer now to four dimensions μ, $\nu = 1, 2, 3, 4$. Recall that $P_\mu = \mathcal{M}_{\mu 5}/R$ and $\Theta_{\mu\nu} = r^2\mathcal{M}_{\mu\nu}$. The above expansion reads

$$\mathcal{A}^\mu = \hat{\kappa}\frac{X^\mu}{R} + \hat{\xi}^\mu + i\Theta^{\mu\nu}\hat{A}_\nu + ir^2RP^\mu\hat{A}_5. \tag{8.342}$$

$$\mathcal{A}^5 = \hat{\kappa}\frac{X^5}{R} + \hat{\xi}^5 - ir^2RP^\mu\hat{A}_\mu. \tag{8.343}$$

We consider the semi-classical limit of this formula. First, at the north pole $X^5 = R$. Second, $\hat{\xi}^a$ is a tangential deformation and hence $\hat{\xi}^5 = 0$. Then, $r^2R \sim L^4_{\text{NC}}/R \longrightarrow 0$. The semi-classical limit of the above expansion becomes then

$$\mathcal{A}^\mu = \hat{\kappa}\frac{x^\mu}{R} + \hat{\xi}^\mu + \theta^{\mu\nu}\hat{A}_\nu. \tag{8.344}$$

$$\mathcal{A}^5 = \hat{\kappa}. \tag{8.345}$$

The expansions of the various fields become (with $\kappa_\mu = 2R\kappa_{\mu 5}$, $\xi^{\mu\nu} = 3R\xi^{\mu\nu 5}$, $A_{\mu\nu} = 2RA_{\mu\nu 5}$)

$$\hat{\kappa} = \kappa + \kappa_\mu P^\mu + \kappa_{\mu\nu}\mathcal{M}^{\mu\nu} + \cdots \tag{8.346}$$

$$\hat{\xi}^\mu = \xi^\mu + \xi^{\mu\nu}P_\nu + \xi^{\mu\nu\rho}\mathcal{M}_{\nu\rho} + \cdots \tag{8.347}$$

$$\hat{A}_\mu = A_\mu + A_{\mu\nu}P^\nu + A_{\mu\nu\rho}\mathcal{M}^{\nu\rho} + \cdots. \tag{8.348}$$

The radial deformation κ is the only mode which modifies the embedding of $\mathbf{S}^4$ in target space. This mode contributes also to the conformal metric. The metric $h_\mu\nu$ is the symmetric part of $A_\mu\nu$, viz

$$A_{\mu\nu} = \frac{1}{2}(h_{\mu\nu} + a_{\mu\nu}). \tag{8.349}$$

The mode $A_\mu\nu\rho$ is antisymmetric in the last two indices. The trace part of this mode is then given by $A_{\mu\nu\rho} = g_{\mu\nu}B_\rho - g_{\mu\rho}B_\nu$. The contribution of this trace part to $\mathcal{A}^\mu$ is given by

$$\mathcal{A}^\mu = \theta^{\mu\nu}A_{\nu\alpha\beta}\mathcal{M}^{\alpha\beta} = \frac{2i}{r^2}g_{\nu\alpha}\theta^{\nu\mu}\theta^{\alpha\beta}B_\beta = 2iR^2(P^TB)^\mu. \tag{8.350}$$

The trace part is then redundant with ξ^{μ}. The modes $A_{\mu}\nu$ and $A_{\mu}\nu\rho$ are tangential in the first index similarly to A_{μ}. The field A_5 leads to a tangential contribution but it drops out. The mode A_{μ} is a $U(1)$ gauge field, the symmetric part $h_{\mu}\nu$ of the mode $A_{\mu}\nu$ is the metric, and $A_{\mu}\nu\rho$ is an $SO(4)$-spin connection. These modes are local degrees of freedom on $\mathbf{S}^4$, i.e. their averages on the fiber $\mathbf{S}^2$ vanish.

8.6.2 Gauge transformations

Finite gauge transformations are given by

$$D^{a}\longrightarrow UD^{a}U^{-1}, \qquad U\in\mathcal{U}(\mathcal{H}). \tag{8.351}$$

Recall that the Hilbert space $\mathcal{H}$ corresponds to the irreducible representation $(n/2, n/2)$ of $SO(5)$. We can write an infinitesimal gauge transformation as

$$U = 1 + i\Lambda + \cdots \tag{8.352}$$

Obviously the gauge parameter is function on fuzzy $\mathbf{CP}^3_N$, i.e.

$$\Lambda\in \mathrm{End}(\mathcal{H}) = \oplus_{0\leqslant k\leqslant l\leqslant n}(l-k,\, 2k)\ \oplus_k\ \Gamma^k\mathbf{S}^4. \tag{8.353}$$

$\Gamma^k\mathbf{S}^4$ can be viewed as a higher-spin tensor bundle over $\mathbf{S}^4$. The local rotation group $SO(4)$ acts on these tensors via $-i[\mathcal{M}^{\mu\nu}, \ldots]$ while the gauge group $\mathcal{U}(\mathcal{H})$ acts non-locally and mixes these tensors together. On the other hand, X^{μ} act on this bundle as derivative operators. For example, they act on functions as

$$[X^{\mu}, \phi] = \theta^{\mu\nu}\partial_{\nu}\phi. \tag{8.354}$$

We explain these things and related issues in detail now.

The infinitesimal gauge transformations are given by

$$D^{a}\longrightarrow D^{a} + i[\Lambda, D^{a}] \Rightarrow \mathcal{A}^{a}\longrightarrow\mathcal{A}^{a} + i[\Lambda, J^{a}] + i[\Lambda, \mathcal{A}^{a}]. \tag{8.355}$$

Again, we expand the gauge parameter keeping only tensors of rank up to 3 as

$$\begin{aligned}\Lambda &= \Lambda_0 + \frac{1}{2}\Lambda_{ab}\mathcal{M}^{ab} + \cdots\\ &= \Lambda_0 + v_{\mu}P^{\mu} + \frac{1}{2}\Lambda_{\mu\nu}\mathcal{M}^{\mu\nu} + \cdots\end{aligned} \tag{8.356}$$

The gauge parameters Λ_0, Λ_{ab}, $v_{\mu} = R\Lambda_{\mu 5}$, etc, are functions on fuzzy $\mathbf{S}^4_N$, i.e. they are $\in C_N(\mathbf{S}^4)$. The complete gauge field $\mathcal{A}_a$ should be decomposed similarly as (8.338) or equivalently as

$$\mathcal{A}^{a} = \mathcal{A}^{a}_{0} + ir^2\mathcal{M}^{ab}\hat{A}_b + \cdots, \tag{8.357}$$

where the definition of $\mathcal{A}^{a}_{0}$ is obvious and the gauge field is given by

$$\hat{A}_b = A_b + A_{bcd}\mathcal{M}^{cd} + \cdots \tag{8.358}$$

The gauge parameter Λ_{ab} generates an $SO(5)$ transformation of the gauge field $\hat{A}_a$ which depends on where we are on $\mathbf{S}^4$. This is therefore a local $SO(5)$ symmetry, and the field $\hat{A}_a$ is a noncommutative $u(1) \times so(5)$-valued gauge field.

We consider the semi-classical limit at the north pole $x^\mu = 0$, $x^5 = R$ where commutators are replaced by Poisson brackets. Around the north pole we can assume that the noncommutativity $\theta_{\mu\nu}(x, \xi) = i\langle x, \xi|\Theta_{\mu\nu}|x, \xi\rangle$ is independent of x, viz $\partial_\mu\theta^{\mu\nu} = 0$. We can approximate the fuzzy $\mathbf{S}^4_N$ here with the Moyal–Weyl noncommutativity $[X^\mu, X^\nu] = i\Theta^{\mu\nu}$ with Θ constant. In particular, derivations $P^\mu \sim i\partial_\mu$ are approximated by

$$P^\mu = (\Theta^{-1})^{\mu\nu}X_\nu \Rightarrow \langle x, \xi|[X_\mu, f]|x, \xi\rangle = \{x_\mu, f\} = \theta_{\mu\nu}\partial^\nu f. \tag{8.359}$$

We have also used the usual star product to compute

$$\langle x, \xi|[f, g]|x, \xi\rangle = \{f, g\} = \theta_{\mu\nu}\partial^\mu f\partial^\nu g. \tag{8.360}$$

We compute then immediately

$$\delta_{\Lambda_0}X^\mu \equiv i[\Lambda_0, X^\mu] \sim -i\theta^{\mu\nu}\partial_\nu\Lambda_0. \tag{8.361}$$

$$\begin{aligned}\delta_\Lambda X^\mu \equiv \frac{i}{2}[\Lambda_{\rho\sigma}\mathcal{M}^{\rho\sigma}, X^\mu] = \frac{i}{2}[\Lambda_{\rho\sigma}, X^\mu]\mathcal{M}^{\rho\sigma}\\ + \frac{i}{2}\Lambda_{\rho\sigma}[\mathcal{M}^{\rho\sigma}, X^\mu] \sim -\frac{i}{2}\theta^{\mu\nu}\partial_\nu\Lambda_{\rho\sigma}\mathcal{M}^{\rho\sigma}.\end{aligned} \tag{8.362}$$

$$\begin{aligned}\delta_v X^\mu \equiv i[v_\nu P^\nu, X^\mu] = -iv^\mu\frac{X_5}{R}\\ + i[v_\nu, X^\mu]P^\nu \sim -iv^\mu - i\theta^{\mu\nu}\partial_\nu v_\lambda P^\lambda.\end{aligned} \tag{8.363}$$

$$\begin{aligned}\delta_v\phi \equiv i[v_\nu P^\nu, \phi] = i[v_\nu, \phi]P^\nu\\ + iv_\nu[P^\nu, \phi] \sim i\theta^{\mu\nu}\partial_\mu v_\rho\partial_\nu\phi P^\rho - v_\nu\partial^\nu\phi.\end{aligned} \tag{8.364}$$

$$\begin{aligned}\delta_\Lambda\phi \equiv \frac{i}{2}[\Lambda_{\rho\sigma}\mathcal{M}^{\rho\sigma}, \phi] = \frac{i}{2}[\Lambda_{\rho\sigma}, \phi]\mathcal{M}^{\rho\sigma}\\ + \frac{i}{2}\Lambda_{\rho\sigma}[\mathcal{M}^{\rho\sigma}, \phi] \sim \frac{i}{2}\theta^{\mu\nu}\partial_\mu\Lambda_{\rho\sigma}\partial_\nu\phi\mathcal{M}^{\rho\sigma}.\end{aligned} \tag{8.365}$$

Thus, δ_Λ corresponds to the action of local $SO(4)$ rotations on tensors, whereas δ_v corresponds to the action of diffeomorphisms. We obtain then the gauge transformations

$$\begin{aligned}\delta\mathcal{A}^\mu = \delta_{\Lambda_0}\mathcal{A}^\mu + \delta_v\mathcal{A}^\mu + \delta_\Lambda\mathcal{A}^\mu\\ -\frac{i}{r}\theta^{\mu\nu}\partial_\nu\left(\Lambda_0 + v_\lambda P^\lambda + \frac{1}{2}\Lambda_{\rho\sigma}\mathcal{M}^{\rho\sigma}\right) - \frac{i}{r}v^\mu.\end{aligned} \tag{8.366}$$

Similarly we have

$$\delta\mathcal{A}^5 = \delta_{\Lambda_0}\hat{\kappa} + \delta_v\hat{\kappa} + \delta_\Lambda\hat{\kappa}. \tag{8.367}$$

Also we compute

$$\delta_{\Lambda_0}(\Theta^{\mu\nu}\hat{A}_\nu) \equiv i[\Lambda_0, \Theta^{\mu\nu}\hat{A}_\nu] \sim -i\theta^{\mu\nu}\delta_{\Lambda_0}\hat{A}_\nu \sim 0. \tag{8.368}$$

$$\begin{aligned}\delta_v(\Theta^{\mu\nu}\hat{A}_\nu) \equiv{}& i[v_\alpha P^\alpha, \Theta^{\mu\nu}\hat{A}_\nu]\sim\\ &- r^2 v_\rho(-g^{\rho\mu}P^\nu + g^{\rho\nu}P^\mu)\hat{A}_\nu - i\theta^{\mu\nu}\delta_v\hat{A}_\nu.\end{aligned} \tag{8.369}$$

The first term will be dropped in the semi-classical limit $r^2 \sim L_{\rm NC}^2/n \longrightarrow 0$, whereas the second term yields $i\theta^{\mu\nu}v^\alpha\partial_\alpha\hat{A}_\nu$. Also we compute

$$\begin{aligned}\delta_\Lambda(\Theta^{\mu\nu}\hat{A}_\nu) &\equiv \frac{i}{2}\left[\Lambda_{\rho\sigma}\mathcal{M}^{\rho\sigma}, \Theta^{\mu\nu}\hat{A}_\nu\right]\\ &= \frac{i}{2}\left[\Lambda_{\rho\sigma}\mathcal{M}^{\rho\sigma}, \Theta^{\mu\nu}\right]\hat{A}_\nu + \Theta^{\mu\nu}\delta_\Lambda\hat{A}_\nu\\ &\sim i\Lambda_{\rho\sigma}(g^{\rho\mu}\theta^{\sigma\nu} - g^{\rho\nu}\theta^{\sigma\mu})\hat{A}_\nu - i\theta^{\mu\nu}\delta_\Lambda\hat{A}_\nu\\ &\sim - i(\Lambda\cdot\theta\hat{A})^\mu + i\theta^{\mu\nu}(\Lambda\cdot\hat{A})_\nu.\end{aligned} \tag{8.370}$$

The action of the local rotation $\Lambda \in so(4)$ on the 4-dimensional gauge field $\hat{A}^\mu$ is defined by $(\Lambda\cdot A)_\mu = -\Lambda_{\mu\nu}A^\nu$. We can then write down the full infinitesimal gauge transformation as

$$\delta\mathcal{A}^\mu = \delta\mathcal{A}_0^\mu + \theta^{\mu\nu}\delta\hat{A}_\nu. \tag{8.371}$$

$$\begin{aligned}\delta\hat{A}_\nu ={}& - \frac{i}{r}\partial_\nu\left(\Lambda_0 + v_\lambda P^\lambda + \frac{1}{2}\Lambda_{\rho\sigma}\mathcal{M}^{\rho\sigma}\right)\\ &- v^\alpha\partial_\alpha\hat{A}_\nu - (\Lambda\cdot\hat{A})_\nu - \frac{4}{L_{\rm NC}^4}\theta_{\nu\alpha}(\Lambda\cdot\theta\hat{A})^\alpha.\end{aligned} \tag{8.372}$$

$$\delta\mathcal{A}_0^\mu = \delta\hat{\xi}^\mu = -\frac{i}{r}v^\mu - v^\alpha\partial_\alpha\xi^\mu. \tag{8.373}$$

These equations are slightly different from those found originally in [127]. In the second equation we have used $g_{\mu\nu}\theta^{\mu\alpha}\theta^{\nu\beta} = \Delta_{x8}^4 g^{\alpha\beta}/4$. We will drop the second term in this equation for the only reason that it is quadratic in the non-commutativity.

Explicitly, the infinitesimal gauge transformation reads then

$$\delta\hat{A}_\nu = \delta A_\nu + \frac{1}{2}(\delta h_{\nu\mu} + \delta a_{\nu\mu})P^\mu + \delta A_{\nu\rho\sigma}\mathcal{M}^{\rho\sigma}, \tag{8.374}$$

where

$$\delta A_\nu = -\frac{i}{r}\partial_\nu\Lambda_0 - v^\alpha\partial_\alpha A_\nu - (\Lambda\cdot A)_\nu. \tag{8.375}$$

$$\delta h_{\nu\mu} = -\frac{i}{r}(\partial_\nu v_\mu + \partial_\mu v_\nu) - v^\alpha \partial_\alpha h_{\nu\mu} - (\Lambda \cdot h)_{\nu\mu}. \quad (8.376)$$

$$\delta a_{\nu\mu} = -\frac{i}{r}(\partial_\nu v_\mu - \partial_\mu v_\nu) - v^\alpha \partial_\alpha a_{\nu\mu} - (\Lambda \cdot a)_{\nu\mu}. \quad (8.377)$$

$$\delta A_{\nu\rho\sigma} = -\frac{i}{2r}\partial_\nu \Lambda_{\rho\sigma} - v^\alpha \partial_\alpha A_{\nu\rho\sigma} - (\Lambda \cdot A)_{\nu\rho\sigma}. \quad (8.378)$$

This is a combined effect of local $SO(4)$ gauge transformations generated by $\Lambda_{\mu\nu}$, local $U(1)$ gauge transformations generated by Λ_0, and diffeomorphism transformations generated by $-v^\rho\partial_\rho$. The $U(1)$ gauge field is A_μ, the $SO(4)$ gauge field is the spin connection $A_{\mu\nu\rho}$, while the gauge fields associated with the diffeomorphisms are $h_\mu\nu$ and $a_\mu\nu$. The metric fluctuation is identified with the symmetric rank two tensor $h_\mu\nu$.

8.6.3 Emergent geometry

We rewrite the metric (8.316) on $\mathbf{S}^4_N$ in the semi-classical limit as

$$\bar{\gamma}^{\mu\nu} = g_{\alpha\beta} e^{\alpha\mu} e^{\beta\nu} = \frac{1}{4}\Delta_x^4 g^{\mu\nu}. \quad (8.379)$$

The vielbein $e^{\alpha\mu}$ is defined by

$$e^{\alpha\mu} = \theta^{\alpha\mu}, \quad e^\alpha = e^{\alpha\mu}\partial_\mu \leftarrow X^\alpha = \Theta^{\alpha\mu}P_\mu. \quad (8.380)$$

This is not a fixed frame on $\mathbf{S}^4$ since it corresponds to the bundle of self-dual tensors $\theta^{\mu\nu}$, which transforms under the local $SO(4)$ in the (1,0) representation along the fiber $\mathbf{S}^2{}_N$, and thus it averages out over the fiber, viz

$$[e^{\alpha\mu}]_{\mathbf{S}^2} = [\theta^{\alpha\mu}]_{\mathbf{S}^2} = 0. \quad (8.381)$$

The metric $\bar{\gamma}$ is however fixed on $\mathbf{S}^4$ and well defined since

$$[\bar{\gamma}^{\mu\nu}]_{\mathbf{S}^2} = \bar{\gamma}^{\mu\nu}. \quad (8.382)$$

As we have discussed, the derivative operators acting on the higher-spin tensor bundle End($\mathcal{H}$) given in equation (8.353) are given by X^a. Indeed, the derivative of a general tensor $\phi \in$ End($\mathcal{H}$) is given by

$$\mathcal{D}^\mu\phi = -i[X^\mu, \phi] = -ie^{\mu\nu}\partial_\nu\phi + \text{non-derivative terms}. \quad (8.383)$$

The non-derivative terms arise from commutators between X, on one hand, and the P and $\mathcal{M}$, which appear in the expansion of ϕ, on the other hand. However, the metric is always obtained from the leading derivative term. Indeed

$$\mathcal{D}^a\phi\mathcal{D}_a\phi = -\bar{\gamma}^{\alpha\beta}\partial_\alpha\phi\partial_\beta\phi = -[X^a, \phi][X_a, \phi]. \quad (8.384)$$

This generalizes when fluctuations are included to

$$\mathcal{D}^a\phi\mathcal{D}_a\phi = -\gamma^{a\beta}\partial_a\phi\partial_\beta\phi = -[Y^a, \phi][Y_a, \phi]. \tag{8.385}$$

The covariant derivative Y is defined by

$$Y^a = rD^a = X^a + r\mathcal{A}^a. \tag{8.386}$$

The metric can then be given by

$$-\mathcal{D}^a x^a \mathcal{D}_a x^\beta = \gamma^{a\beta} = [Y^a, X^a][Y_a, X^\beta]. \tag{8.387}$$

In summary, the curved over-complete basis is now defined by

$$\mathcal{D}^a = -i[Y^a, \ldots] = -ie^a[\mathcal{A}] = -ie^{a\nu}[\mathcal{A}]\partial_\nu. \tag{8.388}$$

After another messy calculation we find the covariant derivative (using $\hat{\phi} = \phi + \phi_\mu P^\mu + \phi_{\rho\sigma}\mathcal{M}^{\rho\sigma}$, $\mathcal{A}^{\mu(2)} = \theta^{\mu\nu}A_{\nu\rho}P^\rho$, $\mathcal{A}^{\mu(3)} = \theta^{\mu\nu}A_{\nu\rho\sigma}\mathcal{M}^{\rho\sigma}$, and setting $x^\mu = 0$ at the north pole, and also neglecting there quadratic terms in θ)

$$\begin{aligned}\mathcal{D}^\mu\hat{\phi} = \partial_\nu\phi&\left[-i\left(1 + r\frac{\hat{\kappa}}{R}\right)\theta^{\mu\nu} - ir\theta^{a\nu}\partial_a\xi^\mu + r\theta^{\mu\lambda}A^{\rho\nu}g_{\lambda\rho}\right]\\ &-\left(1 + r\frac{\hat{\kappa}}{R}\right)\phi^\mu - r\phi_\nu\partial^\nu\hat{\xi}^\mu - ir\left[\mathcal{A}_\mu^{(1)} + \mathcal{A}_\mu^{(3)}, \hat{\phi}\right].\end{aligned} \tag{8.389}$$

Since we must have $\mathcal{D}^\mu\hat{\phi} = -ie^{\mu\nu}[\mathcal{A}]\partial_\nu\phi + \cdots$ we obtain the tangential contribution to the vielbein $e^\mu\nu$ (dropping also higher modes in $\hat{\xi}$)

$$e^{\mu\nu}[\mathcal{A}] = \left(1 + r\frac{\hat{\kappa}}{R}\right)\theta^{\mu\nu} + \delta e^{\mu\nu}, \qquad \delta e^{\mu\nu} = ir\theta^{\mu\lambda}A_{\lambda\rho}g^{\rho\nu} + r\partial_a\xi^\mu\theta^{a\nu}. \tag{8.390}$$

We can drop $\hat{\kappa}$ and re-incorporate it by the replacement $h^{\mu\nu} \longrightarrow \tilde{h}^{\mu\nu} = h^{\mu\nu} + 2r\hat{\kappa}g_{\mu\nu}/R$. In any case the contribution of $\hat{\kappa}$ is subleading.

A very neat calculation gives now the metric fluctuation on fuzzy $\mathbf{S}_N^4$:

$$\gamma^{\mu\nu} = g_{a\beta}e^{a\mu}e^{\beta\nu} = \bar{\gamma}^{\mu\nu} + \delta\gamma^{\mu\nu} \sim [Y^a, X^\mu][Y_a, X^\mu], \tag{8.391}$$

where

$$\begin{aligned}\delta\gamma^{\mu\nu} &= g_{a\beta}(\theta^{a\mu}\delta e^{\beta\nu} + \delta e^{a\mu}\theta^{\beta\nu})\\ &= \frac{ir}{4}\Delta_x^4 h^{\mu\nu} + r\theta^{a\mu}\theta^{\beta\nu}\left(\partial_a\xi_\beta + \partial_\beta\xi_a\right).\end{aligned} \tag{8.392}$$

The crucial observation here is that the metric $h^{\mu\nu}$ arises from the commutators of the P modes in the gauge field $\hat{A}_\mu$, i.e. from the term $A_{\mu\nu}P^\nu$. This crucial property is absent on most other noncommutative spaces as seen as noncommutative branes in the matrix model [48]. However, as we have seen, the ξ^μ is redundant and in fact it can be gauged away, and thus the metric fluctuation does really consist only of the term h. By averaging over the internal sphere $\mathbf{S}_N^2$ (using (8.330)) we obtain the metric

$$[\delta\gamma^{\mu\nu}]_{S^2} = \frac{ir}{4}\Delta_x^4 h^{\mu\nu} + \frac{r\Delta_x^4}{12}(\delta^{\alpha\beta}\delta^{\mu\nu} - \delta^{\mu\beta}\delta^{\alpha\nu})(\partial_\alpha\xi_\beta + \partial_\beta\xi_\alpha). \tag{8.393}$$

We can now write down the action of a scalar field on fuzzy $\mathbf{S}_N^4$ as

$$S = -\frac{2}{g^2}\,\mathrm{Tr}[Y^a, \phi][Y_a, \phi] \sim -\frac{2}{g^2}\frac{\dim \mathcal{H}}{\mathrm{Vol}(\mathcal{M}^4)}\int_{\mathcal{M}} d^4x\gamma^{\mu\nu}\partial_\mu\phi\partial_\nu\phi. \tag{8.394}$$

The covariant effective metric G should be defined by

$$\gamma^{\mu\nu} = \frac{\Delta_x^4}{4}\sqrt{\det G_{\mu\nu}}\,G^{\mu\nu}. \tag{8.395}$$

But we have

$$\begin{aligned}\gamma^{\mu\nu} &= \frac{1}{4}\Delta_x^4 g^{\mu\nu} + \delta\gamma^{\mu\nu}\\ &= \frac{\Delta_x^4}{4}(g^{\mu\nu} + irh^{\mu\nu} + \cdots) \Rightarrow \gamma_{\mu\nu} = \frac{4}{\Delta_x^4}(g_{\mu\nu} - irh_{\mu\nu}).\end{aligned} \tag{8.396}$$

We define

$$G^{\mu\nu} = \alpha\frac{4}{\Delta_x^4}\gamma^{\mu\nu} \Rightarrow \frac{\Delta_x^4}{4}\sqrt{\det G_{\mu\nu}}\,G^{\mu\nu} = \left(\frac{\Delta_x^4}{4}\right)^2\frac{1}{\alpha}\sqrt{\det \gamma_{\mu\nu}}\,\gamma^{\mu\nu}. \tag{8.397}$$

Thus

$$\alpha = (\left(\frac{\Delta_x^4}{4}\right)^4 \det \gamma_{\mu\nu})^{1/2}. \tag{8.398}$$

We compute

$$\det \gamma_{\mu\nu} = \left(\frac{4}{\Delta_x^4}\right)^4(1 - irh) \Rightarrow \alpha = 1 - \frac{ir}{2}h. \tag{8.399}$$

We write

$$G^{\mu\nu} = g^{\mu\nu} + irH^{\mu\nu}. \tag{8.400}$$

We find

$$H^{\mu\nu} = h^{\mu\nu} - \frac{1}{2}hg^{\mu\nu}. \tag{8.401}$$

We will impose the so-called De Donder gauge

$$\partial_\mu h^{\mu\nu} = 0 \Rightarrow \partial_\mu H^{\mu\nu} = \frac{1}{2}\partial^\nu H. \tag{8.402}$$

The action becomes (with $\Phi = \Delta_x^2\phi/2$) given by

$$S = -\frac{2}{g^2}\mathrm{Tr}[Y^a, \phi][Y_a, \phi]\sim - \frac{1}{g^2}\frac{\Delta_x^8}{8}\frac{\dim \mathcal{H}}{\mathrm{Vol}(\mathcal{M}^4)}\int_{\mathcal{M}} d^4x\sqrt{\det G}G^{\mu\nu}\partial_\mu\Phi\partial_\nu\Phi. \tag{8.403}$$

The metric $G^{\mu\nu} = g^{\mu\nu} + irH^{\mu\nu}$ at the linearized level is thus obtained by the replacement $h^{\mu\nu}\longrightarrow H^{\mu\nu} = h^{\mu\nu} - hg^{\mu\nu}/2$. We have

$$H_{\mu\nu} = A_{\mu\nu} + A_{\nu\mu} - \frac{1}{2}hg^{\mu\nu} = A'_{\mu\nu} + A'_{\nu\mu}, \quad A'_{\mu\nu} = A_{\mu\nu} - \frac{1}{4}hg_{\mu\nu}. \tag{8.404}$$

We also introduce the effective vielbeins

$$G^{\mu\nu} = g_{\alpha\beta}\tilde{e}^{\alpha\mu}[\mathcal{A}]\tilde{e}^{\beta\nu}[\mathcal{A}], \tag{8.405}$$

where (using $\det G = 1 + irh/3$)

$$\tilde{e}^{\alpha\nu}[\mathcal{A}] = \frac{2}{\Delta_x^2}\frac{1}{(\det G)^{1/4}}e^{\alpha\nu}[\mathcal{A}] = \frac{2}{\Delta_x^2}\theta^{\alpha\lambda}(\delta_\lambda^\nu + irA'_{\lambda\rho}g^{\rho\nu}). \tag{8.406}$$

This is essentially the open string metric on the noncommutative D-branes in a strong magnetic field in the limit $\tilde{\alpha}'\longrightarrow 0$ considered in [58].

The inverse vielbeins defined by $\tilde{e}^{\alpha\nu}\tilde{e}_{\nu\beta} = \delta_\beta^\alpha$ are given by

$$\tilde{e}_{\alpha\nu}[\mathcal{A}] = \frac{\Delta_x^2}{2}(\theta^{-1})_{\alpha\lambda}(\delta_\lambda^\nu - irA'_{\lambda\rho}g^{\rho\nu}). \tag{8.407}$$

The torsion-free spin connection corresponding to the vielbein $\tilde{e}^{\alpha\mu}$ is given by [48]

$$A_{\mu;\alpha\beta} = \frac{1}{4}(-\partial_\alpha H_{\mu\beta} + \partial_\beta H_{\alpha\mu}). \tag{8.408}$$

We compute further

$$\tilde{e}^{\alpha\nu}[\mathcal{A}] - \tilde{e}^{\nu\alpha}[\mathcal{A}] = \frac{2}{\Delta_x^2}(\theta^{\alpha\nu} + \tilde{\theta}^{\alpha\nu}). \tag{8.409}$$

$$\tilde{\theta}^{\alpha\nu} = \theta^{\alpha\nu} - ir\theta^{\nu\lambda}A'_{\lambda\rho}g^{\rho\alpha} + ir\theta^{\alpha\lambda}A'_{\lambda\rho}g^{\rho\nu}. \tag{8.410}$$

The bit $rh\theta^{\alpha\nu}$ in $r\theta^{\nu\lambda}A'_{\lambda\rho}g^{\rho\alpha}$ can be neglected.

8.6.4 Emergent gauge theory

Recall that the gauge field is given by $\mathcal{A}_\mu = \xi_\mu + \theta_{\mu\nu}\hat{A}^\nu$, $\hat{A}^\lambda = A^\lambda + A^{\lambda\rho}P_\rho + A^{\lambda\rho\sigma}\mathcal{M}_{\rho\sigma}$. The corresponding fluctuations of the flux are given by

$$\begin{aligned} i\Theta_{(Y)}^{\mu\nu} &= [Y^\mu, Y^\nu] \\ &= \theta^{\mu\nu} + r\mathcal{F}^{\mu\nu}, \end{aligned} \tag{8.411}$$

where

$$\begin{aligned}\mathcal{F}_{\mu\nu} &= [X_\mu, \hat{\mathcal{A}}_\nu] - [X_\nu, \hat{\mathcal{A}}_\mu] + r[\hat{\mathcal{A}}_\mu, \hat{\mathcal{A}}_\nu] \\ &= \theta_{\mu\lambda}\partial^\lambda \xi_\nu - \theta_{\nu\lambda}\partial^\lambda \xi_\mu + \theta_{\nu\lambda}\theta_{\mu\rho}(\partial^\rho \hat{A}^\lambda - \partial^\lambda \hat{A}^\rho) \\ &\quad - i\theta_{\nu\lambda}A^{\lambda\mu} + i\theta_{\mu\lambda}A^{\lambda\nu} + r[\hat{\mathcal{A}}_\mu, \hat{\mathcal{A}}_\nu] \\ &= \theta_{\nu\lambda}\theta_{\mu\rho}(\partial^\rho \hat{A}^\lambda - \partial^\lambda \hat{A}^\rho) - i\theta_{\nu\lambda}\tilde{A}^{\lambda\mu} + i\theta_{\mu\lambda}\tilde{A}^{\lambda\nu} + r[\hat{\mathcal{A}}_\mu, \hat{\mathcal{A}}_\nu].\end{aligned} \tag{8.412}$$

The shifted gauge field is invariant under diffeomorphisms and is given by

$$\tilde{A}^{\lambda\nu} = A^{\lambda\nu} - i\partial^\lambda \xi^\nu. \tag{8.413}$$

The most important term in $[\hat{\mathcal{A}}_\mu, \hat{\mathcal{A}}_\nu]$ is $\theta_{\mu\rho}\theta_{\nu\lambda}[\hat{A}_\mu, \hat{A}_\nu]$. We get then the curvature

$$\begin{aligned}\mathcal{F}_{\mu\nu} &= \theta_{\mu\rho}\theta_{\nu\lambda}(\partial^\rho \hat{A}^\lambda - \partial^\lambda \hat{A}^\rho + r[\hat{A}_\mu, \hat{A}_\nu]) \\ &\quad - i\theta_{\nu\lambda}\tilde{A}^{\lambda\mu} + i\theta_{\mu\lambda}\tilde{A}^{\lambda\nu} \\ &= \theta_{\mu\rho}\theta_{\nu\lambda}\hat{F}^{\rho\lambda} - i\theta_{\nu\lambda}\tilde{A}^{\lambda\mu} + i\theta_{\mu\lambda}\tilde{A}^{\lambda\nu}.\end{aligned} \tag{8.414}$$

Obviously, $\hat{F}$ decomposes in terms of the $so(4) \times u(1)$ components as follows

$$\hat{F}^{\rho\lambda} = F^{\rho\lambda} + R^{\rho\lambda} + T^{\rho\lambda}, \tag{8.415}$$

where the $U(1)$ field strength F, the Riemann curvature R of the $SO(4)$ connection $\omega^\mu = A^{\lambda\rho\sigma}\mathcal{M}_{\rho\sigma}$, the linearized spin connection T are given by (with $\alpha_\mu = A_{\mu\lambda}P^\lambda$)

$$F^{\rho\lambda} = \partial^\rho A^\lambda - \partial^\lambda A^\rho. \tag{8.416}$$

$$R^{\rho\lambda} = \partial^\rho \omega^\lambda - \partial^\lambda \omega^\rho + r[\omega^\rho, \omega^\lambda]. \tag{8.417}$$

$$T^{\rho\lambda} = \partial^\rho \alpha^\lambda - \partial^\lambda \alpha^\rho + r[\omega^\rho, \alpha^\lambda] - r[\omega^\lambda, \alpha^\rho]. \tag{8.418}$$

The linearized form of T is precisely given by the spin-connection (8.408), viz (by dropping $a_{\mu\nu}$ and $h = g_{\mu\nu}h^{\mu\nu}$)

$$T_{\mu\nu} = -P^\rho(\partial_\mu A_{\rho\nu} - \partial_\nu A_{\rho\mu}) = 2P^\rho A_{\rho;\mu\nu} + \cdots \tag{8.419}$$

The geometric deformation of the background θ is then given by $\theta \longrightarrow \tilde{\theta}$ where

$$\begin{aligned}i\Theta^{\mu\nu}_{(Y)} &= \theta_{\mu\nu} + r\mathcal{F}_{\mu\nu} \\ &= \tilde{\theta}_{\mu\nu} + r\theta_{\mu\rho}\theta_{\nu\lambda}\hat{F}^{\rho\lambda},\end{aligned} \tag{8.420}$$

where

$$\tilde{\theta}_{\mu\nu} = \theta_{\mu\nu} - ir\theta_{\nu\lambda}\tilde{A}^{\lambda\mu} + ir\theta_{\mu\lambda}\tilde{A}^{\lambda\nu}. \tag{8.421}$$

The mode ξ^α in $\tilde{A}^{\rho\alpha}$ can be dropped since it is unphysical. Thus, $\tilde{\theta}$ can be viewed as deformation of θ in the background Y and this deformation is provided by $\tilde{A}^{\mu\nu}$ or

$A^{\mu\nu}$ whose symmetric part encodes also the metric. The vielbein $\tilde{e}^{\mu\nu}$ given by (8.406) is then viewed as the deformation of the vielbein $e^{\alpha\mu} = \theta^{\alpha\mu}$ (see also (8.409))

We also rewrite the above flux fluctuation as

$$\begin{aligned} i\Theta^{\mu\nu}_{(Y)} &= [Y^\mu, Y^\nu] = \theta^{\mu\rho}\theta^{\nu\lambda}\bar{F}_{\rho\lambda}, \quad \bar{F}_{\rho\lambda} \\ &= (\theta^{-1})_{\rho\rho_1}(\theta^{-1})_{\nu_1\lambda}\tilde{\theta}^{\rho_1\nu_1} + r\hat{F}_{\rho\lambda}. \end{aligned} \tag{8.422}$$

Thus the Yang–Mills action is given by

$$\begin{aligned} -\frac{1}{g^2}\mathrm{Tr}[Y^\mu, Y^\nu][Y_\mu, Y_\nu] &= -\frac{1}{g^2}\left(\frac{\Delta_x^4}{4}\right)^2 \mathrm{Tr}\, g^{\rho\rho_1}g^{\lambda\lambda_1}\bar{F}_{\rho_1\lambda_1}\bar{F}_{\rho\lambda} \\ &= -\frac{1}{g^2}\frac{\dim\mathcal{H}}{\mathrm{Vol}(\mathcal{M}^4)}\int_{\mathcal{M}} d^4x\bar{\gamma}^{\rho\rho_1}\bar{\gamma}^{\lambda\lambda_1}\bar{F}_{\rho_1\lambda_1}\bar{F}_{\rho\lambda}. \end{aligned} \tag{8.423}$$

We concentrate on the gauge field A^λ. Then

$$\begin{aligned} -\frac{1}{g^2}\mathrm{Tr}[Y^\mu, Y^\nu][Y_\mu, Y_\nu] &= -\frac{r^2}{g^2}\frac{\dim\mathcal{H}}{\mathrm{Vol}(\mathcal{M}^4)}\left(\frac{\Delta_x^4}{4}\right)^2 \\ &\times \int_{\mathcal{M}} d^4x g^{\rho\rho_1}g^{\lambda\lambda_1}F_{\rho_1\lambda_1}F_{\rho\lambda}. \end{aligned} \tag{8.424}$$

The covariant form of this action is (which can be shown using the Seiberg–Witten map [49, 58])

$$\begin{aligned} -\frac{1}{g^2}\mathrm{Tr}[Y^\mu, Y^\nu][Y_\mu, Y_\nu] &= -\frac{r^2}{g^2}\frac{\dim\mathcal{H}}{\mathrm{Vol}(\mathcal{M}^4)}\left(\frac{\Delta_x^4}{4}\right)^2 \\ &\times \int_{\mathcal{M}} d^4x\sqrt{\det G}\, G^{\rho\rho_1}G^{\lambda\lambda_1}F_{\rho_1\lambda_1}F_{\rho\lambda}. \end{aligned} \tag{8.425}$$

However, we should insist that the gauge field considered so far is a $U(1)$ gauge field which contributes really to the gravity sector in the matrix theory. We should therefore consider the addition of $SU(n)$ gauge field with the correct scaling $\bar{A}^\mu = \tilde{e}^{\mu\nu}A_\nu$, which mimics $[Y^\mu, \ldots] = e^{\mu\nu}[\mathcal{A}]\partial_\nu$ and thus leads to $i\Theta_{\mu\nu} = r\tilde{e}_{\mu\rho}\tilde{e}_{\nu\lambda}\hat{F}^{\rho\lambda} + \cdots$, to be able to reproduce the above action. The metric G in the above equation is precisely the string metric (8.405).

We compute also the gauge condition

$$\begin{aligned} f(\mathcal{A}) = -i[X^a, \mathcal{A}_a] &= -i[X^5, \mathcal{A}_5] - i[X^\mu, \mathcal{A}_\mu] \\ &= -i\theta^{5\mu}\partial_\mu\hat{\kappa} - i[X^\mu, \mathcal{A}_\mu] \\ &= r^2RP^\mu\partial_\mu\hat{\kappa} - i[X^\mu, \mathcal{A}_\mu] \\ &= \frac{r\Delta_x^2}{2}P^\mu\partial_\mu\hat{\kappa} - i\theta^{\mu\nu}\partial_\nu\xi_\mu + \theta_{\mu\nu}A^{\mu\nu} - \frac{i}{4}\Delta_x^4\partial_\nu\hat{A}^\nu. \end{aligned} \tag{8.426}$$

By inspection we get then the detailed gauge conditions

$$-\frac{i}{4}\Delta_x^4\partial_\nu A^\nu = 0. \tag{8.427}$$

$$-\frac{i}{4}\Delta_x^2\partial_\nu(h^{\nu\rho} + a^{\nu\rho}) + r\partial^\rho\hat{\kappa} = 0. \tag{8.428}$$

$$\theta_{\rho\sigma}(-R^2\partial_\nu A^{\nu\rho\sigma} - i\partial^\rho\xi^\sigma + a^{\rho\sigma}) = 0. \tag{8.429}$$

The first equation gives the Lorentz condition. Since $r\hat{\kappa}/\Delta_x^2 \sim 0$ we have the solution $\partial^\mu h_{\mu\nu} = 0$, $a_\mu \nu = 0$, $\xi = 0$, $\partial_\nu A^{\nu\rho\sigma} = 0$.

8.6.5 Emergent gravity: Einstein equations

By expanding the action (8.334) up to second order in $\mathcal{A}^a$, and using Jacobi identity appropriately, we get [142]

$$\begin{aligned} S[D] = S[J] + \frac{2}{g^2}\,\mathrm{Tr}\Bigg(2\mathcal{A}^a\left(\Box + \frac{\mu^2}{2}\right)J_a \\ + \mathcal{A}_a\left(\Box + \frac{\mu^2}{2}\right)\mathcal{A}^a - 2[J_a, J_b][\mathcal{A}^a, \mathcal{A}^b] - f^2\Bigg). \end{aligned} \tag{8.430}$$

We have redefined the Laplacian on the space of matrix configurations by the formula

$$\Box = [J_a, [J^a, \ldots]]. \tag{8.431}$$

Also f is given by

$$f = i[\mathcal{A}^a, J_a]. \tag{8.432}$$

By assuming that J_a solves the classical equations of motion and also adding a suitable Faddeev–Popov gauge fixing term in the Feynman gauge we get [142, 143]

$$S[D] = S[J] + \frac{2}{g^2}\,\mathrm{Tr}\,\mathcal{A}^a\left(\left(\Box + \frac{\mu^2}{2}\right)\right)g_{ab} + 2i[\mathcal{M}_{ab}, \ldots]\mathcal{A}^b. \tag{8.433}$$

Recall that $i\mathcal{M}^{ab} = [J^a, J^b]$. We define the vector–matrix Laplacian acting on gauge configurations by

$$(\mathcal{D}^2\mathcal{A})_a = \left(\Box + \frac{\mu^2}{2} - M_{rs}^{(A)}[\mathcal{M}^{rs}, \ldots]\right)_{ab}\mathcal{A}^b. \tag{8.434}$$

We have introduced in this last equation the vector representation of $SO(5)$ generators given by

$$(M_{rs}^{(A)})_{ab} = i(\delta_{rb}\delta_{sa} - \delta_{ra}\delta_{sb}). \tag{8.435}$$

The action of the quadratic fluctuations takes then the form

$$S[D] = S[J] + \frac{2}{g^2}\,\mathrm{Tr}\,\mathcal{A}^a(\mathcal{D}^2\mathcal{A})_a. \tag{8.436}$$

We consider now the gravitational ansatz (by setting the redundant mode ξ^μ to zero and dropping the $U(1)$ gauge field for simplicity)

$$\mathcal{A}^\mu_{\rm gr} = \theta^{\mu\nu}(P^\sigma A_{\nu\sigma} + \mathcal{M}^{\rho\sigma}A_{\nu\rho\sigma}), \quad \mathcal{A}^5_{\rm gr} = \kappa. \tag{8.437}$$

We have used that $[P_\mu, g] = i\partial_\mu g$, $[\Theta^{\mu\nu}, g] \sim x\partial_x g \sim 0$. We will also use $[f, g] = \theta^{\mu\nu}\partial_\mu f\,\partial_\nu g$. We compute

$$\begin{aligned}\Box(\theta_{\mu\nu}P_\sigma A^{\nu\sigma}) &= \Box(\theta_{\mu\nu}P_\sigma)A^{\nu\sigma} + (\theta_{\mu\nu}P_\sigma)\Box A^{\nu\sigma} + 2[J^a, \theta_{\mu\nu}P_\sigma][J_a, A^{\nu\sigma}]\\ &\sim 4\theta_{\mu\nu}P_\sigma.\, A^{\nu\sigma} + (\theta_{\mu\nu}P_\sigma)\Box A^{\nu\sigma} - 2i\theta_{\mu\nu}\mathcal{M}_{\sigma\rho}\partial^\sigma A^{\nu\rho}.\end{aligned} \tag{8.438}$$

$$\begin{aligned}\Box(\theta_{\mu\nu}\mathcal{M}_{\rho\sigma}A^{\nu\rho\sigma}) &= \Box(\theta_{\mu\nu}\mathcal{M}_{\rho\sigma})A^{\nu\rho\sigma} + (\theta_{\mu\nu}\mathcal{M}_{\rho\sigma})\Box A^{\nu\rho\sigma}\\ &\quad + 2[J^a, \theta_{\mu\nu}\mathcal{M}_{\rho\sigma}][J_a, A^{\nu\rho\sigma}]\\ &\sim 4\theta_{\mu\nu}\mathcal{M}_{\rho\sigma}.\, A^{\nu\rho\sigma} + (\theta_{\mu\nu}\mathcal{M}_{\rho\sigma})\Box A^{\nu\rho\sigma}.\end{aligned} \tag{8.439}$$

We have used $\Box\theta_{\mu\nu} = 2\theta_{\mu\nu}$ and $\Box P_\sigma = 2P_\sigma$. Thus

$$\left(\Box + \frac{\mu^2}{2}\right)_{\mu\nu}\mathcal{A}^\nu_{\rm gr} = \theta_{\mu\nu}P_\sigma\left(\Box + \frac{1}{2}\mu^2 + 4\right)A^{\nu\sigma} + \theta_{\mu\nu}\mathcal{M}_{\rho\sigma}\left(\Box + \frac{1}{2}\mu^2 + 4\right)A^{\nu\rho\sigma}. \tag{8.440}$$

Also

$$(M^{(A)}_{rs}[\mathcal{M}^{rs}, \ldots])_{\mu\nu}\mathcal{A}^\nu_{\rm grad} = 2i\big([\mathcal{M}_{\nu\mu}, \theta^{\nu\lambda}P^\sigma]A_{\lambda\sigma} + [\mathcal{M}_{\nu\mu}, \theta^{\nu\lambda}\mathcal{M}^{\sigma\rho}]A_{\lambda\sigma\rho}\big). \tag{8.441}$$

We compute (using $\theta^{\mu\nu}P_\nu = 0$ at the point p)

$$[\mathcal{M}_{\nu\mu}, \theta^{\nu\lambda}P^\sigma] = 2i\theta^\lambda_\mu P^\sigma + i\theta^{\sigma\lambda}P_\mu. \tag{8.442}$$

$$[\mathcal{M}_{\nu\mu}, \theta^{\nu\lambda}\mathcal{M}^{\sigma\rho}] = \frac{1}{r^2}\left(2\theta^\lambda_\mu\theta^{\sigma\rho} + \theta^{\sigma\lambda}\theta^\rho_\mu - \theta^{\rho\lambda}\theta^\sigma_\mu + \bar{\gamma}^{\lambda\sigma}g^{\ \rho}_\mu - \bar{\gamma}^{\lambda\rho}g^{\ \sigma}_\mu\right). \tag{8.443}$$

Thus

$$\begin{aligned}(M^{(A)}_{rs}[\mathcal{M}^{rs}, \ldots])_{\mu\nu}\mathcal{A}^\nu_{\rm grad} &= -4\theta_{\mu\nu}P_\sigma A^{\nu\sigma} - 2\theta^{\sigma\lambda}P_\mu A_{\lambda\sigma}\\ &\quad - 4\theta_{\mu\lambda}\mathcal{M}_{\rho\sigma}A^{\lambda\rho\sigma} - 2\mathcal{M}^{\sigma\rho}\theta^{\mu\nu}(A_{\rho\sigma\nu} - A_{\rho\nu\sigma})\\ &\quad + \frac{4i}{r^2}\bar{\gamma}^{\mu\rho}g^{\nu\sigma}A_{\nu\sigma\rho}.\end{aligned} \tag{8.444}$$

Hence

$$\begin{aligned}\mathcal{D}^2_{\mu\nu}\mathcal{A}^\nu_{\rm gr} = \theta_{\mu\nu}P_\sigma\left(\Box + \frac{1}{2}\mu^2 + 8\right)A^{\nu\sigma} + \theta_{\mu\nu}\mathcal{M}_{\rho\sigma}\left(\Box + \frac{1}{2}\mu^2 + 8\right)A^{\nu\rho\sigma}\\ + 2\theta^{\sigma\lambda}P_\mu A_{\lambda\sigma} + 2\mathcal{M}_{\sigma\rho}\theta_{\mu\nu}(A^{\rho\sigma\nu} - A^{\rho\nu\sigma})\\ - \frac{4i}{r^2}\bar{\gamma}_{\mu\rho}g_{\nu\sigma}A^{\nu\sigma\rho} - 2i\theta_{\mu\nu}\mathcal{M}_{\sigma\rho}\partial^\sigma A^{\nu\rho}.\end{aligned} \tag{8.445}$$

We use also $\bar{\gamma}_{\mu\rho}g_{\nu\sigma}A^{\nu\sigma\rho} = -ir^2\theta_{\mu\nu}g^{\nu\sigma}A^{\alpha\beta\rho}g_{\alpha\beta}\mathcal{M}_{\sigma\rho}$. The term $2\theta^{\sigma\lambda}P_\mu A_{\lambda\sigma}$ is a higher mode of $\mathcal{A}^\mu_{\rm gr}$ and thus can be dropped. We get

$$\begin{aligned}\mathcal{D}^2_{\mu\nu}\mathcal{A}^\nu_{\rm gr} = \theta_{\mu\nu}P_\sigma\left(\Box + \frac{1}{2}\mu^2 + 8\right)A^{\nu\sigma} + \theta_{\mu\nu}\mathcal{M}_{\rho\sigma}\left(\Box + \frac{1}{2}\mu^2 + 8\right)A^{\nu\rho\sigma}\\ + 2\mathcal{M}_{\sigma\rho}\theta_{\mu\nu}(A^{\rho\sigma\nu} - A^{\rho\nu\sigma})\\ - 4\theta_{\mu\nu}g^{\nu\sigma}A^{\alpha\beta\rho}g_{\alpha\beta}\mathcal{M}_{\sigma\rho} - 2i\theta_{\mu\nu}\mathcal{M}_{\sigma\rho}\partial^\sigma A^{\nu\rho}.\end{aligned} \tag{8.446}$$

Similarly, we compute

$$(\mathcal{D}^2)_{55}\mathcal{A}^5_{\rm gr} = \left(\Box + \frac{1}{2}\mu^2 + 4\right)\kappa. \tag{8.447}$$

The equations of motion $(\mathcal{D}^2\mathcal{A}_{\rm gr})_a = 0$ give then the detailed equations of motion

$$\left(\Box + \frac{1}{2}\mu^2 + 4\right)\kappa = 0. \tag{8.448}$$

The linear term will drop under averaging over the fiber. Also

$$\left(\Box + \frac{1}{2}\mu^2 + 8\right)A^{\nu\sigma} = 0. \tag{8.449}$$

$$\left(\Box + \frac{1}{2}\mu^2 + 8\right)A^{\nu\rho\sigma} + P_{\rm mix}A^{\nu\rho\sigma} + 2g^{\nu\sigma}A^{\alpha\beta\rho}g_{\alpha\beta} - 2g^{\nu\rho}A^{\alpha\beta\sigma}g_{\alpha\beta} = -i\partial^\sigma A^{\nu\rho} + i\partial^\rho A^{\nu\sigma}, \tag{8.450}$$

where

$$P_{\rm mix}A^{\nu\rho\sigma} = -(A^{\rho\sigma\nu} - A^{\rho\nu\sigma}) + (A^{\sigma\rho\nu} - A^{\sigma\nu\rho}). \tag{8.451}$$

Because of (8.449) a solution of (8.450) is given by a solution of the inhomogeneous equation

$$P_{\rm mix}A^{\nu\rho\sigma} + 2g^{\nu\sigma}A^{\alpha\beta\rho}g_{\alpha\beta} - 2g^{\nu\rho}A^{\alpha\beta\sigma}g_{\alpha\beta} = -i\partial^\sigma A^{\nu\rho} + i\partial^\rho A^{\nu\sigma}, \tag{8.452}$$

We find explicitly [49]

$$A_{\nu\sigma\rho} = \frac{1}{4}(\partial_\rho H_{\nu\sigma} - \partial_\sigma H_{\nu\rho}) = A_{\nu;\sigma\rho}. \tag{8.453}$$

Thus $A_{\nu\sigma\rho}$ is the torsion-free spin connection of $H_{\mu\nu}$. The general solution will be given by this torsion-free spin connection plus a torsion wave solution of the free wave equation, i.e. the equation (8.450) without the $\partial_\mu A_{\alpha\beta}$ terms.

We need to compute the quadratic action for the fluctuation

$$S[\mathcal{A}] = \frac{2}{g^2}\mathrm{Tr}\,\mathcal{A}^a\left(\left(\Box + \frac{\mu^2}{2}\right)\right)g_{ab} + 2i[\mathcal{M}_{ab}, \ldots]\mathcal{A}^b. \tag{8.454}$$

By using the averages (8.330), (8.332) and (8.333) we compute the average over the fiber $\mathbf{S}^2$ of the following term:

$$\begin{aligned}\left[\mathcal{A}^\mu_{\rm gr}\cdot\theta_{\mu\nu}P_\sigma\left(\Box + \frac{\mu^2}{2} + 8\right)\mathcal{A}^{\nu\sigma}\right]_{\mathbf{S}^2} &= \left[\theta^{\mu\lambda}P^\rho A_{\lambda\rho}\cdot\theta_{\mu\nu}P_\sigma\left(\Box + \frac{\mu^2}{2} + 8\right)A^{\nu\sigma}\right]_{\mathbf{S}^2}\\ &= \frac{\Delta_x^2}{4}A^{\lambda\sigma}\left(\Box + \frac{\mu^2}{2} + 8\right)A_{\lambda\sigma}\\ &= \frac{\Delta_x^2}{16}h^{\lambda\sigma}\left(\Box + \frac{\mu^2}{2} + 8\right)h_{\lambda\sigma}\\ &\quad + \frac{\Delta_x^2}{16}a^{\lambda\sigma}\left(\Box + \frac{\mu^2}{2} + 8\right)a_{\lambda\sigma}.\end{aligned} \tag{8.455}$$

$$[\mathcal{A}^5_{\rm gr}(\mathcal{D}^2\mathcal{A}_{\rm gr})_5]_{\mathbf{S}^2} = \kappa\left(\Box + \frac{\mu^2}{2} + 4\right)\kappa. \tag{8.456}$$

$$\begin{aligned}&\left[\mathcal{A}^\mu_{\rm gr}\cdot\theta_{\mu\nu}\mathcal{M}_{\rho\sigma}\left(\left(\Box + \frac{\mu^2}{2} + 8\right)\right)\mathcal{A}^{\nu\rho\sigma} + \cdots\right]_{\mathbf{S}^2}\\ &= \left[\theta^{\mu\lambda}\mathcal{M}^{\alpha\beta}A_{\lambda\alpha\beta}\cdot\theta_{\mu\nu}\mathcal{M}_{\rho\sigma}\left(\left(\Box + \frac{\mu^2}{2} + 8\right)\right)A^{\nu\rho\sigma} + \cdots\right]_{\mathbf{S}^2}\\ &= -\frac{\Delta_x^4}{4r^4}\left[A_{\nu\alpha\beta}\cdot\theta^{\alpha\beta}\theta_{\rho\sigma}\left(\left(\Box + \frac{\mu^2}{2} + 8\right)\right)A^{\nu\rho\sigma} + \cdots\right]_{\mathbf{S}^2}\\ &= -\frac{\Delta_x^8}{48r^4}\left[\epsilon^{\alpha\beta\rho\sigma}g^{\nu\mu}A_{\nu\alpha\beta}\cdot\left(\left(\Box + \frac{\mu^2}{2} + 8\right)\right)A_{\mu\rho\sigma} + \cdots\right]_{\mathbf{S}^2}.\end{aligned} \tag{8.457}$$

We will neglect this last term. The quadratic action for the fluctuation fields $h_{\mu\nu}$, $a_{\mu\nu}$ and κ is given by

$$\begin{aligned}S[\mathcal{A}] = \frac{2}{g^2}\frac{\dim\mathcal{H}}{\mathrm{Vol}(\mathcal{M}^4)}\int_{\mathcal{M}} d^4x\Bigg[&\frac{\Delta_x^2}{16}h^{\lambda\sigma}\left(\Box + \frac{\mu^2}{2} + 8\right)h_{\lambda\sigma}\\ &+ \frac{\Delta_x^2}{16}a^{\lambda\sigma}\left(\Box + \frac{\mu^2}{2} + 8\right)a_{\lambda\sigma} + \kappa\left(\Box + \frac{\mu^2}{2} + 4\right)\kappa\Bigg].\end{aligned} \tag{8.458}$$

Coupling to matter will be of the canonical form (re-incorporating also the contribution of the radial component κ by the replacement $h_{\mu\nu}\longrightarrow\tilde{h}_{\mu\nu}= h_{\mu\nu} + 2r\kappa g_{\mu\nu}/R$)

$$S[\Phi] = \frac{\dim \mathcal{H}}{\mathrm{Vol}(\mathcal{M}^4)}\frac{\Delta_x^8}{8g^2}\int_{\mathcal{M}} d^4x\tilde{h}^{\mu\nu}T_{\mu\nu}[\Phi]. \tag{8.459}$$

The equations of motion are then given by

$$\left(\Box + \frac{\mu^2}{2} + 8\right)h_{\lambda\sigma} = -\frac{\Delta_x^6}{2}T_{\mu\nu}. \tag{8.460}$$

$$\left(\Box + \frac{\mu^2}{2} + 8\right)a_{\lambda\sigma} = 0. \tag{8.461}$$

$$\left(\Box + \frac{\mu^2}{2} + 4\right)\kappa = -\frac{r\Delta_x^8}{16R}T. \tag{8.462}$$

Thus we can see that we can solve with $a_{\mu\nu} = 0$. Also the gauge condition $\partial_\mu h^{\mu\nu} = 0$ is consistent with the equation of motion of $h^\mu\nu$ with a conserved energy–momentum tensor.

We have the metric

$$G^{\mu\nu} = g^{\mu\nu} + irH^{\mu\nu}. \tag{8.400}$$

The corresponding Ricci tensor is given by (see [144] equation (4.4.4))

$$R^{\mu\nu}[G] = R^{\mu\nu}[g] + ir\left(\frac{1}{2}\partial^\mu\partial^\nu H + \frac{1}{2}\partial_\alpha\partial^\alpha H^{\mu\nu} - \frac{1}{2}\partial^\mu\partial_\rho H^{\nu\rho} - \frac{1}{2}\partial^\nu\partial_\rho H^{\mu\rho}\right). \tag{8.464}$$

The Ricci tensor on the sphere $\mathbf{S}^4$ is given by

$$R^{\mu\nu}[g] = \frac{3}{R^2}g_{\mu\nu}. \tag{8.465}$$

In the De Donder gauge (8.402) we find the Ricci tensor and the Ricci scalar

$$R^{\mu\nu}[G] = \frac{3}{R^2}g^{\mu\nu} + \frac{ir}{2}\partial_\alpha\partial^\alpha H^{\mu\nu}. \tag{8.466}$$

$$R[G] = G_{\mu\nu}R^{\mu\nu}[G] = \frac{12}{R^2} + \frac{ir}{2}\partial_\alpha\partial^\alpha H - \frac{3ir}{R^2}H. \tag{8.467}$$

The Einstein tensor is then

$$\begin{aligned}\mathcal{G}^{\mu\nu}[G] &= R^{\mu\nu}[G] - \frac{1}{2}G_{\mu\nu}R[G]\\ &= -\frac{3}{R^2}g^{\mu\nu} + \frac{1}{2}\partial_\alpha\partial^\alpha h^{\mu\nu} + \frac{3ir}{2R^2}(4H_{\mu\nu} + g_{\mu\nu}H).\end{aligned} \tag{8.468}$$

By using $[X^\mu, \phi] = i\theta^{\mu\nu}\partial_\nu\phi$ we can rewrite the equation of motion (8.460) as

$$\left(\frac{1}{4}\partial_\alpha\partial^\alpha - m_h^2\right)h_{\mu\nu} = r^2\Delta_x^2 T_{\mu\nu}. \tag{8.469}$$

The mass m_h^2 is given by

$$m_h^2 = \frac{r^2}{\Delta_x^4}\left(\frac{\mu^2}{2} + 8\right) = \frac{2}{R^2}, \quad \mu^2 \longrightarrow 0. \tag{8.470}$$

By neglecting this mass we obtain the equation of motion

$$\frac{1}{4}\partial_\alpha\partial^\alpha h_{\mu\nu} = r^2\Delta_x^2 T_{\mu\nu}. \tag{8.471}$$

By neglecting also the other terms which are proportional to the background curvature $1/R^2$ we obtain the Einstein tensor

$$\mathcal{G}^{\mu\nu} = 2r^2\Delta_x^2 T_{\mu\nu} = 8\pi G r^2 T_{\mu\nu}. \tag{8.472}$$

The Newton constant is given by

$$G = \frac{\Delta_x^2}{4\pi} = \frac{R^2}{\pi n}. \tag{8.473}$$

The Planck scale is thus suppressed compared to the cosmological scale by the quantization integer n.

8.7 Emergent quantum gravity from multitrace matrix models

In summary, there exists hidden inside any noncommutative $U(1)$ gauge theory a gravitational theory. This is the idea of emergent noncommutative/matrix gravity which is essentially the same idea (but certainly much simpler) as the one found in the AdS/CFT correspondence and gauge/gravity duality in general.

The idea that matrices can capture curvature is also discussed in [145]. This certainly works in the case of quantum gravity in two dimensions which is known to emerge from random Riemannian surfaces [146]. By analogy, a proposal for quantum gravity in higher dimensions emerging from random spaces has recently been put forward in [147]. This works for spaces which are given by finite spectral triples (for example fuzzy spaces) and the underlying matrix models are necessarily multitrace.

8.8 Exercise

Exercise: We consider the IKKT matrix model

$$S_{\text{IKKT}} = \frac{1}{g^2}\,\text{Tr}\left(\frac{1}{4}[X_\mu, X_\nu][X^\mu, X^\nu] + \frac{1}{2}\bar{\Psi}^\alpha\Gamma^\mu_{\alpha\beta}[X_\mu, \Psi^\beta]\right). \tag{8.474}$$

Write a Monte Carlo code for the Euclidean IKKT model in various dimensions (the IKKT matrix model in ten dimensions is also called type IIB matrix model). Consider also adding to the action a mass term for the matrices X_a and a Chern–Simons (Myers) term in the directions 1, 2 and 3.

We consider then the Lorentzian IKKT matrix models with the IR and UV regulators explained in the main text. Write a Monte Carlo. Verify numerically the results discussed in the main text regarding the emergence of (1 + 3)-dimensional Minkowski spacetime, the spontaneous symmetry breaking of rotational symmetry, as well as the expansion (exponential at early times and power law at late times) of the Universe.

Solution: See appendix for the description of various possible algorithms that can be used.

References

[1] Nishimura J, Okubo T and Sugino F 2011 Systematic study of the SO(10) symmetry breaking vacua in the matrix model for type IIB superstrings *J. High Energy Phys.* **1110** 135

[2] Ishibashi N, Kawai H, Kitazawa Y and Tsuchiya A 1997 A Large N reduced model as superstring *Nucl. Phys.* B **498** 467

[3] Aoki H, Iso S, Kawai H, Kitazawa Y and Tada T 1998 Space-time structures from IIB matrix model *Prog. Theor. Phys.* **99** 713

[4] Krauth W, Nicolai H and Staudacher M 1998 Monte Carlo approach to M theory *Phys. Lett.* B **431** 31

[5] Austing P and Wheater J F The Convergence of Yang–Mills integrals *J. High Energy Phys.* **0102** 028

[6] Austing P and Wheater J F 2001 Convergent Yang–Mills matrix theories *J. High Energy Phys.* **0104** 019

[7] Anagnostopoulos K N and Nishimura J 2002 New approach to the complex action problem and its application to a nonperturbative study of superstring theory *Phys. Rev.* D **66** 106008

[8] Nishimura J 2002 Exactly solvable matrix models for the dynamical generation of space-time in superstring theory *Phys. Rev.* D **65** 105012

[9] Nishimura J, Okubo T and Sugino F 2005 Gaussian expansion analysis of a matrix model with the spontaneous breakdown of rotational symmetry *Prog. Theor. Phys.* **114** 487

[10] Anagnostopoulos K N, Azuma T and Nishimura J 2011 A practical solution to the sign problem in a matrix model for dynamical compactification *J. High Energy Phys.* **1110** 126

[11] Ambjorn J, Anagnostopoulos K N, Bietenholz W, Hotta T and Nishimura J 2000 Large N dynamics of dimensionally reduced 4-D SU(N) superYang–Mills theory *J. High Energy Phys.* **0007** 013 (see also [12])

[12] Ambjorn J, Anagnostopoulos K N, Bietenholz W, Hotta T and Nishimura J 2000 Monte Carlo studies of the IIB matrix model at large N *J. High Energy Phys.* **0007** 011

[13] Burda Z, Petersson B and Tabaczek J 2001 Geometry of reduced supersymmetric 4-D Yang–Mills integrals *Nucl. Phys.* B **602** 399

[14] Ambjorn J, Anagnostopoulos K N, Bietenholz W, Hofheinz F and Nishimura J 2002 On the spontaneous breakdown of Lorentz symmetry in matrix models of superstrings *Phys. Rev.* D **65** 086001

[15] Nishimura J and Vernizzi G 2000 Spontaneous breakdown of Lorentz invariance in IIB matrix model *J. High Energy Phys.* **0004** 015

[16] Nishimura J and Vernizzi G 2000 Brane world from IIB matrices *Phys. Rev. Lett.* **85** 4664

[17] Stevenson P M 1981 Optimized perturbation theory *Phys. Rev.* D **23** 2916

[18] Kim S W, Nishimura J and Tsuchiya A 2012 Expanding (3+1)-dimensional universe from a Lorentzian matrix model for superstring theory in (9+1)-dimensions *Phys. Rev. Lett.* **108** 011601

[19] Aoyama T, Nishimura J and Okubo T 2011 Spontaneous breaking of the rotational symmetry in dimensionally reduced super Yang–Mills models *Prog. Theor. Phys.* **125** 537

[20] Kim S W, Nishimura J and Tsuchiya A 2012 Expanding universe as a classical solution in the Lorentzian matrix model for nonperturbative superstring theory *Phys. Rev.* D **86** 027901

[21] Nishimura J and Tsuchiya A 2013 Local field theory from the expanding universe at late times in the IIB matrix model *Progr. Theor. Exp. Phys.* **2013** 043B03

[22] Kim S W, Nishimura J and Tsuchiya A 2012 Late time behaviors of the expanding universe in the IIB matrix model *J. High Energy Phys.* **1210** 147

[23] Nishimura J and Tsuchiya A 2013 Realizing chiral fermions in the type IIB matrix model at finite N *J. High Energy Phys.* **1312** 002

[24] Ito Y, Kim S W, Koizuka Y, Nishimura J and Tsuchiya A 2014 A renormalization group method for studying the early universe in the Lorentzian IIB matrix model *Progr. Theor. Exp. Phys.* **2014**

[25] Aoki H, Nishimura J and Tsuchiya A 2014 Realizing three generations of the Standard Model fermions in the type IIB matrix model *J. High Energy Phys.* **1405** 131

[26] Ito Y, Nishimura J and Tsuchiya A 2015 Power-law expansion of the Universe from the bosonic Lorentzian type IIB matrix model *J. High Energy Phys.* **1511** 070

[27] Ito Y, Nishimura J and Tsuchiya A 2016 Large-scale computation of the exponentially expanding universe in a simplified Lorentzian type IIB matrix model *PoS LATTICE* **2015** 243

[28] Patera J, Sharp R T, Winternitz P and Zassenhaus H 1976 Invariants of real low dimension lie algebras *J. Math. Phys.* **17** 986

[29] Steinacker H 2011 Split noncommutativity and compactified brane solutions in matrix models *Prog. Theor. Phys.* **126** 613

[30] Chatzistavrakidis A 2011 On Lie-algebraic solutions of the type IIB matrix model *Phys. Rev.* D **84** 106010

[31] Vilenkin N and Klimyk A U 1991 *Representation of Lie Groups and Special Functions* vol 1 (Dordrecht: Kluwer)

[32] Stern A 2014 Matrix model cosmology in two space-time dimensions *Phys. Rev.* D **90** 124056

[33] Chaney A, Lu L and Stern A 2015 Lorentzian fuzzy spheres *Phys. Rev.* D **92** 064021

[34] Chaney A and Stern A 2017 Fuzzy CP^2 spacetimes *Phys. Rev.* D **95** 046001

[35] Chaney A, Lu L and Stern A 2016 Matrix model approach to cosmology *Phys. Rev.* D **93** 064074

[36] Freedman D Z, Gibbons G W and Schnabl M 2005 Matrix cosmology *AIP Conf. Proc.* **743** 286

[37] Das S R and Michelson J 2005 pp wave big bangs: Matrix strings and shrinking fuzzy spheres *Phys. Rev.* D **72** 086005

[38] Craps B, Sethi S and Verlinde E P 2005 A matrix big bang *J. High Energy Phys.* **0510** 005

[39] Matsuo T, Tomino D, Wen W Y and Zeze S 2008 Quantum gravity equation in large N Yang–Mills quantum mechanics *J. High Energy Phys.* **0811** 088

[40] Ishino T and Ohta N 2006 Matrix string description of cosmic singularities in a class of time-dependent solutions *Phys. Lett.* B **638** 105

[41] Martinec E J, Robbins D and Sethi S 2006 Toward the end of time *J. High Energy Phys.* **0608** 025

[42] Chen B 2006 The time-dependent supersymmetric configurations in M-theory and matrix models *Phys. Lett.* B **632** 393

[43] Li M 2005 A class of cosmological matrix models *Phys. Lett.* B **626** 202

[44] Klammer D and Steinacker H 2009 Cosmological solutions of emergent noncommutative gravity *Phys. Rev. Lett.* **102** 221301

[45] Lee J and Yang H S 2014 Quantum gravity from noncommutative spacetime *J. Korean Phys. Soc.* **65** 1754

[46] She J H 2006 A matrix model for Misner universe *J. High Energy Phys.* **0601** 002

[47] Rivelles V O 2003 Noncommutative field theories and gravity *Phys. Lett.* B **558** 191

[48] Steinacker H 2010 Emergent geometry and gravity from matrix models: an introduction *Class. Quant. Grav.* **27** 133001

[49] Steinacker H C 2016 Emergent gravity on covariant quantum spaces in the IKKT model *J. High Energy Phys.* **1612** 156

[50] Steinacker H 2007 Emergent gravity from noncommutative gauge theory *J. High Energy Phys.* **0712** 049

[51] Yang H S 2009 Emergent spacetime and the origin of gravity *J. High Energy Phys.* **0905** 012

[52] Yang H S 2009 Emergent gravity from noncommutative spacetime *Int. J. Mod. Phys.* A **24** 4473

[53] Yang H S and Sivakumar M 2010 Emergent gravity from quantized spacetime *Phys. Rev.* D **82** 045004

[54] Kawai H, Kawana K and Sakai K 2017 A note on graviton exchange in the emergent gravity scenario *Progr. Theor. Exp. Phys.* **2017** 043B06

[55] Ydri B 2017 Lectures on matrix field theory *Lecture Notes in Physics* **vol 929** 1

[56] Ydri B 2004 Noncommutative U(1) gauge theory as a non-linear sigma model *Mod. Phys. Lett.* **19** 2205

[57] Yang H S 2009 Noncommutative electromagnetism as a large N gauge theory *Eur. Phys. J.* C **64** 445

[58] Bietenholz W, Hofheinz F and Nishimura J 2004 On the relation between non-commutative field theories at theta = infinity and large N matrix field theories *J. High Energy Phys.* **0405** 047

[59] Seiberg N and Witten E 1999 String theory and noncommutative geometry *J. High Energy Phys.* **9909** 032

[60] Douglas M R and Nekrasov N A 2001 Noncommutative field theory *Rev. Mod. Phys.* **73** 977

[61] Ehlers J and Kundt W 1962 Exact solutions of the gravitational field equations *Gravitation: An Introduction to Current Research* ed L Witten (New York: Wiley) pp 49–101
[62] Balachandran A P, Kurkcuoglu S and Vaidya S 2005 Lectures on fuzzy and fuzzy SUSY physics (arXiv:hep-th/0511114)
[63] O'Connor D 2003 Field theory on low dimensional fuzzy spaces *Mod. Phys. Lett.* A **18** 2423
[64] Ydri B 2001 Fuzzy physics rXiv:hep-th/0110006
[65] Kurkcuoglu S 2004 Explorations in fuzzy physics and non-commutative geometry (UMI-31-60408).
[66] Balachandran A P 2002 Quantum space-times in the year 2002 *Pramana* **59** 359
[67] Steinacker H 2004 Field theoretic models on covariant quantum spaces arXiv:hep-th/0408125
[68] Abe Y 2010 Construction of fuzzy spaces and their applications to matrix models arXiv:1002.4937
[69] Karabali D, Nair V P and Randjbar-Daemi S 2004 Fuzzy spaces, the M-(atrix) model and the quantum Hall effect *From Fields to Strings* vol 1 ed M Shifman *et al* (Singapore: World Scientific) pp 831–75
[70] Grosse H, Klimcik C and Presnajder P 1996 Towards finite quantum field theory in noncommutative geometry *Int. J. Theor. Phys.* **35** 231
[71] Grosse H, Klimcik C and Presnajder P 1996 On finite 4-D quantum field theory in noncommutative geometry *Commun. Math. Phys.* **180** 429
[72] Alekseev A Y, Recknagel A and Schomerus V 1999 Noncommutative world volume geometries: Branes on SU(2) and fuzzy spheres *J. High Energy Phys.* **9909** 023
[73] Hikida Y, Nozaki M and Sugawara Y 2001 Formation of spherical 2D brane from multiple D0 branes *Nucl. Phys.* B **617** 117
[74] Ahluwalia D V 1994 Quantum measurements, gravitation, and locality *Phys. Lett.* B **339** 301
[75] Schwinger J S 1951 The theory of quantized fields. 1 *Phys. Rev.* **82** 914
[76] Snyder H S 1947 Quantized space-time *Phys. Rev.* **71** 38
[77] Yang C N 1947 On quantized space-time *Phys. Rev.* **72** 874
[78] Doplicher S, Fredenhagen K and Roberts J E 1995 The quantum structure of space-time at the Planck scale and quantum fields *Commun. Math. Phys.* **172** 187
[79] Frohlich J and Gawedzki K 1993 Conformal field theory and geometry of strings *Proc. of the Mathematical Quantum Theory, Vancouver 1993* **vol 1** pp 57–97 and Preprint - Gawedzki K (rec.Nov.93) 44 p
[80] Connes A 1994 *Noncommutative Geometry* (London: Academic Press)
[81] Sniatycki J 1980 *Geometric Quantization and Quantum Mechanics Applied Mathematical Sciences*vol 30 (New York: Springer)
[82] Woodhouse N M J 1992 *Geometric Quantization Oxford Mathematical Monographs* (New York: Clarendon)
[83] Nair V P 2005 *Quantum Field Theory: A Modern Perspective* (New York: Springer)
[84] Hu S 2001 *Lecture Notes on Chern-Simons-Witten Theory* (Singapore: World Scientific)
[85] Murray S and Saemann C 2008 Quantization of flag manifolds and their supersymmetric extensions *Adv. Theor. Math. Phys.* **12** 641
[86] Castro-Villarreal P, Delgadillo-Blando R and Ydri B 2005 Quantum effective potential for U(1) fields on S**2(L) x S**2(L) *J. High Energy Phys.* **0509** 066

[87] Azuma T, Bal S, Nagao K and Nishimura J 2005 Perturbative versus nonperturbative dynamics of the fuzzy S**2 x S**2 *J. High Energy Phys.* **0509** 047
[88] Behr W, Meyer F and Steinacker H 2005 Gauge theory on fuzzy S**2 x S**2 and regularization on noncommutative R**4 *J. High Energy Phys.* **0507** 040
[89] Imai T and Takayama Y 2004 Stability of fuzzy S**2 x S**2 geometry in IIB matrix model *Nucl. Phys.* B **686** 248
[90] Vaidya S and Ydri B 2003 On the origin of the UV-IR mixing in noncommutative matrix geometry *Nucl. Phys.* B **671** 401
[91] Kaneko H, Kitazawa Y and Tomino D 2005 Stability of fuzzy S**2 x S**2 x S**2 in IIB type matrix models *Nucl. Phys.* B **725** 93
[92] Alexanian G, Balachandran A P, Immirzi G and Ydri B 2002 Fuzzy CP**2 *J. Geom. Phys.* **42** 28
[93] Azuma T, Bal S, Nagao K and Nishimura J 2006 Dynamical aspects of the fuzzy CP**2 in the large N reduced model with a cubic term *J. High Energy Phys.* **0605** 061
[94] Grosse H and Steinacker H 2005 Finite gauge theory on fuzzy CP**2 *Nucl. Phys.* B **707** 145
[95] Dou D and Ydri B 2007 Topology change from quantum instability of gauge theory on fuzzy CP**2 *Nucl. Phys.* B **771** 167
[96] Grosse H and Strohmaier A 1999 Towards a nonperturbative covariant regularization in 4-D quantum field theory *Lett. Math. Phys.* **48** 163
[97] Kitazawa Y 2002 Matrix models in homogeneous spaces *Nucl. Phys.* B **642** 210
[98] Imai T, Kitazawa Y, Takayama Y and Tomino D 2004 Effective actions of matrix models on homogeneous spaces *Nucl. Phys.* B **679** 143
[99] Dolan B P, Huet I, Murray S and O'Connor D 2007 Noncommutative vector bundles over fuzzy CP**N and their covariant derivatives *J. High Energy Phys.* **0707** 007
[100] Balachandran A P, Dolan B P, Lee J H, Martin X and O'Connor D 2002 Fuzzy complex projective spaces and their star products *J. Geom. Phys.* **43** 184
[101] Karabali D and Nair V P 2002 Quantum Hall effect in higher dimensions *Nucl. Phys.* B **641** 533
[102] Karabali D and Nair V P 2004 The effective action for edge states in higher dimensional quantum Hall systems *Nucl. Phys.* B **679** 427
[103] Karabali D and Nair V P 2004 Edge states for quantum Hall droplets in higher dimensions and a generalized WZW model *Nucl. Phys.* B **697** 513
[104] Anagnostopoulos K N, Azuma T, Nagao K and Nishimura J 2005 Impact of supersymmetry on the nonperturbative dynamics of fuzzy spheres *J. High Energy Phys.* **0509** 046
[105] Valtancoli P 2003 Stability of the fuzzy sphere solution from matrix model *Int. J. Mod. Phys.* A **18** 967
[106] Azuma T, Nagao K and Nishimura J 2005 Perturbative dynamics of fuzzy spheres at large N *J. High Energy Phys.* **0506** 081
[107] Azuma T, Bal S and Nishimura J 2005 Dynamical generation of gauge groups in the massive Yang–Mills-Chern-Simons matrix model *Phys. Rev.* D **72** 066005
[108] Azuma T, Bal S, Nagao K and Nishimura J 2004 Nonperturbative studies of fuzzy spheres in a matrix model with the Chern-Simons term *J. High Energy Phys.* **0405** 005
[109] Ishii T, Ishiki G, Shimasaki S and Tsuchiya A 2008 Fiber bundles and matrix models *Phys. Rev.* D **77** 126015
[110] Ishiki G, Shimasaki S and Tsuchiya A 2009 Large N reduction for Chern-Simons theory on S**3 *Phys. Rev.* D **80** 086004

[111] Ishiki G, Ohta K, Shimasaki S and Tsuchiya A 2009 Two-dimensional gauge theory and matrix model *Phys. Lett.* B **672** 289

[112] Karabali D and Nair V P 2006 Quantum Hall effect in higher dimensions, matrix models and fuzzy geometry *J. Phys.* A **39** 12735

[113] Ho P M and Li M 2001 Fuzzy spheres in AdS/CFT correspondence and holography from noncommutativity *Nucl. Phys.* B **596** 259

[114] Delgadillo-Blando R, O'Connor D and Ydri B 2009 Matrix models, gauge theory and emergent geometry *J. High Energy Phys.* **0905** 049

[115] Delgadillo-Blando R, O'Connor D and Ydri B 2008 *Phys. Rev. Lett.* **100** 201601

[116] O'Connor D and Ydri B 2006 Monte Carlo simulation of a NC gauge theory on the fuzzy sphere *J. High Energy Phys.* **0611** 016

[117] Ydri B, Rouag A and Ramda K 2017 Emergent fuzzy geometry and fuzzy physics in four dimensions *Nucl. Phys.* B **916** 567

[118] Ydri B, Khaled R and Ahlam R 2016 Geometry in transition in four dimensions: A model of emergent geometry in the early universe *Phys. Rev.* D **94** 085020

[119] Ydri B, Ramda K and Rouag A 2016 Phase diagrams of the multitrace quartic matrix models of noncommutative theory *Phys. Rev.* D **93** 065056

[120] Ydri B, Rouag A and Ramda K 2016 Emergent geometry from random multitrace matrix models *Phys. Rev.* D **93** 065055

[121] Szabo R J 2006 Symmetry, gravity and noncommutativity *Class. Quant. Grav.* **23** R199

[122] Szabo R J 2010 Quantum gravity, field theory and signatures of noncommutative spacetime *Gen. Rel. Grav.* **42** 1

[123] Dolan B P and O'Connor D 2003 A Fuzzy three sphere and fuzzy tori *J. High Energy Phys.* **0310** 060

[124] Medina J, Huet I, O'Connor D and Dolan B P 2012 Scalar and spinor field actions on fuzzy S^4: fuzzy CP^3 as a S_F^2 bundle over S_F^4 *J. High Energy Phys.* **1208** 070

[125] Medina J and O'Connor D 2003 Scalar field theory on fuzzy S**4 *J. High Energy Phys.* **0311** 051

[126] Steinacker H C 2016 Emergent 4D gravity on covariant quantum spaces in the IKKT model arXiv:1606.00769 (see also [127])

[127] Steinacker H 2009 Covariant field equations, gauge fields and conservation laws from Yang–Mills matrix models *J. High Energy Phys.* **0902** 044

[128] Steinacker H C 2015 One-loop stabilization of the fuzzy four-sphere via softly broken SUSY *J. High Energy Phys.* **1512** 115

[129] Castelino J, Lee S and Taylor W 1998 Longitudinal five-branes as four spheres in matrix theory *Nucl. Phys.* B **526** 334

[130] Kimura Y 2002 Noncommutative gauge theory on fuzzy four sphere and matrix model *Nucl. Phys.* B **637** 177

[131] Ramgoolam S 2001 On spherical harmonics for fuzzy spheres in diverse dimensions *Nucl. Phys.* B **610** 461

[132] Abe Y 2004 Construction of fuzzy S**4 *Phys. Rev.* D **70** 126004

[133] Valtancoli P 2002 Projective modules over the fuzzy four sphere *Mod. Phys. Lett.* A **17** 2189

[134] Carow-Watamura U, Steinacker H and Watamura S 2005 Monopole bundles over fuzzy complex projective spaces *J. Geom. Phys.* **54** 373

[135] Aoyama S and Masuda T 2003 The fuzzy S4 by quantum deformation *Nucl. Phys.* B **656** 325

[136] Kimura Y 2003 On higher dimensional fuzzy spherical branes *Nucl. Phys.* B **664** 512
[137] Azuma T, Bal S, Nagao K and Nishimura J 2004 Absence of a fuzzy S**4 phase in the dimensionally reduced 5-D Yang–Mills-Chern-Simons model *J. High Energy Phys.* **0407** 066
[138] Ramgoolam S 2002 Higher dimensional geometries related to fuzzy odd dimensional spheres *J. High Energy Phys.* **0210** 064
[139] Fulton W and Harris J 1991 *Representation Theory: A First Course* (Graduate Texts in Mathematics vol 129) (New York: Springer)
[140] Ydri B 2002 Fuzzy S^4, unpublished notes
[141] Ho P M and Ramgoolam S 2002 Higher dimensional geometries from matrix brane constructions *Nucl. Phys.* B **627** 266
[142] Karczmarek J L and Yeh K H C 2015 Noncommutative spaces and matrix embeddings on flat $\mathbb{R}^{2n+1}$ *J. High Energy Phys.* **1511** 146
[143] Castro-Villarreal P, Delgadillo-Blando R and Ydri B 2005 A Gauge-invariant UV-IR mixing and the corresponding phase transition for U(1) fields on the fuzzy sphere *Nucl. Phys.* B **704** 111
[144] Blaschke D N and Steinacker H 2011 On the 1-loop effective action for the IKKT model and non-commutative branes *J. High Energy Phys.* **1110** 120
[145] Wald R M 1984 *General Relativity* (Chicago, IL: Chicago University Press)
[146] Hanada M, Kawai H and Kimura Y 2006 Describing curved spaces by matrices *Prog. Theor. Phys.* **114** 1295
[147] Di Francesco P, Ginsparg P H and Zinn-Justin J 1995 2-D Gravity and random matrices *Phys. Rep.* **254** 1
[148] Ydri B, Soudani C and Rouag A 2017 Quantum gravity as a multitrace matrix model arXiv:1706.07724
[149] Ydri B 2016 The multitrace matrix model: An alternative to Connes NCG and IKKT model in 2 dimensions *Phys. Lett.* B **763** 161

IOP Publishing

Badis Ydri

Appendix A

Algorithms and Monte Carlo codes for the matrix models of string theory

We will show in this appendix how to write Monte Carlo codes for (1) quantum black holes as described by the gauge/gravity duality in one dimension and matrix theory (the BFSS matrix quantum mechanics) and for (2) the expansion of the Universe as described by the Lorentzian type IIB matrix model (the Lorentzian IKKT in 10 dimensions). The IKKT and BFSS matrix models are the only known methods of studying string theory non-perturbatively and rigorously, via exact mathematics, which can be coded precisely in Monte Carlo, allowing us therefore to escape the formalism, and also to test and check explicitly the results, of string theory and of the gauge/gravity duality. In this chapter, I will mostly follow the textbook [1].

A.1 The molecular dynamics

Let us review the main elements of the molecular dynamics algorithm. See for example [1, 2]. We consider a Hamiltonian of the form

$$H = H(q, p) = H_1(p) + H_2(q), \quad H_1(p) = \frac{1}{2}p^2. \tag{A.1}$$

The equations of motion are

$$\frac{\partial H}{\partial q} = -\dot{p}, \quad \frac{\partial H}{\partial p} = \dot{q}. \tag{A.2}$$

doi:10.1088/978-0-7503-1726-9ch9

We need to solve

$$\frac{d}{d\tau}p = -\frac{\partial H_2}{\partial q} = F(\tau)$$
$$\frac{d}{d\tau}q = \frac{\partial H_1}{\partial p} = p(\tau). \quad \text{(A.3)}$$

F is the force exerted on the system. The Stormer–Verlet integrator is equivalent to a Taylor expansion up to the second order, viz

$$p(\tau + \delta\tau) = p(\tau) + \frac{1}{2}\delta\tau(F(\tau + \delta\tau) + F(\tau)) + O(\delta\tau^3)$$
$$q(\tau + \delta\tau) = q(\tau) + \delta\tau \cdot p + \frac{1}{2}\delta\tau^2 \cdot F(\tau) + O(\delta\tau^3). \quad \text{(A.4)}$$

The total integration time will be given by $\tau = \delta\tau \cdot N_\tau$. Thus the total error in this integrator after N_τ steps is $O(\delta\tau^2)$.

Let us define the two Euler steps

$$I_1(\epsilon) \ : \ (q, p) \longrightarrow (q + \epsilon p, p)$$
$$I_2(\epsilon) \ : \ (q, p) \longrightarrow (q, p + \epsilon F). \quad \text{(A.5)}$$

Then it is not difficult to show that the Stormer–Verlet algorithm is given by the iteration (with $\delta\tau = \epsilon$)

$$I_1\left(\frac{\epsilon}{2}\right) \ : \ q\left(t + \frac{\epsilon}{2}\right) = q(t) + \frac{\epsilon}{2}p(t)$$
$$I_2(\epsilon) \ : \ p(t + \epsilon) = p(t) + \epsilon F\left(t + \frac{\epsilon}{2}\right) \quad \text{(A.6)}$$
$$I_1\left(\frac{\epsilon}{2}\right) \ : \ q(t + \epsilon) = q\left(t + \frac{\epsilon}{2}\right) + \frac{\epsilon}{2}p(t + \epsilon).$$

The Stormer–Verlet algorithm (also known as the leap-frog algorithm) is then given by

$$J_{LF}(\epsilon) = \left[I_1\left(\frac{\epsilon}{2}\right)I_2(\epsilon)I_1\left(\frac{\epsilon}{2}\right)\right]^{N_\tau}. \quad \text{(A.7)}$$

Numerically we put the coordinates q in the lattice points $n = 0, 1, \ldots, N_\tau - 1$, whereas we put the momenta p (except the initial and the final momenta) in the mid-interval points $n + 1/2 = 1/2, 3/2, \ldots, N_\tau - 1/2$ and rewrite the above integrator as follows

- Initial step:

$$p(1/2) = p(0) + \epsilon F(0)/2$$
$$q(1) = q(0) + \epsilon p(1/2). \quad \text{(A.8)}$$

- Iteration ($n = 1, \ldots, N_\tau - 1$):

$$\begin{aligned} p(n+1/2) &= p(n-1/2) + \epsilon F(n) \\ q(n+1) &= q(n) + \epsilon p(n+1/2). \end{aligned} \tag{A.9}$$

- Final step:

$$p(N) = p(N-1/2) + \epsilon F(N)/2. \tag{A.10}$$

A more accurate integrator than the leap-frog is the Omelyan integrator given by

$$\begin{aligned} J_{Om}(\epsilon) &= \left[I_1(\zeta\epsilon) I_2\left(\frac{\epsilon}{2}\right) I_1((1-2\zeta)\epsilon) I_2\left(\frac{\epsilon}{2}\right) I_1(\zeta\epsilon) \right]^{N_\tau}, \\ \zeta &= 0.193\,183\,3. \end{aligned} \tag{A.11}$$

The leap-frog and the Omelyan satisfy the crucial properties:

- Reversibility in time.
- Conservation of phase space volume. Thus, although the integrator breaks Hamiltonian conservation by construction, the conservation of the phase space measure implies the identity

$$\langle \exp(-\Delta H) \rangle = 1. \tag{A.12}$$

A.2 The BFSS model

A.2.1 Summary of the BFSS model

We start with D Hermitian $N \times N$ time-dependent bosonic matrices X_i, where $D = 9$, and 16 Hermitian $N \times N$ time-dependent fermionic matrices ψ_α, i.e. ψ is a Majorana–Weyl fermion in $d = D + 1 = 10$ dimensions. The BFSS action is given by

$$\begin{aligned} S = \frac{1}{g^2} \int_0^\beta dt\, \mathrm{Tr} \Bigg[&\frac{1}{2}(D_t X_i)^2 - \frac{1}{4}[X_i, X_j]^2 \\ &+ \frac{1}{2}\psi_\alpha D_t \psi_\alpha - \frac{1}{2}\psi_\alpha (\gamma_i)_{\alpha\beta}[X_i, \psi_\beta] \Bigg]. \end{aligned} \tag{A.13}$$

The fields are periodic with period $\beta = 1/T$ where T is the Hawking temperature as

$$X_i(t+\beta) = X_i(\beta), \quad A(t+\beta) = A(\beta), \quad \psi_\alpha(t) = -\psi_\alpha(t+\beta). \tag{A.14}$$

This is a thermal field theory in one dimension and thus t must be imaginary time, viz $\tilde{t} = -it$ is the real time. The Dirac gamma matrices are defined as usual by the Clifford algebra $\{\gamma_i, \gamma_j\} = 2\delta_{ij}$. The covariant derivative D_t is defined in terms of another Hermitian $N \times N$ time-dependent bosonic matrix A as

$$D_t = \partial_t - i[A, \ldots]. \tag{A.15}$$

Thus, because of periodicity, we should expand the fields as

$$X_i(t) = \sum_{n=-\Lambda}^{+\Lambda} \tilde{X}_i(n)e^{i\omega nt}, \quad \psi_a(t) = \sum_{r=-\Lambda'}^{+\Lambda'} \tilde{\psi}_a(r)e^{i\omega rt}. \tag{A.16}$$

The quantum number n is integer-valued, whereas r is half-integer-valued and we choose $\Lambda' = \Lambda - 1/2$. The frequency ω is given by

$$\omega = \frac{2\pi}{\beta}. \tag{A.17}$$

Of course, we will study the system in the 't Hooft limit given by

$$\lambda = g^2 N. \tag{A.18}$$

There seem to be, therefore, two independent coupling constants λ and T. However, g^2 can always be rescaled away. By dimensional analysis we find that X behaves as inverse length and λ as inverse length cubed, whereas T behaves as inverse length and as a consequence the dimensionless coupling constant must be given by

$$\tilde{\lambda} = \frac{\lambda}{T^3}. \tag{A.19}$$

We will choose $\lambda = 1$.

Next, we gauge-fix non-perturbatively the above action (see next section) and then diagonalize the matrix A as

$$A = \frac{1}{\beta}\text{diag}(\theta_1, \ldots, \theta_N). \tag{A.20}$$

This introduces a Vandermonde determinant from the measure corresponding to a Faddeev–Popov action

$$S_{\text{FP}} = -\frac{1}{2}\sum_{a\neq b} \ln \sin^2 \frac{\theta_a - \theta_b}{2}. \tag{A.21}$$

At this stage we introduce the cutoff Λ shown in (A.16). This shows explicitly why cutoff regularization in one dimension preserves exact gauge invariance. However, the fact that cutoff regularization is consistent with supersymmetry is more subtle since it is really only an approximate statement [3]. For a discussion of this crucial point from a lattice perspective, see for example [4, 5].

We have then the path integral (partition function)

$$Z = \int \mathcal{D}X_i \; \mathcal{D}\psi \prod_{a=1}^{n} d\alpha_a \; \exp(-S_B[X, \alpha] - S_F[X, \psi, \alpha] - S_{\text{FP}}[\alpha]). \tag{A.22}$$

The integration over the fermions can be done exactly to give rise to the Pfaffian, viz

$$\begin{aligned}\int \mathcal{D}\psi \; \exp(-S_F[X, \psi, \alpha]) &= \int \mathcal{D}\psi \; \exp(-\psi \mathcal{O} \psi) \\ &= \text{Pf } \mathcal{O}.\end{aligned} \tag{A.23}$$

The path integral becomes

$$Z = \int \mathcal{D}X_i\ \mathcal{D}\psi \prod_{a=1}^{n} d\alpha_a\ \exp(-S_B[X, \alpha] - S_{\rm FP}[\alpha])\ {\rm Pf}\ \mathcal{O}[X, \alpha]. \tag{A.24}$$

Here, we will employ the most important of all approximations as we will see.

A.2.2 The gauge fixing

The above action is invariant under the $U(N)$ gauge transformations

$$\psi_\alpha \longrightarrow U\psi_\alpha U^\dagger, \quad X_i \longrightarrow UX_iU^\dagger, \quad A \longrightarrow UAU^\dagger - \frac{1}{i}U\partial_t U^\dagger. \tag{A.25}$$

We will now gauge-fix this symmetry non-perturbatively following [6]. First, we need to put the action on a lattice. We define $t = n\delta\, t$, $\beta = \Lambda\delta t$, $\delta\, t = a$, $n = 0, \ldots, \Lambda - 1$. The periodicity becomes $X_i(n + \Lambda) = X_i(n)$. The gauge field A becomes the link variable

$$U_{n,\,n+1} = \mathcal{P}\exp\left(-i\int_{na}^{(n+1)a} dtA(t)\right) = 1 - iaA(n) + O(a^2), \quad a \longrightarrow 0. \tag{A.26}$$

The covariant derivative becomes then

$$\frac{U_{n,\,n+1}X_i(n+1)U_{n+1,\,n} - X_i(n)}{a} = D_tX_i, \quad a \longrightarrow 0. \tag{A.27}$$

The bosonic action is then given by

$$\begin{aligned} S_B = \frac{1}{g^2}\sum_{n=0}^{\Lambda-1} {\rm Tr}\bigg[&\frac{1}{a}X_i^2(n) - \frac{1}{a}U_{n,\,n+1}X_i(n+1) \\ &U_{n+1,\,n}X_i(n) - \frac{a}{4}[X_i(n), X_j(n)]^2\bigg]. \end{aligned} \tag{A.28}$$

We have a local $U(N)$ symmetry at each lattice site. We use this symmetry to rotate the link variables as follows

$$\begin{aligned} X_i'(0) &= X_i(0) \\ X_i'(1) &= U_{0,1}X_i(1)U_{0,\,1}^\dagger \\ X_i'(2) &= U_{0,1}U_{1,2}X_i(2)(U_{0,1}U_{1,2})^\dagger \\ &\vdots \\ X_i'(\Lambda - 1) &= U_{0,1}\cdots U_{\Lambda-2,\,\Lambda-1}X_i(\Lambda - 1)(U_{0,1}\cdots U_{\Lambda-2,\,\Lambda-1})^\dagger. \end{aligned} \tag{A.29}$$

The action becomes then

$$\begin{aligned} S_B = &\frac{1}{g^2}\sum_{n=0}^{\Lambda-1} {\rm Tr}\left[\frac{1}{a}X_i'^2(n) - \frac{a}{4}[X_i'(n), X_j'(n)]^2\right] \\ &- \frac{1}{g^2 a}\left[\sum_{n=0}^{\Lambda-2} {\rm Tr}\ X_i'(n)X_i'(n+1) + X_i'(\Lambda)W^\dagger X_i'(\Lambda - 1)W\right]. \end{aligned} \tag{A.30}$$

The matrix W is defined by

$$W = U_{0,1} \cdots U_{\Lambda-2,\,\Lambda-1} U_{\Lambda-1,\,\Lambda}. \tag{A.31}$$

We can diagonalize the matrix W in the usual way as $W = VDV^\dagger$ where $D = \mathrm{diag}(\exp(i\theta_1), \ldots, \exp(i\theta_N))$. The action becomes (with $\tilde{X}_i(n) = V^\dagger X_i'(n) V$)

$$\begin{aligned} S_B &= \frac{1}{g^2} \sum_{n=0}^{\Lambda-1} \mathrm{Tr}\left[\frac{1}{a}\tilde{X}_i^2(n) - \frac{a}{4}[\tilde{X}_i(n), \tilde{X}_j(n)]^2\right] \\ &\quad - \frac{1}{g^2 a}\left[\sum_{n=0}^{\Lambda-2} \mathrm{Tr}\, \tilde{X}_i(n)\tilde{X}_i(n+1) + \tilde{X}_i(\Lambda) D^\dagger \tilde{X}_i(\Lambda-1) D\right]. \end{aligned} \tag{A.32}$$

This action can also be rewritten in terms of $D_\Lambda = D^{1/\Lambda}$ and $\bar{X}_i(n) = h_n^\dagger \tilde{X}_i(n) h_n$, where $h_n = D_\Lambda^n$, as

$$\begin{aligned} S_B &= \frac{1}{g^2} \sum_{n=0}^{\Lambda-1} \mathrm{Tr}\left[\frac{1}{a}\bar{X}_i^2(n) - \frac{a}{4}[\bar{X}_i(n), \bar{X}_j(n)]^2\right] \\ &\quad - \frac{1}{g^2 a} \sum_{n=0}^{\Lambda-1} \mathrm{Tr}\, \bar{X}_i(n) D_\Lambda \bar{X}_i(n+1) D_\Lambda^\dagger. \end{aligned} \tag{A.33}$$

The measure $\prod_{i,\,n} dX_i(n)$ is invariant under all the above unitary transformations. Finally, we reduce the measure over the transporter fields as follows

$$\begin{aligned} \prod_{n=0}^{\Lambda-1} \mathcal{D}U_{n,\,n+1} &= \prod_{n=1}^{\Lambda-1} \mathcal{D}U_{n,\,n+1} \mathcal{D}U_{0,1} \\ &\sim \prod_{n=1}^{\Lambda} \mathcal{D}U_{n,\,n+1} \mathcal{D}W \\ &\sim \prod_{n=1}^{\Lambda} \mathcal{D}U_{n,\,n+1} \mathcal{D}V \mathcal{D}D \Delta^2(D) \\ &\sim \prod_{i=1}^{N} d\theta_i \cdot \prod_{i>j} |e^{i\theta_i} - e^{i\theta_j}|^2 \\ &\sim \prod_{i=1}^{N} d\theta_i \cdot \prod_{i>j} \sin^2 \frac{\theta_i - \theta_j}{2} \\ &\sim \prod_{i=1}^{N} d\theta_i \cdot \exp(-S_{\mathrm{FP}}). \end{aligned} \tag{A.34}$$

The Faddeev–Popov gauge-fixing action is given explicitly by

$$S_{\mathrm{FP}} = -\frac{1}{2} \sum_{i \neq j} \ln \sin^2 \frac{\theta_i - \theta_j}{2}. \tag{A.35}$$

A.3 The Di Vecchia–Ferrara quantum mechanics

A.3.1 The problem

We can start with the much simpler but very similar Di Vecchia–Ferrara model in one dimension. See [7]. See also [3, 8, 9]. The model is given by the action

$$S = \int dt\left(\frac{1}{2}\dot{\phi}^2 + \frac{1}{2}W'^2 + \bar{\psi}\dot{\psi} + \bar{\psi}W''\psi\right), \quad W' = \frac{dW}{d\phi}. \tag{A.36}$$

This involves a real scalar field ϕ and a one-component Dirac field ψ. This model is invariant for any potential W under the two supersymmetries:

$$\delta^{(1)}\phi = \bar{\epsilon}\psi,\ \delta^{(1)}\psi = 0,\ \delta^{(1)}\bar{\psi} = -\bar{\epsilon}(\dot{\phi} + W'), \tag{A.37}$$

and

$$\delta^{(2)}\phi = \bar{\psi}\epsilon,\ \delta^{(2)}\psi = (\dot{\phi} - W')\epsilon,\ \delta^{(1)}\bar{\psi} = 0, \tag{A.38}$$

where ϵ and $\bar{\epsilon}$ are two anticommuting parameters. We will mostly consider the potential

$$W(\phi) = \frac{m}{2}\phi^2 + \frac{g}{4}\phi^4. \tag{A.39}$$

As a first step we will only consider the bosonic action, viz

$$S_B = \int_0^\beta dt\left(\frac{1}{2}\dot{\phi}^2 + \frac{1}{2}W'^2\right), \tag{A.40}$$

where we have also imposed periodic boundary condition with period $\beta = 1/T$ where T is the temperature. Thus, we have the Fourier expansion

$$\phi(t) = \sum_{n=-\Lambda}^{+\Lambda} \tilde{\phi}(n)\exp(i\omega nt), \quad \omega = \frac{2\pi}{\beta}, \tag{A.41}$$

where we have also introduced a cutoff Λ. The action becomes

$$S_B = \beta\sum_{n=-\Lambda}^{+\Lambda}\frac{1}{2}(n^2\omega^2 + m^2)\tilde{\phi}(n)\tilde{\phi}(-n) + \beta mg(\tilde{\phi}^4)_0 + \frac{\beta}{2}g^2(\tilde{\phi}^6)_0. \tag{A.42}$$

We have introduced the notation

$$(f^{(1)}\cdots f^{(p)})_n = \sum_{n_1}\cdots\sum_{n_p} f^{(1)}(n_1)\cdots f^{(p)}(n_p)|_{n_1+\cdots+n_p=n}. \tag{A.43}$$

A.3.2 Basics of the Fortran code

The goal is to build an operational rational hybrid Monte Carlo code from scratch and we will simply start by coding the above bosonic action.

1. We will use as a programming language *Fortran*. Any other language will do and the choice of the programming language is at the end of the day a

personal choice. We will stress clarity as opposed to efficiency which is always the most beneficial thing for students. We use as an operating system *Linux Ubuntu*, as editor *Emacs*, as compiler *Gfortran* and for graphics *Gnuplot*.

2. The action S_B will be coded either in a **function** or a **subroutine**. We choose subroutine. This will look like:

```
subroutine action(...)
...
...
return
end
```

3. The input of this subroutine consists of all the parameters of the model: m, g, β, Λ, ω as well as of the $2\Lambda + 1$ degrees of freedom $\tilde{\phi}(n)$. The output is precisely the action S_B. In the code these variables will be renamed in an obvious way.
4. The $2\Lambda + 1$ degrees of freedom $\tilde{\phi}(n)$ will be given by an **array** which starts from $-\Lambda$ to $+\Lambda$.
5. All these quantities, with the exception of Λ and $\tilde{\phi}(n)$, are real numbers and thus they will be given as type **double precision**. However, the cutoff Λ is an integer and thus it will be given as type **integer** and the degrees of freedom $\tilde{\phi}(n)$ are complex numbers and they will be given as type **double complex**. Here there is a subtlety due to the fact that $\tilde{\phi}(0)$ is actually real which we will discuss later.

   The subroutine will then look something like:

```
subroutine action(mass,coupling,beta,Lambda,omega,tildephi,actionB)
double precision mass, coupling, beta, omega,actionB
double complex tildephi(-Lambda:+Lambda)
integer Lambda
...
...
return
end
```

6. The first term in S_B is easy to code. It only requires a **do loop**. We initialize the value of the action then we sum via the do loop the terms successively and then multiply by the inverse temperature as:

```
subroutine action(mass,coupling,beta,Lambda,omega,tildephi,actionB)
double precision mass, coupling, beta, omega,actionB
double complex tildephi(-Lambda:+Lambda)
integer Lambda,n

actionB=0.0d0
do n=-Lambda,Lambda
actionB=actionB+0.5d0*(n**2*omega**2+m**2)*tildephi(n)*tildephi(-n)
enddo
actionB=beta*actionB

return
end
```

We remark that the loop variable n is declared as type **integer** as it should be.

7. We need now to add to the action the term proportional to

$$(\tilde{\phi}^4)_0 = \sum_{n_1}\cdots\sum_{n_4}\tilde{\phi}(n_1)\cdots\tilde{\phi}(n_4)|_{n_1+\cdots+n_4=0}. \tag{A.44}$$

This requires an **if statement** because of the conditional $n_1 + \cdots n_4 = 0$. Indeed, the above term is given explicitly by

$$(\tilde{\phi}^4)_0 = \tilde{\phi}(n_1)\tilde{\phi}(n_2)\tilde{\phi}(n_3)\tilde{\phi}(n_4), \quad n_4 = -n_1 - n_2 - n_3. \tag{A.45}$$

The conditional then comes out as the condition that n_4 as given by $-n_1 - n_2 - n_3$ must always lie in the interval between $-\Lambda$ and $+\Lambda$.

In the code this might look like:

```
 phi4=0.0d0
   do n1=-Lambda,Lambda
      do n2=-Lambda,Lambda
         do n3=-Lambda,Lambda
            n4=-n1-n2-n3
            if ((n4.le.Lambda).and.(n4.ge.-Lambda))then
      phi4=phi4+tildephi(n1)*tildephi(n2)*tildephi(n3)*tildephi(n4)
            endif
            enddo
         enddo
      enddo
        phi4=beta*coupling*mass*phi4
```

Again, the variables $n_1,\ldots,n_4$ should be declared as type **integer**, while ϕ_4 is **double complex**.

8. Similarly, the last term

$$(\tilde{\phi}^6)_0 = \tilde{\phi}(n_1)\tilde{\phi}(n_2)\tilde{\phi}(n_3)\tilde{\phi}(n_4)\tilde{\phi}(n_5)\tilde{\phi}(n_6),\qquad n_6 = -n_1 - n_2 - n_3 - n_4 - n_5, \tag{A.46}$$

is coded in the same way, viz:

```
phi6=0.0d0
 do n1=-Lambda,Lambda
    do n2=-Lambda,Lambda
       do n3=-Lambda,Lambda
         do n4=-Lambda,Lambda
           do n5=-Lambda,Lambda
          n6=-n1-n2-n3-n4-n5
          if ((n6.le.Lambda).and.(n6.ge.-Lambda))then
      phi6=phi6+tildephi(n1)*tildephi(n2)*tildephi(n3)
&                *tildephi(n4)*tildephi(n5)*tildephi(n6)
          endif
           enddo
         enddo
       enddo
    enddo
 enddo
      phi6=0.5d0*beta*coupling**2*phi6
```

9. The total action is the sum of the three contributions, viz:

```
actionB=actionB+phi4+phi6
```

10. Next, we create the main body of the code via **program**, give it a name, then close it by **end** as follows:

```
program My_First_Code

return
end
```

11. We call the subroutine which calculates the action from within the main program by using **call**.

    We will need to declare the type of all variables as before. Also, we will need to fix the number of the degrees of freedom Λ by using **parameter** since the array containing the degrees of freedom depends on Λ. However, the other parameters of the model can be fixed in the body of the main program. Also, we need to initialize the field $\tilde{\phi}(n)$. The output which is the value of the action is written to the screen via **write**.

    In the code all this looks as follows:

```
program My_First_Code
double precision mass, coupling, beta, omega,actionB
integer Lambda,n
parameter (Lambda=1)
double complex tildephi(-Lambda:+Lambda)

mass=1.0d0
coupling=1.0d0
beta=1.0d0
omega=1.0d0

do n=-Lambda,+Lambda
    tildephi(n)=1.0d0
enddo

call action(mass,coupling,beta,Lambda,omega,tildephi,actionB)

write(*,*) actionB

return
end
```

12. By putting the main program and the subroutine of the action in a single file we get the final code. The final step is compiling and debugging this code by running the command **gfortran**. This could be the most difficult step.
13. For maximum optimization we compile using gfortran (g77) with the ‘O’ flag as:

```
gfortran -O code.f
```

A.3.3 The molecular dynamics algorithm

We go back now to the anharmonic oscillator problem given by the action

$$S_B = \beta \sum_{n=-\Lambda}^{+\Lambda} \frac{1}{2}(n^2\omega^2 + m^2)\tilde{\phi}(n)\tilde{\phi}(-n) + \beta m g(\tilde{\phi}^4)_0 + \frac{\beta}{2}g^2(\tilde{\phi}^6)_0. \qquad (A.47)$$

The idea of the molecular dynamics algorithm in this context consists in viewing this action S_B as a potential for a pseudo-dynamics with a pseudo-time τ where $\tilde{\phi}(n)$ plays the role of the generalized coordinates, viz

$$\tilde{\phi}(n) = q_n(\tau). \qquad (A.48)$$

The generalized momenta will be denoted by

$$\tilde{\pi}(n) = p_n(\tau), \qquad (A.49)$$

with Hamiltonian given by the usual kinetic energy

$$K = \frac{1}{2} \sum_{n=-\Lambda}^{+\Lambda} \tilde{\pi}(n)\tilde{\pi}(-n), \qquad (A.50)$$

plus the potential S_B, i.e.

$$H = \frac{1}{2} \sum_{n=-\Lambda}^{+\Lambda} \tilde{\pi}(n)\tilde{\pi}(-n) + S_B. \qquad (A.51)$$

Recall the reality conditions

$$\tilde{\phi}(-n) = \tilde{\phi}^*(n), \quad \tilde{\pi}(-n) = \tilde{\pi}^*(n). \qquad (A.52)$$

The kinetic energy and the Hamiltonian will be coded with the action in a single subroutine. A pseudo code is given by the following lines:

```
      Hami=0.0d0
      do n=-Lambda,Lambda
         actionB=actionB+0.5d0*(n**2*omega**2+mass**2)*tildephi(n)
     &         *tildephi(-n)
         Hami=Hami+0.5d0*tildep(n)*tildep(-n)
      enddo
         actionB=beta*actionB
         ...
         ...
         ...
         actionB=actionB+phi4+phi6
         Hami=Hami+actionB
```

The Hamilton equations of motion are

$$\frac{\partial H}{\partial \tilde{\phi}^*(n)} = -\frac{d\tilde{\pi}(n)}{d\tau} \Rightarrow F(n) = -\frac{d\tilde{\pi}(n)}{d\tau}. \tag{A.53}$$

$$\frac{\partial H}{\partial \tilde{\pi}^*(n)} = \frac{d\tilde{\phi}(n)}{d\tau} \Rightarrow \tilde{\pi}(n) = \frac{d\tilde{\phi}(n)}{d\tau}. \tag{A.54}$$

The so-called force $F(n)$ is defined by

$$\begin{aligned} F(n) &= \frac{\partial S_B}{\partial \tilde{\phi}^*(n)} \\ &= \beta(n^2\omega^2 + m^2)\tilde{\phi}(n) + 4\beta m g \sum_{n_1,n_2,n_3} \tilde{\phi}(n_1)\tilde{\phi}(n_2)\tilde{\phi}(n_3)|_{n_1+n_2+n_3=n} \\ &+ 3\beta g^2 \sum_{n_1,n_2,n_3} \tilde{\phi}(n_1)\tilde{\phi}(n_2)\tilde{\phi}(n_3)\tilde{\phi}(n_4)\tilde{\phi}(n_5)|_{n_1+n_2+n_3+n_4+n_5=n}. \end{aligned} \tag{A.55}$$

This will be coded in a similar way as the action. However, remark that the force $F(n)$ is a function of the winding number n, and thus in the code it will appear as a complex **array**. Again, the $n = 0$ case is special since $F(0)$ is actually real.

The actual code is the following subroutine:

```
      subroutine force(mass,coupling,beta,Lambda,omega,tildephi,
     &     forceB)
      double precision mass, coupling, beta, omega
      double complex forceB(-Lambda:Lambda),phi3,phi5
      double complex tildephi(-Lambda:+Lambda)
      integer Lambda,n,n1,n2,n3,n4,n5,n6

      do n=-Lambda,Lambda
         forceB(n)=0.0d0
         forceB(n)=forceB(n)+(n**2*omega**2+mass**2)*tildephi(n)
         forceB(n)=beta*forceB(n)

      phi3=0.0d0
      do n1=-Lambda,Lambda
         do n2=-Lambda,Lambda
               n3=-n1-n2+n
               if ((n3.le.Lambda).and.(n3.ge.-Lambda))then
        phi3=phi3+tildephi(n1)*tildephi(n2)*tildephi(n3)
               endif
         enddo
```

```
      enddo
         phi3=4.0d0*beta*coupling*mass*phi3

         phi5=0.0d0
         do n1=-Lambda,Lambda
            do n2=-Lambda,Lambda
               do n3=-Lambda,Lambda
                  do n4=-Lambda,Lambda
                        n5=-n1-n2-n3-n4+n
                        if ((n5.le.Lambda).and.(n5.ge.-Lambda))then
                     phi5=phi5+tildephi(n1)*tildephi(n2)*tildephi(n3)
     &                           *tildephi(n4)*tildephi(n5)
                        endif
                  enddo
               enddo
            enddo
         enddo

                  phi5=6.0d0*0.5d0*beta*coupling**2*phi5
                  forceB(n)=forceB(n)+phi3+phi5
      enddo

      return
      end
```

The above first order differential Hamilton equations of motion will be solved numerically by means of the leap-frog method which preserves phase space volume and reversibility and only break Hamiltonian conservation. The solution of equations (A.53) and (A.54) using leap-frog is given explicitly by

$$\tilde{\pi}(n)|_{t+\frac{\delta t}{2}} = \tilde{\pi}(n)|_t - \frac{1}{2}\delta t F(n)|_t. \tag{A.56}$$

$$\tilde{\phi}(n)|_{t+\delta t} = \tilde{\phi}(n)|_t + \delta t \tilde{\pi}(n)|_{t+\frac{\delta t}{2}}. \tag{A.57}$$

$$\tilde{\pi}(n)|_{t+\delta t} = \tilde{\pi}(n)|_{t+\frac{\delta t}{2}} - \frac{1}{2}\delta t F(n)|_{t+\delta t}. \tag{A.58}$$

We remark that the force $F(n)$ at the instant t is needed to advance all the degrees of freedom $\tilde{\phi}(n)$ from t to $t + \delta t$. Thus we need to calculate the force $F(n)$ for all n at instant t, then apply equations (A.56) and (A.57) to advance all $\tilde{\phi}(n)$ from t to $t + \delta t$, then calculate again the force $F(n)$ for all n at instant $t + \delta t$, then apply equation (A.58) to advance all $\tilde{\pi}(n)$ from $t + \delta t/2$ to $t + \delta t$.

Also, the leap-frog will need two extra parameters: the step δt and the number of iterations which we call T. The total time of the motion is given by $T\delta t$. The initial values of $\tilde{\phi}(n)$ and $\tilde{\pi}(n)$ at time 0 will be specified and then by applying the above leap-frog algorithm we will obtain the final values of $\tilde{\phi}(n)$ and $\tilde{\pi}(n)$ at time $T\delta t$.

In the code all this might look like the following subroutine:

```
      subroutine molecular(mass,coupling,beta,Lambda,omega,T,dt,
     & tildephi,tildep)
       double precision dt,mass, coupling, beta, omega
       double complex tildephi(-Lambda:+Lambda),
     &  tildep(-Lambda:Lambda),forceB(-Lambda:Lambda)
       integer Lambda,n,T,t_molecular

      do t_molecular=1,T

      call force(mass,coupling,beta,Lambda,omega,tildephi,forceB)
      do n=-Lambda,Lambda
         tildep(n)=tildep(n)-0.5d0*dt*forceB(n)
         tildephi(n)=tildephi(n)+dt*tildep(n)
      enddo

      call force(mass,coupling,beta,Lambda,omega,tildephi,forceB)
      do n=-Lambda,Lambda
         tildep(n)=tildep(n)-0.5d0*dt*forceB(n)
      enddo

      enddo

      return
      end
```

We run a small test of the final code. First, we can easily check that if we set $\omega = g = 0$ we get a harmonic oscillator problem with frequency

$$\Omega = \sqrt{\beta m^2} \Rightarrow \text{period} = \frac{2\pi}{\sqrt{\beta m^2}}. \tag{A.59}$$

Thus the fields $\tilde{\phi}(n)$ and $\tilde{\pi}(n)$ will oscillate with this period, whereas the kinetic energy and the potential/action will oscillate with half this period. The Hamiltonian is supposed to be conserved. But as discussed above, this algorithm contains a systematic error and as such we find that the Hamiltonian which includes necessarily this systematic error oscillates with a very small amplitude with a period given by twice the above value.

A.3.4 The Metropolis and the heat bath algorithms

Random number generator: In this section we will write two algorithms: the Metropolis algorithm and the heat bath algorithm, and in the next section we will write the so-called hybrid Monte Carlo algorithm which is a very powerful algorithm which combines the Metropolis with the heat bath. These are all sampling Monte Carlo algorithms and thus require the use of random numbers in an essential way. As our (pseudo) random generator we will use *Numerical Recipes* **ran2**. This will be included as an independent subroutine in the code. We only need to fix the seed for this random number generator at the beginning of the simulation.

Partition function: In the previous section we have studied by means of molecular dynamics the following action

$$S_B = \beta \sum_{n=-\Lambda}^{+\Lambda} \frac{1}{2}(n^2\omega^2 + m^2)\tilde{\phi}(n)\tilde{\phi}(-n) + \beta mg(\tilde{\phi}^4)_0 + \frac{\beta}{2}g^2(\tilde{\phi}^6)_0. \tag{A.42}$$

Two essential remarks can be stated now:

- The molecular dynamics involves in an obvious way a systematic error.
- The molecular dynamics probes only classical physics.

These two problems can be solved at once via the so-called Metropolis algorithm which is the most general of all Monte Carlo algorithms. Indeed, the Metropolis algorithm allows us to probe quantum physics since it aims at calculating the following partition function (path integral)

$$\begin{aligned} Z &= \int \prod_n d\tilde{\phi}(n)d\tilde{\phi}^*(n)\prod_n d\tilde{\pi}(n)d\tilde{\pi}^*(n) \\ &\quad \exp\left(-\frac{1}{2}\sum_n \tilde{\pi}(n)\tilde{\pi}^*(n) - S_B[\tilde{\phi}, \tilde{\phi}^*]\right) \\ &= \int \prod_n d\tilde{\pi}(n)d\tilde{\pi}^*(n) \exp\left(-\frac{1}{2}\sum_n \tilde{\pi}(n)\tilde{\pi}^*(n)\right) \\ &\quad \int \prod_n d\tilde{\phi}(n)d\tilde{\phi}^*(n) \exp(-S_B[\tilde{\phi}, \tilde{\phi}^*]). \end{aligned} \tag{A.61}$$

The meaning of the measure is precisely given by

$$\int \prod_{n} d\tilde{\phi}(n)d\tilde{\phi}^*(n) = \int d\tilde{\phi}(0) \int \prod_{n=1}^{\Lambda} d\tilde{\phi}(n)d\tilde{\phi}^*(n)$$
$$= \int d\tilde{\phi}(0)d\tilde{\phi}^*(0)\delta(\text{Imag}(\tilde{\phi}(0))) \int \prod_{n=1}^{\Lambda} d \tag{A.62}$$
$$\tilde{\phi}(n)d\tilde{\phi}^*(n).$$

Thus, we will treat $\tilde{\phi}(n)$ as a genuine complex array including $\tilde{\phi}(0)$ with the delta function enforcing the reality of $\tilde{\phi}(0)$. The same thing goes in principle for $\tilde{\pi}(n)$ but this field will be treated differently as we will see.

Metropolis algorithm: The Metropolis algorithm consists of the following steps:

1. We start from some initial configuration $\tilde{\phi}_0$, $\tilde{\pi}_0$.
2. We propose a new configuration $\tilde{\phi}$, $\tilde{\pi}$.
3. We compute the variation of the Hamiltonian

$$\Delta H = H(\tilde{\phi}, \tilde{\pi}) - H(\tilde{\phi}_0, \tilde{\pi}_0). \tag{A.63}$$

4. We accept the new proposal with a probability given by

$$p(\tilde{\phi}, \tilde{p}) = \min(\exp(-\Delta H), 1). \tag{A.64}$$

5. We repeat starting from step 2 until we reach thermalization.
6. We measure thermalized configurations which are used to calculate expectation values of physical observables.

The meaning of equation (A.64) is as follows. After we compute the variation of the Hamiltonian in step 3 we check its sign. If $\Delta\ H < 0$ then we should accept the proposal since it had resulted in a decrease of the Hamiltonian and thus getting us closer to the minimum. Otherwise, if $\Delta\ H > 0$ we accept the proposal only with a probability given by the Boltzmann distribution $\exp(-\Delta H)$. This is the part which is simulating quantum mechanics and it is implemented numerically via the Von Neumann method. In other words, choose a uniform random number r between 0 and 1 and compare it to the Boltzmann distribution $\exp(-\Delta H)$. If $r < \Delta H$ then accept the proposal otherwise reject the proposal.

Explicitly, in the code this will look as follows:

```
c1........initial configuration.............

      tildephi0=tildephi
```

```
      tildep0=tildep

c2.........initial value of Hamiltonian.......

      call action(mass,coupling,beta,Lambda,omega,tildephi,tildep,
     &     actionB,Hami)
      VariationH=Hami
      .
      .
c3.........new proposal.......................

     tildephi
     tildep
      .
      .
c4.........variation of Hamiltonian...........

      call action(mass,coupling,beta,Lambda,omega,tildephi,tildep,
     &     actionB,Hami)
      VariationH=VariationH-Hami
      VariationH=-VariationH

c5.........metropolis step....................

         if (variationH.lt.0.0d0)then
            accept=accept+1.0d0
         else
            probabilityB=dexp(-VariationH)
            r=ran2(idum)
            if (r.lt.probabilityB)then
               accept=accept+1.0d0
         else
            reject=reject+1.0d0
            tildephi=tildephi0
            tildep=tildep0
         endif
      endif
```

Heat bath algorithm: The hybrid Monte Carlo algorithm is an algorithm in which two crucial extra steps are added to the Metropolis algorithm:

1. The step number 2 in the Metropolis algorithm is implemented via the **molecular dynamics algorithm**. In other words, the new configuration $\tilde{\phi}$, $\tilde{\pi}$ is

given by the solution of the molecular dynamics problem with $\tilde{\phi}_0$, $\tilde{\pi}_0$ as the initial condition.

2. As it turns out the first integral over the conjugate momentum field in equation (A.61) should be sampled using the so-called **heat bath algorithm** in order to avoid the ergodic problem, i.e. to be able to reach every point in phase space.

   This is also very natural since the integral is Gaussian, namely

$$\begin{aligned} Z &= \int \prod_n d\tilde{\pi}(n) d\tilde{\pi}^*(n) \exp\left(-\frac{1}{2} \sum_{n=-\Lambda}^{+\Lambda} \tilde{\pi}(n)\tilde{\pi}(-n) \right) \\ &= \int d\tilde{\pi}(0) \exp\left(-\frac{1}{2} \tilde{\pi}(0)\tilde{\pi}(0) \right) \\ &\quad \prod_{n=1}^{\Lambda} \int d\tilde{\pi}(n) d\tilde{\pi}^*(n) \exp\left(- \sum_{n=-\Lambda}^{+\Lambda} \tilde{\pi}(n)\tilde{\pi}(-n) \right) \end{aligned} \tag{A.65}$$

All these integrals are Gaussian of the form

$$\frac{a}{\pi} \int dz dz^* \exp(-azz^*) = \int_0^1 dv_1 \int_0^1 dv_2. \tag{A.66}$$

$$v_1 = \exp(-ar^2), \quad v_2 = \frac{\theta}{2\pi}. \tag{A.67}$$

These two equations show that v_1 and v_2 are uniform random numbers between 0 and 1 and thus z is a complex random number given by the formula

$$z = \sqrt{-\frac{1}{a}\ln v_1} (\cos 2\pi v_2 + i \sin 2\pi v_2). \tag{A.68}$$

All $\tilde{\pi}(n)$, with $n = 1, \ldots, \Lambda$, are given by this formula with $a = 1$. However, $\tilde{\pi}(0)$ is a real random number corresponding to $a = 1/2$ given by the formula

$$\tilde{\pi}(0) = \sqrt{-2\ln v_1} \cos 2\pi v_2. \tag{A.69}$$

We will mostly initialize the configuration from a Gaussian distribution with $a = \beta\, m2$ for $n = 1, \ldots, \Lambda$ and with $a = \beta$ m2 /2 for $n = 0$.

The heat bath algorithm will be coded in a subroutine which might take the form:

```
subroutine gaussian(Lambda,tildep,idum)
integer lambda,n,idum
double precision pi,theta,r,ran2
double complex ii, tildep(-Lambda:+Lambda)

pi=dacos(-1.0d0)
ii=cmplx(0.0d0,1.0d0)

do n=1,Lambda
   theta=2.0d0*pi*ran2(idum)
   r=dsqrt(-1.0d0*dlog(1.0d0-ran2(idum)))
   tildep(n)=r*dcos(theta)+ii*r*dsin(theta)
   tildep(-n)=conjg(tildep(n))
enddo

theta=2.0d0*pi*ran2(idum)
r=dsqrt(-2.0d0*dlog(1.0d0-ran2(idum)))
tildep(0)=r*dcos(theta)

return
end
```

A.3.5 Hybrid Monte Carlo algorithm

In summary, the hybrid Monte Carlo algorithm is a Metropolis algorithm with updating of the field $\tilde{\phi}$ provided by the molecular dynamics algorithm, whereas updating of the conjugate field $\tilde{\pi}$ is provided by heat bath algorithm.

Thus, in the third step in the above pseudo code we will have to call the molecular dynamics subroutine we wrote in the previous section to give a new configuration $\tilde{\phi}$ as:

```
c3.........new proposal.......................

      call molecular(mass,coupling,beta,Lambda,omega,T,dt,
     &     tildephi,tildep)
```

Whereas, in the first step in the above pseudo code we will have to call the heat bath subroutine to initialize the conjugate momentum field as:

```
c1.........initial configuration.............

      call gaussian(Lambda,tildep,idum)
      tildephi0=tildephi
      tildep0=tildep

c3.........new proposal......................

      call molecular(mass,coupling,beta,Lambda,omega,T,dt,
     &     tildephi,tildep)
```

The hybrid Monte Carlo algorithm will then look like in the code as perhaps the following subroutine:

```
      subroutine metropolis(mass,coupling,beta,Lambda,omega,T,dt,idum,
     &     accept,reject,tildephi,variationH)
      integer Lambda,T,idum
      double precision mass,coupling,beta,omega,dt,ran2,r,actionB,Hami,
     &     VariationS,VariationH,probabilityB,accept,reject
      double complex tildephi(-Lambda:+Lambda),
     &     tildephi0(-Lambda:+Lambda), tildep(-Lambda:+Lambda),
     &     tildep0(-Lambda:+Lambda)

      call gaussian(Lambda,tildep,idum)

      tildephi0=tildephi
      tildep0=tildep

      call action(mass,coupling,beta,Lambda,omega,tildephi,tildep,
     &     actionB,Hami)
      VariationS=actionB
```

```
      VariationH=Hami

      call molecular(mass,coupling,beta,Lambda,omega,T,dt,
     &     tildephi,tildep)

      call gaussian(Lambda,tildep,idum)

      call action(mass,coupling,beta,Lambda,omega,tildephi,tildep,
     &     actionB,Hami)

      VariationS=VariationS-actionB
      VariationH=VariationH-Hami
      VariationS=-VariationS
      VariationH=-VariationH

c.........metropolis step....
         if (variationH.lt.0.0d0)then
            accept=accept+1.0d0
         else
            probabilityB=dexp(-VariationH)
            r=ran2(idum)
            if (r.lt.probabilityB)then
               accept=accept+1.0d0
         else
            reject=reject+1.0d0
            tildephi=tildephi0
            tildep=tildep0
         endif
      endif

      return
      end
```

Test: As a test of the proper running of the final code we consider again the harmonic oscillator problem corresponding to $\omega = g = 0$. The Hamiltonian reads

$$H = \frac{1}{2}\sum_{n=-\Lambda}^{+\Lambda} \tilde{\pi}(n)\tilde{\pi}^*(n) + \sum_{n=-\Lambda}^{+\Lambda} \frac{1}{2}\beta m^2 \tilde{\phi}(n)\tilde{\phi}^*(n). \tag{A.70}$$

The expectation value of the Hamiltonian is given by the formula

$$\langle H \rangle = -\frac{\partial}{\partial a} \ln Z_a|_{a=1}. \tag{A.71}$$

The partition function Z_a can be calculated quite easily to find

$$
\begin{aligned}
Z_a &= \int \prod_n d\tilde{\pi}(n) d\tilde{\pi}^*(n) \exp\left(-\frac{a}{2}\sum_n \tilde{\pi}(n)\tilde{\pi}^*(n)\right) \\
&\int \prod_n d\tilde{\phi}(n) d\tilde{\phi}^*(n) \exp\left(-\frac{a\beta m^2}{2}\sum_n \tilde{\phi}(n)\tilde{\phi}^*(n)\right) \\
&= \frac{2\mathcal{N}}{(a^2\beta m^2)^{\Lambda+\frac{1}{2}}},
\end{aligned}
\tag{A.72}
$$

where $\mathcal{N}$ is some unimportant constant. We get immediately

$$
\langle H \rangle = 2\Lambda + 1. \tag{A.73}
$$

This is independent of β and m. We remark also that this result is only correct for $m \neq 0$. For $m = 0$ we should get half this prediction.

A.3.6 Remez algorithm and conjugate gradient method: how one deals with the fermionic determinant?

We add now to the bosonic action S_B of the previous sections the fermionic action (with $\bar{\psi} = \psi^\dagger$)

$$
S_F = \int dt(\bar{\psi}\dot{\psi} + \bar{\psi}W''\psi). \tag{A.74}
$$

The one-component Dirac field ψ is a complex Grassmann number and thus it cannot be simulated directly in Monte Carlo sampling. It is assumed to obey periodic boundary condition similarly to the real scalar field ϕ and thus it can be expanded as

$$
\psi(t) = \sum_{n=-\Lambda}^{+\Lambda} \tilde{\psi}(n)\exp(i\omega n t), \quad \omega = \frac{2\pi}{\beta}. \tag{A.75}
$$

The action S_F in terms of the Grassmannian Fourier modes $\tilde{\psi}(n)$ takes the form

$$
\begin{aligned}
S_F &= \sum_{nk} \tilde{\psi}^*(n)\mathcal{M}_{nk}\tilde{\psi}(k), \\
\mathcal{M}_{nk} &= \beta\left[(in\omega + m)\delta_{nk} + 3g\sum_l (\tilde{\phi}^2)_l \delta_{n,\,k+l}\right]
\end{aligned}
\tag{A.76}
$$

The partition function of interest is now given by

$$Z = \int \prod_n d\tilde{\phi}(n)d\tilde{\phi}^*(n)\exp(-S_B(\tilde{\phi},\tilde{\phi}^*)) \int \prod_n d\tilde{\psi}(n)d\tilde{\psi}^*(n)\exp(-\tilde{\psi}^\dagger \mathcal{M}\tilde{\psi}). \tag{A.77}$$

The first integral is the bosonic part which we have already studied in previous sections, whereas the second integral is the new contribution coming from fermions. This second integral can be done exactly to give the determinant of the matrix $\mathcal{M}$, viz

$$Z = \int \prod_n d\tilde{\phi}(n)d\tilde{\phi}^*(n)\exp(-S_B(\tilde{\phi},\tilde{\phi}^*))\det\ \mathcal{M}. \tag{A.78}$$

This determinant is positive definite for positive g and m. We rewrite this as

$$Z = \int \prod_n d\tilde{\phi}(n)d\tilde{\phi}^*(n)\exp(-S_B(\tilde{\phi},\tilde{\phi}^*))\det\ \mathcal{O}, \quad \mathcal{O} = \Delta^{1/2}, \quad \Delta = \mathcal{M}^\dagger\mathcal{M}. \tag{A.79}$$

This determinant will not be computed directly in Monte Carlo, because it is obviously a very expensive calculation, but rather it will be sampled indirectly via the so-called pseudo fermions. We note the identity

$$\int \prod_n d\tilde{\Phi}(n)d\tilde{\Phi}^*(n)\exp(-\tilde{\Phi}^\dagger\mathcal{O}^{-1}\tilde{\Phi}) = \det\ \mathcal{O}. \tag{A.80}$$

The new scalar fields $\tilde{\Phi}(n)$ are the pseudo fermions because they are in fact bosons, which carries the same indices as the original fermions $\tilde{\psi}(n)$, but are designed to give exactly the fermion contribution.

We are led then to the partition function

$$Z = \int \prod_n d\tilde{\phi}(n)d\tilde{\phi}^*(n)\int \prod_n d\tilde{\Phi}(n)d\tilde{\Phi}^*(n) \exp(-S_B(\tilde{\phi},\tilde{\phi}^*) - \tilde{\Phi}^\dagger r(\Delta)\tilde{\Phi}). \tag{A.81}$$

$$r(\Delta) = \mathcal{O}^{-1} = \Delta^{-1/2}. \tag{A.82}$$

We define the pseudo fermions potential by

$$V = \tilde{\Phi}^\dagger r(\Delta)\tilde{\Phi}. \tag{A.83}$$

Next, we employ the rational approximation of functions by a ratio of two polynomials given by

$$r(x) = x^{-1/2} \simeq a_0 + \sum_{\sigma=1}^{M} \frac{a_\sigma}{x + b_\sigma}. \tag{A.84}$$

The constants M, a_0, a_σ and b_σ are determined by the requirement that this rational approximation is close to the actual function by some prescribed tolerance. This is done via the so-called **Remez algorithm**.

The potential V becomes

$$\begin{aligned} V &= a_0\tilde{\Phi}^\dagger\tilde{\Phi}^\dagger + \sum_{\sigma=1}^{M} a_\sigma\tilde{\Phi}^\dagger(x + b_\sigma)^{-1}\tilde{\Phi} \\ &= a_0\tilde{\Phi}^\dagger\tilde{\Phi}^\dagger + \sum_{\sigma=1}^{M} a_\sigma\tilde{\Phi}^\dagger G_\sigma = \tilde{\Phi}^\dagger W \\ &= a_0\tilde{\Phi}^\dagger\tilde{\Phi}^\dagger + \sum_{\sigma=1}^{M} a_\sigma G_\sigma^\dagger\tilde{\Phi} = W^\dagger\tilde{\Phi}. \end{aligned} \tag{A.85}$$

The objects G_σ and W are defined by

$$G_\sigma = (x + b_\sigma)^{-1}\tilde{\Phi}, \qquad W = a_0\tilde{\Phi} + \sum_{\sigma=1}^{M} a_\sigma G_\sigma. \tag{A.86}$$

The evaluation of G_σ involves the inversion of the matrix Δ with M different values of the mass parameter b_σ. This is clearly another very expensive and demanding calculation which will be done indirectly via the use of the so-called **conjugate gradient method**.

In the molecular dynamics part we will introduce in addition to the momenta $\tilde{\pi}(n)$ associated with the boson fields $\tilde{\phi}(n)$ new momenta $\tilde{\Pi}(n)$ associated with the pseudo fermion fields $\tilde{\Phi}(n)$. Thus, we should deal with the Hamiltonian

$$H = \frac{1}{2}\sum_{n=-\Lambda}^{+\Lambda} \tilde{\pi}(n)\tilde{\pi}^*(n) + \frac{1}{2}\sum_{n=-\Lambda}^{+\Lambda} \tilde{\Pi}(n)\tilde{\Pi}^*(n) + S_B + V. \tag{A.87}$$

The Hamilton equations of motion of the fields $\tilde{\Phi}$ and $\tilde{\Pi}$ are given by

$$\frac{d\tilde{\Phi}(n)}{d\tau} = \tilde{\Pi}(n), \quad \frac{d\tilde{\Pi}(n)}{d\tau} = -W(n). \tag{A.88}$$

These are the new equations of motion. The Hamilton equations of motion of the fields $\tilde{\phi}$ and $\tilde{\pi}$ are given by

$$\frac{d\tilde{\phi}(n)}{d\tau} = \tilde{\pi}(n), \quad \frac{d\tilde{\pi}(n)}{d\tau} = -F(n) - \frac{\partial V}{\partial\tilde{\phi}^*(n)}. \tag{A.89}$$

Clearly, $F(n)$ is the bosonic force introduced in a previous section. Thus, the second equation is modified by the introduction of the fermion force given by

$$\begin{aligned} f(n) &= \frac{\partial V}{\partial \tilde{\phi}^*(n)} \\ &= \tilde{\Phi}^\dagger \frac{\partial r}{\partial \tilde{\phi}^*(n)} \tilde{\Phi} \\ &= -\sum_{\sigma=1}^{M} a_\sigma G_\sigma^\dagger \frac{\partial \Delta}{\partial \tilde{\phi}^*(n)} G_\sigma \\ &= -\sum_{\sigma=1}^{M} a_\sigma F_\sigma^\dagger \frac{\partial \mathcal{M}}{\partial \tilde{\phi}^*(n)} G_\sigma - \left(\sum_{\sigma=1}^{M} a_\sigma F_\sigma^\dagger \frac{\partial \mathcal{M}}{\partial \tilde{\phi}^*(n)} G_\sigma \right)^* . \end{aligned} \tag{A.90}$$

The object F_σ is defined by

$$F_\sigma = \mathcal{M} G_\sigma. \tag{A.91}$$

In the Metropolis step we will generate, before and after the molecular dynamics, the fields $\tilde{\pi}(n)$ as well as the fields $\tilde{\Pi}(n)$ by means of the heat bath algorithm in order to avoid ergodic problems. Similarly, the fields $\tilde{\Phi}(n)$ should be generated from a Gaussian number ξ by the formula

$$\tilde{\Phi} = \bar{r}(\Delta)\xi, \quad \bar{r}(\Delta) = \Delta^{1/4}. \tag{A.92}$$

Again, we will employ a different rational approximation

$$\bar{r}(x) = x^{1/4} \simeq a_0 + \sum_{\sigma=1}^{M} \frac{a_\sigma}{x + b_\sigma}, \tag{A.93}$$

by calling the Remez algorithm.

A.3.7 How one really deals with the fermionic determinant?

Remez algorithm

We will need to compute the operator Δ^δ, for some fractional power δ, via the minimax approximation

$$\Delta^\delta \simeq r(\Delta) = a_0 + \sum_{\sigma=1}^{M} \frac{a_\sigma}{x + b_\sigma}. \tag{A.94}$$

In the actual code we will use the Remez algorithm written by Clark and Kennedy to compute a_0, a_σ, b_σ and M.

The steps are:

1. Install mpfr.
2. Install gmp.
3. Install g++.
4. Download AlgRemez and unzip.

5. Run ‘make’.
6. We choose the function to approximate $f = x^{y/z}$, the degrees n and d of the rational approximation $r(x) = P_n(x)/P_d(x)$ where $P_n(x)$ and $P_d(x)$ are polynomials of degrees n and d, respectively, the range over which the rational approximation is to be calculated $[\lambda_{\mathrm{low}}, \lambda_{\mathrm{high}}]$, and the precision used.

   Of course, we can only take $n = d$ (limitation of this code) and $M = n = d$.
7. Run a calculation as:

```
./test y z n d lambda_low lambda_high precision
```

8. Results are found in the file ‘approx.dat’ while errors are found in the file ‘error.dat’.
9. Modify the file ‘main.C’ so that the poles and residues of the rational approximations appear as a column. The approximations for $x^{y/z}$ and $x^{-y/z}$ appear in separate files ‘approx.dat’ and ‘approx1.dat’, respectively.
10. Add the error function $r - f$ to the file error.dat, and compute the uniform norm $|r - f|_\infty = \max|r - f|$ in a different file ‘error1.dat’. Do not forget to run ‘make’ each time you modify one of the files.
11. Read the uniform norm and test whether or not it is smaller than some tolerance. If it is then stop, otherwise go back and change the degrees of the approximation n and d and repeat.
12. Create a Fortran file which automates this task. The command ./test is implemented in the Fortran file by *call system(command)*, whereas the *inquire statement* is used to test whether or not the file error1.dat is ready to use.

A Fortran code which does all these things may look like:

```
c.........this code should be compiled in the directory where "test", which is
..........obtained in the make of AlgRemez, is found.....

      program my_remez
      implicit none
      integer y,z,n,d,precision,i,counter,j,n0
      parameter(n0=100)
      doubleprecision lambda_low, lambda_high,e,tolerance
      doubleprecision a0,a(n0),b(n0),c0,c(n0),dd(n0),coefficient(n0)
      parameter (tolerance=1.0d-5)
      character*100 degree, com
```

```
      character*50 h1
      LOGICAL THERE

c........we choose the function to approximate, the range over which the rational
.........approximation is to be calculated, and the precision used....

      y=1
      z=2
      lambda_low=0.0004d0
      lambda_high=100.0d0
      precision=40
      print*, "Approximating the functions x^{y/z} and x^{-y/z}:"
     &     , "y=",y,"z=",z
      print*, "Approximation bounds:", lambda_low,lambda_high
      print*, "Precision of arithmetic:", precision
      write(*,*)".................."

c.... we start the iteration on the degree of approximation at n=d=2....

      counter=0
      i=1
 14   i=i+1
      counter=counter+1
      print*, "ITERATION:",counter
      write(degree,â ("", i7)â)i
      read(degree,â (i7)â)n
      read(degree,â (i7)â)d
      write(*,*)"degrees of approximation", n,d

c..we call AlgRemez by the command="./test y z n d lambda_low lambda_high precision".....

      write(com,â (a,i5," ",i5," ",i7," ",i7," ",F10.5," ",F10.5," "
     &,i5," ",a)â) "./test ",y,z,d,n,lambda_low,lambda_high
     &     ,precision,""
      print*, "command:", com
      call system(com)

c........we check whether or not the uniform norm is found.......................

      inquire(file=âerror1.datâ, exist=THERE)
 11   if ( THERE) then
         write(*,*) "file exists!"
      else
```

```
          go to  11
       end if

c......we read the uniform norm and test whether or not it is smaller than some
.......tolerance, if it is not, we go back and repeat with  increased degrees of
.......approximation n=n+1 and d=d+1............

       open(unit=50+i,file=â error1.datâ ,status=â oldâ )
       read(50+i,555) e
       write(*,*)"uniform norm", e
       write(*,*)".................."
 555   format(1F20.10)
       close(50+i)
       if (e.gt.tolerance) go to 14

c.............the solution for x^{y/z}...........

       write(*,*)"rational approximation of x^{y/z}"
       open(unit=60,file=â approx.datâ ,status=â oldâ )

       do j=1,2*n+1
          read(60,*)coefficient(j)
       enddo

       c0=coefficient(1)
       write(*,*)"c0=",c0
       do i=2,n+1
          c(i-1)=coefficient(i)
          dd(i-1)=coefficient(i+n)
          write(*,*)"i-1=",i-1,"c(i-1)=", c(i-1),"d(i-1)=",dd(i-1)
       enddo

c.................the solution for x^{-y/z}..........

       write(*,*)"rational approximation of x^{-y/z}"
       open(unit=61,file=â approx1.datâ ,status=â oldâ )

       do j=1,2*n+1
          read(61,*)coefficient(j)
       enddo

       a0=coefficient(1)
       write(*,*)"a0=",a0
```

```
do i=2,n+1
   a(i-1)=coefficient(i)
   b(i-1)=coefficient(i+n)
   write(*,*)"i-1=",i-1,"a(i-1)=", a(i-1),"b(i-1)=",b(i-1)
enddo



return
end
```

Conjugate gradient method

We will also need to compute for all values of b_σ generated by the Remez algorithm the vector x given by the equation

$$(A + b_\sigma)x = v, \tag{A.95}$$

where $A = \Delta$ and $v = \tilde{\Phi}$. We will define the residue by

$$r = v - Ax. \tag{A.96}$$

We use here the conjugate gradient method with multi-mass Krylov solver as discussed by Jegerlehner. This consists in solving the multi-mass system iteratively using only a single set of vector–matrix operations. In other words, we will generate the solutions for $b_\sigma \neq 0$ from the solution for $b_\sigma = 0$ and this latter linear system will only involve a single set of vector–matrix operations. The errors are generated directly from the errors of the no-sigma problem.

Starting at some point x_n, the iterative solution x_{n+1} of the no-sigma problem is obtained from the minimum of the function $f(x) = (xAx - 2xv)/2$ in some direction p_n. The initial point is $x_0 = 0$ and the initial searching direction is given by $p_0 = v$ while the next searching direction p_n is determined by the requirement that it must be A-orthogonal to the previous searching direction p_{n-1}.

Without any proof [1] the main ingredients of this algorithm are given by:

1. We start from

$$x = x_0^\sigma = 0, \quad r_0 = r_0^\sigma = v, \quad p = p_0^\sigma = v. \tag{A.97}$$

We must also start from

$$\alpha_0 = \alpha_0^\sigma = 0, \beta_{-1} = \beta_{-1}^\sigma = 1, \quad \zeta_0^\sigma = \zeta_{-1}^\sigma = 1. \tag{A.98}$$

2. We solve the no-sigma problem (we start from $n = 0$):

$$q_n = Ap_n. \tag{A.99}$$

$$\beta_n = -\frac{r_n r_n}{p_n q_n}$$
$$x_{n+1} = x_n - \beta_n p_n. \tag{A.100}$$

$$r_{n+1} = r_n + \beta_n q_n. \tag{A.101}$$

$$\alpha_{n+1} = \frac{r_{n+1} r_{n+1}}{r_n r_n}$$
$$p_{n+1} = r_{n+1} + \alpha_{n+1} p_n. \tag{A.102}$$

3. We generate solutions of the sigma problems by the relations (we start from $n = 0$):

$$\zeta_{n+1}^{\sigma} = \frac{\zeta_n^{\sigma} \zeta_{n-1}^{\sigma} \beta_{n-1}}{\alpha_n \beta_n (\zeta_{n-1}^{\sigma} - \zeta_n^{\sigma}) + \zeta_{n-1}^{\sigma} \beta_{n-1} (1 - \sigma \beta_n)}. \tag{A.103}$$

$$\beta_n^{\sigma} = \beta_n \frac{\zeta_{n+1}^{\sigma}}{\zeta_n^{\sigma}}. \tag{A.104}$$

$$x_{n+1}^{\sigma} = x_n^{\sigma} - \beta_n^{\sigma} p_n^{\sigma}. \tag{A.105}$$

$$r_{n+1}^{\sigma} = \zeta_{n+1}^{\sigma} r_{n+1}. \tag{A.106}$$

$$\alpha_{n+1}^{\sigma} = \alpha_{n+1} \frac{\zeta_{n+1}^{\sigma} \beta_n^{\sigma}}{\zeta_n^{\sigma} \beta_n}. \tag{A.107}$$

$$p_{n+1}^{\sigma} = r_{n+1}^{\sigma} + \alpha_{n+1}^{\sigma} p_n^{\sigma}. \tag{A.108}$$

   Note how the residues are generated directly from the residues of the no-sigma problem.

4. The above procedure continues as long as $|r| \geqslant \epsilon$ where ϵ is some tolerance, otherwise stop. Thus

$$|r| \geqslant \epsilon, \quad \text{continue.} \tag{A.109}$$

A Fortran code which does all this may look like:

```
      program my_conjugate_gradient
      implicit none
      integer N,M,i,j,counter,sig
      parameter (N=3,M=2)
      double precision A(1:N,1:N),v(1:N),sigma(1:M)
      double precision x(1:N),r(1:N),p(1:N),q(1:N),product,product1,
     & product2,residue,tolerance
      double precision  alpha,beta,alpha_previous,beta_previous,xii,
     & xii0
      double precision beta_sigma(1:M),alpha_sigma(1:M),xi(1:M),
     & xi_previous(1:M)
      double precision x_sigma(1:N,1:M),p_sigma(1:N,1:M),
     & r_sigma(1:N,1:M)
      parameter(tolerance=1.0d-10)

c............example input..........................

      call input(N,M,A,v,sigma)

c.............initialization........................

      do i=1,N
         x(i)=0.0d0
         r(i)=v(i)
         do sig=1,M
            x_sigma(i,sig)=0.0d0
         enddo
      enddo

c.............we start with alpha(0)=0, beta(-1)=1, xi^sigma(-1)=xi^sigma(0)=1,
..............alpha^sigma(0)=0 and beta^sigma(-1)=1...

      alpha=0.0d0
      beta=1.0d0
      do sig=1,M
         xi_previous(sig)=1.0d0
         xi(sig)=1.0d0
         alpha_sigma(sig)=0.0d0
         beta_sigma(sig)=1.0d0
      enddo

c.............starting iteration.........
```

```
      counter=0

c..............choosing search directions...............

 13   do i=1,N
         p(i)=r(i)+alpha*p(i)
         do sig=1,M
            p_sigma(i,sig)=xi(sig)*r(i)
     &           +alpha_sigma(sig)*p_sigma(i,sig)
         enddo
      enddo
       write(*,*)counter,p

c.......solving the sigma=0 problem.........

      product=0.0d0
      product1=0.0d0
c.......the only matrix-vector multiplication in the problem..........
      do i=1,N
         q(i)=0.0d0
         do j=1,N
            q(i)=q(i)+A(i,j)*p(j)
         enddo
         product=product+p(i)*q(i)
         product1=product1+r(i)*r(i)
      enddo
      beta_previous=beta
      beta=-product1/product

      product2=0.0d0
      do i=1,N
         x(i)=x(i)-beta*p(i)
         r(i)=r(i)+beta*q(i)
         product2=product2+r(i)*r(i)
      enddo
      alpha_previous=alpha
      alpha=product2/product1

c.......solving the sigma problems.............

      do sig=1,M
c......the xi coefficients..........
```

```
          xii0=alpha_previous*beta*(xi_previous(sig)-xi(sig))
     &        +xi_previous(sig)*beta_previous*(1.0d0-sigma(sig)*beta)
          xii=xi(sig)*xi_previous(sig)*beta_previous/xii0
          xi_previous(sig)=xi(sig)
          xi(sig)=xii
c........the beta coefficients......
          beta_sigma(sig)=beta*xi(sig)/xi_previous(sig)
c.........the solutions and residues...........
          do i=1,N
             x_sigma(i,sig)=x_sigma(i,sig)-beta_sigma(sig)*p_sigma(i,sig)
             r_sigma(i,sig)=xi(sig)*r(i)
          enddo
c.......the alpha coefficients.......
          alpha_sigma(sig)=alpha
          alpha_sigma(sig)= alpha_sigma(sig)*xi(sig)*beta_sigma(sig)
          alpha_sigma(sig)=alpha_sigma(sig)/(xi_previous(sig)*beta)
       enddo

c......testing whether or not the interation should be continued........

       counter=counter+1
       residue=0.0d0
       do i=1,N
          residue=residue+r(i)*r(i)
       enddo
       residue=dsqrt(residue)
c       write(*,*)counter,residue,tolerance
       if(residue.ge.tolerance)  go to 13

c........verification 1: if we set sigma=0 then xi must be equal 1 whereas
.........the other pairs must be equal.........
       write(*,*)"verification 1"
       write(*,*)counter,xi(1),xi_previous(1)
       write(*,*)counter,beta,beta_sigma(1)
       write(*,*)counter,alpha,alpha_sigma(1)

c............verification 2.....
       write(*,*)"verification 2"
       do i=1,N
          q(i)=0.0d0
          do j=1,N
             q(i)=q(i)+A(i,j)*x(j)
          enddo
```

```
      enddo
      write(*,*)"v",v
      write(*,*)"q",q

c...........verification 3.....
      write(*,*)"verification 3"
      sig=1
      do i=1,N
         q(i)=sigma(sig)*x_sigma(i,sig)
         do j=1,N
            q(i)=q(i)+A(i,j)*x_sigma(j,sig)
         enddo
      enddo
      write(*,*)"v",v
      write(*,*)"q",q

      return
      end
```

In the code of the supersymmetric anharmonic oscillator this program should be turned into a subroutine.

Finally we can include the fermions
Fermion matrix: We will need to code the fermion matrix $\mathcal{M}_{nk}$, the fermion force $\partial\mathcal{M}_{nk}/\partial\tilde{\phi}^*(i)$ and the fermion Laplacian $\Delta = \mathcal{M}^\dagger\mathcal{M}$. Recall that the fermion matrix is given by

$$\mathcal{M}_{nk} = \beta\left[(in\omega + m)\delta_{nk} + 3g\sum_l(\tilde{\phi}^2)_l\delta_{n,\,k+l}\right]. \tag{A.110}$$

We need also need to compute the pseudo-fermion potential V, the fermion force f and the pseudo-fermion kinetic energy K_f given by

$$V = \tilde{\Phi}^\dagger W. \tag{A.111}$$

$$f(i) = -2\cdot\mathrm{Real}\left(\sum_{\sigma=1}^{M} a_\sigma G_\sigma^\dagger\mathcal{M}^\dagger\frac{\partial\mathcal{M}}{\partial\tilde{\phi}^*(i)}G_\sigma\right). \tag{A.112}$$

$$K_f = \frac{1}{2}\sum_{n=-\Lambda}^{+\Lambda}\tilde{\Pi}(n)\tilde{\Pi}^*(n). \tag{A.113}$$

Note that V requires the calculation of $W = a_0\tilde{\Phi}+\sum_{\sigma=1}^{M} a_\sigma G_\sigma$ which requires $G_\sigma = (x + b_\sigma)^{-1}\tilde{\Phi}$. These are both given by the conjugate gradient subroutine.

This part is straightforward to code.

Molecular dynamics: The molecular dynamics should also be modified according to the equations

$$\frac{d\tilde{\Phi}(n)}{d\tau} = \tilde{\Pi}(n),\ \frac{d\tilde{\Pi}(n)}{d\tau} = -W(n). \tag{A.88}$$

$$\frac{d\tilde{\phi}(n)}{d\tau} = \tilde{\pi}(n),\ \frac{d\tilde{\pi}(n)}{d\tau} = -F(n) - f(n). \tag{A.115}$$

Thus we will call here the fermion matrix subroutine which will give us the fermion force $f(n)$. This part is also straightforward.

Metropolis step: Now the Metropolis should be implemented with the full Hamiltonian

$$\begin{aligned} H =& \frac{1}{2}\sum_{n=-\Lambda}^{+\Lambda} \tilde{\pi}(n)\tilde{\pi}^*(n) \\ &+ \frac{1}{2}\sum_{n=-\Lambda}^{+\Lambda} \tilde{\Pi}(n)\tilde{\Pi}^*(n) + S_B[\tilde{\phi}, \tilde{\phi}^*] + V[\tilde{\Phi}, \tilde{\Phi}^*]. \end{aligned} \tag{A.116}$$

This is also straightforward.

Heat bath: The fields $\tilde{\pi}$ and $\tilde{\Pi}$ are generated by a Gaussian distribution before each molecular dynamics evolution.

Similarly, the pseudo-fermions $\tilde{\Phi}$ should be generated by a Gaussian distribution as follows

$$\tilde{\Phi} = \Delta^{1/4}\xi. \tag{A.117}$$

$$\Delta^{1/4} \simeq c_0 + \sum_{\sigma=1}^{M_0} \frac{c_\sigma}{\Delta + d_\sigma}. \tag{A.118}$$

Again M_0, c_0, c_σ and d_σ are generated by calling the Remez algorithm for the function $x^{1/4}$. Clearly, the calculation of this rational approximation involves again the conjugate gradient method.

This part can be coded using the following subroutine:

```
subroutine pseudo_gaussian(Lambda,idum,M0,c0,c,d,pseudo_ph)
integer M0,idum,Lambda,n,sig
double precision c0,c(1:M0),d(1:M0),aa
double complex Delta(-Lambda:+Lambda,-Lambda:+Lambda)
doublecomplex pseudo_ph(-Lambda:Lambda)
double complex G(1:M0,-Lambda:Lambda),W(-Lambda:Lambda)
double complex xi(-Lambda:Lambda)

aa=2.0d0
call  gaussian(Lambda,aa,xi,idum)
call conjugate_gradient(Lambda,M0,c0,c,d,Delta,xi,G,W)

do n=-Lambda,Lambda
   pseudo_ph(n)=c0*xi(n)
   do sig=1,M0
      pseudo_ph(n)=pseudo_ph(n)+c(sig)*G(sig,n)
   enddo
enddo

return
end
```

A.4 Hybrid Monte Carlo and heat bath algorithms for the BFSS matrix model

A.4.1 Coding BFSS quantum mechanics in Fourier space

We code now the BFSS model following the above same steps. We start as before from the gauge-fixed bosonic action (gauge fixing in the static diagonal gauge $\beta A = \text{diag}(\alpha_1, \ldots, \alpha_N)$) given by

$$S_B = \frac{1}{g^2}\int_0^\beta dt\, \text{Tr}\left[\frac{1}{2}(D_t X_a)^2 - \frac{1}{4}[X_a, X_b]^2\right] - \frac{1}{2}\sum_{i\neq j} \ln \sin^2 \frac{\theta_i - \theta_j}{2}. \tag{A.119}$$

In Fourier space this action reads

$$S_B = N\beta\left[\frac{1}{2}\sum_{n=-\Lambda}^{n=+\Lambda}(n\omega - \frac{\alpha_i - \alpha_j}{\beta})^2 \tilde{X}_a^{ji}(-n)\tilde{X}_a^{ij}(n) - \frac{1}{4}(\text{Tr}[\tilde{X}_a, \tilde{X}_b]^2)_0\right]. \tag{A.120}$$

Recall that $\omega = 2\pi/\beta$ and $\beta = 1/T$ where T is the temperature. The definition of the object $(...)_0$ is as in the previous section. Thus, the only difference with the anharmonic oscillator problem of the previous section is the increase of the number of degrees of freedom as

$$\tilde{\phi}(n) \longrightarrow \tilde{X}_a^{ij}(n). \tag{A.121}$$

In other words, the number of degrees of freedom goes from $2\Lambda + 1$ to $(2\Lambda + 1)DN^2$. The coding of this action is therefore straightforward.

Let us consider now the fermionic part of the BFSS action given by

$$S_F = \frac{1}{g^2}\int_0^\beta dt\, \mathrm{Tr}\left[\frac{1}{2}\psi_\alpha D_t \psi_\alpha - \frac{1}{2}\psi_\alpha (\gamma_i)_{\alpha\beta}[X_i, \psi_\beta]\right]. \tag{A.122}$$

In Fourier space this reads

$$\begin{aligned} S_F = \frac{N\beta}{2}\Bigg[&\sum_{r=-\Lambda'}^{r=+\Lambda'} i(r\omega - \frac{\alpha_i - \alpha_j}{\beta})\tilde{\psi}_\alpha^{ji}(-r)\tilde{\psi}_\alpha^{ij}(r) \\ &- (\gamma_a)_{\alpha\beta}\, \mathrm{Tr}\, \tilde{\psi}_\alpha(-r)\Big([\tilde{X}_a, \tilde{\psi}_\beta]\Big)_r\Bigg]. \end{aligned} \tag{A.123}$$

To simplify even further, we consider only the first term in the above action, viz

$$S_F = \frac{N\beta}{2}\left[\sum_{r=-\Lambda'}^{r=+\Lambda'} i(r\omega)\sum_i\sum_j \tilde{\psi}_\alpha^{ji}(-r)\tilde{\psi}_\alpha^{ij}(r).....\right]. \tag{A.124}$$

We will introduce the $N \times N$ matrices T^A defined by

$$(T^A)_{ij} = \delta_{i_A i}\delta_{jj_A}, \quad A = N(i_A - 1) + j_A, \quad \bar{A} = N(j_A - 1) + i_A, \tag{A.125}$$

and expand the Majorana–Weyl spinors $\tilde{\psi}_\alpha(r)$ as

$$\tilde{\psi}_\alpha(r) = \sum_{A=1}^{N^2} \tilde{\psi}_\alpha^A(r) T^A. \tag{A.126}$$

We can then rewrite the above fermion action as

$$\begin{aligned} S_F &= \frac{N\beta}{2}\left[\sum_{r=-\Lambda'}^{r=+\Lambda'} i(r\omega)\sum_A\sum_B \tilde{\psi}_\alpha^A(-r)\tilde{\psi}_\alpha^B(r)\delta_{\bar{A}B}.....\right] \\ &= \frac{1}{2}\sum_A\sum_B \tilde{\psi}_\alpha^A(s)\mathcal{M}_{A\alpha s;B\beta r}\tilde{\psi}_\beta^B(r), \end{aligned} \tag{A.127}$$

where the matrix $\mathcal{M}$ is given by

$$\mathcal{M}_{A\alpha s;B\beta r} = N\beta(ir\omega)\delta_{\bar{A}B}\delta_{\alpha,\beta}\delta_{s,-r}. \tag{A.128}$$

The integration over the fermions gives the pfaffian, viz

$$\int \mathcal{D}\bar{\psi} \ \exp(-\bar{\psi}\mathcal{M}\bar{\psi}) = \text{Pf } \mathcal{M}. \tag{A.129}$$

This is in general a complex number which cannot be handled in Monte Carlo and thus an approximation here is needed to be able to proceed. We will replace the pfaffian by its modulus which can be converted into a determinant as follows

$$\int \mathcal{D}\bar{\psi} \ \exp(-\bar{\psi}\mathcal{M}\bar{\psi}) \simeq |\text{Pf } \mathcal{M}| = (\det \mathcal{D})^{1/4}, \quad \mathcal{D} = \mathcal{M}^{+}\mathcal{M}. \tag{A.130}$$

This approximation will be justified numerically later.

A.4.2 Coding BFSS quantum mechanics on the beloved lattice

In [6, 10] a lattice approach instead of a Fourier space regularization is used to code the BFSS quantum mechanics.

In the free code [11] the author in his users' manual in response to his own question '(W)hy am I using lattice now?', he states: 'The reason is that, while the momentum cutoff method has various advantages such as a fast convergence to the continuum limit and quick restoration of supersymmetry, it is not adequate for a large-scale parallelization, because the action is highly nonlocal in momentum space'.

Thus for our purposes we go back to lattice regularization in this section. The gauge-fixed bosonic BFSS action is given by (with $1/g^2 = N$)

$$\begin{aligned} S_B = N\sum_{n=0}^{\Lambda-1} \text{Tr}\left[\frac{1}{a}X_i^2(n) - \frac{a}{4}[X_i(n), X_j(n)]^2\right] \\ -\frac{N}{a}\left[\sum_{n=0}^{\Lambda-2} \text{Tr } X_i(n)X_i(n+1) + X_i(\Lambda)D^{\dagger}X_i(\Lambda-1)D\right]. \end{aligned} \tag{A.131}$$

The holonomy matrix D is given by

$$D = \text{diag}(\exp(i\theta_1), \dots, \exp(i\theta_N)). \tag{A.132}$$

The fourth term in the action reads explicitly

$$-\frac{N}{a}\exp(i(\theta_a - \theta_b))X_i^{ab}(\Lambda)X_i^{ba}(\Lambda - 1). \tag{A.133}$$

The Hamiltonian of the system is then given by

$$H = \frac{1}{2}\sum_{a=1}^{N} p_a^2 + \frac{1}{2}\sum_{n=0}^{\Lambda-1} \text{Tr } P_i^2(n) + S_B + S_{\text{FP}}. \tag{A.134}$$

The Faddeev–Popov action is given by

$$S_{\rm FP} = -\frac{1}{2}\sum_{a\neq b} \ln \sin^2 \frac{\theta_a - \theta_b}{2}. \qquad \text{(A.21)}$$

Obviously, the trace parts Tr $X_i(n)$ of the matrices $X_i(n)$ decouple from the remaining degrees of freedom. Similarly, the center of mass $\sum_a \theta_a / N$ drops from the action. Thus, in the path integral we should integrate with the tracelessness constraints

$$\mathrm{Tr}\, X_i(n) = 0, \quad \sum_a \theta_a = 0. \qquad \text{(A.136)}$$

The first numerical test we can perform is to verify the exact identity

$$\langle \exp(-\Delta H)\rangle = 1, \qquad \text{(A.137)}$$

where ΔH is the variation of the Hamiltonian used in the Metropolis step. Another test is to verify that this variation scales as $\delta\tau^2$ in the leap-frog algorithm used in solving the Hamiltonian equations. See for example [2].

Another very simple test is to study the harmonic oscillator problem given by the Hamiltonian

$$S_B = \frac{1}{2}\sum_{n=0}^{\Lambda-1} \mathrm{Tr}\, P_i^2(n) + \frac{N}{a}\sum_{n=0}^{\Lambda-1} \mathrm{Tr}\, X_i^2(n). \qquad \text{(A.138)}$$

In this case the physics is completely known given by a periodic motion with a known period and the average value of the Hamiltonian is exactly equal to the number of degrees of freedom, viz $\langle H\rangle = dN^2\Lambda$.

In the molecular dynamics part we need the force in a crucial way. The force acting on the θ variables is

$$\begin{aligned} F_a = \frac{\delta S_B}{\delta\theta_a} &= -\frac{iN}{a}\exp(i(\theta_a - \theta_b))X_i^{ab}(0)X_i^{ba}(\Lambda - 1) \\ &+ \frac{iN}{a}\exp(-i(\theta_a - \theta_b))X_i^{ab}(\Lambda - 1)X_i^{ba}(0) - \sum_b \tan^{-1}\frac{\theta_a - \theta_b}{2}. \end{aligned} \qquad \text{(A.139)}$$

The force acting on the X variables is

$$\begin{aligned} F_i^{ab}(n) = \frac{\delta S_B}{\delta X_i^{ab}(n)} &= -aN[X_j(n), [X_i(n), X_j(n)]]^{ba} + \frac{2N}{a}X_i^{ba}(n) \\ &- \frac{N}{a}(X_i^{ba}(n-1) + X_i^{ba}(n+1)). \\ n = 1, \ldots, \Lambda - 2. \end{aligned} \qquad \text{(A.140)}$$

$$
\begin{aligned}
F_i^{ab}(0) = \frac{\delta S_B}{\delta X_i^{ab}(0)} &= -aN[X_j(0), [X_i(0), X_j(0)]]^{ba} + \frac{2N}{a}X_i^{ba}(0) \\
&- \frac{N}{a}X_i^{ba}(1) - \frac{N}{a}\exp(i(\theta_a - \theta_b))X_i^{ba}(\Lambda - 1).
\end{aligned} \tag{A.141}
$$

$$
\begin{aligned}
F_i^{ab}(n) = \frac{\delta S_B}{\delta X_i^{ab}(n)} &= -aN[X_j(n), [X_i(n), X_j(n)]]^{ba} + \frac{2N}{a}X_i^{ba}(n) \\
&- \frac{N}{a}X_i^{ba}(\Lambda - 2) - \frac{N}{a}\exp(-i(\theta_a - \theta_b))X_i^{ba}(0). \\
n &= \Lambda - 1.
\end{aligned} \tag{A.142}
$$

The molecular dynamics equations are Hamilton equations, viz

$$
\frac{\partial H}{\partial q} = -\dot{p}, \quad \frac{\partial H}{\partial p} = \dot{q}. \tag{A.2}
$$

We denote the molecular dynamics (pseudo) time by τ. The molecular dynamics equations, in the Stormer–Verlet or leap-frog approximation, for θ are

$$
p_a\left(\tau + \frac{1}{2}\right) = p_a(\tau) - \frac{\delta\tau}{2}F_a(\tau). \tag{A.144}
$$

$$
\theta_a(\tau + 1) = \tau_a(\tau) + \delta\tau p_a\left(\tau + \frac{1}{2}\right). \tag{A.145}
$$

$$
p_a(\tau + 1) = p_a\left(\tau + \frac{1}{2}\right) - \frac{\delta\tau}{2}F_a(\tau + 1). \tag{A.146}
$$

The molecular dynamics equations, in the leap-frog approximation, for X are

$$
P_i^{ab}\left(n, \tau + \frac{1}{2}\right) = P_i^{ab}(n, \tau) - \frac{\delta\tau}{2}F_i^{ba}(n, \tau). \tag{A.147}
$$

$$
X_i^{ab}(n, \tau + 1) = X_i^{ab}(n, \tau) + \delta\tau P_i^{ab}\left(n, \tau + \frac{1}{2}\right). \tag{A.148}
$$

$$
P_i^{ab}(n, \tau + 1) = P_i^{ab}\left(n, \tau + \frac{1}{2}\right) - \frac{\delta\tau}{2}F_i^{ba}(n, \tau + 1). \tag{A.149}
$$

A.4.3 The heat bath algorithm for the bosonic BFSS quantum mechanics

We go back to the lattice action (with a symmetric static diagonal gauge $V = D^{1/\Lambda}$)

$$S_B = N\sum_{n=0}^{\Lambda-1} \mathrm{Tr}\left[\frac{1}{a}X_i^2(n) - \frac{a}{4}[X_i(n), X_j(n)]^2\right] - \frac{N}{a}\left[\sum_{n=0}^{\Lambda-1} \mathrm{Tr}\, X_i(n+1)V^\dagger X_i(n)V\right]. \tag{A.150}$$

We focus first on the IKKT term

$$S_1 = \frac{N}{a}\sum_{n=0}^{\Lambda-1} \mathrm{Tr}\, X_i^2(n) - \frac{Na}{4}\sum_{n=0}^{\Lambda-1} \mathrm{Tr}[X_i(n), X_j(n)]^2. \tag{A.151}$$

We rewrite this as

$$S_1' = \frac{N}{a}\sum_{n=0}^{\Lambda-1} \mathrm{Tr}\, X_i^2(n) - \frac{Na}{2}\sum_{n=0}^{\Lambda-1} \mathrm{Tr} \sum_{i>j}\Big(-Q_{ij}^2(n) + 2Q_{ij}(n)\{X_i(n), X_j(n)\} - 4X_i^2(n)X_j^2(n)\Big). \tag{A.152}$$

We fix the lattice site to be m and the vector index to be k. Then the relevant part of the action depending on $X_k(m)$ is

$$\frac{N}{a}\mathrm{Tr}\, X_k^2(m) - \frac{Na}{2}\mathrm{Tr}\left(-4X_k^2(m)\sum_{j\neq k} X_j^2(m) + 2X_k(m)\sum_{j\neq k}\{X_j(m), Q_{kj}(m)\}\right). \tag{A.153}$$

Define the matrices

$$M_k(m) = 2aN\sum_{j\neq k} X_j^2(m). \tag{A.154}$$

$$N_k(m) = -aN\sum_{j\neq k}\{X_j(m), Q_{kj}(m)\}. \tag{A.155}$$

The diagonal elements of $X_k(m)^{aa}$ will appear with the action (a fixed)

$$(X_k(m)^{aa})^2\left(\frac{N}{a} + M_k(m)^{aa}\right) + X_k(m)^{aa} \times\left(\sum_{b\neq a} X_k(m)^{ab}M_k(m)^{ba} + \sum_{b\neq a} X_k(m)^{ba}M_k(m)^{ab} + N_k(m)^{aa}\right). \tag{A.156}$$

The off diagonal elements of $X_k(m)^{ab}$ will appear with the action ($a \neq b$ fixed)

$$
\begin{aligned}
& X_k(m)^{ab} X_k(m)^{ba}\left(\frac{2N}{a} + M_k(m)^{aa} + M_k(m)^{bb}\right) \\
& + X_k(m)^{ba}\left(N_k(m)^{ab} + \sum_{c\neq a} M_k(m)^{ac} X_k(m)^{cb}\right. \\
& \left. + \sum_{c\neq b} X_k(m)^{ac} M_k(m)^{cb}\right) + X_k(m)^{ab} \\
& \left(N_k(m)^{ab} + \sum_{c\neq a} M_k(m)^{ac} X_k(m)^{cb} + \sum_{c\neq b} X_k(m)^{ac} M_k(m)^{cb}\right)^* .
\end{aligned}
\tag{A.157}
$$

We include now the effect of the Hopping term

$$
\begin{aligned}
S_2 &= -\frac{N}{a}\left[\sum_{n=0}^{\Lambda-1} \mathrm{Tr}\, X_i(n+1) V^{\dagger} X_i(n) V\right] \\
&= -\frac{N}{a}\sum_{n=0}^{\Lambda-1} \exp\left(i\frac{\theta_a - \theta_b}{\Lambda}\right) X_i^{ab}(n+1) X_i^{ba}(n).
\end{aligned}
\tag{A.158}
$$

We have immediately the following results:

- The diagonal elements will appear as

$$
(X_k(m)^{aa})^2 l_k(m)^{aa} + X_k(m)^{aa} h_k(m)^{aa} = l\left(X + \frac{h}{2l}\right)^2 + \cdots. \tag{A.159}
$$

$$
l_k(m)^{aa} = \frac{N}{a} + M_k(m)^{aa}. \tag{A.160}
$$

$$
\begin{aligned}
h_k(m)^{aa} &= N_k(m)^{aa} + \sum_{c\neq a} M_k(m)^{ac} X_k(m)^{ca} \\
&\quad + \sum_{c\neq a} X_k(m)^{ac} M_k(m)^{ca} - \frac{N}{a} X_k(m+1)^{aa} \\
&\quad - \frac{N}{a} X_k(m-1)^{aa}.
\end{aligned}
\tag{A.161}
$$

Thus the diagonal elements are given in terms of a Gaussian real random number x with width $1/a$ where $a = l_k(m)^{aa}$ by the following rule

$$
X_k(m)^{aa} = x - \frac{h_k(m)^{aa}}{2 l_k(m)^{aa}}, \quad x = \sqrt{-\frac{1}{a}\ln v_1}\cos 2\pi v_2. \tag{A.162}
$$

- On the other hand, the off diagonal elements will appear as

$$2X_k(m)^{ab}X_k(m)^{ba}l_k(m)^{ab} + X_k(m)^{ba}h_k(m)^{ab} + X_k(m)^{ab}h_k(m)^{ba} = 2l\left(X + \frac{h}{2l}\right)\left(X^* + \frac{h^*}{2l}\right) + \cdots. \quad (A.163)$$

$$2l_k(m)^{ab} = \frac{2N}{a} + M_k(m)^{aa} + M_k(m)^{bb}. \quad (A.164)$$

$$h_k(m)^{ab} = N_k(m)^{ab} + \sum_{c \neq a} M_k(m)^{ac}X_k(m)^{cb} + \sum_{c \neq b} X_k(m)^{ac}M_k(m)^{cb} - \frac{N}{a}e^{i\frac{\theta_a - \theta_b}{\Lambda}}X_k(m+1)^{ab} - \frac{N}{a}e^{-i\frac{\theta_a - \theta_b}{\Lambda}}X_k(m-1)^{ab}. \quad (A.165)$$

Thus, the off diagonal elements are given in terms of a Gaussian complex random number z with width $1/a$ where $a = 2l_k(m)^{ab}$ by the following rule

$$X_k(m)^{ab} = z - \frac{h_k(m)^{ab}}{2l_k(m)^{ab}}, \quad z = \sqrt{-\frac{1}{a}\ln v_1}(\cos 2\pi v_2 + i \sin 2\pi v_2). \quad (A.166)$$

In the above formulas v_1 an v_2 are uniform random numbers between 0 and 1.

Thus, the updating of the diagonal elements for each vector index and for each lattice site is done simultaneously. Also, the updating of the off diagonal elements for each vector index and for each lattice site is done simultaneously [13]. However, here one should be careful since only the elements $A_{\mu_1,\mu_2}, A_{\mu_3\mu_4}, \ldots, A_{\mu_{N-1}\mu_N}$ for completely different indices $\mu_1, \ldots \mu_N$ are really independent. In other words, we should update off diagonal elements in $N - 1$ blocks each block containing $N/2$ independent elements with completely different indices [12]. The trace part (trace over matrix and lattice) should then be subtracted after this update.

The updating of $Q_{ij}(m)$ is controlled by the piece

$$\frac{Na}{2}\mathrm{Tr}(Q_{ij}(m) - \{X_i(m), X_j(m)\})^2 + \cdots \quad (A.167)$$

These are also shifted Gaussian random numbers as above.

Finally, the updating of the holonomy matrix is preformed via the ordinary Metropolis–Hastings algorithm.

The most time consuming piece of the heat bath algorithm as explained above is the part concerning the update of the off diagonal components which goes at each lattice site as $O(N^3)$. The diagonal components and the auxiliary fields cost $O(N^2)$.

The main observables we will measure are:

- The energy is defined by

$$E = -\frac{1}{Z(\beta)}\frac{Z(\beta^{'}) - Z(\beta)}{\Delta\beta}, \quad \Delta\beta = \beta^{'} - \beta \longrightarrow 0. \tag{A.168}$$

We compute [13]

$$\frac{E}{N^2} = \frac{3T}{N^2}\langle \text{commu} \rangle, \quad \text{commu} = -\frac{Na}{4}\sum_{n=0}^{\Lambda-1} \text{Tr}[X_i(n), X_j(n)]^2. \tag{A.169}$$

The proof goes as follows. We relate the partition functions $Z(\beta^{'})$ and $Z(\beta)$ by the following scalings

$$\frac{t'}{t} = \frac{\beta^{'}}{\beta}, \quad \frac{A'}{A} = \frac{\beta}{\beta^{'}}, \quad \frac{X'}{X} = \sqrt{\frac{\beta^{'}}{\beta}}. \tag{A.170}$$

This guarantees that the kinetic term is fully invariant. We assume that the measures over $X_i(n)$ and $A(t)$ are also invariant under these scalings. Then from the non invariance of the Yang–Mills term we can derive the above formula of the energy.

- The radius of space defined by

$$R^2 = \frac{a}{\Lambda N^2}\langle \text{radius} \rangle, \quad \text{radius} = \frac{N}{a}\sum_{n=0}^{\Lambda-1} \text{Tr } X_i^2(n). \tag{A.171}$$

- The Polyakov line which is the order parameter defined by

$$\langle |P| \rangle = \frac{1}{N}\langle |\text{Tr } U| \rangle. \tag{A.172}$$

The holonomy matrix U is defined by the path ordered exponential

$$U = \mathcal{P}\exp\left(i\int_0^{\beta} dtA(t)\right). \tag{A.173}$$

On the lattice U is given by

$$U = V(0)V(1).... V(\Lambda - 1) = \text{diag}(e^{i\theta_1}, e^{i\theta_2}, \dots, e^{i\theta_N}). \tag{A.174}$$

- The eigenvalue distribution of the holonomy matrix defined by

$$\rho(\theta) = \frac{1}{N}\sum_i \langle \delta(\theta - \theta_i) \rangle. \tag{A.175}$$

- Here some general guidance on implementing the above algorithm:

- We start off the holonomy angles θ_a from zero to guarantee that we remain around the group identity and do not wander off to other central group elements while we start off the matrices $X_i(n)$ from a Gaussian distribution with a narrow peak.
- It is imperative in the heat bath update to organize the update as explained above. First we calculate the matrices h and l and update all the diagonal elements at once, then recalculate the matrices h and l and update the first block of $N/2$ independent off diagonal elements, then recalculate h and l and update the second block of $N/2$ independent off diagonal elements, and so on until all the $N - 1$ blocks of off diagonal elements get updated.
- In the heat bath after each update we subtract the global (lattice and matrix) traces of $X_i(n)$ for every i or after all i's are updated (preferable).
- The angles θ_a are subjected to the Metropolis step with action

$$S_\theta = -\frac{1}{2}\sum_{a\neq b} \ln \sin^2 \frac{\theta_a - \theta_b}{2} - \frac{N}{a} \sum_{n=0}^{\Lambda-1}\sum_{a,b} \exp\left(-i\frac{\theta_a - \theta_b}{\Lambda}\right) X_i^{ba}(m+1) X_i^{ab}(m). \tag{A.176}$$

- The removal of the trace of θ_a is done as follows [11]. The traceless $\tilde{\theta}_a$ are defined by

$$\tilde{\theta}_a = \theta_a - \frac{1}{N}\sum_a \theta_a. \tag{A.177}$$

The integral over θ_a can be replaced with the integral over $\tilde{\theta}_a$ plus the integral over the center of mass which does not appear in the action S_θ. The angles θ_a are between $-\pi$ and π while the angles $\tilde{\theta}_a$ are between $\tilde{\theta}_{\rm mi}$ and $\tilde{\theta}_{\rm ma}$. The center of mass is therefore between $-\pi - \tilde{\theta}_{\rm ma}$ and $+\pi - \tilde{\theta}_{\rm mi}$. The Boltzmann should then be multiplied by the factor

$$\omega = 2\pi - \mu, \quad \mu < 2\pi, \quad \omega = 0, \quad \mu \geqslant 2\pi. \tag{A.178}$$

In actual simulation we may avoid the zero by adopting the prescription of [11] given by (with $\epsilon = 1/g$ and $g = 100$)

$$\omega = 2\pi - \mu + \epsilon, \quad \mu < 2\pi, \quad \omega = \epsilon \exp(-g(\mu - 2\pi)), \quad \mu \geqslant 2\pi. \tag{A.179}$$

The interval from which we draw proposals for the holonomy angles θ_a is $[-\pi, \pi]$. The acceptance rate is typically quite low and thus in order to accumulate a good sample we repeat the Metropolis step. Generally, we repeat this step N times which is the number of degrees of freedom contained in the holonomy matrix.

- The errors are estimated using the theoretical formula which includes he effect of the auto-correlation time. We have also tried the Jackknife method.

  The random number generator we use is Ran2. We strongly advise to use the Mersenne twister.

  We fix the lattice size a and the matrix size N and run the code for some number T_t of thermalization steps and some number T_m of measurement steps. The range of temperature is [0,3] where the number of lattice iterations is given by $\Lambda = 1/aT$. We measure the energy per matrix degrees of freedom, the extent of space, and the Polyakov line.

We will observe in the numerical results a phase transition at the temperature $T \sim 1$ at which the order parameter (the Polyakov line) goes from 1 to zero. This is a second order phase transition between confined and deconfined phases associated with the spontaneous breakdown of the $U(1)$ symmetry $A(t) \longrightarrow A(t) + \alpha\mathbf{1}$ and the emergence of a gap in the eigenvalue distribution or from a non-inform distribution to a uniform distribution.

In fact, there also exists another phase transition, which is third order, in this model since the deconfined phase is actually divided into a non-uniform and gapped phases. The confined phase is the uniform phase. The confined/deconfined phase transition is closely related to the black string/black holes transition in the two-dimensional supersymmetric extension of this theory, i.e. it is related to the Gregory–Laflamme transition [15, 16]. There are also intimate relations with the Hagedorn transition in string theory [17–19]. A beautiful discussion of these issues can be found in [14]. See also [6].

The fit used for the eigenvalue density can be taken to be the Gross–Witten (GW) gapped eigenvalue distribution. Because of the very narrow critical regime in the deconfined non-uniform phase it is very hard to observe the GW non-uniform gapless distribution. These are given respectively by

$$\rho(\theta) = \frac{2}{\pi\kappa}\cos\frac{\theta}{2}\sqrt{\frac{\kappa}{2} - \sin^2\frac{\theta}{2}}, \quad |\theta| \leqslant 2\sin^{-1}\sqrt{\frac{\kappa}{2}}, \quad \kappa < 2, \quad \text{gapped.} \tag{A.180}$$

$$\rho(\theta) = \frac{1}{2\pi}\left(1 + \frac{2}{\kappa}\cos\theta\right), \quad \kappa > 2, \quad \text{gapless or non uniform.} \tag{A.181}$$

As we increase the temperature, the transition between the confined uniform phase to the deconfined non-uniform phase is third order at some T_{c2}, the range of temperature in the non-uniform gapless phase is very narrow, then a second order transition to the deconfined gapped phase occurs at a slightly larger temperature T_{c1}. The measurement of [14] is $T_{c2} = 0.8761(3)$ and $T_{c1} = 0.905(2)$.

A.4.4 The $D = \infty$ approximation

It was shown in [6] that in the large $1/D$ expansion of the bosonic part of the BFSS model we can effectively replace the commutator term, through an indirect saddle point approximation, by a mass term. The action becomes the Gaussian quantum mechanics model

$$S = N\int_0^\beta dt \,\mathrm{Tr}\left[\frac{1}{2}(D_tX_i)^2 + \frac{1}{2}mX_i^2\right], \quad m = D^{2/3}. \tag{A.182}$$

This model enjoys a phase transition at

$$T_c = \frac{m}{\ln D} = 0.9467. \tag{A.183}$$

The eigenvalue distribution of any of the matrices X_i is a Wigner semicircle law with a radius

$$R = \left(\frac{8}{D}\right)^{1/6} = 0.98. \tag{A.184}$$

On the lattice this becomes

$$S_B = \left(\frac{N}{a} + \frac{Nam}{2}\right)\sum_{n=0}^{\Lambda-1} \mathrm{Tr}\, X_i^2(n) - \frac{N}{a}\left[\sum_{n=0}^{\Lambda-1} \mathrm{Tr}\, X_i(n+1)V^\dagger X_i(n)V\right]. \tag{A.185}$$

The Polyakov line is defined as before. The energy in this model is given by the extent of space, viz

$$\frac{E}{N^2} = \frac{a^2Tm}{N^2}\langle\text{radius}\rangle. \tag{A.186}$$

The adoption of the heat bath algorithm to this problem is straightforward. For a given fixed vector index k and lattice site m we have

$$\begin{aligned}&\left(\frac{N}{a} + \frac{Nam}{2}\right)X_k^{ab}(m)X_k^{ba}(m) - \frac{N}{a}X_k^{ab}(m)\\ &\times\left[e^{-\frac{i}{\Lambda}(\theta_a-\theta_b)}X_k^{ba}(m+1) + e^{\frac{i}{\Lambda}(\theta_a-\theta_b)}X_k^{ba}(m-1)\right].\end{aligned} \tag{A.187}$$

The diagonal elements will be given in terms of a real Gaussian number x with width

$$\Delta = \frac{N}{a} + \frac{m_x}{2}, \quad m_x = Nam \tag{A.188}$$

by the formula

$$X_k^{aa}(m) = x + \frac{q}{2}[X_k^{aa}(m+1) + X_k^{aa}(m-1)], \quad q = \frac{1}{1 + \frac{am_x}{2N}}. \tag{A.189}$$

The off diagonal elements will be given in terms of a complex Gaussian number z with width 2Δ by the formula

$$X_k^{ab}(m) = z + \frac{q}{2}\left[\exp\left(\frac{i}{\Lambda}(\theta_a - \theta_b)\right)X_k^{ab}(m+1) + \exp\left(-\frac{i}{\Lambda}(\theta_a - \theta_b)\right)X_k^{ab}(m-1)\right]. \quad (A.190)$$

The updating the holonomy is done as before through standard Metropolis. A subroutine which implements the updating of X is given as follows:

```
      subroutine heat_bath_reduce(N,dim,Lambda,LM,lattice,idum,X,thet,Q,
     &     accept,reject,pa,inn,a1,a2,a3,af,a0,mx)
      implicit none
      integer N,dim,LM,Lambda,nn1,a,b,c,j,k,l,idum,i1,i2,i3,i4
      doubleprecision lattice,thet(1:N),thet0(1:N),thett(1:N),pp(1:N)
      double complex X(1:dim,0:LM-1,1:N,1:N),X0(1:dim,0:LM-1,1:N,1:N)
      doublecomplex  P(1:dim,0:LM-1,1:N,1:N)
      double complex Q(1:dim,1:dim,0:LM-1,1:N,1:N)
      double complex Q0(1:dim,1:dim,0:LM-1,1:N,1:N)
      double precision LL(1:dim,0:LM-1,1:N,1:N)
      double complex HH(1:dim,0:LM-1,1:N,1:N),hh1,hh2,hh3,ii,hh4
      double precision kinetic, commu,radius,actionB,lp,SFP,Ham,a1,a2,a3
     & ,af,a0,mt,mx,variationS,accept,reject,pa,probabilityB,pi,ran2,aa
     & ,a11,theta,r,actionB_HB,inn,lp2,qq,fab,tr,tr0

      ii=cmplx(0.0d0,1.0d0)
      pi=dacos(-1.0d0)

      qq=mx*lattice/(2.0d0*N)
      qq=qq+1.0d0
      qq=1.0d0/qq
c..............UPDATING OF X_i(n)...................

      do k=1,dim
         do nn1=0,Lambda-1
c................diagonal elements.............
            do a=1,N
               aa=0.5d0*mx+(N/lattice)
               a11=1.0d0/aa
               theta=2.0d0*pi*ran2(idum)
               r=dsqrt(-a11*dlog(1.0d0-ran2(idum)))
```

```
            if ((nn1.gt.0).and.(nn1.lt.(Lambda-1))) then
               X(k,nn1,a,a)=r*dcos(theta)
     &             +0.5d0*qq*X(k,nn1+1,a,a)+0.5d0*qq*X(k,nn1-1,a,a)
            elseif (nn1.eq.0) then
               X(k,nn1,a,a)=r*dcos(theta)
     &             +0.5d0*qq*X(k,nn1+1,a,a)+0.5d0*qq*X(k,Lambda-1,a,a)
            else
               X(k,nn1,a,a)=r*dcos(theta)
     &             +0.5d0*qq*X(k,0,a,a)+0.5d0*qq*X(k,Lambda-2,a,a)
            endif
          enddo
c...............off diagonal elements...........
          do a=1,N
            do b=a+1,N
               aa=2.0d0*(0.5d0*mx+(N/lattice))
               a11=1.0d0/aa
               theta=2.0d0*pi*ran2(idum)
               r=dsqrt(-a11*dlog(1.0d0-ran2(idum)))
               fab=(thet(a)-thet(b))/Lambda
               hh3=dcos(fab)+ii*dsin(fab)
               hh4=dcos(fab)-ii*dsin(fab)
               if ((nn1.gt.0).and.(nn1.lt.(Lambda-1))) then
                  X(k,nn1,a,b)=r*dcos(theta)+ii*r*dsin(theta)
     &       +0.5d0*qq*hh4*X(k,nn1+1,b,a)+0.5d0*qq*hh3*X(k,nn1-1,b,a)
               elseif (nn1.eq.0) then
                  X(k,nn1,a,b)=r*dcos(theta)+ii*r*dsin(theta)
     &       +0.5d0*qq*hh4*X(k,nn1+1,b,a)+0.5d0*qq*hh3*X(k,Lambda-1,b,a)
               else
                  X(k,nn1,a,b)=r*dcos(theta)+ii*r*dsin(theta)
     &        +0.5d0*qq*hh4*X(k,0,b,a)+0.5d0*qq*hh3*X(k,nn1-1,b,a)
               endif
               X(k,nn1,b,a)=conjg(X(k,nn1,a,b))
            enddo
          enddo
        enddo
      enddo
c............removing the global trace................
      call  tracelessX(N,dim,Lambda,LM,X)

      return
      end
```

A.4.5 Coding the BMN matrix model

The extension of the above algorithms and Monte Carlo codes to the BMN model is quite trivial. However, the resulting physics of the BMN model is expected to be far from trivial and quite different from the physics of the BFSS model.

A.5 Algorithms for Yang–Mills matrix models

A.5.1 Summary of the Lorentzian IKKT model

The model of interest here is the Lorentzian IKKT matrix model (properly regularized in the IR and UV) given by the Yang–Mills matrix models

$$\begin{aligned}V_{\text{pot}} &= \frac{1}{2}\gamma_C\left(\frac{1}{N}\,\text{Tr}\,F_{\mu\nu}F^{\mu\nu}\right)^2 + \frac{1}{2}\gamma_L\left(\frac{1}{N}\,\text{Tr}\,A_a^2 - L^2\right)^2 \\ &+ \frac{1}{2}\gamma_\kappa\left(\kappa L^2 - \frac{1}{N}\,\text{Tr}\,A_0^2\right)^2\theta\left(\frac{1}{N}\,\text{Tr}\,A_0^2 - \kappa L^2\right).\end{aligned} \tag{A.191}$$

A.5.2 Metropolis algorithm

The crucial ingredient in the Metropolis algorithm is the variation of the action which is computed in the following.

Variation of A_k: The variation of the element ij of the matrix A_k is given by

$$A_k' = A_k + \hat{\epsilon}, \quad \hat{\epsilon}_{nm} = \delta_{ni}\delta_{mj}\epsilon + \delta_{nj}\delta_{mi}\epsilon^*. \tag{A.192}$$

For the diagonal part we define $\tilde{\epsilon} = \epsilon + \epsilon^*$. We define the matrices

$$F = 2\,\text{Tr}[A_0, A_a]^2 - \text{Tr}[A_a, A_b]^2, \quad \Phi = \frac{1}{N}\,\text{Tr}\,A_a^2 - L^2. \tag{A.193}$$

$$\delta\Phi = \frac{2}{N}(\epsilon(A_k)_{ji} + \epsilon^*(A_K)_{ij}) + \frac{2|\epsilon|^2}{N} + \frac{1}{N}(\epsilon^2 + \epsilon^{*2})\delta_{ij}. \tag{A.194}$$

$$\begin{aligned}\delta F &= -2(\lambda_n - \lambda_m)^2(2(A_k)_{nm}\hat{\epsilon}_{mn} + \hat{\epsilon}_{nm}\hat{\epsilon}_{mn}) \\ &\quad -2\text{Tr}[\hat{\epsilon}, A_j]^2 - 4\text{Tr}[A_k, A_j][\hat{\epsilon}, A_j] \\ &= (-4\epsilon(\lambda_i - \lambda_j)^2(A_k)_{ji} - 2|\epsilon|^2(\lambda_i - \lambda_j)^2 \\ &\quad -4\epsilon^2(A_a)_{ji}(A_a)_{ji} - 4|\epsilon|^2(A_a)_{ii}(A_a)_{jj} \\ &\quad +4\epsilon^2\delta_{ij}(A_a^2)_{ji} + 2|\epsilon|^2((A_a^2)_{ii} + (A_a^2)_{jj}) - 4\epsilon[A_a, [A_k, A_a]]_{ji}) \\ &\quad + \text{h. c.}\end{aligned} \tag{A.195}$$

$$\Delta V_{\text{pot}} = \frac{1}{2N^2}\gamma_C \delta F(\delta F + 2F) + \frac{1}{2}\gamma_L \delta\Phi(\delta\Phi + 2\Phi). \tag{A.196}$$

Variation of A_0: For the variation of the eigenvalue $\lambda_k \longrightarrow \lambda_k + \epsilon$ we define the two expressions

$$\Phi_0 = \frac{1}{N}\,\mathrm{Tr}\,A_0^2 - \kappa L^2, \qquad \delta\Phi_0 = \frac{\epsilon}{N}(2\lambda_k + \epsilon). \tag{A.197}$$

We have three kinds of variations:

- **Term proportional to the step function**: The variation is given by

$$\Delta V_1 = \frac{1}{2}\gamma_\kappa \delta\Phi_0(2\Phi_0 + \delta\Phi_0), \quad \Phi_0 \geqslant 0, \quad \Phi_0 \geqslant -\delta\Phi_0. \tag{A.198}$$

$$\Delta V_1 = -\frac{1}{2}\gamma_\kappa \delta\Phi_0^2, \quad \Phi_0 \geqslant 0, \quad \Phi_0 < -\delta\Phi_0. \tag{A.199}$$

$$\Delta V_1 = \frac{1}{2}\gamma_\kappa(\Phi_0 + \delta\Phi_0)^2, \quad \Phi_0 < 0, \quad \Phi_0 \geqslant -\delta\Phi_0. \tag{A.200}$$

$$\Delta V_1 = 0, \quad \Phi_0 < 0, \quad \Phi_0 < -\delta\Phi_0. \tag{A.201}$$

- **Term proportional to the Yang–Mills action**: The variation is given by

$$\Delta V_2 = \frac{1}{2N^2}\gamma_C \delta F(2F + \delta F). \tag{A.202}$$

$$\delta F = -8\epsilon(\lambda_k - \lambda_j)(A_a)_{kj}(A_a)_{jk}. \tag{A.203}$$

- **Term proportional to the Vandermonde and Pfaffian**: The variation is given by

$$\Delta V_3 = -(D-1)\sum_{j\neq k} \ln\left(1 + \frac{\epsilon}{\lambda(k) - \lambda(j)}\right)^2. \tag{A.204}$$

Metropolis setup: The update is accepted with a probability given by

$$\text{probability} = \min(1, \exp(-\Delta V_{\text{pot}})). \tag{A.205}$$

A.5.3 Hybrid Monte Carlo

The actions and the Hamiltonian and the equations of motion of the Lorentzian type IIB matrix model are given by

$$\begin{aligned} V_{\text{pot}} = \frac{1}{2}\gamma_C\left(\frac{1}{N}\,\text{Tr}\,F_{\mu\nu}F^{\mu\nu}\right)^2 + \frac{1}{2}\gamma_L\left(\frac{1}{N}\,\text{Tr}\,A_a^2 - L^2\right)^2 \\ + \frac{1}{2}\gamma_\kappa\left(\kappa L^2 - \frac{1}{N}\,\text{Tr}\,A_0^2\right)^2 \theta\left(\frac{1}{N}\,\text{Tr}\,A_0^2 - \kappa L^2\right). \end{aligned} \tag{A.206}$$

$$V_{\text{eff}} = V_{\text{pot}} - \frac{1}{2}\sum_{k\neq l}\ln(\lambda_k - \lambda_l)^2. \tag{A.207}$$

$$H = \frac{1}{2}\sum_i p_i^2 + \frac{1}{2}\,\text{Tr}\,\Pi_a^2 + V_{\text{eff}}. \tag{A.208}$$

$$\begin{aligned} \dot{p}_i &= f_i \\ \dot{\lambda}_i &= p_i \\ (\dot{\Pi}_k)_{ji} &= (F_k)_{ji} \\ (\dot{A}_k)_{ji} &= (\Pi_k)_{ji}. \end{aligned} \tag{A.209}$$

The numerical solution of these equations is given by the leap-frog algorithm (A.56), (A.57) and (A.58). The forces are defines as follows. First we define

$$\Phi_0 = \kappa L^2 - \frac{1}{N}\,\text{Tr}\,A_0^2. \tag{A.210}$$

$$\Phi = \frac{1}{N}\,\text{Tr}\,A_a^2 - L^2 \Rightarrow \frac{\partial\Phi}{\partial(A_k)_{ij}} = \frac{2}{N}(A_k)_{ji}. \tag{A.211}$$

$$\begin{aligned} F &= -2\,\text{Tr}[A_0, A_a]^2 + \text{Tr}[A_a, A_b]^2 \\ &= 2(\lambda_n - \lambda_m)^2(A_a)_{mn}(A_a)_{nm} + \text{Tr}[A_i, A_j]^2 \Rightarrow \\ \frac{\partial F}{\partial(A_k)_{ij}} &= 4(\lambda_i - \lambda_j)^2(A_k)_{ji} + 4[A_a, [A_k, A_a]]_{ji} \\ \frac{\partial F}{\partial\lambda_i} &= 8(\lambda_i - \lambda_j)(A_a)_{ji}(A_a)_{ij}. \end{aligned} \tag{A.212}$$

The forces are

$$
\begin{aligned}
(F_k)_{ji} &= -\frac{\delta V_{\text{eff}}}{\delta (A_k)_{ij}} \\
&= -\frac{1}{N^2}\gamma_C F \frac{\partial F}{\partial (A_k)_{ij}} - \gamma_L \Phi \frac{\partial \Phi}{\partial (A_k)_{ij}}.
\end{aligned} \tag{A.213}
$$

$$
\begin{aligned}
f_i &= -\frac{\delta V_{\text{eff}}}{\delta \lambda_i} \\
&= -\frac{1}{N^2}\gamma_C F \frac{\partial F}{\partial \lambda_i} + 2\sum_{j\neq i}\frac{1}{\lambda_i - \lambda_j} + \frac{2}{N}\gamma_\kappa \lambda_i \Phi_0 \theta(\Phi_0).
\end{aligned} \tag{A.214}
$$

We will also need to fix the shift symmetry $A_\mu \longrightarrow A_\mu + \alpha_\mu \mathbf{1}$ by adding the action

$$
V_S = \frac{1}{2}\gamma_S \left(\frac{1}{N}\,\mathrm{Tr}\, A_a^2|_L - \frac{1}{N}\,\mathrm{Tr}\, A_a^2|_R\right)^2 \equiv \frac{1}{2}\gamma_S \Phi_1^2, \tag{A.215}
$$

where

$$
\begin{aligned}
\mathrm{Tr}\, A_a^2|_L &= (A_a)_{ij}(A_a)_{ji}, \quad i + j < N + 1, \\
\mathrm{Tr}\, A_a^2|_R &= (A_a)_{ij}(A_a)_{ji}, \quad i + j > N + 1.
\end{aligned} \tag{A.216}
$$

The extra contribution to the force is given by

$$
(F_k)_{ji} = -\frac{\partial V_S}{\partial (A_k)_{ij}} = -\frac{2}{N}\gamma_S \Phi_1 (A_k)_{ji}, \quad \text{if } i + j < N + 1. \tag{A.217}
$$

$$
(F_k)_{ji} = -\frac{\partial V_S}{\partial (A_k)_{ij}} = +\frac{2}{N}\gamma_S \Phi_1 (A_k)_{ji}, \quad \text{if } i + j > N + 1. \tag{A.218}
$$

The rest of the steps are similar to the BFSS matrix model and in fact they are technically much simpler because there is no lattice here to worry about.

References

[1] Ydri B 2017 *Computational Physics: An Introduction to Monte Carlo Simulations of Matrix Field Theory* (Singapore: World Scientific)

[2] Schaefer S, Simulations with the Hybrid Monte Carlo algorithm: implementation and data analysis (downloaded from author website)

[3] Hanada M, Nishimura J and Takeuchi S 2007 Non-lattice simulation for supersymmetric gauge theories in one dimension *Phys. Rev. Lett.* **99** 161602

[4] Catterall S and Wiseman T 2007 Towards lattice simulation of the gauge theory duals to black holes and hot strings *J. High Energy Phys.* **0712** 104

[5] Catterall S and Karamov S 2002 Testing a Fourier accelerated hybrid Monte Carlo algorithm *Phys. Lett.* B **528** 301

[6] Filev V G and O'Connor D 2016 The BFSS model on the lattice *J. High Energy Phys.* **1605** 167

[7] Di Vecchia P and Ferrara S 1977 Classical solutions in two-dimensional supersymmetric field theories *Nucl. Phys.* B **130** 93

[8] Volkholz J and Bietenholz W 2007 Simulations of a supersymmetry inspired model on a fuzzy sphere *PoS LAT* **2007** 283

[9] Bergner G, Kristen T, Uhlmann S and Wipf A 2008 Low-dimensional supersymmetric lattice models *Ann. Phys.* **323** 946

[10] O'Connor D and Filev V G 2016 Membrane matrix models and non-perturbative checks of gauge/gravity duality *PoS CORFU* **2015** 111

[11] Masanori H, Users Manual of Monte Carlo Code for the Matrix Model of M-theory, https://sites.google.com/site/hanadamasanori/home/mmmm.

[12] Bal Subrata, private communication.

[13] Hotta T, Nishimura J and Tsuchiya A 1999 Dynamical aspects of large N reduced models *Nucl. Phys.* B **545** 543

[14] Kawahara N, Nishimura J and Takeuchi S 2007 Phase structure of matrix quantum mechanics at finite temperature *J. High Energy Phys.* **0710** 097

[15] Aharony O, Marsano J, Minwalla S and Wiseman T 2004 Black hole-black string phase transitions in thermal 1.1 dimensional supersymmetric Yang–Mills theory on a circle *Class. Quant. Grav.* **21** 5169

[16] Harmark T and Obers N A 2004 New phases of near-extremal branes on a circle *J. High Energy Phys.* **0409** 022

[17] Aharony O, Marsano J, Minwalla S, Papadodimas K and Van Raamsdonk M 2004 The Hagedorn–deconfinement phase transition in weakly coupled large N gauge theories *Adv. Theor. Math. Phys.* **8** 603

[18] Aharony O, Marsano J, Minwalla S, Papadodimas K and Van Raamsdonk M 2005 A first order deconfinement transition in large N Yang-Mills theory on a small S**3 *Phys. Rev.* D **71** 125018

[19] Atick J J and Witten E 1988 The Hagedorn transition and the number of degrees of freedom of string theory *Nucl. Phys.* B **310** 291

www.ingramcontent.com/pod-product-compliance
Lightning Source LLC
Chambersburg PA
CBHW082126210326
41599CB00031B/5887
* 9 7 8 0 7 5 0 3 1 7 6 6 5 *